BIOMEMBRANOMICS

Structure, Techniques, and Applications

edited by

Hongda Wang

JENNY STANFORD
PUBLISHING

Published by

Jenny Stanford Publishing Pte. Ltd.
101 Thomson Road
#06-01, United Square
Singapore 307591

Email: editorial@jennystanford.com
Web: www.jennystanford.com

British Library Cataloguing-in-Publication Data
A catalogue record for this book is available from the British Library.

Biomembranomics: Structure, Techniques, and Applications

ISBN 978-981-4968-61-4 (Hardcover)
ISBN 978-1-003-45635-3 (eBook)

Contents

Preface

The basic functional unit of life is molecules, which work together in an orderly manner to form a "molecular society." Cells that construct the structure of life are defined by the biomembrane system. Thousands of molecules, including phospholipids, proteins, and sugars, combine in different ways to form a complex, diverse, and dynamic membrane structure. Biomembranes form an essential barrier between living cells and the surrounding world, as well as serve to compartmentalize intracellular organelles within eukaryotes. Cell communication and signaling, material transport, energy conversion, and viral and bacterial infections are a few examples of processes involving biomembranes. The abnormal expression and distribution of some membrane proteins are even closely related to genetic diseases, neurodegenerative diseases, and malignant tumors.

Although the importance of membranes has been widely recognized and studied, there is still a lack of comprehensive and systematic research on membranes, such as genomics, proteomics, and metabolomics. Moreover, the broad sense of biomembranes is not limited to the plasma membrane, but organelle membranes and endocytosis or exocytosis vesicle membranes are also derived from the biomembrane system. Therefore, we introduce the concept of "omics" into membranes in this book and propose "biomembranomics," which regards the structure, function, and intermolecular interaction of the membrane system as an organic whole. More importantly, we pay more attention to the fine structure and dynamic function of membranes at the three-dimensional and single-molecule levels.

The book has 12 chapters. Chapter 1 introduces the structure, properties, and functions of membranes and proposes the concept of "biomembranomics." Chapter 2 introduces mass spectrometry, an ensemble technology for studying omics, and summarizes the research progress of mass spectrometry in exploring the types and contents of membrane proteins, phospholipids, and glycans.

From Chapter 3 to Chapter 12, we update the latest development in structural analysis techniques of biomembranes, including atomic force microscopy, super-resolution fluorescence imaging, single-molecule fluorescence techniques, cryo-transmission and cryo-scanning electron microscopy, infrared and Raman spectrum, single-molecule force spectroscopy and force probes, and computer simulations. These 10 chapters focus on the organization and function of biomembranes at the single-molecule level, beginning with simple erythrocyte membranes and extending to complex somatic cell membranes and organelle membranes. The properties, localizations, and interactions of membrane components during the dynamic processes of membrane transport and vesicle traffic are also summarized and discussed. Meanwhile, experimental methods currently used in membrane research are presented in both text and illustration in every chapter.

The purpose of this book is to reach a consensus on the organizational and functional properties of biomembranes at the three-dimensional and single-molecule levels by integrating different methods/disciplines, that is, the concept of "biomembranomics." It holds significant potential for the future development of targeted drugs for membrane-related diseases. We hope the book will be of interest to the readers and extend their studies in this field since its view of biomembranomics and new approaches set new priorities for a deep understanding of biomembrane organization at the single-molecule level and future study.

We are deeply grateful to all the authors for voluntarily contributing to the efforts that brought this work to fruition. We also want to express our gratitude to the editors for their constant support and guidance during the editing process.

This book is dedicated to Harry and Memary at the new starting point of life science research.

Hongda Wang

Summer 2023

Acknowledgment

We dedicate this book to Prof. Erkang Wang on his 90th birthday and thank him for his strong support for biomembrane research.

Prof. Erkang Wang was born in Jiangsu, PRC, in 1933. He graduated from the University of Shanghai in 1952 and obtained a candidate degree (PhD) under the guidance of Nobel Prize laureate Prof. J. Heyrovsky in the Czechoslovak Academy of Sciences in 1959. He is a member of Chinese Academy of Sciences (CAS) since 1991 and of the Academy of Sciences for the Developing World (formerly called the Third World Academy of Sciences, TWAS) since 1993. He is an honorary member of the Japanese Analytical Chemistry Society since 2006. He was a visiting professor at the University of Houston (USA), Dijon University (France), Kyoto and Yamanashi Universities (Japan), and the Hong Kong University of Science and Technology. He is the former head of the Changchun Institute of Applied Chemistry (CIAC), CAS. He has been the editor-in-chief of the *Chinese Journal of Analytical Chemistry* and an editorial board member of nine international SCI journals, including *Analytica Chimica Acta*, *Talanta*, *Electroanalysis*, and *Chemistry: An Asian Journal*. He has been the chair of numerous international and domestic conferences in analytical chemistry and electroanalytical chemistry and has presented over 100 plenary, keynote, and invited lectures in international symposia as well as over 200 seminars in 27 countries and regions. He has directed 15 postdocs and over 100 graduates among which 3 have received national hundred excellent doctorate thesis awards, 5 special President Awards, and 4 excellent doctorate thesis awards from CAS.

He is the winner of the National Science Congress Award, 7 CAS and ministerial Awards, the first Special Outstanding Award in Jilin Province, 4 National Natural Science Awards, many CAS Excellent Graduate Supervisor Awards, and the World Intellectual Property Organization Award. He is also the first winner of the 10th Kharazmi International Festival. He is known for his advancements in bioelectrochemistry, electroanalytical chemistry, sensors, biosensors, hyphenated interface with separation technics, microfluidics, and nanozymes. He has published over 1000 papers in international SCI journals that have been cited over 62,000 times with h index 120. He also has more than 40 invention patents and published more than 60 monographs and translations. He has been selected as the global "highly cited researcher" 9 times over 19 years (2002–2020) by the ISI Web of Science. He is a recipient of the Lifetime Achievement Awards for electrochemistry at the 19th National Electrochemistry Conference (2017), for electroanalytical chemistry at the 13th National Electroanalytical Chemistry Conference (2017), for chemical sensors at the 15th National Chemical Sensor Conference (2021). These achievements, patents, papers, monographs, and translations are the crystallization of his hard work for more than half a century, which is of far-reaching significance to the development of electroanalytical chemistry.

Prof. Wang is the founder of the State Key Laboratory of Electroanalytical Chemistry, Changchun Institute of Applied Chemistry, PRC. Over the past few decades, he has devoted himself to science and made fundamental contributions to the innovation and development of the laboratory. Throughout the year, he takes the lab as his home, without weekends and holidays. He fully embodies the lofty quality and scientific spirit of loving science, hard work, determined innovation, selfless dedication, and self-improvement. Today, he still actively participates in various scientific research and academic activities, consults the latest literature and reports every day, and personally guides students' experiments and thesis writing. With his rigorous, realistic, and tireless pursuit of science, he has set an example for scientific researchers and young students and is a model for us to learn all our life.

On the occasion of Prof. Wang Erkang's 90th birthday, we sincerely wish him health and happiness and his tree of scientific research evergreen!

Chapter 1

Biomembranomics: New Concept

Jing Gao and Hongda Wang
State Key Laboratory of Electroanalytical Chemistry, Research Center of Biomembranomics, Changchun Institute of Applied Chemistry, Chinese Academy of Sciences, Changchun, Jilin, China
gaojing@ciac.ac.cn, hdwang@ciac.ac.cn

The membrane is the basis for cells to perform a variety of life activities. It is of great biological significance to study the structural characteristics and functional mechanisms of membranes. In this chapter, we summarize the composition, structure, and mechanical properties of membranes, as well as the relationship between membrane function and disease, and discuss the strengths and weaknesses of current research methods. Based on this, we propose a new concept of "biomembranomics," which regards the structure, function, and intermolecular interaction of membrane systems as an organic whole, to establish a new and real membrane structure at the molecular level by integrating different methods/disciplines. We hope to provide guidance for the accurate and effective development of targeted drugs for membrane-related diseases.

Biomembranomics: Structure, Techniques, and Applications
Edited by Hongda Wang

ISBN 978-981-4968-61-4 (Hardcover), 978-1-003-45635-3 (eBook)
www.jennystanford.com

1.1 Introduction

Biological membranes are unique and complex physical systems. They form an essential barrier between living cells and their external environments, as well as serve to compartmentalize intracellular organelles within eukaryotes. The membranes are also fundamental building blocks regulating an extensive repertoire of biological functions, such as cell communication and signaling, material transport, energy conversion, embryo and histogenesis, tumor homeostasis, and viral and bacterial infections [1–3]. The membranes comprise a mixture of lipids and proteins that arrange in a lipid bilayer [4]. Lipids are the basic scaffold of the membranes, and their composition determines membrane curvature, stiffness, and fluidity [5]. Aside from the structural role, lipids function mainly as solvents and provide the proper microenvironment for membrane proteins to attach [6]. Membrane proteins are the main executors of membrane functions, and they play a variety of roles related to material transport, receptors, enzymes, immunity, and interactions with extracellular matrix [7].

The compositional, structural, and functional complexity of membranes often catches researchers by surprise. In terms of composition, both lipids and proteins have hundreds or even thousands of species, and about one-third of the genome code for membrane proteins [8]. For a static structure, phospholipids, which have hydrophobic tails and hydrophilic heads, spontaneously assemble into a ~5 nm wide bilayer, and proteins span or attach to the periphery of the bilayer. However, lipid movement, including lateral diffusion, rotation, and flip-flop, and the interactions between proteins and between proteins and lipids are formidably challenging. Functionally, quantitative changes in the main groups of lipids and the loss of lipid asymmetry are observed in all neoplastic cells as compared to non-malignant cells [9]. Therefore, lipid profile can be considered a crucial prognostic factor in cancer patients [10]. As to membrane proteins, identification of these proteins with aberrant properties can result in the discovery of novel therapeutic targets [11]. Actually, they represent 60% of the known existing and future drug targets [12].

Due to the vital roles of biological membranes in a variety of activities, scientists have been trying intensively to understand and reveal all the knowledge related to membranes. From the discovery of the lipid bilayer membrane by Gorter and Grendel in 1925 [13] and the classical fluid mosaic model proposed by Singer and Nicolson in 1972 [14] to modified fluid mosaic model [15] and protein layer-lipid-protein island (PLLPI) model [16] in recent years, the research on membranes has a history of 100 years, and the techniques applied and developed during this period are becoming more and more advanced and accurate. Nevertheless, it is still a great challenge to completely uncover the mystery of membrane structure and functional mechanism. One is due to the physical properties of the membrane. The diversity of composition, the fluidity of membrane, and the limitations of membrane proteins in dissolution, separation, and identification make it difficult to study in depth. The other is due to the lack or immaturity of technology and methods. Because single-molecule, in situ, three-dimensional, and dynamic investigation of membranes requires not only experimental approaches to maintain the integrity and activity of membranes, but also analytical instruments with high sensitivity and spatiotemporal resolution. Apparently, a single technique can hardly meet these requirements simultaneously. Thus, the combination of two or more techniques is essential to tackle these issues and bridge the biophysical observations to the biological function.

In this part, we will summarize the research status of membranes and discuss the strengths and weaknesses of related techniques by reviewing the molecular composition, dynamics and compartmentalization, and mechanical properties of membranes and the relationship between membrane functions and diseases. On this basis, we will propose a new concept of "biomembranomics," a molecular level research system, in order to obtain systematic information such as single-molecule structure, molecular interactions, molecular dynamic function, overall membrane structure, and different membrane system structures. This, we hope, will unify the previous scattered research directions and establish a new system and overall framework for membrane research, so as to realize clinical diagnosis and treatment guidance.

1.2 Composition of Membranes

The mammalian membrane consists of a mixture of lipids and proteins that form the outer and inner leaflets of an asymmetric bilayer (Fig. 1.1) [17]. This lipid bilayer is afforded mostly structural and barrier roles and provides a platform for sensing the extracellular environment. The remainder portion of the membrane is proteins. Some of them are embedded on the membrane surface, and some are cross the lipid bilayer. They are randomly distributed and aggregated, forming a mosaic-like arrangement. The types of proteins, the asymmetry of protein distribution, and the synergy with lipids endow biomembranes with their respective characteristics and functions.

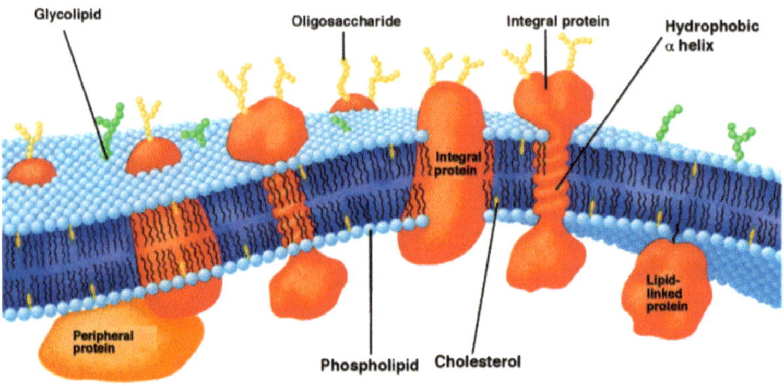

Figure 1.1 Schematic representation of the bilayer fluid mosaic model of the cell membrane. Integral proteins are embedded in the bilayer composed of phospholipids and cholesterol. Reprinted with permission from Ref. [17]. Copyright 2002 American Chemical Society.

1.2.1 Lipids

Lipids are the basic components of biomembranes. There are about 10^9 lipid molecules in mammalian cell membrane and about 5×10^6 lipid molecules per square micron of plasma membrane. The lipid fraction of the membrane could be divided into phosphoglycerides, sphingolipids, glycolipids, and cholesterol [18].

1.2.1.1 Phosphoglycerides

Phosphoglycerides constitute the basic components of membrane lipids, accounting for more than 50% of the whole membrane lipids. A typical phosphoglyceride molecule has a polar head and a nonpolar tail formed by two fatty acid chains. The carbon chains of fatty acids are even, and most of them are composed of 16, 18, or 20 carbon atoms. According to the different head groups, phosphoglycerides can be mainly classified into four types: phosphatidylcholines (PC), phosphatidylethanolamines (PE), phosphatidylserines (PS), and phosphatidylinositols (PI) (Fig. 1.2) [19].

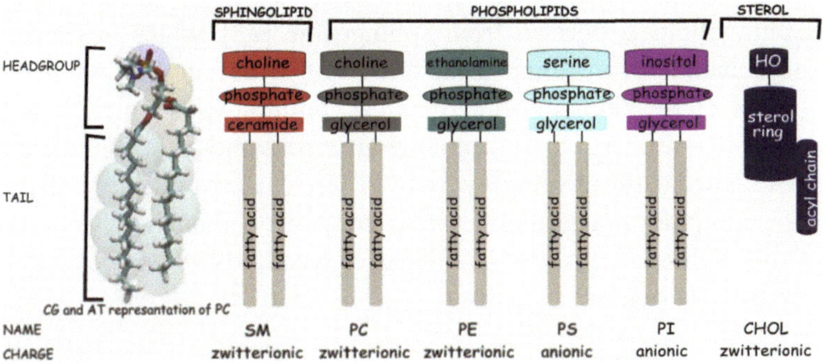

Figure 1.2 Coarse-grained and all-atom representation of a phosphatidylcholine lipid and the major lipids comprising the plasma membrane. SM: sphingomyelin; PC: phosphatidylcholine; PE: phosphatidylethanolamine; PS: phosphatidylserine; PI: phosphatidylinositol; CHOL: cholesterol. Reprinted from Ref. [19], Copyright 2021, with permission from Elsevier.

PC is the most abundant compound of the lipid bilayer [20], mostly present in the outer leaflet of the plasma membrane and higher ester/ether form ratio [20]. Although it is primarily a neutral lipid, it can also be referred to as a zwitterion, as it consists of a positively charged choline head group and a negatively charged phosphate group substituent [21].

PE is the second most common membrane phospholipid, occurring in ester or ether form [20]. It is present on both sides of the plasma membrane but exhibits the greatest concentration in the cytosolic leaflet of the lipid bilayer [21]. Due to its wedge shape, PE is responsible for plasma membrane curvature modulation [22].

PS under physiological pH is defined as negatively charged lipid, present in the intracellular layer of the cell membrane [21]. Like PE, PS is mostly composed of palmitic, oleic, and stearic acids attached at the C1′ glycerol position and residues of oleic acid in the C2′ position [20].

The PI family is a heterogeneous group of membrane lipids that differ from each other with regard to fatty acid moieties composition and the number of phosphate groups attached to polar myoinositol head group [21]. It is an anionic lipid, present in the outer leaflet of the membrane [23]. Due to the variety of phosphoinositol metabolism, the membrane level of PI is volatile [20, 24].

1.2.1.2 Sphingolipids

Sphingolipid is derived from sphingosine [23], which has a long acyl chain with an amino group and two hydroxyl groups at one end (Fig. 1.2). In sphingomyelin, the most common sphingolipid, a fatty acid tail is attached to the amino group, and a phosphocholine group is attached to the terminal hydroxyl group. Thus, the overall structure of sphingomyelin is quite similar to that of phosphatidylcholine. Other sphingolipids are amphipathic glycolipids whose polar head groups are sugars.

1.2.1.3 Glycolipids

Glycolipids are mostly composed of ceramide or phosphatidic acid conjugated with glycan group [21]. The simplest glycolipid is cerebroside, with only one glucose or galactose residue connected to ceramide. More complex gangliosides can contain up to 7 sugar residues, including different numbers of sialic acids. Although glycolipids account for less than 5% of the total membrane lipids, they are ubiquitous on the plasma membrane. And the types of glycolipids in different cells are different. For example, neuron cells have gangliosides and human erythrocytes have ABO blood group glycolipids.

1.2.1.4 Cholesterol

Cholesterol is a steroid compound. It has a sterol ring, a hydroxyl substituent, and an acyl chain (Fig. 1.2). As its hydroxyl group can interact with water, cholesterol is amphipathic. In lipid bilayer, cholesterol molecules are oriented by their hydroxyl groups close to the polar head groups of adjacent phospholipids. The level of cholesterol generally does not exceed one-third of the membrane lipids [25]. Due to its small molecular volume and excessive hydrophobic interactions with fatty acids, cholesterol plays an important role in regulating membrane fluidity, increasing membrane stability, and reducing the permeability of water-soluble substances [26, 27]. Moreover, it is also the basic structural component of lipid raft.

1.2.2 Membrane Proteins

In addition to lipids, proteins are also loaded on the membrane and account for ~50% mass ratio of most cell plasma membrane. This means that there are 50 lipid molecules to one protein molecule, because lipids are relatively small and proteins are large in comparison. Depending on their mode of association with lipids, membrane proteins can be broadly classified into integral, peripheral, and lipid-anchored membrane proteins (Fig. 1.3) [28].

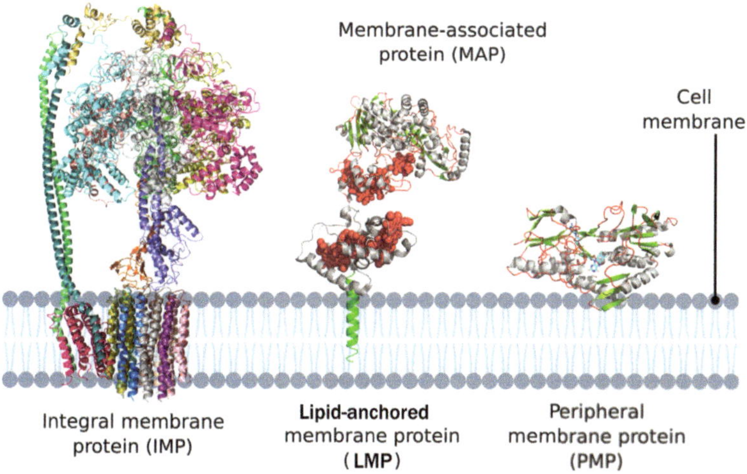

Figure 1.3 Different types of membrane proteins shown in the cell membrane. Integral membrane proteins are shown to span the entire membrane. Peripheral membrane proteins only associate with the membrane via electrostatic or hydrophobic interactions, and lipid-anchored membrane proteins anchor themselves in the lipid bilayer using a hydrophobic segment that does not span the entire membrane. Reproduced from Ref. [28] under Creative Commons Attribution (CC BY) license. Copyright 2021, MDPI (http://www.mdpi.org).

1.2.2.1 Integral membrane proteins

Integral membrane proteins comprise single or multiple transmembrane (TM) domains. These can be classified as type 1, type 2, or multipass proteins (Fig. 1.4) [29] and are estimated to encompass 10, 10, and 11–12% of the genome, respectively [30]. Type 1 proteins are oriented with their N-terminus facing the luminal or extracellular space in an organelle, while type 2 proteins have the opposite orientation. Multipass proteins can have either or both of their N- and C-terminus facing either the luminal or extracellular space.

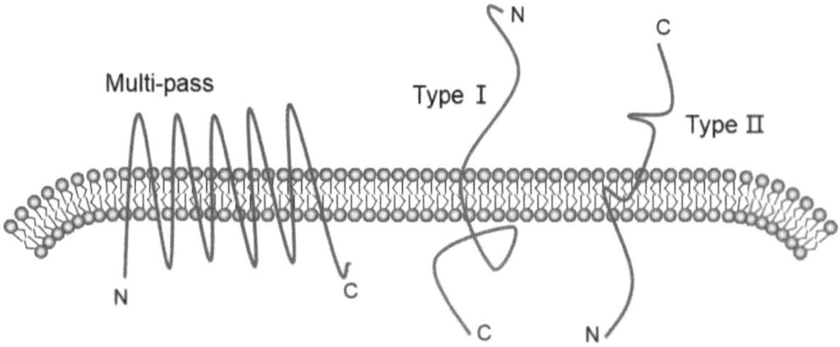

Figure 1.4 Different orientations of integral membrane proteins that embedded in the lipid bilayer.

There are three main binding modes between internal membrane proteins and lipids: (1) TM domain forms a hydrophobic interaction with the lipid bilayer; (2) positively charged amino acid residues at both ends of the TM domain form ionic bonds with negatively charged polar heads of phospholipids; (3) some membrane proteins can covalently bind fatty acid through cysteine residues on the cytoplasmic side and insert into the lipid bilayer.

TM domain is an important site for binding to lipids. These membrane-spanning regions are formed as hydrophobic stretches of nonpolar amino acids that transverse the bilayer as α-helices. The hydrophobicity increases with the number of membrane-spanning α-helices. Yet other TM domains form β-barrels as pores. These β-barrels consist of alternate polar and nonpolar amino acids facing the aqueous channels or lipid bilayer, respectively. The hydrophobicity is considerably lower than several membrane-spanning α-helices.

1.2.2.2 Peripheral membrane proteins

Peripheral membrane proteins are those that do not transverse the membrane but associate with the membrane surface to varying extents and with various moieties. These proteins are usually bound to the membrane indirectly by interactions with integral membrane proteins or directly by interactions with lipid polar head groups.

There are various forms in which proteins can directly interact with the phospholipids. One of the common anchors is via a covalent linkage such as glycosyl-phosphatidylinositol (GPI). GPI-anchored proteins lack a TM domain, have no cytoplasmic tail, and are, therefore, located exclusively on the extracellular side of the plasma membrane. These proteins include membrane-associated enzymes, adhesion molecules, activation antigens, and differentiation markers.

Another class of peripheral membrane proteins are attached to the cytosolic face of membranes by a hydrocarbon moiety, such as prenyl, farnesyl, and geranylgeranyl groups, covalently attached to a cysteine near the protein C-terminus.

1.2.2.3 Lipid-anchored membrane proteins

Lipid-anchored membrane proteins are anchored on the cell membrane by covalently linking to lipid molecules (such as fatty acids or glycolipids). Those bound to fatty acids are mostly distributed on the inner side of the plasma membrane, whereas those bound to glycolipids are mostly distributed on the outer side of the plasma membrane. These anchors are necessary for protein function.

1.3 Membrane Compartmentalization and Dynamics

Biological membranes are not homogenous mixtures of lipids and proteins; rather, they are segregated into different physical and/or biochemical domains. It is now well recognized that membranes contain a variety of such domains and that these domains impose a local structure that prevents free mixing of the lipid and protein components of the membrane [31]. Dynamic compartmentalization is ubiquitous for regulating the spatiotemporal organization of the living cell membrane from nanoscale to micron-scale [32, 33]. This non-random organization is intricately linked to cell function.

1.3.1 Lipid Rafts-Dynamic Domains

One of the most widely studied subdomains of biological membranes is lipid rafts. According to the lipid raft hypothesis [34], these rafts are dynamic domains enriched with cholesterol and sphingolipids as well as a variety of proteins, including lipid-linked proteins such as glycosylphosphatidylinositol (GPI)-anchored proteins and signaling molecules (Fig. 1.5) [35]. These domains form tightly packed, more ordered and thicker assemblies, coexisting within the less tightly packed, non-raft membrane regions. Their function can be described as a dynamic platform for lateral protein sorting, which is capable of facilitating or inhibiting interactions with other biomolecules [36]. On living cell membranes, lipid rafts are thought to be essential for various physiological roles, such as signal transduction and membrane trafficking [2, 37].

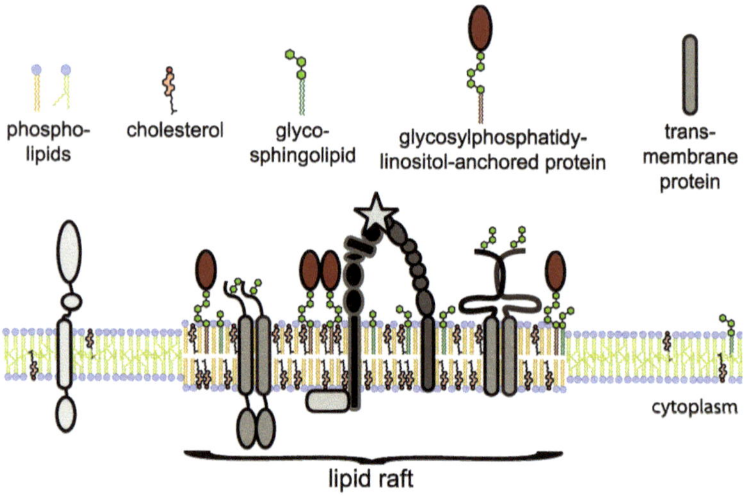

Figure 1.5 Schematic of lipid raft organization. The asymmetric plasma membrane contains phospholipids, glycosphingolipids, cholesterol, and protein receptors that are organized in the thicker lipid microdomains. Lipid rafts float freely in the plasma membrane while being packed tighter and more ordered compared to non-raft regions. Reproduced from Ref. [35] under Creative Commons Attribution (CC BY) license. Copyright 2020, MDPI (http://www.mdpi.org).

With the introduction of the liquid-ordered (Lo) and liquid-disordered (Ld) domains [38], lateral membrane compartmentalization was supported by model lipid membrane [39]. Sterols and saturated lipids associate to a Lo phase, forming domains of macroscopic as well as nanoscopic sizes, whereas domains of the Ld phase contain highly unsaturated and very short lipids and thus stabilize phase separation. This mechanism was further found to govern the lipid membrane organization in simulations [40], on mimetic membranes [41], on isolated cell membranes [42], and on giant plasma membrane vesicles [43]. The membrane raft concept was profound in 1997 [34]. However, undoubted nanoscale visualization of membrane rafts in the unperturbed living cell membranes is still uncompleted, increasing the skepticism about their existence.

1.3.2 Approaches to Resolve Membrane Dynamics

Given the notorious difficulty in detecting membrane rafts in cells, the raft model has also driven advances in technology to develop techniques capable of imaging biomolecules with increasing spatial and temporal resolution. We here summarize three main approaches for resolving membrane dynamics (Fig. 1.6) [44], especially for observing lipid rafts: (1) imaging by super-resolution fluorescence microscopy (SRFM), (2) single-particle tracking and correlation techniques, and (3) imaging mass spectrometry to determine the composition.

1.3.2.1 Imaging by SRFM

Membrane domains such as rafts have long been accepted to have an average size of 10–200 nm, and they are dynamic structures with sub-millisecond lifetimes. Traditional optical techniques such as confocal microscopy, Föerster resonance energy transfer (FRET) [45], or fluorescence recovery after photobleaching (FRAP) [46] are limited by the diffraction limit of light to 200–300 nm. Fortunately, a wealth of super-resolution fluorescence techniques have emerged that readily allow the study of membranes at the nanometer level.

The SRFM usually exploits the photophysical properties of fluorophores together with different illumination schemes. (The main principles and applications of these techniques are explained in Chapter 4.) They have achieved a lateral resolution of ~50 nm for saturated structured illumination microscopy (SSIM), ~35 nm for stimulated emission depletion (STED) microscopy [47], and ~20 nm for photoactivatable localization microscopy (PALM), stochastic optical reconstruction microscopy (STORM), and points accumulation for imaging in nanoscale topography (PAINT) [48], and an unparalleled lateral resolution of 2–6 nm in living cells for minimal emission fluxes (MINFLUX) [49]. As such, they have provided insights into nanoscale regions of the membrane with much greater spatial resolution. For example, direct visualization of cholesterol-enriched nanodomains in living cells has been obtained with STED nanoscopy, providing direct evidence of the existence of lipid raft in living cells [50]. Additionally, SRFM techniques have also been used to determine the heterogeneity of the membrane structure and dynamics [51].

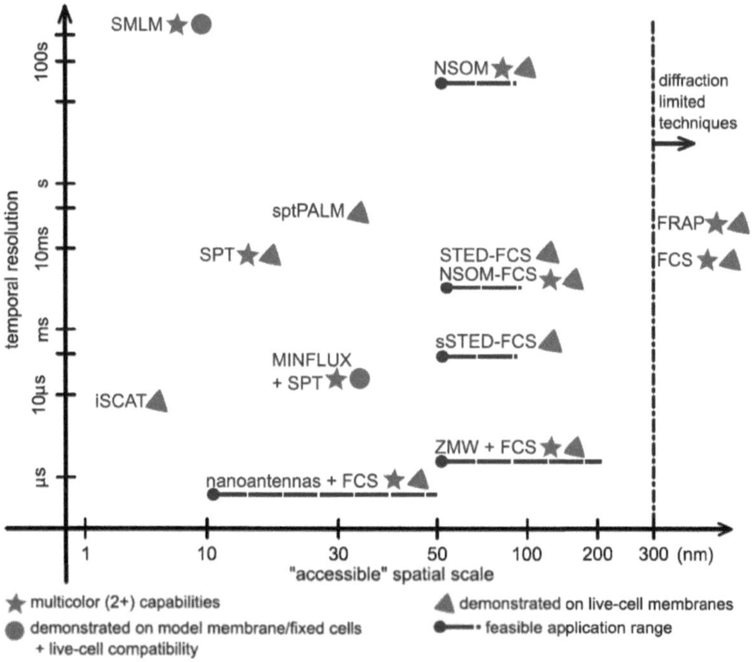

Figure 1.6 Approaches for membrane dynamics studies with respect to the accessible spatial scale and temporal resolution. FCS: fluorescence correlation spectroscopy, FRAP: fluorescence recovery after photobleaching, SMLM: single-molecule localization microscopy, NSOM: near-field scanning optical microscopy, SPT: single particle tracking, sptPALM: single particle tracking photo-activated localization microscopy, (s)STED: (scanning) stimulated emission microscopy, iSCAT: interferometric scattering microscopy, ZMW: zero-mode waveguide. Reprinted from Ref. [44], an open-access article published by Portland Press. Copyright 2021 Winkler, P.M. and Garcia-Parajo, M.F.

It is noteworthy that, to our knowledge, no definitive direct nanoscale imaging of all the lipid components forming membrane rafts and their dynamics in living cells has been put forward yet. First, the acquisition of images with high enough temporal resolution (<1 ms) for imaging the highly dynamic membrane rafts is quite a difficult task. Because SRFM techniques usually require long acquisition times, samples have to be fixed. Second, it needs highly stable, small-sized fluorescent probes that are suitable for SRFM and that do not perturb the natural properties of the membrane rafts, which is also challenging.

1.3.2.2 Single-particle tracking and correlation techniques

To observe the highly dynamic membrane rafts, SRFM can be combined with single-particle tracking and correlation techniques to obtain information about the kinetics of membranes. Some of the most common approaches include fluorescence correlation spectroscopy (FCS), raster image correlation spectroscopy (RICS), single-particle tracking (SPT), and the tracking mode in MINFLUX.

FCS is an autocorrelation-based analysis that provides information about the kinetic parameters of molecules. Several FCS-based derivative techniques have been developed and applied to the field of membrane research, such as fluorescence cross-correlation spectroscopy (FCCS) [52], scanning FCS [53], two-focus FCS [54], spot variation FCS [55], or z-scan FCS [55]. Moreover, the combination of STED with FCS has been a major milestone. STED-FCS resolved the transient trapping (<10–20 ms) of sphingolipids and GPI-anchored proteins (GPI-APs) into cholesterol-dependent nanoscopic domains as small as 30 nm in size [56]. RICS makes use of the time information contained in the movement of the scanning laser beam to extract information about the dynamics of the studied molecules. When combined with STED, RICS can be used in living cells to obtain information of the molecule dynamics with high spatial resolution in a wide dynamic temporal range [57].

An alternative approach is SPT, which can track individual molecules as they diffuse on the plasma membrane with nanometer localization precision (see Chapter 6 for details). SPT has provided enormous information on the nanoscale dynamics of raft-associated GPI-APs on the cell membrane. It has revealed that GPI-APs continuously assemble in transient (~200 ms) homodimers likely constituting a possible basic unit of raft domains (~3–15 nm in size) [58]. Subsequently, using dual-color SPT, it was shown that gangliosides dynamically interact on the timescale of ~10 ms with monomers and dimers of GPI-AP in a cholesterol-dependent manner [59]. In addition, with live-cell SPT, the rapid exchange of sphingomyelin (SM) between GPI-AP assemblies and the membrane has been shown [60]. These results further confirm the dynamic character of membrane rafts at the millisecond to second timescale. Recently, live-cell SPT has been coupled to PALM, i.e., sptPALM. The approach generated high-density single-molecule maps of the Gag and VSVG membrane proteins and showed different diffusion and clustering behavior [61]. It also resolved the interaction dynamics between membrane and water-soluble proteins on the living cell membrane [62].

Finally, MINFLUX can also be applied to SPT with a temporal resolution below 1 ms [63]. Indeed, MINFLUX tracking mode has been used to track the diffusion of lipid probes in model membranes with a temporal resolution of ~100 μs [64].

1.3.2.3 Imaging mass spectrometry

The lipid composition of membranes can also be studied using secondary ion mass spectrometry (SIMS) and its nanoscale variant NanoSIMS with lateral resolutions of up to ~50 nm. NanoSIMS has been used to study the association of GM1 with cholesterol and SM [65], to identify non-raft sphingolipid-enriched domains [66], and to determine the cholesterol distribution in the plasma membrane of cells [67]. Nevertheless, this technique requires previous fixation of the cells and can only be performed under ultra-high vacuum, limiting its use in membrane dynamic studies in living cells.

1.3.3 Organizers of Membrane Compartmentalization

With the deepening of the research on cell membrane, the compartmentalization and dynamics of membrane have attracted more and more attention as this spatiotemporal organization reflects the native state of cell membrane in vivo and is closely related to its functions. Until now, several local organizers that dynamically organize and are responsible for the plasma membrane compartmentalization have been investigated. First, the lipid and cholesterol content of the membrane is crucial for its lateral organization. Second, proteins play an essential role in regulating, localizing, and templating the lipid environment. Third, the cortical actin cytoskeleton and the extracellular matrix modulate the membrane with a hierarchical order from the meso-down to the nanoscale [68, 69]. Taken together, the cell membrane appears to be highly heterogeneous with multiple components actively and passively interacting with each other at multiple spatiotemporal scales.

1.4 Mechanobiology of the Cell Membrane

As the boundary between a living cell and its microenvironment, cell membrane is also the first sensor for various external stimuli. It forms a key interface for transmitting mechanical stimuli between intra- and extracellular compartments and converting them into intracellular signals. Its molecular composition and physical state enable it to change cell shape and mechanical properties to accommodate any variation in the environmental mechanical conditions. Recently, much attention has been directed to the mechanics of the cell membrane, and in particular membrane tension, due to its critical roles in diverse cellular processes ranging from membrane trafficking [70], cell spreading [71], phagocytosis [72], and cell polarity [73].

1.4.1 Membrane Tension

The membrane can be characterized by its tension, which is defined as the force per unit length across the membrane section. In lipid vesicles, in-plane membrane tension can propagate globally because a lipid bilayer has the properties of a 2D liquid, allowing rapid diffusion of its components. However, in cells, plasma membrane tension can be greater and more local than in pure lipid vesicles because of the following factors: peripheral protein binding, the presence of transmembrane proteins, and interactions with the actomyosin cortex, as well as adhesions between the cell surface. These factors provide additional resistance to membrane area changes and can constrain such changes to a localized area of the cell. The resting tension of the plasma membrane varies depending on the cell type, and the values are within a range of 0.01–0.04 mN/m [74].

Since membrane tension is determined by a balance of mechanical forces, any variation from the environment would lead to a change in tension distribution, leading to a change in the mechanical behavior of the membrane. Recent reports have shown that tension is not equal over the cellular membrane leading to direct physiological consequences. Tension gradients have been shown to be important mechanical regulators for both mechanosensitive ion channel activation [75] or motile cell behavior [76]. Collectively, membrane tension plays a direct role in establishing mechanical properties of the cell.

1.4.2 Methods to Access Membrane Mechanics

Advances in nanotechnology have led to the development of new, innovative solutions to analyze membrane mechanics (Table 1.1) [77]. For example, developments in the field of microfabrication led to the creation of micrometer-sized cantilevers and other mechanical parts that offer greater spatial resolution for probing single cells with forces and displacements down to the pico-scale.

In general, techniques used for measuring the mechanical properties of cells can be subdivided into two categories as summarized in Fig. 1.7 [77]. The first category contains techniques that apply force to the whole cell while simultaneously recording its mechanical response (force-application techniques). The second category refers to techniques used for measurements of forces generated by cells during physiological processes, including development, contraction, and migration (force-sensing techniques) [78].

Techniques and methods used to apply controlled forces on cells include optical tweezers (laser traps) [79], magnetic tweezers [80], atomic force microscopy (AFM) [81], microneedle manipulation [82], cell stretchers, and flow rheometry [83]. To improve on experimental throughput and the span of force-loading rates, a new technique was developed, called acoustic force spectroscopy (AFS) [84]. Several techniques have also been developed to monitor the ability of cells to interact with and to generate forces on their local environment. These techniques include tracking force microscopy (TFM) [85] and micropillar arrays [86]. Another promising technique is particle-tracking microrheology, used to measure the viscoelastic properties of the cell at high spatiotemporal resolution [87].

Table 1.1 Techniques used for measurements of membrane mechanics

Tool	Force Application	Force Sensing	Spatial Sens.	Cell State	Force Directionality
Micropipette aspiration	1.1–10^3 nN	—	1–100 μm	Suspended	Pull
Microneedle manipulation	1 pN	—	1–10^3 μm	Suspended	Push?
AFM	10–10^7 pN	10–10^5 pN	1–10^5 nm	Adherent	Push/pull
Optical tweezer	0.01–10^3 pN	1–100 pN	10–10^5 nm	Adherent	Push/pull
Optical stretcher	10–10^3 pN	—	10–10^6 nm	Suspended	Pull
Magnetic tweezer	0.1–10^4 pN	10–10^3 pN	0.1100 μm	Adherent	Pull
Magnetic twisting cytometry	1–100 pN	—	1–10^3 μm	Adherent	Angular
Acoustic force spectroscopy	10^{-4}–10^3 pN/s	0.01–10^2 pN	—	Adherent	Pull
Traction force microscopy	—	10–10^6 pN	—	Adherent	Push/pull
Micropillar arrays	—	0.01–10^3 nN	1–10^3 μm	Adherent	Push/pull

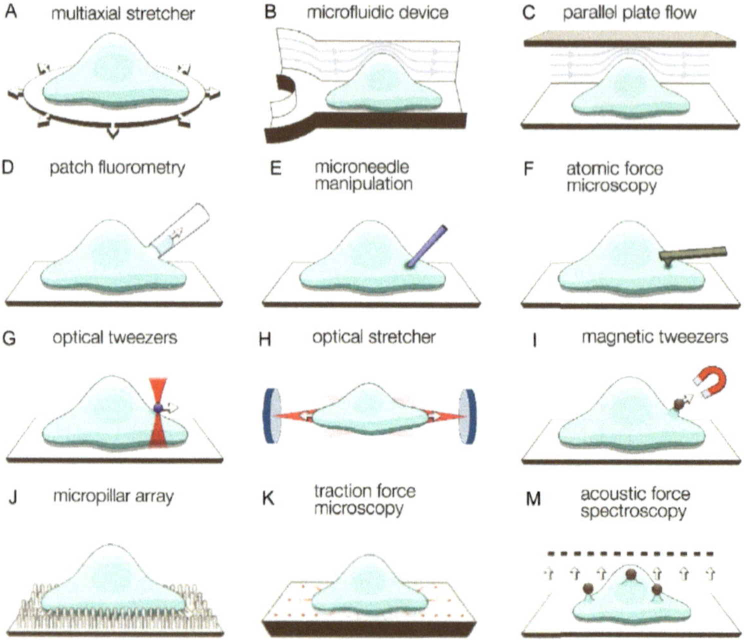

Figure 1.7 Force-application and -sensing techniques. Reprinted from Ref. [77], Copyright 2020, with permission from Elsevier.

Thanks to these sophisticated experimental techniques, our ability to study fundamental biological processes has been revolutionized over the last 20 years. However, due to various force directionalities, magnitudes, loading rates, and modes of operation, different techniques stimulate different mechanical responses in cells. It is, therefore, important to evaluate the factors that can influence these measurements to choose the most appropriate experimental tool.

1.5 Membrane Functions and Diseases

As mentioned above, plasma membranes serve as barriers and gatekeepers that possess crucial functions and associates with many physiological and pathological processes. All the components that make up the cell membrane play an important role in this process. Membrane proteins encompass a wide variety of functions required for normal development and physiology and when disrupted lead to a range of diseases. Identification of membrane proteins with aberrant properties can result in the discovery of novel therapeutic targets [11]. Membrane lipids and lipid structures participate directly as messengers or regulators of signal transduction. In addition, protein–lipid interactions participate in the localization of signaling protein partners to specific membrane microdomains. Thus, lipid alterations change cell signaling that are associated with a variety of diseases, including cancer, obesity, neurodegenerative disorders, cardiovascular pathologies [19], etc.

1.5.1 Membrane Proteins as Biomarkers and Drug Targets

Change in membrane proteins is one of the first events that take place under pathological conditions, making plasma membrane proteins a likely source of potential disease biomarkers with prognostic or diagnostic potential. Membrane proteins are also potential targets for monoclonal antibodies and other drugs that block receptors or inhibit enzymes essential to the disease progress. Approximately, 38% of all proteins encoded by the mammalian genome and more than one-third of the current list of biomarker candidates are membrane proteins.

1.5.1.1 Membrane receptors

Membrane receptors are the best known and characterized members and mostly represent integral membrane-protein families. Receptors in the plasma membrane have one or more membrane-spanning hydrophobic domain and ligand-binding regions for recognizing mostly extracellular ligands.

The receptor tyrosine kinases (RTKs) have single transmembrane domains and are activated in most cases by dimerization, resulting in an activated intracellular tyrosine kinase subunit. The human genome contains almost 60 genes coding for RTKs, which participate in a great deal of cellular functions—regulating cell survival, metabolism, proliferation, or differentiation. Dysfunctions of RTKs are involved in developmental diseases, diabetes, atherosclerosis, and cancer development [88].

The G-protein-coupled receptors (GPCRs) comprise another large and well-characterized family of membrane receptors—represented by approximately 800 genes in the human genome. In addition to their key role in major signaling pathways, such as sensing (e.g., smell, light, hormones, or neurotransmitters), numerous drugs and toxins also act through GPCR proteins. Most of the currently available medicines, whether agonists, antagonists, inverse agonists, or allosteric modulators, interact with GPCRs [89].

Channel proteins are key modulators of cellular ion and water balance, execute information transfer, and regulate membrane potential in a variety of cells. Neurotransmitter ion channels are important drug targets and thus relevant biomarkers in major brain diseases [90]. The potassium channels in cardiac tissues, HERG and KCNQ1, in most cases responsible for long QT syndrome and related drug side effects, are exceptionally important personalized biomarkers in pharmaceutical development projects [91]. An important ion channel complex, formed by the ABCC8 protein and the Kir6.1 potassium channel, senses cellular ATP levels and regulates insulin secretion in pancreatic β-cells [92].

1.5.1.2 Membrane transporter proteins

Membrane transporter proteins compose another large group of potential biomarkers and therapeutic targets, as approximately 10% of all human genes are transporter related.

Active, ATP-dependent transporters include the P-type ATPases, responsible for the energy-dependent transport of various ions. The Na-K ATPase family has a key role in the establishment of unequal ion distribution, that is, the high cellular K and low Na content. The synergistic function of the plasma membrane Ca-ATPases and the endoplasmic reticulum Ca-ATPases provide the physiologically low intracellular calcium concentration, the H-K ATPase is a basic component of acid production in the gastric epithelia. Although these transporters are essential components of cell survival, partial malfunctions are involved in various disease conditions, and thus they may serve as important biomarkers. The role of Na-K ATPase mutations was indicated in migraines, in renal diseases and tumors; the mutations in PMCA cause heritable deafness and balance disorder [93].

The ATP-binding cassette (ABC) transporters have basic structural similarities, but numerous divergent functions [94]. Primary ATP-dependent active transporters, called multidrug resistance (MDR)-ABC transporters, are involved in the active cellular extrusion of hydrophobic xenobiotics and toxic agents. Because of their important function in cancer drug resistance and general drug and toxin metabolism, we emphasize the potential personalized biomarker properties of these ABC proteins. The expression of MDR-ABC transporters in cancer cells, especially in cancer stem cells, causes multiple drug resistance; thus, the inclusion of these proteins as cancer biomarkers may predict drug response and modify disease prognosis and treatment [95].

Based on evolutionary homology, many membrane transporters are grouped as solute carrier (SLC) proteins, and this huge group includes ion-coupled transporters, exchangers, and so-called passive transporters. Surveying some of the most relevant SLC groups as potential biomarkers, the SLC1A1–7 proteins include Na-dependent high-affinity glutamate and neutral amino acid transporters, involved in neurodegenerative diseases. The SLC2A1–14 proteins are facilitated glucose or other sugar/anion transporters, some of which (SLC2A1–4 [GLUT1–4]) have important alterations in diabetes and in other metabolic diseases.

1.5.1.3 Changes in PTMs of membrane proteins

Post-translational modifications (PTMs) of plasma membrane proteins are frequently altered under pathological conditions, and identification of differences in PTMs has been recognized as a potential way for identifying disease biomarkers [96].

A number of biomarker candidates for cancer, for example, are associated with quantitative and qualitative changes in the glycosylation of plasma membrane proteins [97]. Changes in glycosylation also provide markers for stages in tissue development [98], for cell signaling events [99] as well as for early recognition of potentially harmful reactions after organ transplantation [100]. In Fig. 1.8 [96], different glycoprotein glycan structures associated with cancer are presented. These structures are normally not found in glycans derived from membrane glycoproteins in healthy tissues.

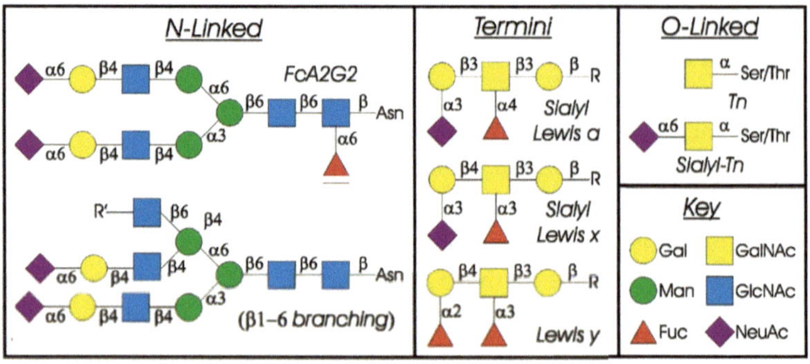

Figure 1.8 Glycoprotein glycan structures associated with cancers. The terminal Lewis structures are often highly expressed in cancer on N- or O-linked glycans as well as on glycolipids. FcA2G2 is not specific to cancer but has been found as a unique glycoform associated with several proteins in hepatocellular carcinoma. The β1– 6 branching glycans are highly expressed in cancers. Tn and sialyl-Tn are aberrant glycans found on mucins as a result of underglycosylation. Gal, galactose; GalNac, N-acetylgalactosamine; Man, mannose; Glc-NAc, N-acetylglucosamine; Fuc, fucose; NeuAc, neuraminic acid (sialic acid). Reprinted from Ref. [96], Copyright 2006, with permission from Elsevier.

Alterations in protein phosphorylation during malignant transformation, or other cellular malfunctions, were a topic of recent investigations [101]. Mausbacher et al. [102] combined glycoproteome enrichment by immobilized lectin affinity chromatography with SILAC-based quantitative phosphoproteomics for the analysis of these PTMs in SCC-9 squamous cell carcinoma cells. Zarling et al. [103] identified phosphopeptides restricted to the surface of solid tumors that were not found in normal tissues. These phosphopeptides were recognized by specific T lymphocytes and were seen as potential targets for new drugs.

1.5.2 Membrane-Lipid Therapy

The type and abundance of lipids in membranes influence their organization and properties, so it is not surprising that disorders in lipid metabolism and transport have a role in human disease. Thus, a new therapeutic approach called membrane-lipid therapy (MLT) has emerged. The molecular targets of MLT are the lipids and the structures they form. The aim of MLT is to regulate cell functions by modulation of membrane-protein localization and activity. Studies have now demonstrated the potential of MLT in the treatment of various human pathologies.

Alterations in membrane-lipid composition and structure is related to various cardiovascular diseases, such as hypertension, atherosclerosis, coronary heart disease, and sudden cardiac death [104, 105]. Oleic acid plays an important role in regulating membrane properties and the activity of GPCRs, G-proteins, and effectors that control blood pressure [106]. This effect, a prototype of the MLT, is highly structurally specific because elaidic and stearic acids (oleic acid congeners) do not induce similar effects. Recently, a synthetic monounsaturated FA that is structurally analogous to oleic acid (2-hydro-xyoleic acid) was shown to exert a potent hypotensive effect [107].

The levels of membrane lipids are altered in cell membranes from patients with cancer and from cancer cells that are resistant to chemotherapy [108]. Therefore, the activity of various anticancer drugs is associated with their ability to alter the membrane-lipid structure. Indeed, daunorubicin, hexamethylene bisacetamide, and minerval all alter the non-lamellar-phase propensity of membranes and PKC activity [109]. Edelfosine and miltefosine are also targeted to membranes [110]. Lipid analogs of conventional antitumor drugs have recently been developed. One example is NEO6002, a less toxic citarabine derivative with a cardiolipin moiety that binds to membranes and overcomes resistance to citarabine in cancer cells [111].

Obesity, a disorder that is reaching epidemic proportions all around the world, is associated with an increase in the incidence of cardiovascular pathologies, cancer, and diabetes. Numerous reports have also shown alterations in the membrane-lipid composition of different cell types in obese subjects. The intake of cis-unsaturated FAs is associated with loss of body weight, whereas the intake of saturated or trans-unsaturated FAs has the opposite effect [112]. Therefore, lipids can regulate appetite and satiety through mechanisms that involve membrane signaling, and they can be used in the treatment of obesity, diabetes, and related metabolic diseases [113]. In addition to these mechanisms, FAs regulate body weight and metabolism through binding to peroxisome proliferator-activated receptors (PPARs).

After adipose tissue, the central nervous system (CNS) represents the largest location of lipids in the body. Certain neurodegenerative disorders are associated with demyelinization or significant alterations in membrane lipids. These facts and the capacity of lipid drugs to interact with myelin and to cross the blood–brain barrier suggest that MLT may play a major role in the treatment of CNS disorders. Alzheimer's disease is the most common neurodegenerative disease resulting in progressive dementia in the elderly. Changes in the lipid composition of membranes have been shown to regulate lipid–protein interactions and the toxicity of β-amyloid peptides [114]. Indeed, it is noteworthy that some treatments and diets supplemented with phospholipids and FAs have beneficial effects on the mental status of patients with Alzheimer's disease and animal models of aging.

In addition, membrane lipids have been implicated in mechanotransduction and related respiratory signaling processes, and lipid-binding molecules can be used in the treatment of respiratory pathologies [115]. Therefore, MLT gives rise to new possibilities for the treatment of diseases that currently still require therapies or alternative or improved therapeutic solutions.

1.5.3 Membrane Rafts in Diseases

Lipids in membrane rafts impart lateral segregation of membrane proteins, thus facilitating the spatial organization and regulation of membrane proteins involved in many cellular processes, such as cell proliferation, apoptosis, and cell signaling. Therefore, abnormalities in membrane rafts and associated molecules have been found in various diseases, including atherosclerosis, diabetes, cancer, and neurodegenerative disorders such as Alzheimer's disease (Table 1.2) [116]. Furthermore, increasing data show that membrane rafts function as the entry and/or budding site in host cells for various bacteria, viruses, and parasites. Here, we describe some examples of how lipid rafts are involved in diseases. It should be noted that caveolae belong to another prominent class of membrane rafts [117]. The major distinction between lipid rafts and caveolae is the presence of caveolin, a coat protein that gives caveolae the characteristic flask-shaped appearance [118].

Table 1.2 Diseases in which lipid rafts are involved

Disease	Viral Infections	Bacterial Infections
Alzheimer's disease	Influenza virus	*Escherichia coli*
Parkinson's disease	HIV-1	*Vibrio cholerae*
Atherosclerosis	Herpes simplex virus 1	*Mycobacterium tuberculosis*
Diabetes	Marburg virus	
Pulmonary hypertension	Epstein–Barr virus	*Clostridium tetani*
Systemic lupus erythematosus	Ebola virus	*Salmonella typhimurium*
Asthma	SV-40	*Campylobacter jejuni*
Niemann–Pick disease	Rotavirus	
Gaucher, Fabry, and Tay-Sachs diseases	Echovirus type 1	
	Enterovirus	
Muscular dystrophy		
Neoplasis		

1.5.3.1 Atherosclerosis

The pathogenesis of atherosclerosis is most often initiated by injury to the vascular endothelium resulting in the accumulation of macrophages and low-density lipoprotein (LDL) at the site of injury. Studies have shown that caveolae are involved in regulating cholesterol transport and the activity of endothelial nitric oxide synthase (eNOS), an important mediator of vasorelaxation, angiogenesis, and vasodilation [119]. Moreover, lipid rafts have also been shown to be involved in platelet-mediated thrombosis at the site of injury and atherosclerotic plaque rupture in the endothelium [120]. These observations suggest that perturbations in caveolae-mediated cholesterol homeostasis, regulation of eNOS activity, and thrombosis may contribute to the development of cardiovascular disease and morbidity.

1.5.3.2 Diabetes

It is known that an impairment of insulin signaling is closely related to the onset of type II diabetes. Lipid rafts have been implicated as critical for the proper compartmentalization of insulin signaling in adipocytes [117]. The localization of insulin receptor (IR) in caveolae has been suggested by the data showing colocalization and the direct interaction of caveolin-1 with IR. Cholesterol depletion, a treatment that disrupts lipid rafts and caveolae, leads to a significantly reduced response to insulin stimulation [121], suggesting a close relationship between lipid rafts and the metabolic signaling of insulin.

1.5.3.3 Cancer

Perturbations in cell signaling are a hallmark of cancer; therefore, it is not surprising that lipid rafts and caveolae are implicated in this disease. Cancer-related proteins and receptors have been identified in lipid rafts and caveolae. Regulation of cancer-related signaling proteins, such as Ras, hetero-trimeric G-proteins, eNOS, src-family kinases, protein kinase C (PKC) alpha, and Neu, can be directly mediated through binding to the scaffolding domain of the caveolin protein [122]. In addition, lipid–lipid interactions can also regulate lipid raft resident proteins. Acylation of proteins by palmitoylation or myristoylation facilitates the integration of these proteins into the rigid lipid environment of lipid rafts.

1.5.3.4 Viral infection

A number of bacteria and viruses utilize lipid raft-associated molecules as receptors for entering host cells (Table 1.2). For example, the receptors (CD4, CXCR4, and CCR5) of human immunodeficiency virus (HIV)-1, an enveloped virus, are associated with lipid rafts in T lymphocytes [123]. Cholesterol depletion decreases the infection efficiency of the virus [123]. Lipid rafts also play a crucial role in assembly and budding processes. When enveloped viruses bud from host cells, they acquire their membranes from the host cell. Lipid analyses of several enveloped viruses, such as influenza viruses and HIV-1, have shown that their envelopes are enriched in raft lipids, suggesting that the viruses bud from raft domains [124, 125].

1.6 Why Should Biomembranomics Be Proposed?

The concept of omics, which was put forward as early as the 1980s, is an important research field to reveal the law of life and human health. At present, omics mainly includes genomics, proteomics, metabolomics, lipidomics, glycomics, and so on. As mentioned above, the complexity and importance of biological membranes have been recognized, but there is still a lack of comprehensive and systematic research on membranes like various omics. Therefore, we introduce the concept of omics into biological membranes, including cell membrane and intracellular membranes, and regard the structure, function, and intermolecular interaction of membrane system as an organic whole.

1.6.1 Significance of Biomembranomics

At present, omics mainly studies the composition and content of the target system by using ensemble technologies such as mass spectrometry and gene sequencing. Although the research systems of various omics have been gradually established, they are essentially large-scale characteristic research at the overall level and cannot be directly used for disease detection and treatment. For example, many diseases involve gene mutations, but gene sequencing cannot be fully applied to clinical treatment. Proteomics provides a variety of mass spectrometry information, but we still cannot use complex data for the clinical diagnosis of diseases. As we all know, the protein is the smallest unit performing various biological functions. Defects and abnormal expression of a single protein will lead to pathological changes in life. Therefore, it is necessary to establish a new omics system to study the structure, localization, and dynamic function of membrane proteins at the molecular level, which will provide accurate design scheme for the development of targeted drugs.

In addition, the previous research on cell membrane is relatively scattered and the method is single. The experimental results obtained from different techniques have thus far not reached a consensus view for the organizational and functional properties of the membrane. Therefore, it is necessary to unify these research directions and establish a new system and overall framework for membrane research. For these reasons, we propose biomembranomics to solve this scientific problem.

1.6.2 Progresses and Limitations of Membrane Research

Cell membrane is an important part of cells. Its research has a history of nearly 100 years. The development of analytical technology has contributed to our understanding of cell membrane. At first, Gorter et al. found that the cell membrane was composed of phospholipid bilayer membrane by extraction technique [13]. With the appearance of electron microscopy, Singer et al. found that membrane proteins were embedded in phospholipid bilayers and proposed the classical flow mosaic model (FMM) [14]. With the development of ultracentrifuges, Simons et al. found that some membrane proteins exist in cholesterol aggregates and proposed the lipid raft model [34]. In fact, different membrane structure models have their own advantages and disadvantages. For example, FMM reveals the fluidity and mosaic of cell membrane but cannot explain the synergistic functions between membrane proteins. Another example is the lipid raft model, which explains the synergistic functions of membrane proteins but does not involve the overall protein distribution. However, many important problems such as the overall distribution of membrane proteins, the relationship between proteins, and the differences of membrane structure in different cells or organelles have not been systematically studied.

The research on the structure of cell membrane is still in the stage of model hypothesis. The main reasons are as follows. One is the complex components of cell membrane, and each component will change dynamically in the process of endocytosis and signal transmission, which is difficult to analyze in situ and in real time. Second, the three-dimensional structure information of membrane proteins cannot be fully obtained. Most of the studies on the synergy between membrane proteins are from the perspective of two-dimensional and specific domains, lacking spatial three-dimensional and in situ analysis. Third, different techniques or methods have limitations, and it is difficult to comprehensively clarify the overall structure and functional mechanism of cell membrane with each method in isolation. For example, although the traditional electron microscopy has a high resolution, the samples need to be stained and fixed before imaging, which destroys the real structure of the cell membrane. For another example, although optical microscopy can realize real-time observation without damaging the sample, its resolution is low and cannot reveal the structure of cell membrane at the molecular level.

1.6.3 Combination of Methods and Interdisciplinary Collaborations

Different from traditional proteomics, biomembranomics not only provides the information of species of membrane proteins, but also focuses on single-molecule localization and interaction. It is a three-dimensional proteomic concept at the molecular level, which specifically covers the interactions and dynamic functions between individual membrane proteins. Therefore, in situ investigations in living cell membranes are important to elucidate the dynamic behaviors of membrane proteins in complex biomembrane environments.

Elucidating the complex coupling of multiple spatial and temporal scales of molecules on the cell membrane requires the application of different techniques, with a particular need for simultaneous nanometer spatial precision and microsecond temporal resolution. The relevant spatiotemporal scales for the fine structure of the cell membrane are in the range of a few tens of nanometers to micrometers, and of tens of microseconds to seconds [31, 126]. However, few experimental approaches can access this combined scale range simultaneously. For example, electron microscopy has superior sub-nanometer image resolution and can image tens to hundreds of microns of specimen area, but only at fixed time points with very limited specific molecular staining options. In contrast, spectroscopy techniques, such as electron spin resonance (ESR) and nuclear magnetic resonance (NMR), can quantify short-range lateral motion of lipids at length scales of a few nanometers at millisecond timescales but cannot be extended to longer spatiotemporal scales. Finally, there exist a range of optical microscopy techniques, including FRAP, various FCS techniques, and SPT. These techniques have moderate to low phototoxicity and high temporal resolution. Nevertheless, their spatial resolution is limited by the diffraction limit of light to 200–350 nm, preventing the visualization of membranes at smaller spatial scales.

With the continuous progress of modern science and technology, new analytical techniques and means are constantly updated and improved. The rapid development of cryo-electron microscopy (cryo-ET), SRFM, AFM, mass spectrometry (MS), NMR, infrared and Raman spectroscopy, machine learning, and molecular dynamics simulation provides a powerful tool for the study of biomembranomics. Moreover, the emergence of some combined techniques and instruments also opens a new way for solving the problem of temporal or/and spatial resolution of membrane structure. Several examples are listed below.

- **Combination of methods:** SPT and the combination of FCS techniques with the super-resolution imaging method STED can directly investigate lateral dynamics at diffraction-unlimited spatial scales ranging from tens to either hundreds of nanometers (FCS) or micrometers (SPT) while simultaneously accessing timescales ranging from a few or tens of microseconds to seconds.
- **Combination of instruments:** AFM and SRFM can provide cell membrane morphology images and single-molecule localization images for the same sample and analyze the composition and distribution of membrane-protein aggregates and the interactions between membrane molecules.
- **Interdisciplinary collaborations:** The fixed-point labeling technology of gene fusion is introduced into cryo-ET imaging of membrane proteins, and the machine learning method is used to solve the high-throughput processing of cryo-ET data. Based on the real in situ structure and interaction of membrane proteins, a new membrane model at the molecular level that breaks through the traditional spatiotemporal scale will be established by molecular dynamics simulation, and the dynamic mechanism will be revealed as well. The whole process involves different experimental methods, new modern analytical techniques, model establishment, structural analysis, functional simulation, and so on. Therefore, interdisciplinary collaborations such as chemistry, biology, physics, and computer science will bypass the disadvantages of using each discipline insolation and provide new ideas for biomembranomics research.

1.7 Conclusion

As the most basic unit of life, a "cell" is defined by a membrane system. The appearance of a cell membrane marks the origin and formation of life and determines how cell life performs and realizes its function through the membrane. Studying and understanding the essentials of cell membranes is of great significance to the origin, evolution, and function of cells and even life. The research on membranes has a history of nearly 100 years. However, due to its complex structure and technical limitations, the cell membrane structure is still in the stage of model hypothesis. Moreover, it is impossible to understand the membrane system in essence only from one aspect or using a certain method/discipline in isolation. Therefore, we put forward the new concept of "biomembranomics," which regards the structure, function, and intermolecular interaction of the membrane system as an organic whole, and pay more attention to the fine structure and dynamic function of membranes at three-dimensional and single-molecule levels. This is also the biggest difference from traditional omics that study the overall level and characteristics of the target system. We hope to reach a consensus on the organizational and functional properties of biomembranes by integrating experimental results obtained from different methods/disciplines, which provides guidance for the accurate and effective development of targeted drugs for membrane-related diseases.

References

1. Hobson, J.P., Liu, S.H., Rono, B., Leppla, S.H., and Bugge, T.H. (2006). Imaging specific cell-surface proteolytic activity in single living cells, *Nat. Methods*, **3**, pp. 259–261.
2. Simons, K. and Toomre, D. (2000). Lipid rafts and signal transduction, *Nat. Rev. Mol. Cell Biol.*, **1**, pp. 31–39.
3. Schlessinger, J., Axelrod, D., Koppel, D.E., Webb, W.W., and Elson, E.L. (1977). Lateral transport of a lipid probe and labeled proteins on a cell membrane, *Science*, **195**, pp. 307–309.
4. Escriba, P.V., Gonzalez-Ros, J.M., Goni, F.M., Kinnunen, P.K.J., Vigh, L., Sanchez-Magraner, L., Fernandez, A.M., Busquets, X., Horvath, I., and Barcelo-Coblijn, G. (2008). Membranes: A meeting point for lipids, proteins and therapies, *J. Cell. Mol. Med.*, **12**, pp. 829–875.

5. van Meer, G., Voelker, D.R., and Feigenson, G.W. (2008). Membrane lipids: Where they are and how they behave? *Nat. Rev. Mol. Cell Biol.*, **9**, pp. 112–124.

6. Buschiazzo, J., Ialy-Radio, C., Auer, J., Wolf, J.-P., Serres, C., Lefevre, B., and Ziyyat, A. (2013). Cholesterol depletion disorganizes oocyte membrane rafts altering mouse fertilization, *PLoS One*, **8**, pp. e62919.

7. Ng, D.P., Poulsen, B.E., and Deber, C.M. (2012). Membrane protein misassembly in disease, *Biochim. Biophys. Acta-Biomembr.*, **1818**, pp. 1115–1122.

8. Stevens, T.J. and Arkin, I.T. (2000). The effect of nucleotide bias upon the composition and prediction of transmembrane helices, *Protein Sci.*, **9**, pp. 505–511.

9. Szlasa, W., Zendran, I., Zalesinska, A., Tarek, M., and Kulbacka, J. (2020). Lipid composition of the cancer cell membrane, *J. Bioenerg. Biomembr.*, **52**, pp. 321–342.

10. Luo, X., Zhao, X., Cheng, C., Li, N., Liu, Y., and Cao, Y. (2018). The implications of signaling lipids in cancer metastasis, *Exp. Mol. Med.*, **50**, pp. 1–10.

11. Carter, P., Smith, L., and Ryan, M. (2004). Identification and validation of cell surface antigens for antibody targeting in oncology, *Endocr. Relat. Cancer*, **11**, pp. 659–687.

12. Hopkins, A.L. and Groom, C.R. (2002). The druggable genome, *Nat. Rev. Drug Discovery*, **1**, pp. 727–730.

13. Gorter, E. and Grendel, F. (1925). On bimolecular layers of lipoids on the chromocytes of the blood, *J. Exp. Med.*, **41**, pp. 439–443.

14. Singer, S.J. and Nicolson, G.L. (1972). The fluid mosaic model of the structure of cell membranes, *Science*, **175**, pp. 720–731.

15. Goni, F.M. (2014). The basic structure and dynamics of cell membranes: An update of the Singer–Nicolson model, *Biochim. Biophys. Acta-Biomembr.*, **1838**, pp. 1467–1476.

16. Zhao, W., Tian, Y., Cai, M., Wang, F., Wu, J., Gao, J., Liu, S., Jiang, J., Jiang, S., and Wang, H. (2014). Studying the nucleated mammalian cell membrane by single molecule approaches, *PLoS One*, **9**, pp. e91595.

17. Percec, V. and Bera, T.K. (2002). Cell membrane as a model for the design of ion-active nanostructured supramolecular systems, *Biomacromolecules*, **3**, pp. 167–181.

18. Fahy, E., Cotter, D., Sud, M., and Subramaniam, S. (2011). Lipid classification, structures and tools, *Biochim. Biophys. Acta-Mol. Cell Biol. Lipids*, **1811**, pp. 637–647.

19. Soteriou, C., Kalli, A.C., Connell, S.D., Tyler, A.I.I., and Thorne, J.L. (2021). Advances in understanding and in multi-disciplinary methodology used to assess lipid regulation of signalling cascades from the cancer cell plasma membrane, *Prog. Lipid Res.*, **81**, pp. 101080.

20. Zech, T., Ejsing, C.S., Gaus, K., de Wet, B., Shevchenko, A., Simons, K., and Harder, T. (2009). Accumulation of raft lipids in T-cell plasma membrane domains engaged in TCR signalling, *Embo J.*, **28**, pp. 466–476.

21. Kennedy, E.P. (1973). Biogenesis and function of membrane lipids, *Proc. Aust. Biochem. Soc.*, **6**, pp. P2–P3.

22. Llado, V., Lopez, D.J., Ibarguren, M., Alonso, M., Soriano, J.B., Escriba, P.V., and Busquets, X. (2014). Regulation of the cancer cell membrane lipid composition by NaCHOleate Effects on cell signaling and therapeutical relevance in glioma, *Biochim. Biophys. Acta-Biomembr.*, **1838**, pp. 1619–1627.

23. van Meer, G. (2005). Cellular lipidomics, *Embo J.*, **24**, pp. 3159–3165.

24. Hoejholt, K.L., Muzic, T., Jensen, S.D., Dalgaard, L.T., Bilgin, M., Nylandsted, J., Heimburg, T., Frandsen, S.K., and Gehl, J. (2019). Calcium electroporation and electrochemotherapy for cancer treatment: Importance of cell membrane composition investigated by lipidomics, calorimetry and in vitro efficacy, *Sci. Rep.*, **9**, pp. 4758.

25. Brown, M.S. and Goldstein, J.L. (1999). A proteolytic pathway that controls the cholesterol content of membranes, cells, and blood, *Proc. Natl. Acad. Sci. U. S. A.*, **96**, pp. 11041–11048.

26. Grouleff, J., Irudayam, S.J., Skeby, K.K., and Schiott, B. (2015). The influence of cholesterol on membrane protein structure, function, and dynamics studied by molecular dynamics simulations, *Biochim. Biophys. Acta-Biomembr.*, **1848**, pp. 1783–1795.

27. Brown, D.A. and London, E. (1998). Structure and origin of ordered lipid domains in biological membranes, *J. Membr. Biol.*, **164**, pp. 103–114.

28. Boes, D.M., Godoy-Hernandez, A., and McMillan, D.G.G. (2021). Peripheral membrane proteins: Promising therapeutic targets across domains of life, *Membranes*, **11**, pp. 346.

29. Tan, S., Tan, H.T., and Chung, M.C.M. (2008). Membrane proteins and membrane proteomics, *Proteomics*, **8**, pp. 3924–3932.

30. Kanapin, A., Batalov, S., Davis, M.J., Gough, J., Grimmond, S., Kawaji, H., Magrane, M., Matsuda, H., Schonbach, C., Teasdale, R.D., Yuan, Z., Grp, R.G., and Memebers, G.S.L. (2003). Mouse proteome analysis, *Genome Res.*, **13**, pp. 1335–1344.

31. Kusumi, A., Fujiwara, T.K., Chadda, R., Xie, M., Tsunoyama, T.A., Kalay, Z., Kasai, R.S., and Suzuki, K.G.N. (2012). Dynamic organizing principles of the plasma membrane that regulate signal transduction: Commemorating the fortieth anniversary of Singer and Nicolson's fluid-mosaic model, in *Annual Review of Cell and Developmental Biology,* Vol 28, R. Schekman (Ed), pp. 215–250.

32. de la Serna, J.B., Schuetz, G.J., Eggeling, C., and Cebecauer, M. (2016). There is no simple model of the plasma membrane organization, *Frontiers Cell Dev. Biol.*, **4**, pp. 106.

33. Kusumi, A., Nakada, C., Ritchie, K., Murase, K., Suzuki, K., Murakoshi, H., Kasai, R.S., Kondo, J., and Fujiwara, T. (2005). Paradigm shift of the plasma membrane concept from the two-dimensional continuum fluid to the partitioned fluid: High-speed single-molecule tracking of membrane molecules, in *Annual Review of Biophysics and Biomolecular Structure*, pp. 351–378.

34. Simons, K. and Ikonen, E. (1997). Functional rafts in cell membranes, *Nature*, **387**, pp. 569–572.

35. Lietha, D. and Izard, T. (2020). Roles of membrane domains in integrin-mediated cell adhesion, *Int. J. Mol. Sci.*, **21**, pp. 5531.

36. Jacobson, K. and Dietrich, C. (1999). Looking at lipid rafts? *Trends Cell Biol.*, **9**, pp. 87–91.

37. Brown, D.A. and London, E. (1998). Functions of lipid rafts in biological membranes, *Annu. Rev. Cell Dev. Biol.*, **14**, pp. 111–136.

38. Ipsen, J.H., Karlstrom, G., Mouritsen, O.G., Wennerstrom, H., and Zuckermann, M.J. (1987). Phase-equilibria in the phosphatidylcholine-cholesterol system, *Biochim. Biophys. Acta*, **905**, pp. 162–172.

39. Brown, D.A. and London, E. (1997). Structure of detergent-resistant membrane domains: Does phase separation occur in biological membranes? *Biochem. Biophys. Res. Commun.*, **240**, pp. 1–7.

40. Lin, X., Lorent, J.H., Skinlde, A.D., Levental, K.R., Waxham, M.N., Gorfe, A.A., and Levental, I. (2016). Domain stability in biomimetic membranes driven by lipid polyunsaturation, *J. Phys. Chem. B*, **120**, pp. 11930–11941.

41. Heberle, F.A., Petruzielo, R.S., Pan, J., Drazba, P., Kucerka, N., Standaert, R.F., Feigenson, G.W., and Katsaras, J. (2013). Bilayer thickness mismatch controls domain size in model membranes, *J. Am. Chem. Soc.*, **135**, pp. 6853–6859.

42. Sezgin, E., Levental, I., Grzybek, M., Schwarzmann, G., Mueller, V., Honigmann, A., Belov, V.N., Eggeling, C., Coskun, U., Simons, K., and

Schwille, P. (2012). Partitioning, diffusion, and ligand binding of raft lipid analogs in model and cellular plasma membranes, *Biochim. Biophys. Acta-Biomembr.*, **1818**, pp. 1777–1784.

43. Levental, I., Grzybek, M., and Simons, K. (2011). Raft domains of variable properties and compositions in plasma membrane vesicles, *Proc. Natl. Acad. Sci. U. S. A.*, **108**, pp. 11411–11416.

44. Winkler, P.M. and Garcia-Parajo, M.F. (2021). Correlative nanophotonic approaches to enlighten the nanoscale dynamics of living cell membranes, *Biochem. Soc. Trans.*, **49**, pp. 2357–2369.

45. Goswami, D., Gowrishankar, K., Bilgrami, S., Ghosh, S., Raghupathy, R., Chadda, R., Vishwakarma, R., Rao, M., and Mayor, S. (2008). Nanoclusters of GPI-anchored proteins are formed by cortical actin-driven activity, *Cell*, **135**, pp. 1085–1097.

46. Kenworthy, A.K. (2007). Fluorescence recovery after photobleaching studies of lipid rafts, *Methods Mol. Biol.*, **398**, pp. 179–192.

47. Sreedharan, S., Gill, M.R., Garcia, E., Saeed, H.K., Robinson, D., Byrne, A., Cadby, A., Keyes, T.E., Smythe, C., Pellett, P., de la Serna, J.B., and Thomas, J.A. (2017). Multimodal super-resolution optical microscopy using a transition-metal-based probe provides unprecedented capabilities for imaging both nuclear chromatin and mitochondria, *J. Am. Chem. Soc.*, **139**, pp. 15907–15913.

48. Owen, D.M., Magenau, A., Williamson, D., and Gaus, K. (2012). The lipid raft hypothesis revisited: New insights on raft composition and function from super-resolution fluorescence microscopy, *Bioessays*, **34**, pp. 739–747.

49. Gwosch, K.C., Pape, J.K., Balzarotti, F., Hoess, P., Ellenberg, J., Ries, J., and Hell, S.W. (2020). MINFLUX nanoscopy delivers 3D multicolor nanometer resolution in cells, *Nat. Methods*, **17**, pp. 217–224.

50. Lorizate, M., Terrones, O., Nieto-Garai, J.A., Rojo-Bartolome, I., Ciceri, D., Morana, O., Olazar-Intxausti, J., Arboleya, A., Martin, A., Szynkiewicz, M., Calleja-Felipe, M., Bernardino de la Serna, J., and Contreras, F.X. (2021). Super-resolution microscopy using a bioorthogonal-based cholesterol probe provides unprecedented capabilities for imaging nanoscale lipid heterogeneity in living cells, *Small Methods*, **5**, pp. 2100430.

51. Danylchuk, D.I., Moon, S., Xu, K., and Klymchenko, A.S. (2019). Switchable solvatochromic probes for live-cell super-resolution imaging of plasma membrane organization, *Angew. Chem. Int. Ed.*, **58**, pp. 14920–14924.

52. Schwille, P., MeyerAlmes, F.J., and Rigler, R. (1997). Dual-color fluorescence cross-correlation spectroscopy for multicomponent diffusional analysis in solution, *Biophys. J.*, **72**, pp. 1878–1886.

53. Petrasek, Z. and Schwille, P. (2008). Precise measurement of diffusion coefficients using scanning fluorescence correlation spectroscopy, *Biophys. J.*, **94**, pp. 1437–1448.

54. Dertinger, T., Pacheco, V., von der Hocht, I., Hartmann, R., Gregor, I., and Enderlein, J. (2007). Two-focus fluorescence correlation spectroscopy: A new tool for accurate and absolute diffusion measurements, *Chemphyschem*, **8**, pp. 433–443.

55. Wawrezinieck, L., Rigneault, H., Marguet, D., and Lenne, P.F. (2005). Fluorescence correlation spectroscopy diffusion laws to probe the submicron cell membrane organization, *Biophys. J.*, **89**, pp. 4029–4042.

56. Eggeling, C., Ringemann, C., Medda, R., Schwarzmann, G., Sandhoff, K., Polyakova, S., Belov, V.N., Hein, B., von Middendorff, C., Schoenle, A., and Hell, S.W. (2009). Direct observation of the nanoscale dynamics of membrane lipids in a living cell, *Nature*, **457**, pp. 1159–1162.

57. Hedde, P.N., Doerlich, R.M., Blomley, R., Gradl, D., Oppong, E., Cato, A.C.B., and Nienhaus, G.U. (2013). Stimulated emission depletion-based raster image correlation spectroscopy reveals biomolecular dynamics in live cells, *Nat. Commun.*, **4**, pp. 2093.

58. Suzuki, K.G.N., Kasai, R.S., Hirosawa, K.M., Nemoto, Y.L., Ishibashi, M., Miwa, Y., Fujiwara, T.K., and Kusumi, A. (2012). Transient GPI-anchored protein homodimers are units for raft organization and function, *Nat. Chem. Biol.*, **8**, pp. 774–783.

59. Komura, N., Suzuki, K.G.N., Ando, H., Konishi, M., Koikeda, M., Imamura, A., Chadda, R., Fujiwara, T.K., Tsuboi, H., Sheng, R., Cho, W., Furukawa, K., Furukawa, K., Yamauchi, Y., Ishida, H., Kusumi, A., and Kiso, M. (2016). Raft-based interactions of gangliosides with a GPI-anchored receptor, *Nat. Chem. Biol.*, **12**, pp. 402–410.

60. Kinoshita, M., Suzuki, K.G.N., Matsumori, N., Takada, M., Ano, H., Morigaki, K., Abe, M., Makino, A., Kobayashi, T., Hirosawa, K.M., Fujiwara, T.K., Kusumi, A., and Murata, M. (2017). Raft-based sphingomyelin interactions revealed by new fluorescent sphingomyelin analogs, *J. Cell Biol.*, **216**, pp. 1183–1204.

61. Manley, S., Gillette, J.M., Patterson, G.H., Shroff, H., Hess, H.F., Betzig, E., and Lippincott-Schwartz, J. (2008). High-density mapping of single-molecule trajectories with photoactivated localization microscopy, *Nat. Methods*, **5**, pp. 155–157.

62. Kim, D.-H., Zhou, K., Kim, D.-K., Park, S., Noh, J., Kwon, Y., Kim, D., Song, N.W., Lee, J.-B., Suh, P.-G., Lee, N.K., and Ryu, S.H. (2015). Analysis of interactions between the epidermal growth factor receptor and soluble ligands on the basis of single-molecule diffusivity in the membrane of living cells, *Angew. Chem. Int. Ed.*, **54**, pp. 7028–7032.

63. Eilers, Y., Ta, H., Gwosch, K.C., Balzarotti, F., and Hell, S.W. (2018). MINFLUX monitors rapid molecular jumps with superior spatiotemporal resolution, *Proc. Natl. Acad. Sci. U. S. A.*, **115**, pp. 6117–6122.

64. Schmidt, R., Weihs, T., Wurm, C.A., Jansen, I., Rehman, J., Sahl, S.J., and Hell, S.W. (2021). MINFLUX nanometer-scale 3D imaging and microsecond-range tracking on a common fluorescence microscope, *Nat. Commun.*, **12**, pp. 1478.

65. Lozano, M.M., Liu, Z., Sunnick, E., Janshoff, A., Kumar, K., and Boxer, S.G. (2013). Colocalization of the ganglioside G(M1) and cholesterol detected by secondary ion mass spectrometry, *J. Am. Chem. Soc.*, **135**, pp. 5620–5630.

66. Frisz, J.F., Lou, K., Klitzing, H.A., Hanafin, W.P., Lizunov, V., Wilson, R.L., Carpenter, K.J., Kim, R., Hutcheon, I.D., Zimmerberg, J., Weber, P.K., and Kraft, M.L. (2013). Direct chemical evidence for sphingolipid domains in the plasma membranes of fibroblasts, *Proc. Natl. Acad. Sci. U. S. A.*, **110**, pp. E613–E622.

67. He, C., Hu, X., Jung, R.S., Weston, T.A., Sandoval, N.P., Tontonoz, P., Kilburn, M.R., Fong, L.G., Young, S.G., and Jiang, H. (2017). High-resolution imaging and quantification of plasma membrane cholesterol by NanoSIMS, *Proc. Natl. Acad. Sci. U. S. A.*, **114**, pp. 2000–2005.

68. Sezgin, E., Levental, I., Mayor, S., and Eggeling, C. (2017). The mystery of membrane organization: Composition, regulation and roles of lipid rafts, *Nat. Rev. Mol. Cell Biol.*, **18**, pp. 361–374.

69. Kalappurakkal, J.M., Sil, P., and Mayor, S. (2020). Toward a new picture of the living plasma membrane, *Protein Sci.*, **29**, pp. 1355–1365.

70. Apodaca, G. (2002). Modulation of membrane traffic by mechanical stimuli, *Am. J. Physiol. Renal Physiol.*, **282**, pp. F179–F190.

71. Raucher, D. and Sheetz, M.P. (2000). Cell spreading and lamellipodial extension rate is regulated by membrane tension, *J. Cell Biol.*, **148**, pp. 127–136.

72. Masters, T.A., Pontes, B., Viasnoff, V., Li, Y., and Gauthier, N.C. (2013). Plasma membrane tension orchestrates membrane trafficking, cytoskeletal remodeling, and biochemical signaling during phagocytosis, *Proc. Natl. Acad. Sci. U. S. A.*, **110**, pp. 11875–11880.

73. Houk, A.R., Jilkine, A., Mejean, C.O., Boltyanskiy, R., Dufresne, E.R., Angenent, S.B., Altschuler, S.J., Wu, L.F., and Weiner, O.D. (2012). Membrane tension maintains cell polarity by confining signals to the leading edge during neutrophil migration, *Cell*, **148**, pp. 175–188.

74. Gauthier, N.C., Masters, T.A., and Sheetz, M.P. (2012). Mechanical feedback between membrane tension and dynamics, *Trends Cell Biol.*, **22**, pp. 527–535.

75. Cox, C.D., Bavi, N., and Martinac, B. (2019). Biophysical principles of ion-channel-mediated mechanosensory transduction, *Cell Rep.*, **29**, pp. 1–12.

76. Lieber, A.D., Schweitzer, Y., Kozlov, M.M., and Keren, K. (2015). Front-to-rear membrane tension gradient in rapidly moving cells, *Biophys. J.*, **108**, pp. 1599–1603.

77. Martinac, B., Nikolaev, Y.A., Silvani, G., Bavi, N., Romanov, V., Nakayama, Y., Martinac, A.D., Rohde, P., Bavi, O., and Cox, C.D. (2020). Cell membrane mechanics and mechanosensory transduction, in *Membrane Biomechanics*, I. Levitan and A. Trache (Eds), pp. 83–141.

78. Rodriguez, M.L., McGarry, P.J., and Sniadecki, N.J. (2013). Review on cell mechanics: Experimental and modeling approaches, *ApMRv*, **65**, pp. 060801.

79. Mio, C., Gong, T., Terray, A., and Marr, D.W.M. (2000). Design of a scanning laser optical trap for multiparticle manipulation, *Rev. Sci. Instrum.*, **71**, pp. 2196–2200.

80. Hosu, B.G., Jakab, K., Banki, P., Toth, F.I., and Forgacs, G. (2003). Magnetic tweezers for intracellular applications, *Rev. Sci. Instrum.*, **74**, pp. 4158–4163.

81. Mueller, D.J. and Dufrene, Y.F. (2008). Atomic force microscopy as a multifunctional molecular toolbox in nanobiotechnology, *Nat. Nanotechnol.*, **3**, pp. 261–269.

82. Baldi, A., Fass, J.N., De Silva, M.N., Odde, D.J., and Ziaie, B. (2003). A micro-tool for mechanical manipulation of in vitro cell arrays, *Biomed. Microdevices*, **5**, pp. 291–295.

83. Gossett, D.R., Tse, H.T.K., Lee, S.A., Ying, Y., Lindgren, A.G., Yang, O.O., Rao, J., Clark, A.T., and Di Carlo, D. (2012). Hydrodynamic stretching of single cells for large population mechanical phenotyping, *Proc. Natl. Acad. Sci. U. S. A.*, **109**, pp. 7630–7635.

84. Sitters, G., Kamsma, D., Thalhammer, G., Ritsch-Marte, M., Peterman, E.J.G., and Wuite, G.J.L. (2015). Acoustic force spectroscopy, *Nat. Methods*, **12**, pp. 47–50.

85. Style, R.W., Boltyanskiy, R., German, G.K., Hyland, C., MacMinn, C.W., Mertz, A.F., Wilen, L.A., Xu, Y., and Dufresne, E.R. (2014). Traction force microscopy in physics and biology, *Soft Matter*, **10**, pp. 4047–4055.

86. Schoen, I., Hu, W., Klotzsch, E., and Vogel, V. (2010). Probing cellular traction forces by micropillar arrays: Contribution of substrate warping to pillar deflection, *Nano Lett.*, **10**, pp. 1823–1830.

87. Wirtz, D. (2009). Particle-tracking microrheology of living cells: Principles and applications, *Annu. Rev. Biophys.*, **38**, pp. 301–326.

88. Lemmon, M.A. and Schlessinger, J. (2010). Cell signaling by receptor tyrosine kinases, *Cell*, **141**, pp. 1117–1134.

89. Congreve, M. and Marshall, F. (2010). The impact of GPCR structures on pharmacology and structure-based drug design, *Br. J. Pharmacol.*, **159**, pp. 986–996.

90. Zamponi, G.W. (2010). Welcome to "ion channels: key therapeutic targets" foreword, *Future Med. Chem.*, **2**, pp. 689–690.

91. Morita, H., Wu, J., and Zipes, D.P. (2008). The QT syndromes: Long and short, *Lancet*, **372**, pp. 750–763.

92. Gottesman, M.M. and Ambudkar, S.V. (2001). Overview: ABC transporters and human disease, *J. Bioenerg. Biomembr.*, **33**, pp. 453–458.

93. Palmgren, M.G. and Nissen, P. (2011). P-type ATPases, in *Annual Review of Biophysics*, Vol 40, D.C. Rees, K.A. Dill, and J.R. Williamson (Eds), pp. 243–266.

94. Borst, P. and Elferink, R.O. (2002). Mammalian ABC transporters in health and disease, *Annu. Rev. Biochem.*, **71**, pp. 537–592.

95. Sarkadi, B., Homolya, L., Szakacs, G., and Varadi, A. (2006). Human multidrug resistance ABCB and ABCG transporters: Participation in a chemoimmunity defense system, *Physiol. Rev.*, **86**, pp. 1179–1236.

96. Krueger, K.E. and Srivastava, S. (2006). Posttranslational protein modifications: Current implications for cancer detection, prevention, and therapeutics, *Mol. Cell. Proteomics*, **5**, pp. 1799–1810.

97. Block, T.M., Comunale, M.A., Lowman, M., Steel, L.F., Romano, P.R., Fimmel, C., Tennant, B.C., London, W.T., Evans, A.A., Blumberg, B.S., Dwek, R.A., Mattu, T.S., and Mehta, A.S. (2005). Use of targeted glycoproteomics to identify serum glycoproteins that correlate with liver cancer in woodchucks and humans, *Proc. Natl. Acad. Sci. U. S. A.*, **102**, pp. 779–784.

98. Gundry, R.L., Raginski, K., Tarasova, Y., Tchernyshyov, I., Bausch-Fluck, D., Elliott, S.T., Boheler, K.R., Van Eyk, J.E., and Wollscheid, B. (2009).

The mouse C2C12 myoblast cell surface N-linked glycoproteome, *Mol. Cell. Proteomics*, **8**, pp. 2555–2569.

99. Schiess, R., Mueller, L.N., Schmidt, A., Mueller, M., Wollscheid, B., and Aebersold, R. (2009). Analysis of cell surface proteome changes via label-free, quantitative mass spectrometry, *Mol. Cell. Proteomics*, **8**, pp. 624–638.

100. Uematsu, R., Shinohara, Y., Nakagawa, H., Kurogochi, M., Furukawa, J.-I., Miura, Y., Akiyama, M., Shimizu, H., and Nishimura, S.-I. (2009). Glycosylation specific for adhesion molecules in epidermis and its receptor revealed by glycoform-focused reverse genomics, *Mol. Cell. Proteomics*, **8**, pp. 232–244.

101. Brill, L.M., Salomon, A.R., Ficarro, S.B., Mukherji, M., Stettler-Gill, M., and Peters, E.C. (2004). Robust phosphoproteomic profiling of tyrosine phosphorylation sites from human T cells using immobilized metal affinity chromatography and tandem mass spectrometry, *Anal. Chem.*, **76**, pp. 2763–2772.

102. Maeusbacher, N., Schreiber, T.B., and Daub, H. (2010). Glycoprotein capture and quantitative phosphoproteomics indicate coordinated regulation of cell migration upon lysophosphatidic acid stimulation, *Mol. Cell. Proteomics*, **9**, pp. 2337–2353.

103. Zarling, A.L., Polefrone, J.M., Evans, A.M., Mikesh, L.M., Shabanowitz, J., Lewis, S.T., Engelhardt, V.H., and Hunt, D.F. (2006). Identification of class I MHC-associated phosphopeptides as targets for cancer immunotherapy, *Proc. Natl. Acad. Sci. U. S. A.*, **103**, pp. 14889–14894.

104. Escriba, P.V., Sanchez-Dominguez, J.M., Alemany, R., Perona, J.S., and Ruiz-Gutierrez, V. (2003). Alteration of lipids, G proteins, and PKC in cell membranes of elderly hypertensives, *Hypertension*, **41**, pp. 176–182.

105. Luostarinen, R., Boberg, M., and Saldeen, T. (1993). Fatty-acid composition in total phospholipids of human coronary-arteries in sudden cardiac death, *Atherosclerosis*, **99**, pp. 187–193.

106. Yang, Q., Alemany, R., Casas, J., Kitajka, K., Lanier, S.M., and Escriba, P.V. (2005). Influence of the membrane lipid structure on signal processing via G protein-coupled receptors, *Mol. Pharmacol.*, **68**, pp. 210–217.

107. Alemany, R., Teres, S., Baamonde, C., Benet, M., Vogler, O., and Escriba, P.V. (2004). 2-Hydroxyoleic acid: A new hypotensive molecule, *Hypertension*, **43**, pp. 249–254.

108. Mikirova, N., Riordan, H.D., Jackson, J.A., Wong, K., Miranda-Massari, J.R., and Gonzalez, M.J. (2004). Erythrocyte membrane fatty acid composition in cancer patients, *P. R. Health Sci. J.*, **23**, pp. 107–113.

109. Martinez, J., Vogler, O., Casas, J., Barcelo, F., Alemany, R., Prades, J., Nagy, T., Baamonde, C., Kasprzyk, P.G., Teres, S., Saus, C., and Escriba, P.V. (2005). Membrane structure modulation, protein kinase C alpha activation, and anticancer activity of Minerval, *Mol. Pharmacol.*, **67**, pp. 531–540.

110. Jendrossek, V. and Handrick, R. (2003). Membrane targeted anticancer drugs: Potent inducers of apoptosis and putative radiosensitisers, *Curr. Med. Chem. Anticancer Agents*, **3**, pp. 343–353.

111. Ali, S.M., Khan, A.R., Ahmad, M.U., Chen, P., Sheikh, S., and Ahmad, I. (2005). Synthesis and biological evaluation of gemcitabine-lipid conjugate (NE06002), *Bioorg. Med. Chem. Lett.*, **15**, pp. 2571–2574.

112. Schroder, H., Marrugat, J., Vila, J., Covas, M.I., and Elosua, R. (2004). Adherence to the traditional Mediterranean diet is inversely associated with body mass index and obesity in a Spanish population, *J. Nutr.*, **134**, pp. 3355–3361.

113. St-Onge, M.P. (2005). Dietary fats, teas, dairy, and nuts: Potential functional foods for weight control? *Am. J. Clin. Nutr.*, **81**, pp. 7–15.

114. Arispe, N. and Doh, M. (2002). Plasma membrane cholesterol controls the cytotoxicity of Alzheimer's disease A beta P (1-40) and (1-42) peptides, *FASEB J.*, **16**, pp. 1526–1536.

115. Palestini, P., Calvi, C., Conforti, E., Daffara, R., Botto, L., and Miserocchi, G. (2003). Compositional changes in lipid microdomains of air-blood barrier plasma membranes in pulmonary interstitial edema, *J. Appl. Physiol.*, **95**, pp. 1446–1452.

116. Ohno-Iwashita, Y., Shimada, Y., Hayashi, M., and Inomata, M. (2010). Plasma membrane microdomains in aging and disease, *Geriatr. Gerontol. Int.*, **10**, pp. S41–S52.

117. Cohen, A.W., Hnasko, R., Schubert, W., and Lisanti, M.P. (2004). Role of caveolae and caveolins in health and disease, *Physiol. Rev.*, **84**, pp. 1341–1379.

118. Parton, R.G., Hanzal-Bayer, M., and Hancock, J.F. (2006). Biogenesis of caveolae: A structural model for caveolin-induced domain formation, *J. Cell Sci.*, **119**, pp. 787–796.

119. Li, X.A., Everson, W.V., and Smart, E.J. (2005). Caveolae, lipid rafts, and vascular disease, *Trends Cardiovasc. Med.*, **15**, pp. 92–96.

120. Lopez, J.A., Del Conde, I., and Shrimpton, C.N. (2005). Receptors, rafts, and microvesicles in thrombosis and inflammation, *J. Thromb. Haemost.*, **3**, pp. 1737–1744.

121. Bickel, P.E. (2002). Lipid rafts and insulin signaling, *American Am. J. Physiol.-Endocrinol. Metab.*, **282**, pp. E1–E10.

122. Ma, D.W.L. (2007). Lipid mediators in membrane rafts are important determinants of human health and disease, *Appl. Physiol. Nutr. Metab.*, **32**, pp. 341–350.

123. Manes, S., del Real, G., Lacalle, R.A., Lucas, P., Gomez-Mouton, C., Sanchez-Palomino, S., Delgado, R., Alcami, J., Mira, E., and Martinez, C. (2000). Membrane raft microdomains mediate lateral assemblies required for HIV-1 infection, *EMBO Rep.*, **1**, pp. 190–196.

124. Zhang, J., Pekosz, A., and Lamb, R.A. (2000). Influenza virus assembly and lipid raft microdomains: A role for the cytoplasmic tails of the spike glycoproteins, *J. Virol.*, **74**, pp. 4634–4644.

125. Brugger, B., Glass, B., Haberkant, P., Leibrecht, I., Wieland, F.T., and Krausslich, H.G. (2006). The HIV lipidome: A raft with an unusual composition, *Proc. Natl. Acad. Sci. U. S. A.*, **103**, pp. 2641–2646.

126. Kraft, M.L. (2013). Plasma membrane organization and function: Moving past lipid rafts, *Mol. Biol. Cell*, **24**, pp. 2765–2768.

Chapter 2

Mass Spectrometry for Studying Ensemble Biomembranomics

Fangjun Wang, Zheyi Liu, and Jing Liu
Dalian Institute of Chemical Physics, Chinese Academy of Science, 457 Zhongshan Road, Dalian, Liaoning 116023, China
wangfj@dicp.ac.cn

Mass spectrometry (MS) is an unbiased profiling technology with high throughput. The application of MS in biomembranomics is aimed to comprehensively profile all biomolecules within biomembranes, which can provide a global view of the molecular constitutions of biomembranes as well as their dynamic changes at different conditions. In the last two decades, advanced MS-based omics strategies such as proteomics, glycoproteomics, and lipidomics have been greatly developed, promoting the in-depth profiling of all the proteins, glycans, lipids, as well as their interaction complexes in biomembranes. MS has been demonstrated as a powerful tool for the global characterization of thousands of membrane proteins as well as their glycosylation sites, providing crucial molecular-level information in the membrane's physiological

Biomembranomics: Structure, Techniques, and Applications
Edited by Hongda Wang

ISBN 978-981-4968-61-4 (Hardcover), 978-1-003-45635-3 (eBook)
www.jennystanford.com

and pathological processes. We believe that the rapid development of biomembranomics will continue in the future and novel MS methods such as native MS can bring biomembranomics to a new period of interaction and function exploration. MS-based biomembranomics analysis will become an important inter-discipline and play an essential role in fundamental biological study and drug development.

2.1 Introduction

In the last two decades, the technologies of various types of MS-based omics have been greatly developed for the large-scale investigation of proteins, peptides, lipids, and glycans. Biomembranes are mainly composed of lipids, proteins, and glycans. The application of MS in biomembranomics is aimed to comprehensively characterize all biomolecules mainly including membrane proteins, lipids and carbohydrates, and their interactions within biomembranes at the overall level. Currently, liquid chromatography coupled with tandem mass spectrometry (LC-MS/MS, Fig. 2.1) is the core platform for comprehensive omics-level profiling of different types of biomolecules except for nucleic acids.

Membrane proteins, accounting for 20–30% of the eukaryote genome encodes, are essential macro-biomolecules penetrating or embedding in the phospholipid bilayers of biomembranes. About one-third of human proteins are membrane proteins. Membrane proteins play crucial roles in fundamental physiological processes, including intercellular communication as well as ion and molecule transportation into and out of cells. The dysregulation of membrane proteins in expression, post-translational modification (PTM), and functional conformation usually leads to serious diseases such as cancer. More than half of the drug targets are membrane proteins, including the famous G-protein coupled receptors (GPCRs). The membrane proteome can be defined as the entire membrane proteins with different spatial localization, sizes, hydrophobicity, abundances, PTMs, and proteoforms in a cell. Therefore, it is a challenging task to realize global membrane proteomic profiling by MS.

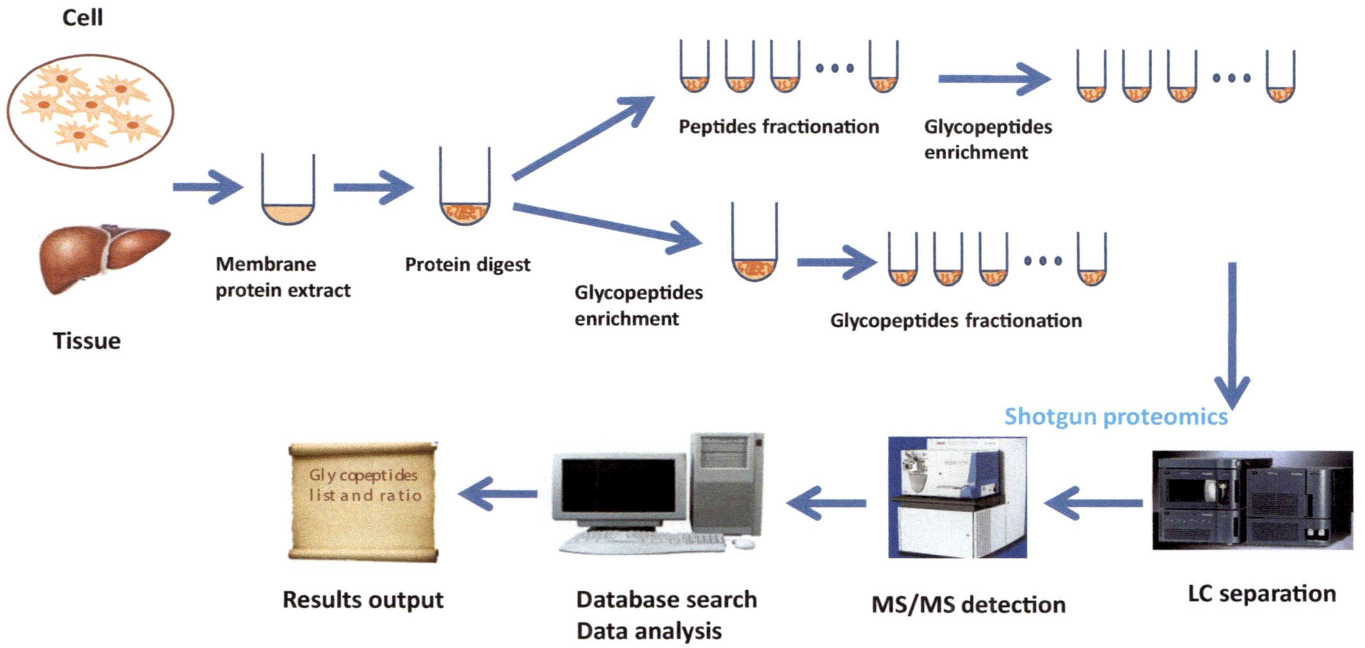

Figure 2.1 LC-MS/MS shogun strategy for membrane proteomics and glycoproteomics analysis.

The extracellular domains of most of the membrane proteins are often covalently modified with glycan chains at multiple Asn residues locating in conserved Asn-x-Ser/Thr sequence motifs (N-link) or Ser, Thr residues (O-link). N-glycosylation and O-glycosylation are closely related with the regulation of structures, interactions, and functions of membrane proteins in signaling and adhesion. Glycosylation is a non-template-based process, resulting in high heterogeneity in glycoforms, which are mainly constructed by monosaccharide components fucose, mannose, galactose, N-acetylglucosamine, N-acetylgalactosamine, N-acetylneuraminic acid, and N-glycolylneuraminic acid. Thus, membrane glycoproteomics has been rapidly developed for elucidating the glycosylation sites, glycoforms, and related biofunctions. On the other hand, all the N-link glycans of a complex membrane protein sample can be released by PNGase F cleavage to realize global glycan profiling, which is termed membrane glycomics. Therefore, MS-based membrane glycoproteomics is devoted to characterizing intact glycoproteins and glycopeptides, while glycomics is focused on only glycan chains.

Lipids are the main constituents of biomembranes. The arrangements of lipids and membrane proteins usually act as crucial membrane receptors and channel pores with important biofunctions. Membrane lipidomics is aimed to profile the whole membrane lipids with their constitutions, proportions, and structures. Furthermore, the interactions of biomolecules within biomembranes are crucial to the regulation of many bioprocesses, such as the modulation of surrounding lipids on membrane protein structures and functions. Therefore, advanced MS technologies turn to elucidate the endogenous membrane protein–ligand and protein–lipid interactions in recent years and uncover the molecular details underlining membrane recognition and signaling.

2.2 Membrane Proteomics

Integral membrane proteins usually have an extracellular segment, one or multiple transmembrane domains or lipid anchors and a cytoplasmic region. The most challenging task of membrane proteomic profiling is the high-recovery sample preparation due to the relatively high hydrophobicity of membrane proteins, mainly including the procedures of membrane protein extraction, proteolytic digestion, and sample purification. Then, the purified peptide mixture is analyzed by LC-MS/MS, followed with database matching and bioinformatic analysis to identify the membrane proteins and predict their functions, which is also termed shotgun proteomics (Fig. 2.1).

2.2.1 Development of Shotgun Proteomics

Proteomics is aimed at profiling all the proteins expressed within a cell, tissue, or organism at a specific state, which is more complex than genomics because proteins are constituted of 20 types of amino acids and can be dynamically modified by more than 300 types of PTMs such as phosphorylation, glycosylation, acetylation, and methylation. In the current bottom-up proteomic strategy established at the end of the last century, the protein samples are usually digested at first, and then the complex peptides mixture is analyzed by LC-MS/MS (Fig. 2.1) [1, 2]. Thousands of peptides and proteins can be identified and quantified in just one LC-MS/MS experiment due to the high separation resolution of LC and high identification capability of MS. However, it is still impossible to detect all the expressed proteins within a complex biological system due to the extremely high complexity and high dynamic range of biological protein samples, especially for the low-abundant ones. Therefore, high-resolution LC separation is essential to increase the detection coverage, sensitivity, and throughput of proteomic analysis.

Online multidimensional LC separation exhibits low sample loss and contamination in combination with MS detection, and the proteomic analysis sensitivity and throughput can be greatly improved. The most famous online multidimensional LC system is multidimensional protein identification technology (MudPIT) developed by Yates's lab in 2001 (Fig. 2.2) [3, 4]. The automated MudPIT is achieved by using a biphasic capillary column packed with strong cation exchange (SCX) and reversed phase (RP) stationary phases in sequence. Peptides loaded onto the first segment of SCX adsorbent are stepwise eluted to the RP segment by flushing salt step, and a nanoflow RPLC-MS/MS analysis is flowed with each salt step. This online 2DLC-MS/MS system can be fully automated and exhibits advantages such as minimal sample loss, no vial contamination and sample dilution, and is widely applied to various biological sample analysis at the initial stage of shotgun proteomics.

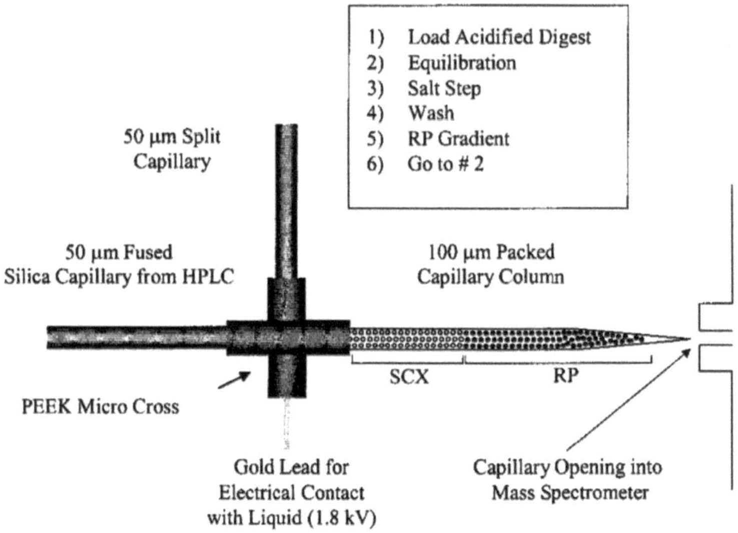

Figure 2.2 The schematic diagram of MudPIT 2DLC-MS/MS system. Reprinted with the permission from Ref. [4]. Copyright 2001 American Chemical Society.

2.2.2 Practical Examples of Membrane Proteomics

In the traditional proteomic studies of more than 20 years ago, MS detection was usually coupled with gel electrophoresis separation. However, the sample preparation procedures of gel-separated proteins are tedious and hydrophobic proteins, including membrane proteins, are difficult to analyze. Therefore, the shotgun proteomics strategy has been rapidly developed to analyze the proteolytic peptides of extracted membrane proteins by using LC-MS/MS coupled with database matching in the early 21st century.

The pioneering works of Yates's lab greatly promoted the in-depth profiling of membrane proteins in shotgun proteomics [3–6]. Compared with conventional gel electrophoresis, LC-MS/MS approaches exhibit higher detection sensitivity and are capable of discovering PTMs on membrane proteins. Yates et al. established a method using high pH and proteinase K (hpPK) for membrane proteomic analysis (Fig. 2.3) [7]. High-pH buffer could result in the formation of membrane sheets, while the non-specific proteinase K would cleave the exposed regions of membrane proteins efficiently. Then, the membrane was solubilized, and the hydrophobic transmembrane protein domains were re-digested with CNBr. Both peptide samples were analyzed by online MudPIT 2DLC-MS/MS to identify the exposed hydrophilic sequences and transmembrane hydrophobic sequences. Therefore, global membrane protein identification as well as topology and localization characterization can be achieved.

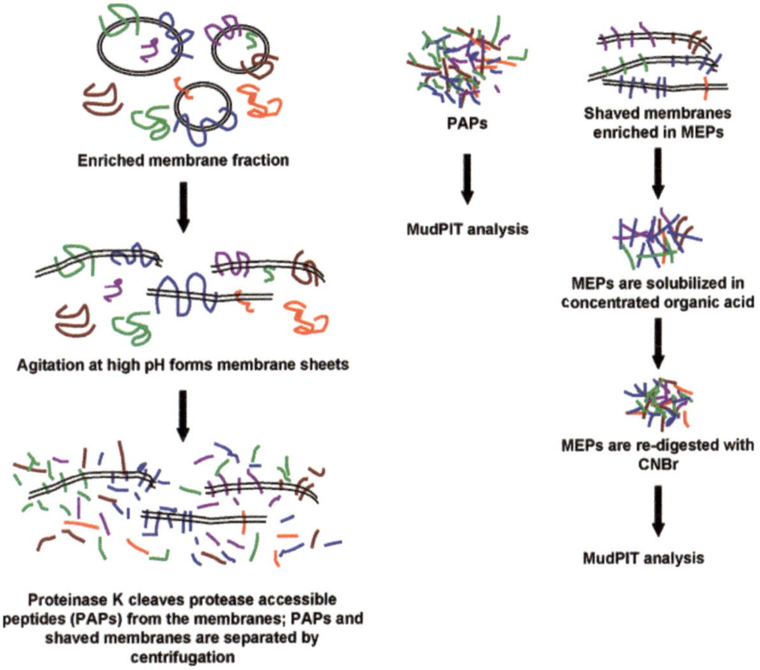

Figure 2.3 Workflow for the preparation and proteomic analysis of protease-accessible peptides (PAPs) and membrane embedded peptides (MEPs). Reprinted with the permission from Ref. [7]. Copyright 2008 American Chemical Society.

Moreover, further efforts in membrane proteomics have been made to improve the extraction efficiency of membrane proteins with integrated sample preparation procedures to reduce sample loss. As conventional chaotropic agents such as urea and guanidine hydrochloride have low solubilization efficiency for membrane proteins, a sodium dodecyl sulfate (SDS) detergent has been successfully applied to extract hydrophobic membrane proteins. However, SDS will denature the active proteins and is not compatible with protease digestion. Therefore, the extracted membrane protein samples must be purified by organic solvent precipitation or ultrafiltration and re-dissolved into solutions without a detergent, resulting in high sample loss. Other detergents such as Nonidet P-40 (NP-40), Triton X-100, and n-dodecyl-β-D-maltoside (DDM) can also be utilized in membrane protein extraction and proteomic analysis. There is still a need for a reagent with high efficiency in hydrophobic membrane protein extraction as well as high compatibility to both enzyme digestion and LC-MS/MS analysis. Zhang's lab adopted ionic liquids (ILs) in membrane protein sample extraction with performance superior to SDS, Rapigest, and methanol [8–12]. They systematically studied the effect of ionic liquid structure on membrane protein solubilization and enzyme biocompatibility, based on which 1-dodecyl-3-methylimidazolium chloride (C12Im-Cl) was selected for the sample preparation of integral membrane proteins. Compared with SDS, 1.4-times and 3.5-times improvements on the numbers of identified membrane proteins (251 versus 178) and hydrophobic peptides (982 versus 279) were achieved (Fig. 2.4) [8].

As described above, the sample preparation procedures of membrane proteomic analysis are tedious and easily induce high sample loss, limiting the detection sensitivity especially for low-abundant membrane proteins. A promising technology for solving this problem is the development of a proteomic reactor that can integrate all the sample preparation procedures into a single microfluidic device, including sample pre-concentration, purification, reduction, digestion, and fractionation [13]. Figeys's lab established a novel approach for the identification of plasma membrane proteins and intracellular microsomal membrane proteins. They combined membrane isolation and a centrifugal proteomic reactor for streamlined protein extraction, protein digestion, and fractionation by centrifugation (Fig. 2.5) [14]. This approach was demonstrated for the proteomic profiling of ER and Golgi microsomal membranes of rat hepatic cells. The centrifugal proteomic reactor identified 945 plasma membrane proteins and 955 microsomal membrane proteins, of which 63% and 47% were predicted as bona fide membrane proteins, respectively. Among these proteins, more than 800 proteins were undetectable by the conventional in-gel digestion strategy.

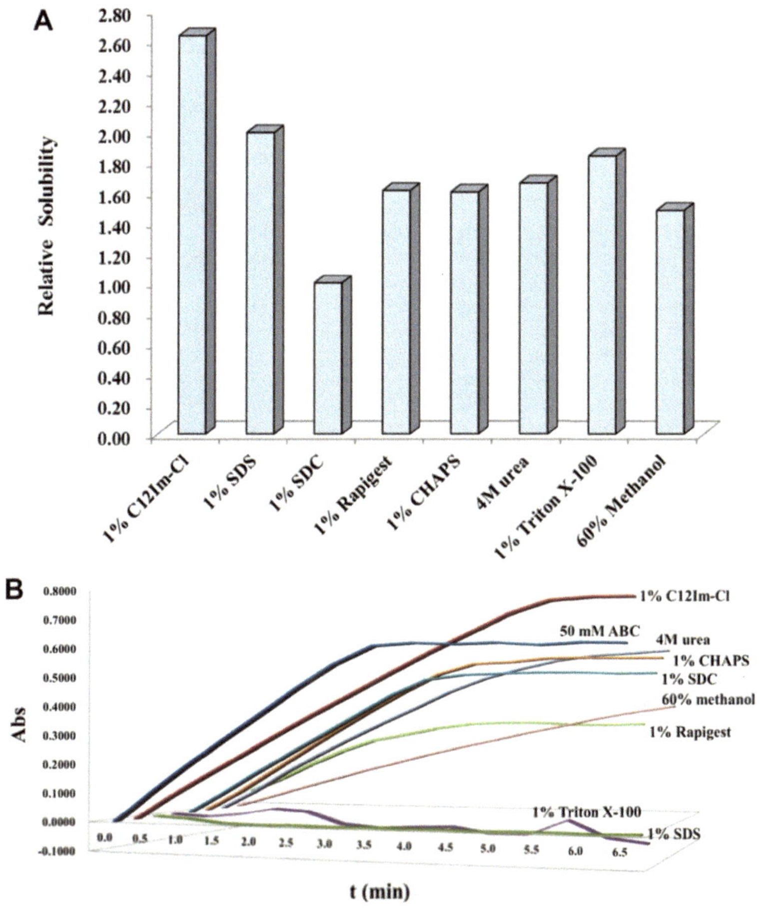

Figure 2.4 (A) Relative solubility of membrane proteins and (B) trypsin activity with different additives. Reprinted with the permission from Ref. [8]. Copyright 2014 American Chemical Society.

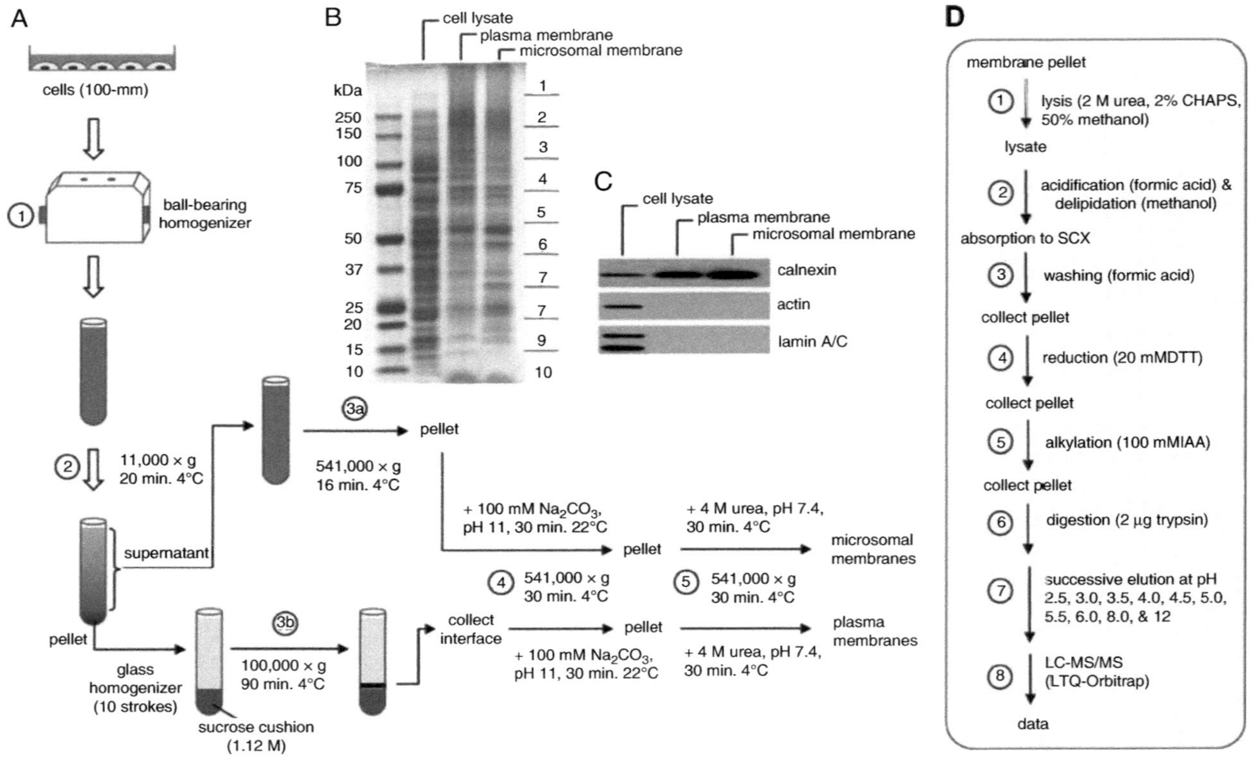

Figure 2.5 Flow charts of microsomal and plasma membrane isolation and the centrifugal proteomic reactor for proteomic analysis of membrane proteins. Reprinted from Ref. [14]. Copyright 2011, with permission from Elsevier.

Conventional proteases utilized in shotgun proteomics usually cleave at basic, acidic, or aromatic amino acids in protein sequences. Trypsin, Arg-C, Lys-C, and LysargiNase cleave at basic amino acids (Lys or Arg). Pepsin and Glu-C cleave at acidic amino acids (Asp or Glu). Chymotrypsin cleaves mainly at aromatic amino acids (Phe, Tyr, and Trp). Trypsin is the golden standard of common proteolytic digestion in proteomic analysis due to its stringent cleavage specificity and high efficiency. However, the cleavage of most of these proteases in proteomics is targeted at the hydrophilic amino acids, usually exhibiting poor efficiency in the cleavage of hydrophobic domains of membrane proteins. Although chymotrypsin prefers cleavage at the large aromatic amino acids, its specificity is poor. Further, the aromatic amino acid residues with high hydrophobicity always hide into the hydrophobic core of protein structures, which are hard to be accessed and cleaved in aqueous solutions.

On the other hand, the hydrophobic short-chain aliphatic amino acids, including Val, Ala, Ile, and Leu, are widely present in all prokaryotic and eukaryotic proteomes. Therefore, a specific protease for aliphatic amino acid cleavage will be beneficial for probing the hydrophobic membrane protein domains. Wang et al. reported a cyanobacterial protease VAILase, which can specifically cleave at the short-chain aliphatic amino acids Val, Ala, Ile, Leu, and Thr with cleavage specificity up to 92% for proteomics analysis [15–17]. They utilized cyanobacterial photosystem II membrane protein complex (PSII, dimer, 700 kDa) to demonstrate the performance of VAILase in sequencing super membrane protein complex. Nearly all PSII subunits have at least one transmembrane domain. The sequence coverages of most of the subunits obtained by VAILase were higher than trypsin as the transmembrane regions lack Lys and Arg sites (Fig. 2.6). The total sequence coverages of the transmembrane regions obtained by trypsin and VAILase were 8% and 35% (Fig. 2.6). For example, the PsbD transmembrane region coverages were 0 and 46.4%, respectively [16].

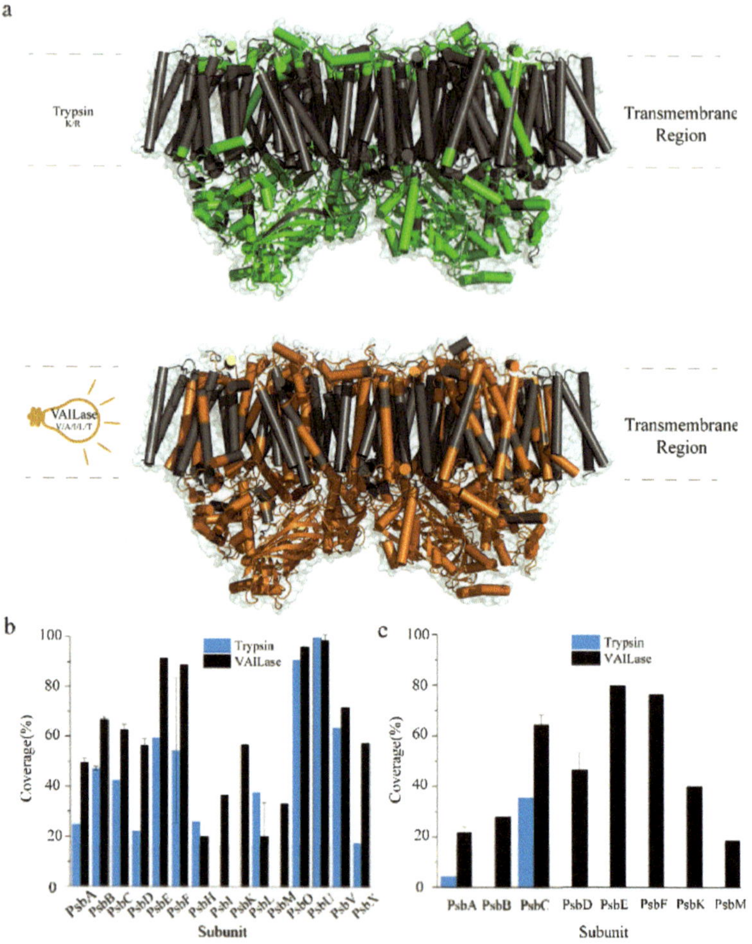

Figure 2.6 (a) The proteomics covered sequence regions of PSII subunits by using trypsin and VAILase (PDB: 3WU2). (b) Comparison of the whole sequence (left) and transmembrane sequence (right) coverages of PSII subunits obtained by using trypsin and VAILase. Reprinted with the permission from Ref. [16]. Copyright 2020 American Chemical Society.

2.3 Membrane Glycoproteomics

The extracellular domains of most of the membrane proteins are modified with glycans in highly heterogenous glycoforms. The alterations on glycosylation sites and glycan structures are both highly correlative with cancer and other serious diseases; thus, membrane glycoproteomics is a crucial tool to globally profile the cell surface glycosylation in biomarker discovery. Highly efficient enrichment of glycoproteins or glycopeptides before LC-MS/MS detection is usually the key step of membrane glycoproteomic analysis due to the much lower stoichiometry of glycoproteins compared with other unmodified proteins in the complex biological samples (Fig. 2.1). In the last 20 years, the development of glycoproteomics has been mainly driven by the specific glycopeptide enrichment methods as well as by the software for glycosylation and glycan structure elucidation.

Although the traditional lectin affinity chromatography can capture glycans and glycopeptides, the specificity and efficiency are relatively low. Further, one type of lectin usually cannot cover different types of glycans. Therefore, new glycoprotein or glycopeptide enrichment methods are urgently needed when shotgun proteomics has been developed around 2000. In 2003, Aebersold's lab described the hydrazide chemistry strategy for glycoprotein enrichment [18]. It was based on the chemical conjugation of glycoproteins to hydrazide resins by using hydrazide chemistry and the release of N-glycopeptides was achieved via specific PNGase F cleavage. The recovered peptides were then identified by LC-MS/MS for accurate characterization of the glycosylation sites due to a 0.98 Da mass difference induced during PNGase F cleavage. Stable isotope labeling of glycopeptides could be further incorporated for comparative glycoproteomic analysis of different samples (Fig. 2.7). They used a crude membrane fraction from the prostate cancer epithelial cell line for membrane glycoproteomic analysis and identified a total of 64 glycoproteins, among which 45 (70%) were bona fide or predicted transmembrane proteins. Membrane proteins that are difficult to be detected by gel electrophoresis were also identified, such as the acidic transmembrane protein signal sequence receptor α subunit (SSRA, pI = 4.39). From here, hydrazide chemistry became one of the most popular strategies for both glycoprotein and glycopeptide enrichment with high specificity. The combination of hydrazide chemistry and multiple enzyme digestion further improved the coverage of glycosylation sites significantly [19].

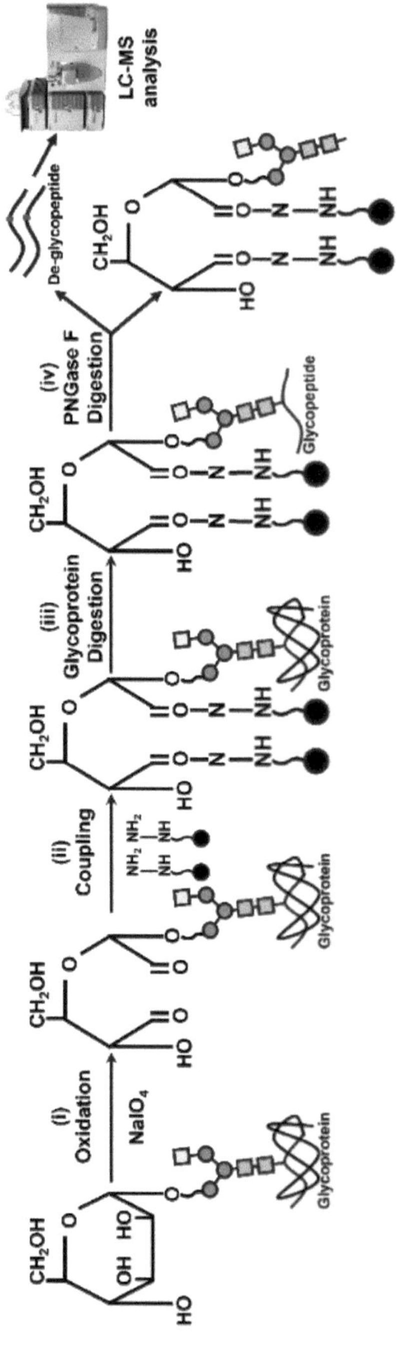

Figure 2.7 Schematic diagram of quantitative analysis of N-glycopeptides by hydrazide chemistry.

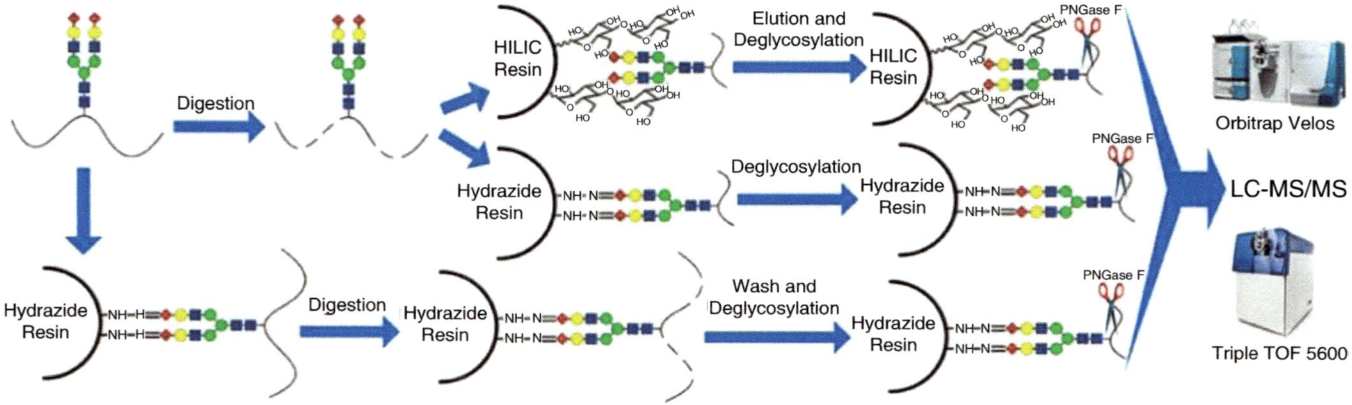

Figure 2.8 Workflow for comprehensive mapping of N-glycosylation of human liver proteins. Reprinted with the permission from Ref. [22]. Copyright 2014 American Chemical Society.

Although the hydrazide chemistry strategy can capture glycoproteins with high specificity, its first step of oxidation will break the glycan structure; thus, it is not suitable for intact glycopeptide analysis. In 2004, Roepstorff et al. developed a hydrophilic interaction chromatography (HILIC) strategy for intact glycopeptide enrichment without damaging their glycan structure [20]. They identified 62 glycosylation sites in 37 glycoproteins from the human plasma sample. Then, a lot of efforts have been made to improve the efficiency of HILIC in glycopeptide enrichment, including the preparation of novel adsorbents with higher hydrophilicity and the optimization of enrichment conditions, especially the buffer systems. Zou's lab demonstrated that the click maltose–HILIC adsorbents had good efficiency and selectivity in glycopeptide enrichment [21]. They further mapped the N-glycosylation sites of human liver proteins by combining click maltose–HILIC and hydrazide chemistry enrichment strategies (Fig. 2.8). The glycopeptide enrichment specificity was as high as 90% for hydrazide chemistry and 80% for HILIC. Finally, they identified 14480 glycopeptides, corresponding to 2210 glycoproteins and 4783 glycosylation sites [22]. About 60% of the identified glycoproteins were membrane proteins, including transforming growth factor beta receptor and 29 members of GPCRs. Most of the protein receptors on cellular membranes of the early and activated hepatic stellate cells (HSCs) were glycosylated, including platelet-derived growth factor receptor (PDGFR), leptin receptor (LEPR), tumor necrosis factor receptor (TNFR), transforming growth factor beta receptor (TGFβR), epidermal growth factor receptor (EGFR), and interleukin-4 receptor (IL4R), demonstrating the crucial role of protein glycosylation in HSC activation.

The sample preparation procedures for membrane glycoproteomic analysis are more complex than membrane proteomic analysis due to the addition of enrichment steps. Therefore, highly sensitive integrated platforms are beneficial to highly sensitive glycoproteomic profiling, especially for the low-abundant glycoproteins [21]. Wang et al. developed a monolithic capillary-column-based glycoproteomic reactor enabling high-sensitive mapping of N-glycosylation sites from minute amounts of protein sample. This novel glycoproteomic reactor was composed of an 8 cm long C12 hydrophobic monolithic capillary column for protein digestion and a 6 cm long organic–silica hybrid hydrophilic monolithic capillary column for glycopeptide enrichment and deglycosylation, which could perform all sample preparation procedures, including protein purification/desalting, tryptic digestion, HILIC enrichment, and deglycosylation of glycopeptides in about 3 h (Fig. 2.9). Finally, 486 unique N-glycosylation sites and 279 glycoproteins were reliably mapped from $\sim 10^4$ HeLa cells [23]. About 73% (204) of the identified glycoproteins were membrane glycoproteins, which were involved in cell metabolism, localization, differentiation, stimulus response, and regulation of multiple biological process. They further applied a nonionic detergent N-dodecyl β-D-maltoside (DDM), which is compatible with trypsin digestion and LC-MS/MS analysis in membrane protein extraction, to avoid tedious detergent removal steps. The combination of DDM-assistant protein extraction with the capillary-based glycoproteomic reactor further improved the sensitivity of membrane glycoprotein identification significantly [24].

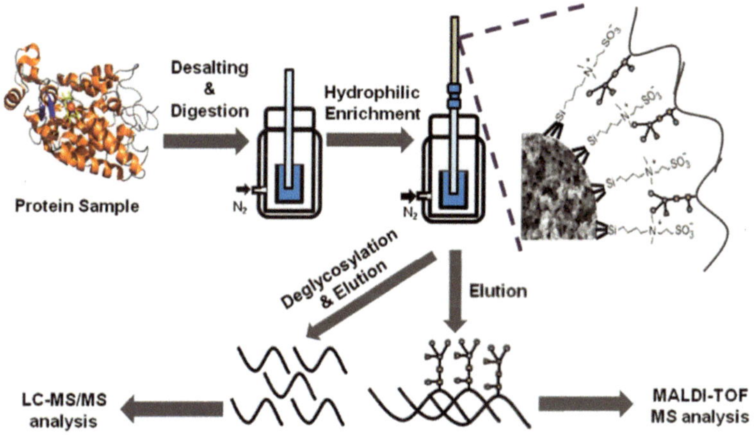

Figure 2.9 Workflow for glycoproteomic analysis with a monolithic glycoproteomic reactor. Reprinted with the permission from Ref. [23]. Copyright 2013 American Chemical Society.

The glycoprotein heterogeneity information of diverse glycan structures is lost in the glycoproteomic analysis at site level. In the recent years, the developments of tandem mass spectrometry (MS/MS) [25] and sophisticated data interpretation software [26–30] have greatly promoted the capability of glycoproteomics to characterize the site-specific glycoforms of intact glycopeptides without cleavage of glycan chains. Sufficient fragment information of both peptide backbones and attached glycans is required to comprehensively elucidate the intact glycopeptides. MS/MS dissociation methods, including collision-induced dissociation (CID), high-energy collision dissociation (HCD), electron-transfer dissociation (ETD), and electron-transfer/higher-energy collision dissociation (EThcD), have been utilized in glycopeptide identification. As CID favors cleavage of glycosidic linkages in the carbohydrate moiety to produce carbohydrate Y series ions, it is hard to confidently identify the location of the glycosylation site and peptide sequences. ETD generates abundant c/z-ions from peptide backbone fragments by primarily breaking N-Cα bonds but often suffers from incomplete fragmentation with a large amount of residual precursor ions [31–33]. HCD is a newly developed dissociation approach unique to Orbitrap MS with b- and y-type fragment ions. HCD with lower NCE values prefers glycan fragmentation, while HCD with higher NCE values prefers peptide backbone fragmentation. By applying HCD with alternating low and high NCE values, comprehensive structure information of both the glycan chain and the backbone sequence can be collected [34]. The combination of HCD with CID or ETD for complementary fragmentation further increases the identification confidence of intact glycopeptides [35, 36].

Data interpretation software have been usually designed for specific MS/MS fragmentation modes and now most commonly for HCD. However, the HCD fragmentation of peptide backbones is typically suppressed due to the labile nature of glycosidic bonds, which greatly hinders the structure elucidation of intact glycopeptides. To improve the identification of intact glycopeptides with unusual glycans or poor peptide backbone fragmentation, Ye et al. presented a platform called Glyco-Decipher to interpret the glycoproteomics data of N-glycopeptides (Fig. 2.10) [26]. Glyco-Decipher conducted a glycan database-independent peptide matching and exploited the fragmentation pattern of shared peptide backbones in glycopeptides to improve the spectrum interpretation. Glyco-Decipher identified more peptide-spectrum matches than other similar software, including Byonic, MSFragger-Glyco, StrucGP, and pGlyco 3.0, with a 33.5–178.5% increase in identified glycopeptide spectra and discovered 164 modified glycans in mouse tissues. However, current MS/MS fragmentation methods including HCD and EThcD are still disabled to provide enough informative fragment ions to fully elucidate the peptide backbone and diverse glycan structures. Highly efficient fragmentation methods such as ultraviolet laser dissociation (UVPD) coupled with specially designed powerful software are still an important development area of membrane glycoproteomics in the future [15, 37, 38].

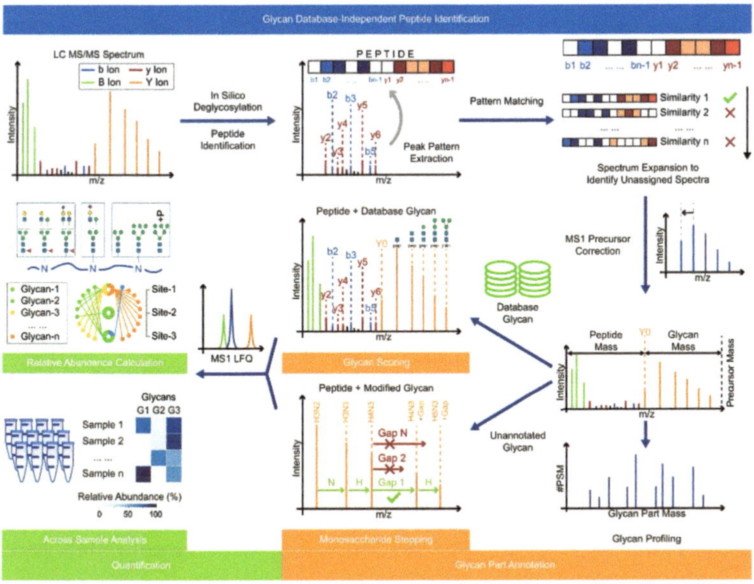

Figure 2.10 Workflow of Glyco-Decipher. Reprinted by permission from Macmillan Publishers Ltd: Ref. [26], Copyright 2022.

The novel severe acute respiratory syndrome coronavirus (SARS-CoV-2) caused a worldwide pandemic of coronavirus disease in 2019 (COVID-19). The SARS-CoV-2 is decorated by many highly glycosylated proteins, and its glycosylation (both N-link and O-link) extensively affects the host recognition, penetration, binding, recycling, and pathogenesis. Spike protein (S) is the most important surface protein of SARS-CoV-2 with about 40% of the protein surface in S trimer shields with glycans. S protein binding with ACE2 receptor, a glycoprotein expressed on the membrane of human cells, primarily mediates virus infection. Watanabe et al. first reported the site-specific glycosylation profiling of S protein

[39]. By combining multienzyme digestion, LC-MS/MS with HCD fragmentation, and interpretation by ByonicTM (Version 3.5) and ByologicTM software, 22 N-glycosylation sites with a variety of glycan structures and trace levels of O-linked glycosylation at T323/S325 were identified. Of the 22 N-glycosylation sites on the S protein, eight contain significant populations of oligomannose-type glycans, while the remaining 14 sites are dominated by complex-type glycans. Then, more glycoforms are identified by combining multienzyme digestion and Zic-HILIC enrichment (Fig. 2.11) [40]. Also 157 N-glycans attached on 22 potential N-glycosylation sites were identified from S protein. Similarly, site-specific glycosylation profiling of hACE2 with seven N-glycosylation sites and one O-site is obtained by the glycoproteomic approach [41]. The combined usage of pre-deglycosylation and O-protease OpeRATOR digestion further improved the O-glycosylation profiling with 27, 11 O-glycosylation sites attaching 23, 11 O-glycoforms identified for S and ACE2 protein, respectively [42].

Recently, metabolic labeling using unnatural sugar analogs has emerged as a powerful method to study glycosylated proteins due to the compatibility of site-specific profiling and dynamically identifying changes in glycoproteins. Lebrilla et al. investigated the incorporation efficiencies and kinetics of ^{13}C-labeled glucose into specific cell-membrane glycans and glycoproteins. By analyzing the metabolic-derivatized glycoproteins with isotope labels by LC-MS/MS, incorporation efficiencies and kinetics of common monosaccharides into specific glycans and glycoproteins on cell membrane of human intestinal and hepatic cells were obtained. Time-course studies revealed both glycan-specific and protein-dependent features of the incorporation rates. Comparative studies using different dietary saccharides and multiple cell lines revealed the variance of monosaccharide utilization and interconversion in different tissues and organisms [43]. The azide-derived monosaccharide is another widely used sugar analogs in the metabolic labeling of glycans. Their subsequent modification by click-chemistry CuAAC (copper-catalyzed azydealkyne cycloaddition) allows the formation of reactive sugars "chemical handles," which can be used selectively to increase the enrichment of glycoproteins [44, 45]. Lebrilla et al. [46] developed a glycan–protein cross-linking method to characterize the sialic-acid-mediated protein networks

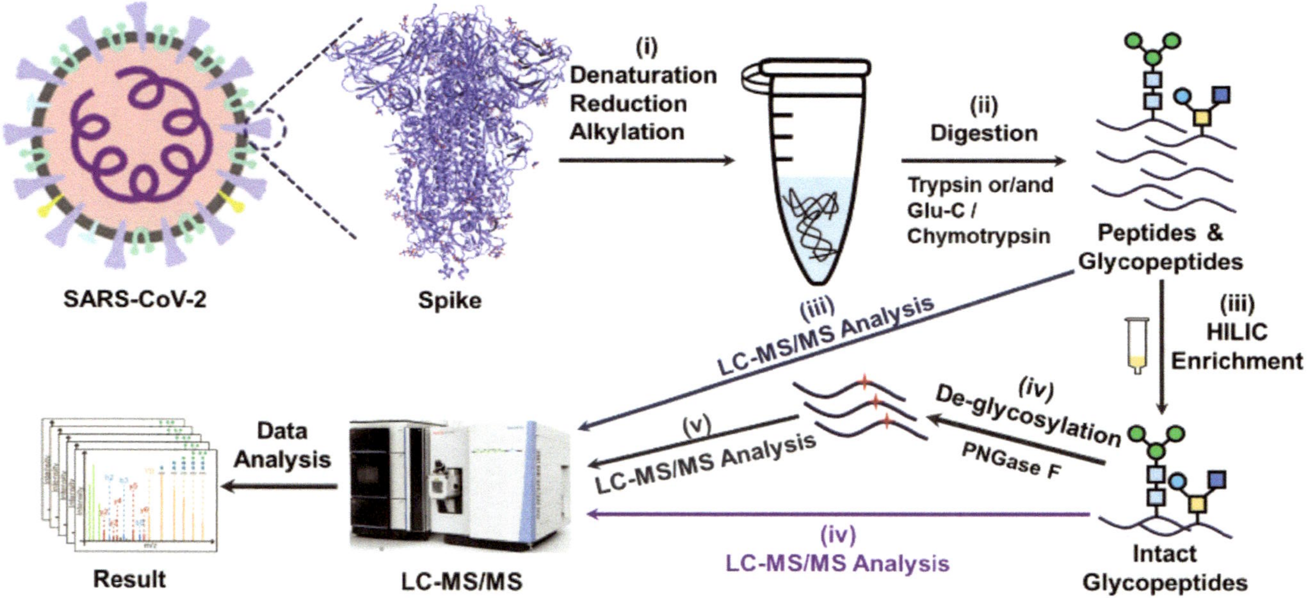

Figure 2.11 Workflow for site-specific N-glycosylation characterization of recombinant SARS-CoV-2 S proteins by using multienzyme digestion and an integrated N-glycoproteomic analysis.

on cell membranes (Fig. 2.12). In the absence of the sialic acid substrate, the analog sugar N-azidoacetylmannosamine (ManNAz) is taken to be metabolically incorporated into cell surface sialic acid-containing glycoproteins through the biosynthetic pathway. Then, cell culture is treated with a synthesized N-hydroxysuccinimide-cyclooctyne reagent, which further reacted with primary amines on the lysine side chains of nearby proteins. The peptide–glycan–peptide cross-linked pairs are analyzed via LC-MS/MS to provide a wealth of information on the site-specific glycoforms, the source glycopeptide, and the non-glycosylated peptide of the cross-linked pairs. As a result, more than 300 unique glycoprotein–protein pairs were found in the PNT2 cell membrane with over 40 glycoproteins identified as the source proteins. Nearly 50% of the source proteins were involved in cell adhesion, while only 12% of the target proteins were found with similar functions, suggesting the unique role of sialic acid in cell adhesion and migration.

2.4 Membrane Lipidomics

Although biomembranes are mainly composed of bilayers of lipids, the membrane lipid constituents are not as complex as membrane proteins and glycans. The main classes of lipids found in eukaryotic cells are glycerophospholipids (GPL), sphingolipids (SPL), and sterols (ST), and fatty acyls are building blocks in membrane lipids. Alterations of membrane lipids are linked to various diseases, while many of their functions remain unknown. Therefore, membrane lipidomics can provide an unbiased and global view of the whole membrane lipids with their constitutions, proportions, and structures, which might be extremely helpful for the discovery of functional lipids [47–50].

The membrane lipid sample is usually prepared by liquid–liquid extraction to yield an organic phase containing a complex mixture of lipids. Then, LC separation coupled with a combination of single-stage MS and tandem MS (MS/MS)—such as product ion analysis, neutral loss, and precursor scanning—is used for lipidomic profiling. Wenk et al. established an in-depth lipidomic profiling strategy and provided a large-scale semi-quantification of lipids during the life cycle of *Drosophila melanogaster*, including 230

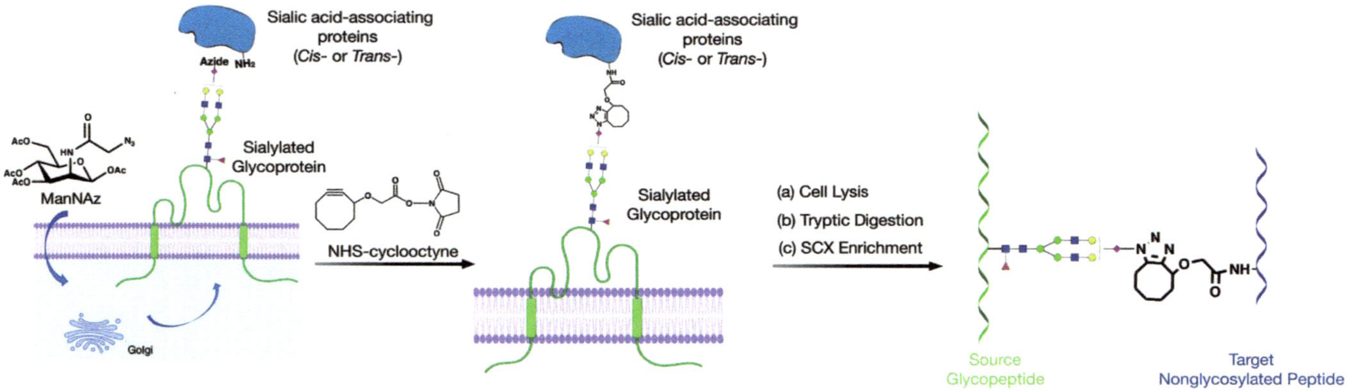

Figure 2.12 Workflow for producing and implementing glycan–protein cross-linking on the cell membranes. The metabolic incorporation of the azido groups into cell surface glycoproteins as SiaNAz. The cross-linking reaction of NHS–cyclooctyne to the azido groups and primary amines. The cross-linked glycopeptide–peptide products were obtained with subsequent cell lysis, tryptic digestion, and SCX enrichment. Reproduced from Ref. [46] by permission of The Royal Society of Chemistry.

glycerophospholipids, 210 sphingolipids, 6 sterols and sterol esters, and 60 glycerolipids (Fig. 2.13) [51]. This work was a comprehensive, systematic description of membrane lipids found in Drosophila with an emphasis on SPL, enhancing our understanding of the dynamics of lipid metabolism during the natural process of organismal development (Fig. 2.14).

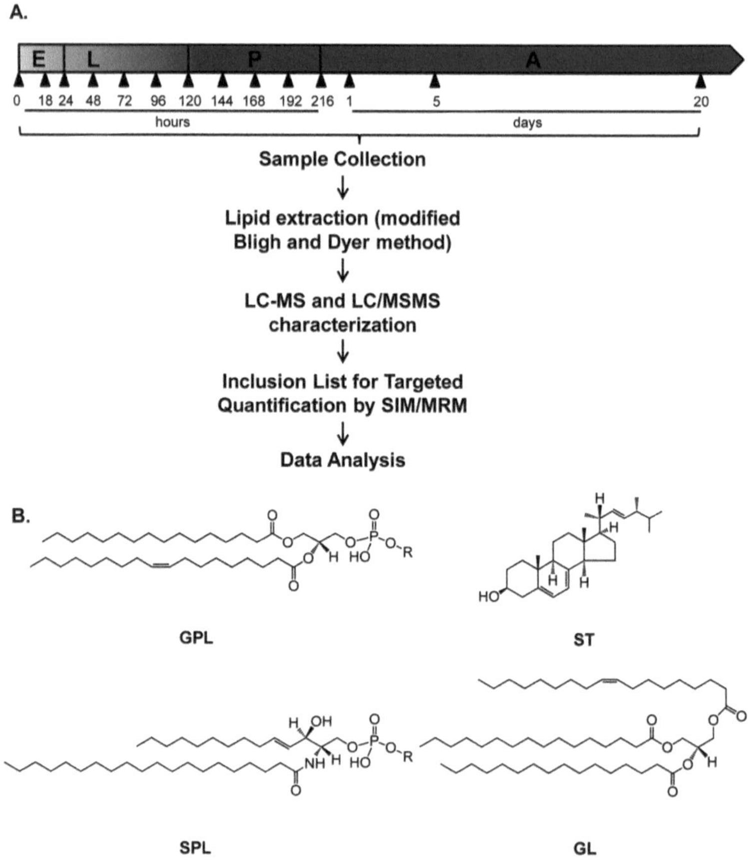

Figure 2.13 Workflow of membrane lipidomic profiling (A) and lipid classes identified (B) in Drosophila melanogaster development. Abbreviations: GPL, glycerophospholipids; ST, sterols; SPL, sphingolipids; GL, glycerolipids. Reprinted from Ref. [51]. Copyright 2013, with permission from Elsevier.

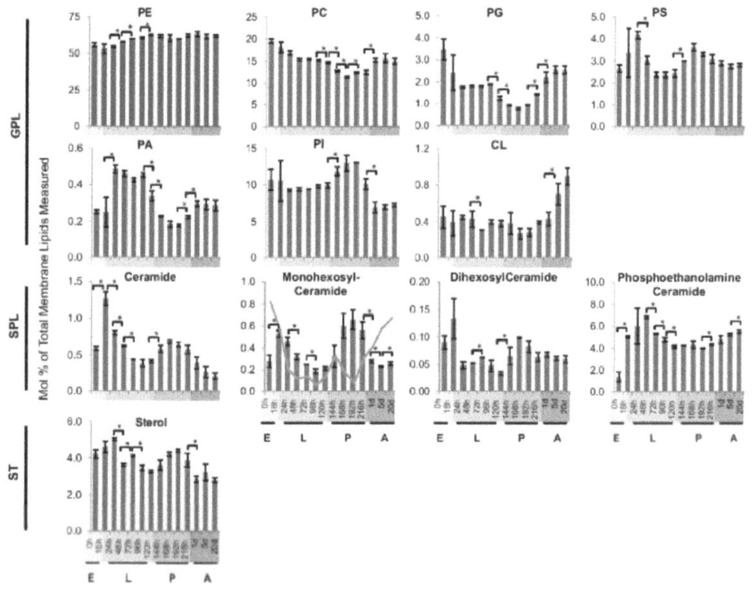

Figure 2.14 Temporal changes in membrane lipids during Drosophila development. Twelve membrane lipid classes are analyzed by using multiple LC-MS/MS approaches. Significance of difference between two adjacent time points is calculated by Student's t test, and *$p < 0.05$.

Abbreviations: A, adult; CL, cardiolipin; E, embryo; L, larva; PA, phosphatidic acid; PC, phosphatidylcholine; PE, phosphatidylethanolamine; PI, phosphatidylinositol, PG, phosphatidylglycerol, PS, phosphatidylserine. P, pupa. Reprinted from Ref. [51]. Copyright 2013, with permission from Elsevier.

2.5 Interactions between Membrane Molecules

The biomolecules within the membrane system can interact with each other to form functional complexes. For example, the lipids surrounding membrane proteins can modulate their structures and biological functions. Further, the exogenous ligand bound to membrane proteins is critical to the modulation of their biological functions and targeted drug development. However, it is still extremely challenging to characterize these essential interactions occurring on biomembranes. Recent advances in detection sensitivity and resolving power in native mass spectrometry (nMS) provide powerful tools for exploring the dynamic membrane protein–lipid interactions [52–54]. nMS performs electrospray ionization (ESI)-MS in non-denature conditions to preserve and characterize native structure, composition, interaction, and stoichiometry of biomolecules and their complexes (Fig. 2.15) [55, 56].

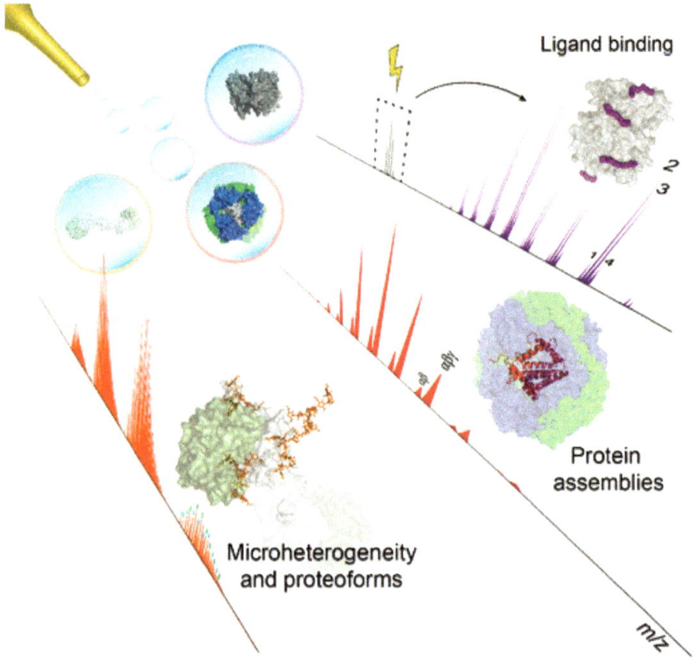

Figure 2.15 Schematic diagram of high-resolution nMS to characterize native biomolecules, including proteins, protein assemblies, and interactions between proteins and small-molecule ligands, including lipids, drugs, and carbohydrates. Reprinted with the permission from Ref. [56]. Copyright 2022 American Chemical Society.

The pioneering works of Robinson's lab established an nMS platform for investigating how lipids regulate the membrane protein associations [57]. They demonstrated that interfacial lipids are essential to the oligomeric stability of α-helical oligomeric membrane proteins by using nMS combined with HCD dissociation (Fig. 2.16) [58]. For example, a cohort of lipids were found within the dimer interface of eukaryotic biogenic transporters LeuT, and mutation of the lipid-binding sites abrogated the dimer formation. Molecular dynamics (MD) simulation demonstrated that cardiolipin acts as a bidentate ligand for the dimer formation. Cardiolipin was further observed to shift the equilibrium of the sugar transporter SemiSWEET from monomer to functional dimer. Therefore, a rational understanding of the roles of lipids in the interactions with membrane proteins could be achieved by nMS characterization.

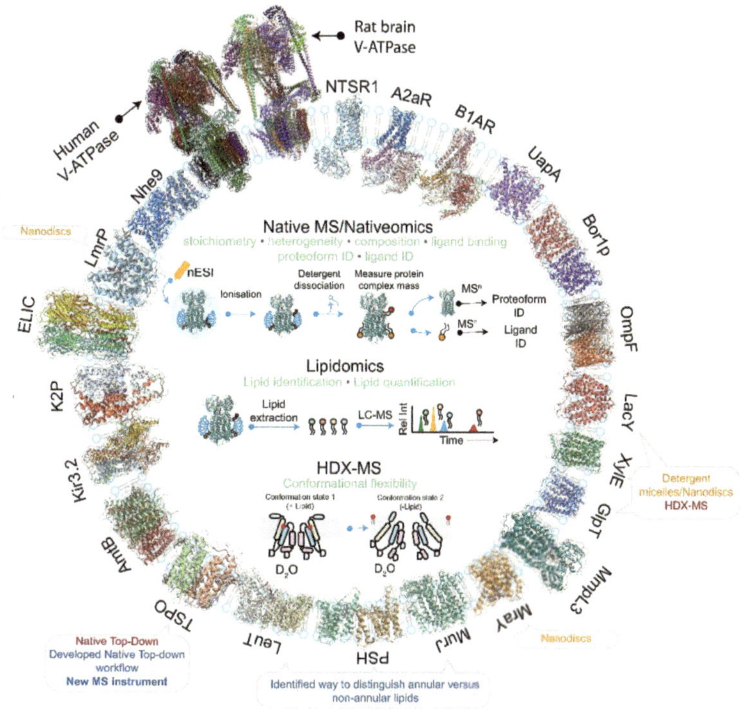

Figure 2.16 Recent membrane proteins studied by mass spectrometry. Notable experiments performed in different membrane mimetics (orange), mass spectrometry methods (red), or mass spectrometry milestones (blue) have additional descriptions found near protein structures. The interior of the vesicle shows the major mass spectrometry approaches used to interrogate different complexes, including native mass spectrometry and native top-down (nativeomics), lipidomics, and hydrogen–deuterium exchange coupled to mass spectrometry (HDX-MS). Reprinted from Ref. [59]. Copyright 2021, with permission from Elsevier.

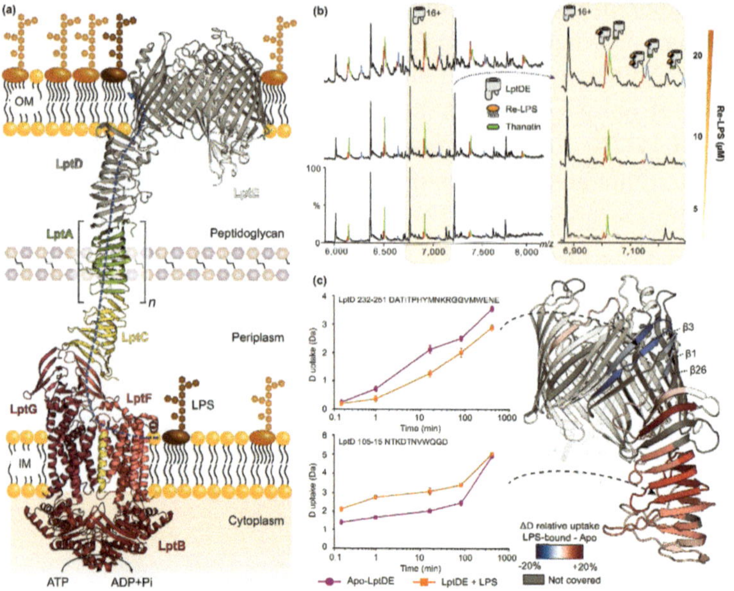

Figure 2.17 Schematic of the LPS transport system. (a) LPS is transported from the IM to the OM through a multiprotein complex comprising seven proteins. (b) Native MS analysis of LptDE with Re-LPS and thanatin. (c) Conformational dynamics of LPS-bound LptDE obtained from HDX-MS (PDB ID: 5IV9). Reprinted from Ref. [60]. Copyright 2021, with permission from Elsevier.

Lipopolysaccharide (LPS) is composed of lipid A, core oligosaccharide, and an O-antigen glycan. LPS transport to the outer membrane (OM) is crucial for the microbial surface defenses, yet the molecular details of LPS insertion by LPS transport OM protein LptDE are still unknown. Robinson et al. integrated nMS, hydrogen–deuterium exchange (HDX)-MS, and MD simulation to investigate the conformational dynamics of LPS–LptDE interactions. They demonstrated that LPS induces the LptD β-taco domain in an open state, coupled with conformational changes on β-strands adjacent to the putative lateral exit gate, and antimicrobial peptide thanatin could stabilize the β-taco, preventing the LPS transport [61].

Then, Robinson et al. further established a top-down method, "nativeomics," by integrating lipidomics, proteomics, and metabolomics analysis with nMS to identify ligands bound to membrane protein assemblies [62]. To realize nMS analysis of protein complexes with the ability of MS^n multiple-step dissociation, they modified the tribrid Orbitrap MS instrument at first. By preserving protein–ligand complexes, they defined the lipidome/metabolome interacting with membrane porins and a mitochondrial translocator to discover potential regulators of protein function.

2.6 Protocols for Mass-Spectrometry-Based Biomembranomics Study

2.6.1 Protocols for Membrane Proteomics Study

The method consists of the following steps [24, 63]:

Sample preparation of membrane proteins
Cells were washed and resuspended in 50 mM PBS (pH 7.4).

They underwent lysis by pressure-assisted cell lysis method. The cell samples were filled in a French press cell and passed dropwise at a pressure of 20,000 psi for three times.

The lysis samples were centrifuged at 1200 g and 4°C for 10 min to remove unbroken cells, and then subjected to ultracentrifugation at 120,000 g and 4°C for 90 min. The supernatant was discarded, and the pellets were collected.

The pellets were resuspended in digestion buffer containing 50 mM NH_4HCO_3 and 1% DDM. The samples were reduced with 5 mM tris(2-carboxyethyl)phosphine (TCEP) at room temperature for 20 min, followed by alkylation with 10 mM iodoacetamide (IAA) in dark at room temperature for 25 min.

Before digestion, high-frequency sonication was performed. The TPCK-treated trypsin was added at a ratio of 1:50. The digestion was conducted at 37°C for 16~18 h. High-frequency sonication could be repeated during the digestion. The digestion was quenched by adjusting the pH to about 2 using 10% formic acid.

The membrane protein digest was desalted by C18 column or fractionated by high-pH RPLC strategy.

LC-MS/MS analysis

The membrane digest was resuspended in buffer A (0.1% formic acid in H_2O). The peptide concentration was determined by NanoDrop, and 1 µg peptides were subjected to LC-MS/MS analysis.

The peptides were first loaded onto a 3 cm long trap column (150 um i.d., 3 µm 120 Å C18 beads) at a flowrate of 5 µL/min.

The loaded peptides were separated by a 50 cm long analytical column (75 i.d., 1.9 µm 120 Å C18 beads) at a flowrate of 300 nL/min. The column was kept at 50°C to improve the separation and recovery. The peptides were eluted by a gradient from 6% to 45% Buffer B (0.1% formic acid in 80% acetonitrile) in 90 min.

The MS datasets were collected by a Thermo Fusion Lumos in a data-dependent acquisition (DDA) mode. The MS was conducted at positive mode. The full scan was recorded with an Orbitrap analyzer with a resolution of 120,000 and a scan range of 350–1800 m/z. Precursor ions with a charge state of 2+ to 8+ and a minimum intensity of 5E4 were cumulated with an isolation width of 1.6 m/z and subjected to HCD with a normalized energy of 28%. The total cycle time of DDA was limited to 3 s. The MS2 spectra were recorded by an Orbitrap analyzer with a resolution of 15,000. The dynamic exclusion was enabled with an exclusion time of 60 s.

Data processing and bioinformatic analysis

The collected MS data could be analyzed by the MaxQuant software package with a false detection rate of 1% at the peptide level. The raw data were directly loaded into the software. The database in FASTA filetype could be downloaded from UniProt. The digestion rules were set to trypsin with a maximum missed cleavage of 2. The oxidation of methionine was set as variable modification, and the carbamidomethyl of cystine was set as fixed modification. The function "match between runs" was enabled with the default setting (0.7 min match window and 20 min alignment time). The other parameters were set as default.

The identified proteins with a minimum of two unique peptides were filtered as confident identifications. The proteins with a relationship to the lipid bilayer were classified as "membrane protein" based on the UniProt knowledge base.

2.6.2 Protocols for Membrane Glycoproteomics Study

Glycopeptide enrichment and hydrazide chemistry enrichment:

The method consists of six steps (Fig. 2.7) [18].

1. **Glycoprotein oxidation:** Periodate oxidation converts the cis-diol groups of carbohydrates to aldehydes (Fig. 2.7b).
2. **Coupling:** The aldehydes react with hydrazide groups immobilized on a solid support to form covalent hydrazone bonds (Fig. 2.7b). Non-glycosylated proteins are removed.
3. **Proteolysis:** The immobilized glycoproteins are proteolyzed on the solid support. The non-glycosylated peptides are removed by washing and can be collected, if desired, for further analysis, whereas the glycosylated peptides remain on the solid support.
4. **Isotope labeling:** The α-amino groups of the immobilized glycopeptides are labeled with isotopically light (d0, contains no deuteriums) or heavy (d4, contains four deuteriums) forms of succinic anhydride after the ε-amino groups of lysine are converted to homoarginine.
5. **Release:** Formerly N-linked glycopeptides are released from the solid phase by PNGase F treatment.
6. **Analysis:** The isolated peptides are identified and quantified using microcapillary high-performance liquid chromatography electrospray ionization (μLC-ESI) MS/MS or μLC separation followed by matrix-assisted laser desorption/ionization (MALDI) MS/MS.

Integrated glycoproteomic reactor

The method consists of five steps (Fig. 2.9) [23].

1. **Protein preparation:** Protein solutions (1 μg/μL) were acidified by 0.2% FA, mixed with trypsin at a mass ratio of 2.5:1, and loaded onto the C12 monolithic column by nitrogen pressure. Protein was desalted, and the column was alkalifying by ammonium bicarbonate solution.

2. **Proteolysis:** Protein digestion was initiated through filling the microreactor with 80% ACN/6 mM NH_4HCO_3 buffer at 37°C for 2 h.
3. **Enrichment:** C12 monolithic column was joined with the hydrophilic column, and the peptides were eluted to the HILIC monolithic column by using 0.2% TFA/80% ACN solution.
4. **Elution and deglycosylation:** The enriched glycopeptides were eluted with 0.1% FA for intact glycopeptide analysis or 20 mM NH_4HCO_3 containing PNGase F for deglycosylation.
5. **Analysis:** The isolated peptides are identified using microcapillary high-performance liquid chromatography electrospray ionization (μLC-ESI) MS/MS or matrix-assisted laser desorption/ionization (MALDI) MS.

Data interpretation of intact glycopeptides by Glyco-Decipher

The method consists of three steps (Fig. 2.10) [26].

1. **Glycan database-independent peptide matching:** The in silico deglycosylated spectra are searched against the protein database without setting any glycans as modifications, which determines the peptide backbone for glycopeptide spectra with rich peptide fragment ions. Then, the fragmentation patterns of the peptide backbones are extracted and utilized to match spectra that remain unannotated. This step, termed "spectrum expansion," enables the identification of peptide backbones of glycopeptide spectra with poor peptide fragmentation.
2. **Glycan annotation:** The mass of the glycan part is precisely derived by the mass difference between the precursor and the peptide backbone. The mass profile of glycans in a system is constructed without the use of glycan databases. For glycan annotation, the experimental B/Y ions in glycopeptide spectra are first matched to their theoretical fragment ions of database glycans to identify glycans. For glycans that do not match any database entries, Glyco-Decipher performs monosaccharide stepping to reveal the composition of modified glycans and potential modification moiety on them.
3. **Quantification:** The quantification module based on the elution profiles of glycopeptides is embedded in Glyco-

Decipher and allows the computation of the abundance distributions of site-specific glycans.

Site-specific glycosylation profiling of SARS-CoV-2 Spike protein

The method consists of five steps (Fig. 2.11) [40].

1. **Proteolysis:** After reduction by DTT and alkylating with IAA, recombinant S proteins are proteolyzed using an in-solution protease digestion protocol with trypsin, Glu-C, and a combination of trypsin and Glu-C added to the tube and incubated for 16 h at 37°C.
2. **Desalting:** The peptides are desalted using a pipette tip packed with a C18 membrane and lyophilized for further analysis.
3. **Enrichment:** Peptides are resuspended in 80% ACN/0.2% TFA solution after adding 2 mg of processed Zic-HILIC to the peptide solution, and rotated for 2 h at 37°C. Then, the mixture is transferred to a 200 μL pipette tip packed with a C8 membrane and washed twice with 80% ACN/0.2% TFA. Finally, intact N-glycopeptides are eluted three times with 0.1% TFA and lyophilized for further analysis.
4. **Deglycosylation:** Enriched intact N-glycopeptides were digested with PNGase F in 50 mM NH_4HCO_3 for 2 h at 37°C.
5. **Analysis:** Deglycosylated peptides and intact glycopeptides before and after HILIC enrichment are identified using microcapillary high-performance liquid chromatography electrospray ionization (μLC-ESI) MS/MS.

Metabolic labeling strategy to characterize sialic-acid-mediated protein networks on cell membranes

The method consists of six steps (Fig. 2.12) [46].

1. **PNT2 cell culture:** PNT2 cells are grown in 1640 medium supplemented with 10% (v/v) fetal bovine serum (FBS), 100 U/ml penicillin, and 100 mg/ml streptomycin at 37°C in a humidified incubator with 5% CO_2.
2. **ManNAz treatment in PNT2 cells:** At 60% cell confluency, the cells are treated with 100 mM ManNAz in cell culture media and let to stand in the incubator for 72 h.

3. **Cross-linking:** Cells are treated with a 100 mM synthesized cross-linker in PBS for 60 min at 37°C. The reaction was quenched with 0.1 M tris-buffer for 5 min at 37°C and washed with PBS three times.
4. **Cell lysis:** The cells are harvested with scrapers and resuspended in a homogenization buffer, and then lysed at 4°C using a probe sonicator.
5. **Proteolysis:** After reduction by DTT and alkylating with IAA, proteins are proteolyzed by trypsin in solution at 37°C for 18 h.
6. **Analysis:** Glycan–protein cross-links (GPX) are identified using microcapillary high-performance liquid chromatography electrospray ionization (μLC-ESI) MS/MS.

2.6.3 Protocols for Membrane Lipidomics Study

The method consists of the following steps [51, 64]:

Membrane lipid extraction

1. Cells were washed and resuspended in 100 μL 50 mM PBS (pH 7.4).
2. The cells were homogenized by high-frequency sonication in ice bath.
3. Preparation of internal standard (IS) solution containing 8 lipid standards: LPC 19:0, SM (d18:1/12:0), Cer (d18:1/17:0), PE (17:0/17:0), PC (19:0/19:0), TAG (15:0/15:0/15:0), FFA 16:0-d3, and FFA 18:0-d3 in methanol.
4. 400 μL IS solution and 200 μL chloroform were added to the cell homogenates to extraction the lipids.
5. The samples were vortexed for 1 min and subjected to shaking at 1000 rpm for 2 h at 4°C.
6. 200 μL chloroform and 250 μL H_2O were added to samples for phase separation. The samples were vortexed for 1 min and centrifugated at 10,000 g for 5 min at 4°C. The lower organic phase was collected.
7. An additional 400 μL of chloroform was added to the collected water phase and vortexed for 2 min, followed by centrifugation at 10,000 g for 5 min at 4°C. The lower organic phase was also

collected and combined with the previous collected organic phase.

8. The samples were lyophilized at 4°C and kept at −20°C until use.

LC-MS/MS analysis

9. The lyophilized samples were reconstituted in loading buffer (acetonitrile\isopropanol\H_2O, 60\35\5, 5 mM ammonium acetate).
10. Non-targeted lipidomics was performed by an ultra-performance liquid chromatography (UPLC) – AB Sciex tripleTOF 5600 plus mass spectrometry system. The extracted lipids were separated by an ACQUITY UPLC BEH C8 column (2.1 × 100 mm, 1.7 μm). The mobile phases consisted of acetonitrile\H_2O (60\40, 5 mM ammonium acetate) and acetonitrile\isopropanol (10/90, 5 mM ammonium acetate) for buffer A and buffer B, respectively. The elution gradient was set from 32% to 97% buffer B in 15 min. The column oven was applied with a temperature of 55°C.
11. The samples were first analyzed by UPLC-MS/MS in the positive mode. The gases of ion source were optimized to increase the signal and obtain a stable electrospray. The source temperature was set at 500°C, and the spray voltage was set at 5.5 kV. Both full MS and MS2 spectra were collected with high resolution. The precursor ions with top 20 intensity were isolated and subjected to fragmentation with a collision energy of 35 V and a collision energy spread (CES) of ±15 V.
12. The same samples were analyzed by UPLC-MS/MS in the negative mode again. The gases of ion source were optimized to increase the signal and obtain a stable electrospray. The source temperature was set at 600°C, and the spray voltage was set at −4.5 kV. Both full MS and MS2 spectra were collected with high resolution. The precursor ions with top 20 intensity were isolated and subjected to fragmentation with a collision energy of −45 V and a CES of ±15 V.

Data analysis

13. Both the positive and negative MS/MS data were processed via the AB Sciex LipidView software. The identification of lipids

was based on the combined MS information, including m/z, retention time, and fragmentation pattern. The detected lipids were quantified by normalization to the internal standards.

2.6.4 Experimental Details of Membrane Interactomics by nMS

The method consists of following steps [62, 65]:

Setup and optimization of mass spectrometry

1. The nMS was conducted on a Synapt G2 HDMS mass spectrometer (Waters). The MS was calibrated with CSI calibration solution containing 10 mg/mL CSI in water.
2. The parameters of MS, including ESI voltage, source pressure, trap gas, et al., were optimized to maximize the transmission of the lipid–protein complex ions.

Sample preparation

3. The expressed protein–lipid complexes were buffer exchanged into MS capable solution containing 200 mM NH4Ac and 2 × critical micelle concentration (CMC) detergent (C8E4, DDM, or β-OG) using a Micro-BioSpin P6 column (BioRad).
4. The samples were further purified by SEC to remove excess detergent and concentrated to 2–10 μM using ultrafiltration with a proper MWCO, which should be higher than the detergent micelle size to avoid the concentration of the detergent.

Native MS analysis

5. The sample was loaded onto a borosilicate capillary made nano-ESI emitter.
6. The protein–lipid complex was ionized and introduced into the MS by the static nanoflow ESI.
7. The collision energy increased to release the protein–lipid complex from the detergent micelle.
8. The full MS spectrum of the protein–lipid complex was recorded, and the peaks corresponding to the complex were identified.

9. Deconvolution of the full MS was done to obtain the accurate mass of the protein–lipid complex.
10. The peaks corresponding to the protein–lipid complex were isolated by a quadrupole.
11. Dissociation of the protein–lipid complex was observed by stepped increase in the collision energy, which could induce the departure of the lipids from the complexes.
12. The spectrum for the newly generated peaks by dissociation was monitored, which may belong to the peaks of the bound lipids. The collision energy should be carefully tuned to dissociate the protein–lipid complex, while avoiding the dissociation of the proteins and lipids.

Data analysis

13. Deconvolution of the full MS was done to obtain the accurate mass of the protein–lipid complex.
14. Based on the masses of the protein–lipid complex, proteins, and lipids, the stoichiometric ratio could be calculated.
15. Based on the recoded MS2 spectra with stepped collision energy, the binding strength between lipids and proteins could be semi-quantified.

References

1. Yates, I. and John R. (1998). Mass spectrometry and the age of the proteome, *J. Mass Spectrom.*, **33**, pp. 1–19.
2. Aebersold, R. and Mann, M. (2003). Mass spectrometry-based proteomics, *Nature*, **422**, pp. 198–207.
3. Washburn, M. P., Wolters, D., and Yates, J. R. (2001). Large-scale analysis of the yeast proteome by multidimensional protein identification technology, *Nat. Biotechnol.*, **19**, pp. 242–247.
4. Wolters, D. A., Washburn, M. P., and Yates, J. R. (2001). An automated multidimensional protein identification technology for shotgun proteomics, *Anal. Chem.*, **73**, pp. 5683–5690.
5. Wu, C. C. and Yates, J. R. (2003). The application of mass spectrometry to membrane proteomics, *Nat. Biotechnol.*, **21**, pp. 262–267.
6. Wu, C. C., MacCoss, M. J., Howell, K. E., and Yates, J. R. (2003). A method for the comprehensive proteomic analysis of membrane proteins, *Nat. Biotechnol.*, **21**, pp. 532–538.

7. Blackler, A. R., Speers, A. E., Ladinsky, M. S., and Wu, C. C. (2008). A shotgun proteomic method for the identification of membrane-embedded proteins and peptides, *J. Proteome Res.*, **7**, pp. 3028–3034.

8. Zhao, Q., Fang, F., Liang, Y., Yuan, H., Yang, K., Wu, Q., Liang, Z., Zhang, L., and Zhang, Y. (2014). 1-Dodecyl-3-methylimidazolium chloride-assisted sample preparation method for efficient integral membrane proteome analysis, *Anal. Chem.*, **86**, pp. 7544–7550.

9. Fang, F., Zhao, Q., Li, X., Liang, Z., Zhang, L., and Zhang, Y. (2016). Dissolving capability difference based sequential extraction: A versatile tool for in-depth membrane proteome analysis, *Anal. Chim. Acta*, **945**, pp. 39–46.

10. Zhao, Q., Fang, F., Shan, Y., Sui, Z., Zhao, B., Liang, Z., Zhang, L., and Zhang, Y. (2017). In-depth proteome coverage by improving efficiency for membrane proteome analysis, *Anal. Chem.*, **89**, pp. 5179–5185.

11. Fang, F., Zhao, Q., Chu, H., Liu, M., Zhao, B., Liang, Z., Zhang, L., Li, G., Wang, L., Qin, J., and Zhang, Y. (2020). Molecular dynamics simulation-assisted ionic liquid screening for deep coverage proteome analysis, *Mol. Cell Proteomics*, **19**, pp. 1724–1737.

12. Chu, H., Zhao, Q., Liu, J., Yang, K., Wang, Y., Liu, J., Zhang, K., Zhao, B., He, H., Zheng, Y., Zhong, S., Liang, Z., Zhang, L., and Zhang, Y. (2022). Ionic liquid-based extraction system for in-depth analysis of membrane protein complexes, *Anal. Chem.*, **94**, pp. 758–767.

13. Zhou, H., Ning, Z., Wang, F., Seebun, D., and Figeys, D. (2011). Proteomic reactors and their applications in biology, *FEBS J.*, **278**, pp. 3796–3806.

14. Zhou, H., Wang, F., Wang, Y., Ning, Z., Hou, W., Wright, T. G., Sundaram, M., Zhong, S., Yao, Z., and Figeys, D. (2011). Improved recovery and identification of membrane proteins from rat hepatic cells using a centrifugal proteomic reactor, *Mol. Cell Proteomics*, **10**, pp. 1–11.

15. Sun, B., Liu, Z., Fang, X., Wang, X., Lai, C., Liu, L., Xiao, C., Jiang, Y., and Wang, F. (2021). Improving the performance of proteomic analysis via VAILase cleavage and 193-nm ultraviolet photodissociation, *Anal. Chim. Acta*, **1155**, pp. 338340–338345.

16. Sun, B., Liu, Z., Fang, Z., Dong, W., Yu, Y., Ye, M., Liu, L., Wang, H., and Wang, F. (2020). Probing the proteomics dark regions by VAILase cleavage at aliphatic amino acids, *Anal. Chem.* **92**, pp. 2770–2777.

17. Sun, B., Lv, J., Chen, J., Liu, Z., Zhou, Y., Liu, L., Jin, Y., and Wang, F. (2021). Size-selective VAILase proteolysis provides dynamic insights into protein structures, *Anal Chem*, **93**, pp. 10653–10660.

18. Zhang, H., Li, X. J., Martin, D. B., and Aebersold, R. (2003). Identification and quantification of N-linked glycoproteins using hydrazide chemistry,

stable isotope labeling and mass spectrometry, *Nat. Biotechnol.*, **21**, pp. 660–666.

19. Chen, R., Jiang, X., Sun, D., Han, G., Wang, F., Ye, M., Wang, L., and Zou, H. (2009). Glycoproteomics analysis of human liver tissue by combination of multiple enzyme digestion and hydrazide chemistry, *J. Proteome Res.* **8**, pp. 651–661.

20. Hagglund, P., Bunkenborg, J., Elortza, F., Jensen, O. N., and Roepstorff, P. (2004). A new strategy for identification of N-glycosylated proteins and unambiguous assignment of their glycosylation sites using HILIC enrichment and partial deglycosylation, *J. Proteome Res.*, **3**, pp. 556–566.

21. Zhu, J., Wang, F., Chen, R., Cheng, K., Xu, B., Guo, Z., Liang, X., Ye, M., and Zou, H. (2012). Centrifugation assisted microreactor enables facile integration of trypsin digestion, hydrophilic interaction chromatography enrichment, and on-column deglycosylation for rapid and sensitive N-glycoproteome analysis, *Anal. Chem.*, **84**, pp. 5146–5153.

22. Zhu, J., Sun, Z., Cheng, K., Chen, R., Ye, M., Xu, B., Sun, D., Wang, L., Liu, J., Wang, F., and Zou, H. (2014). Comprehensive mapping of protein N-glycosylation in human liver by combining hydrophilic interaction chromatography and hydrazide chemistry, *J. Proteome Res.*, **13**, pp. 1713–1721.

23. Liu, J., Wang, F., Lin, H., Zhu, J., Bian, Y., Cheng, K., and Zou, H. (2013). Monolithic capillary column based glycoproteomic reactor for high-sensitive analysis of N-glycoproteome, *Anal. Chem.*, **85**, pp. 2847–2852.

24. Liu, J., Wang, F., Mao, J., Zhang, Z., Liu, Z., Huang, G., Cheng, K., and Zou, H. (2015). High-sensitivity N-glycoproteomic analysis of mouse brain tissue by protein extraction with a mild detergent of N-dodecyl beta-D-maltoside, *Anal. Chem.*, **87**, pp. 2054–2057.

25. Chen, Z. W., Huang, J. F., and Li, L. J. (2019). Recent advances in mass spectrometry (MS)-based glycoproteomics in complex biological samples, *Trends Anal. Chem.*, **118**, pp. 880–892.

26. Fang, Z., Qin, H., Mao, J., Wang, Z., Zhang, N., Wang, Y., Liu, L., Nie, Y., Dong, M., and Ye, M. (2022). Glyco-Decipher enables glycan database-independent peptide matching and in-depth characterization of site-specific N-glycosylation, *Nat. Commun.*, **13**, pp. 1900–1914.

27. Lu, L., Riley, N. M., Shortreed, M. R., Bertozzi, C. R., and Smith, L. M. (2020). O-Pair Search with MetaMorpheus for O-glycopeptide characterization, *Nat. Methods*, **17**, pp. 1133–1138.

28. Polasky, D. A., Yu, F. C., Teo, G. C., and Nesvizhskii, A. I. (2020). Fast and comprehensive N- and O-glycoproteomics analysis with MSFragger-Glyco, *Nat. Methods*, **17**, pp. 1125–1132.

29. Shen, J., Jia, L., Dang, L., Su, Y., Zhang, J., Xu, Y., Zhu, B., Chen, Z., Wu, J., Lan, R., Hao, Z., Ma, C., Zhao, T., Gao, N., Bai, J., Zhi, Y., Li, J., Zhang, J., and Sun, S (2021). StrucGP: De novo structural sequencing of site-specific N-glycan on glycoproteins using a modularization strategy, *Nat. Methods*, **18**, pp. 921–929.

30. Zeng, W.-F., Cao, W.-Q., Liu, M.-Q., He, S.-M., and Yang, P.-Y. (2021). Precise, fast and comprehensive analysis of intact glycopeptides and modified glycans with pGlyco3, *Nat. Methods*. **18**, pp. 1515–1523.

31. Riley, N. M., Hebert, A. S., Westphall, M. S., and Coon, J. J. (2019). Capturing site-specific heterogeneity with large-scale N-glycoproteome analysis, *Nat. Commun.*, **10**, pp. 1311–1323.

32. Li, S., Zhou, Y., Xiao, K., Li, J., and Tian, Z. (2018). Selective fragmentation of the N-glycan moiety and protein backbone of ribonuclease B on an Orbitrap Fusion Lumos Tribrid mass spectrometer, *Rapid. Commun. Mass Spectrom.*, **32**, pp. 2031–2039.

33. Yu, Q., Wang, B., Chen, Z., Urabe, G., Glover, M. S., Shi, X., Guo, L. W., Kent, K. C., and Li, L. (2017). Electron-transfer/higher-energy collision dissociation (EThcD)-enabled intact glycopeptide/glycoproteome characterization, *J. Am. Soc. Mass Spectrom.*, **28**, pp. 1751–1764.

34. Cao, L., Tolic, N., Qu, Y., Meng, D., Zhao, R., Zhang, Q., Moore, R. J., Zink, E. M., Lipton, M. S., Pasa-Tolic, L., and Wu, S. (2014). Characterization of intact N- and O-linked glycopeptides using higher energy collisional dissociation, *Anal. Biochem.*, **452**, pp. 96–102.

35. Scott, N. E., Parker, B. L., Connolly, A. M., Paulech, J., Edwards, A. V., Crossett, B., Falconer, L., Kolarich, D., Djordjevic, S. P., Hojrup, P., Packer, N. H., Larsen, M. R., and Cordwell, S. J. (2011). Simultaneous glycan-peptide characterization using hydrophilic interaction chromatography and parallel fragmentation by CID, higher energy collisional dissociation, and electron transfer dissociation MS applied to the N-linked glycoproteome of Campylobacter jejuni, *Mol. Cell Proteomics*, **10**, pp. M000031–MCP201.

36. Singh, C., Zampronio, C. G., Creese, A. J., and Cooper, H. J. (2012). Higher energy collision dissociation (HCD) product ion-triggered electron transfer dissociation (ETD) mass spectrometry for the analysis of N-linked glycoproteins, *J. Proteome Res.*, **11**, pp. 4517–4525.

37. Liu, Z., Jiang, Y., Xiao, C., Xiong, X., Wang, T., Li, ., Wang, F., Fang, X., and Yang, X. (2018). Elucidating the various multi-phosphorylation statuses

of protein functional regions by 193-nm ultraviolet photodissociation, *Chinese Chem. Lett.*, **29**, pp. 694–698.

38. Brodbelt, J. S., Morrison, L. J., and Santos, I. (2020). Ultraviolet photodissociation mass spectrometry for analysis of biological molecules, *Chem Rev*, **120**, pp. 3328–3380.
39. Watanabe, Y., Allen, J. D., Wrapp, D., McLellan, J. S., and Crispin, M. (2020). Site-specific glycan analysis of the SARS-CoV-2 spike, *Science*, pp. 330–333.
40. Zhang, Y., Zhao, W., Mao, Y., Chen, Y., Wang, S., Zhong, Y., Su, T., Gong, M., Du, D., Lu, X., Cheng, J., and Yang, H. (2020). Site-specific N-glycosylation characterization of recombinant SARS-CoV-2 spike proteins, *Mol. Cell Proteomics*, pp. 100058–100069.
41. Shajahan, A., Archer-Hartmann, S., Supekar, N. T., Gleinich, A. S., Heiss, C., and Azadi, P. (2021). Comprehensive characterization of N- and O-glycosylation of SARS-CoV-2 human receptor angiotensin converting enzyme 2, *Glycobiology*, **31**, pp. 410–424.
42. Zhao, P., Praissman, J. L., Grant, O. C., Cai, Y., Xiao, T., Rosenbalm, K. E., Aoki, K., Kellman, B. P., Bridger, R., Barouch, D. H., Brindley, M. A., Lewis, N. E., Tiemeyer, M., Chen, B., Woods, R. J., and Wells, L. (2020). Virus-receptor interactions of glycosylated SARS-CoV-2 spike and human ACE2 receptor, *Cell Host Microbe*, **28**, pp. 586–601 e6.
43. Xu, G., Wong, M., Li, Q., Park, D., Cheng, Z., and Lebrilla, C. B. (2019). Unveiling the metabolic fate of monosaccharides in cell membranes with glycomic and glycoproteomic analyses, *Chem Sci*, **10**, pp. 6992–7002.
44. Herber, J., Njavro, J., Feederle, R., Schepers, U., Muller, U. C., Brase, S., Muller, S. A., and Lichtenthaler, S. F. (2018). Click chemistry-mediated biotinylation reveals a function for the protease BACE1 in modulating the neuronal surface glycoproteome, *Mol. Cell Proteomics*, **17**, pp. 1487–1501.
45. Kim, E. and Koo, H. (2019). Biomedical applications of copper-free click chemistry: In vitro, in vivo, and ex vivo, *Chem. Sci.*, **10**, pp. 7835–7851.
46. Xie, Y., Chen, S., Li, Q., Sheng, Y., Alvarez, M. R., Reyes, J., Xu, G., Solakyildirim, K., and Lebrilla, C. B. (2021). Glycan-protein cross-linking mass spectrometry reveals sialic acid-mediated protein networks on cell surfaces, *Chem. Sci.*, **12**, pp. 8767–8777.
47. van Meer, G., Voelker, D. R., and Feigenson, G. W. (2008). Membrane lipids: Where they are and how they behave, *Nat. Rev. Mol. Cell Biol.*, **9**, pp. 112–124.

48. Ferreri, C., Sansone, A., Ferreri, R., Amézaga, J., and Tueros, I. (2020). Fatty acids and membrane lipidomics in oncology: A cross-road of nutritional, signaling and metabolic pathways, *Metabolites*, **10**, pp. 345–370.

49. Harayama, T. and Riezman, H. (2018). Understanding the diversity of membrane lipid composition, *Nat. Rev. Mol. Cell Biol.*, **19**, pp. 281–296.

50. Tessier, C., Sweers, K., Frajerman, A., Bergaoui, H., Ferreri, F., Delva, C., Lapidus, N., Lamaziere, A., Roiser, J. P., De Hert, M., and Nuss, P. (2016). Membrane lipidomics in schizophrenia patients: A correlational study with clinical and cognitive manifestations, *Transl. Psychiatry*, **6**, pp. e906–e906.

51. Guan, X. L., Cestra, G., Shui, G., Kuhrs, A., Schittenhelm, R. B, Hafen, E., van der Goot, F. G., Robinett, C. C, Gatti, M., Gonzalez-Gaitan, M., and Wenk, M. R. (2013). Biochemical membrane lipidomics during Drosophila development, *Dev. Cell*, **24**, pp. 98–111.

52. Agasid, M. T. and Robinson, C. V. (2021). Probing membrane protein–lipid interactions, *Curr. Opin. Struc. Biol.*, **69**, pp. 78–85.

53. Bolla, J. R., Fiorentino, F., and Robinson, C. V. (2021). Mass spectrometry informs the structure and dynamics of membrane proteins involved in lipid and drug transport, *Curr. Opin. Struc. Biol.*, **70**, pp. 53–60.

54. Chorev, D. S., Tang, H., Rouse, S. L., Bolla, J. R., von Kügelgen, A., Baker, L. A., Wu, D., Gault, J., Grünewald, K., Bharat, T. A. M., Matthews, S. J., and Robinson, C. V. (2020). The use of sonicated lipid vesicles for mass spectrometry of membrane protein complexes, *Nat. Protoc.*, **15**, pp. 1690–1706.

55. Chorev, D. S., and Robinson, C. V. (2020). The importance of the membrane for biophysical measurements, *Nat. Chem. Biol.*, **16**, pp. 1285–1292.

56. Tamara, S., den Boer, M. A., and Heck, A. J. R. (2022). High-resolution native mass spectrometry, *Chem. Rev.*, **122**, pp. 7269–7326.

57. Laganowsky, A., Reading, E., Allison, T. M., Ulmschneider, M B., Degiacomi, M. T., Baldwin, A. J., and Robinson, C. V. (2014). Membrane proteins bind lipids selectively to modulate their structure and function, *Nature*, **510**, pp. 172–175.

58. Gupta, K., Donlan, J. A. C., Hopper, J. T. S., Uzdavinys, P., Landreh, M., Struwe, W. B., Drew, D., Baldwin, A. J., Stansfeld, P. J., and Robinson, C. V. (2017). The role of interfacial lipids in stabilizing membrane protein oligomers, *Nature*, **541**, pp. 421–424.

59. Agasid, M. T. and Robinson, C. V. (2021). Probing membrane protein–lipid interactions, *Curr. Opin. Struc. Biol.*, **69**, pp. 78–85.

60. Bolla, J. R., Fiorentino, F., and Robinson, C. V. (2021). Mass spectrometry informs the structure and dynamics of membrane proteins involved in lipid and drug transport, *Curr. Opin. Struc. Biol.*, **70**, pp. 53–60.

61. Fiorentino, F., Sauer, J. B., Qiu, X., Corey, R. A., Cassidy, C. K., Mynors-Wallis, B., Mehmood, S., Bolla, J. R., Stansfeld, P. J., and Robinson, C. V. (2021). Dynamics of an LPS translocon induced by substrate and an antimicrobial peptide, *Nat. Chem. Biol.*, **17**, pp. 187–195.

62. Gault, J., Liko, I., Landreh, M., Shutin, D., Bolla, J. R., Jefferies, D., Agasid, M., Yen, H.-Y., Ladds, M. J. G. W., Lane, D. P., Khalid, S., Mullen, C., Remes, P. M., Huguet, R., McAlister, G., Goodwin, M., Viner, R., Syka, J. E. P., and Robinson, C. V. (2020). Combining native and 'omics' mass spectrometry to identify endogenous ligands bound to membrane proteins, *Nat. Methods*, **17**, pp. 505–508.

63. Moore, S. M., Hess, S. M., and Jorgenson, J. W. (2016). Extraction, enrichment, solubilization, and digestion techniques for membrane proteomics, *J. Proteome Res.*, **15**, pp. 1243–1252.

64. Hoene, M., Li, J., Häring, H-U., Weigert, C., Xu, G., and Lehmann, R. (2014). The lipid profile of brown adipose tissue is sex-specific in mice, *Biochim. Biophys. Acta Mol. Cell Biol. Lipids*, **1841**, pp. 1563–1570.

65. Gupta, K., Li, J., Liko, I., Gault, J., Bechara, C., Wu, D., Hopper, J. T. S., Giles, K., Benesch, J. L. P., and Robinson, C. V. (2018). Identifying key membrane protein lipid interactions using mass spectrometry, *Nat. Protoc.*, **13**, pp. 1106–1120.

Chapter 3

Mapping Membrane by High-Resolution Atomic Force Microscopy

Mingjun Cai, Jing Gao, Yan Shi, and Hongda Wang
State Key Laboratory of Electroanalytical Chemistry,
Research Center of Biomembranomics,
Changchun Institute of Applied Chemistry,
Chinese Academy of Sciences, Changchun, Jilin, China
sherry@ciac.ac.cn, hdwang@ciac.ac.cn

In this chapter, we will introduce the basic principles of atomic force microscopy (AFM) and its application in the structural information of cell membranes. High-resolution AFM imaging of membranes has shown that membrane proteins are asymmetrically distributed in a smooth lipid bilayer, and the membrane structure is diverse according to the cell type. Based on these, new cell membrane models have also been proposed. We hope these discoveries about new characteristics of cell membrane structure obtained by AFM will lay the foundation for a deep understanding of biomembranomics.

Biomembranomics: Structure, Techniques, and Applications
Edited by Hongda Wang

ISBN 978-981-4968-61-4 (Hardcover), 978-1-003-45635-3 (eBook)
www.jennystanford.com

3.1 Introduction

The cell membrane (also named the plasma membrane) encloses the entire cell, compartmentalizing the cell from the environment and maintaining the essential differences between the cytosol and the extracellular environment [1]. All cell membranes have a general structure: a very thin film consisting of lipids and proteins linked together mainly by noncovalent interactions (see Chapter 1). The lipid molecules are arranged as a continuous double layer. This lipid bilayer provides the basic fluid structure of the membrane and serves as a relatively impermeable barrier to the passage of most water-soluble molecules. Most membrane proteins span the lipid bilayer and mediate nearly all of the other functions of the membrane, including the transport of specific molecules and the catalysis of membrane-associated reactions such as ATP synthesis [2]. The cell membrane is involved in various functions of the cell, including cell communication, molecular import and export, cell growth and motility, which are all essential for cell survival. It is the complex structure of the cell membrane that endows it with so many important functions. Therefore, the structure of cell membranes at the molecular level is a fundamental and crucial question in cellular biology.

The advantage of AFM is that it can image samples in solution under physiological conditions, which is very important for biological samples, especially for the imaging analysis of cell membrane structure. AFM can image cell membranes at nanoscale spatial resolution and obtain 3D structural characteristics of cell membranes.

Here, we summarize the progress of AFM in the study of membrane structure, which is of great significance in studying the structure and function of cell membrane from the perspective of biomembranomics.

3.2 Principles of AFM and Its Related Technologies

Atomic force microscopy, which was invented in 1986 by Binnig et al., is an important member of the family of scanning probe microscopy techniques and provides molecular-scale imaging of nonconducting materials [3]. AFM has been widely applied to biological samples because of the high resolution and the ability to work under physiological conditions. Living cells and cell membranes have been extensively imaged by AFM, promoting the understanding of cell membranes [4, 5].

3.2.1 Basics of AFM

The device and the working principle of AFM are shown in Fig. 3.1 [6]. The basic components include a piezoelectric scanner, flexible cantilever containing a sharp probe, laser, photodiode detector, and feedback electronics. The principle of AFM is simple whereby the movements of a flexible cantilever with a tiny tip detect the weak interactions between atoms on the tip and the sample surface. As the probe approaches the sample surface, the attraction force between the sample and the tip induces either a deflection of the cantilever probe (contact mode) or an amplitude change (tapping mode). A photodetector detects the shift or amplitude change and converts this signal to an electrical signal read by a computer; as a result, the surface information is recorded.

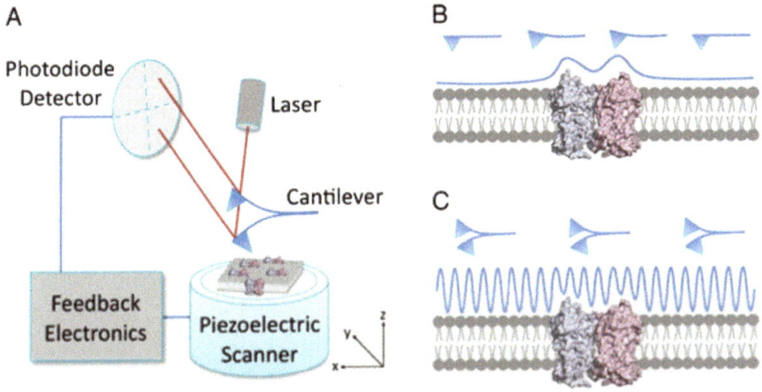

Figure 3.1 Schematic diagram of atomic force microscopy. (A) Components of an atomic force microscope. (B) Contact mode imaging. (C) Tapping mode or MAC mode imaging [6]. Reprinted from Ref. [6], Copyright 2014, with permission from Elsevier.

3.2.2 AFM Imaging Mode

The imaging modes of AFM are divided into the following major categories according to whether the tip contacts the sample:

contact mode AFM and the tapping mode (acoustic AC (AAC) mode) AFM [6, 7]. There is another type of tapping mode AFM in which the cantilever is attached to a magnetic particle and then the cantilever is driven directly by a magnetic field, named magnetic AC mode (MAC mode) AFM. A better signal-to-noise ratio means that cantilevers with much lower force constants and much smaller vibrating amplitudes can be used. The MAC mode reduces damage to samples and greatly improves the resolution of AFM imaging in liquid environments [8, 9].

The contact mode was developed first. It has the highest resolution and is commonly used. In this mode, the AFM tip contacts the sample surface and scans using a piezoelectric scanner (Fig. 3.1B). The feedback electronics adjust the vertical position of the piezoelectric scanner to maintain a constant laser deflection of the cantilever at a predetermined set-point value, thereby maintaining constant force on the sample. The adjustment in the vertical position required to maintain a constant deflection is recorded and used to generate a topographical image of the sample. The contact mode can obtain high-resolution images in either air or liquid. However, due to the existence of surface shear stress, this mode would deform the soft sample, so it is not a best imaging mode for the soft sample.

The tapping mode uses a constant driving force (acoustic, magnetic) to vibrate the cantilever at a certain frequency [10, 11]. The oscillation amplitude is monitored, and the feedback electronics adjusts the vertical position of the piezoelectric scanner to maintain a constant amplitude. The amplitude of the cantilever changes as the height of the sample changes (Fig. 3.1C). The tapping mode is more suitable for softer or loosely adsorbed samples. The resolution of the tapping mode is similar to that of the contact mode. The advantage of tapping mode AFM is the short contact time and the minimum contact force, which effectively prevents the samples from damage during scanning. Tapping mode AFM (or MAC mode) in liquid is the most widely used AFM mode in biological fields [12–14].

3.2.3 Principle of Topography and Recognition (TREC) Imaging

Atomic force microscopy provides incomparable advantages for obtaining topography information regarding the surface of a sample.

However, for a complex sample with multiple components, the height measurements of AFM are unable to differentiate between the species (e.g., for a mixture of two proteins, we cannot distinguish them from their shape and height). Molecular recognition (TREC) imaging is a technique for the identification of a particular molecule in a complex [15, 16]. In principle, the AFM tip is functionalized with specific molecules (e.g., antibodies) to recognize the counterpart molecules on the surface (e.g., antigens). High-resolution topography and recognition images are simultaneously obtained. As shown in Fig. 3.2A, an AFM tip tethered to an antibody scans the surface. When there is a specific interaction between the sample and the antibody on the AFM tip, the amplitude of the AFM cantilever will be reduced, and feedback systems adjust the cantilever position (ΔA) to maintain a constant amplitude (Fig. 3.2B,C). This peak change corresponds to the molecular recognition signal from the specific interactions. Molecular recognition imaging has excellent specificity and reproducibility, and it further expands the biological applications of AFM. AFM and TREC have excellent properties for the study of cell membrane structure [17, 18].

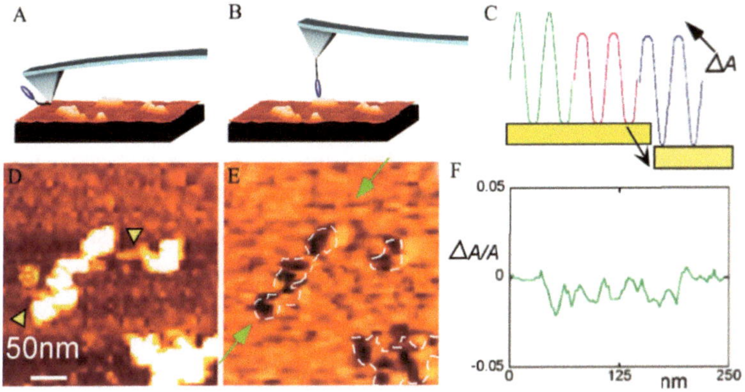

Figure 3.2 Principle of topography and recognition imaging. An AFM tip tethered to an antibody scans the sample (A, B). Antigen–antibody binding causes a transient reduction in the oscillation amplitude of the tip (C). The AFM feedback system converts this change into a recognition signal. (D, E) A topographic image of MMTV chromatin arrays and the corresponding recognition image. (F) A plot of the peak signal, ΔA, for the portion of the recognition image between the green arrows in (E). Reprinted from Ref. [18]. Copyright 2018, with permission from Elsevier.

3.3 Imaging the Cell Membrane

Atomic force microscopy has advantages over other analytical methods in terms of working environment, resolution, and sample preparation, such as easy sample preparation without special treatment (fixation or staining). As a result, an accurate and quick image of the cell membrane's surface and structure of the lipid bilayer and the membrane protein can be obtained with nanometer resolution under most physiological conditions. For that reason, the AFM imaging technology has been used in the study of the cell membrane proteins and the cell membrane structure for a long time [19–21].

3.3.1 Imaging the Lipid/Lipid Bilayer

Cell membranes are complex quasi-two-dimensional, supramolecular assemblies of a variety of lipids, proteins, and carbohydrates. Cell membranes contain thousands of diverse lipid types. Membranes provide an identity to the cell and its organelles and represent an ideal milieu for the proper functioning of membrane proteins. Lipids are the main component of cellular membranes and have key roles in many essential cellular processes, as well as structural and signaling functions [22, 23].

The complexity of the structure and function of natural cell membranes makes it very difficult to study them. Therefore, researchers first used artificial membranes to mimic the cell membrane. Under near physiological conditions, AFM shows the nanostructures and mechanical properties of artificial membranes, the biomembrane functions, and the interactions between membranes and nanoparticles. For example, using AFM, Karim El Kirat et al. explored the nanostructured information of artificial lipid bilayer membranes composed of dioleoylphosphatidylcholine/dipalmitoylphosphatidylcholine (DOPC/ DPPC) and the effect of Triton X-100 on these membranes (Fig. 3.3). The addition of a cholesterol molecule to the artificial membrane significantly increased the resistance of the bilayer to Triton X-100 (Fig. 3.4). AFM image indicated that cholesterol molecules play a key role in forming domains of resistant to Triton X-100 [24–26].

Figure 3.3 Triton X-100 interaction with supported lipid membranes on an AFM height image (20 × 20 μm; Z-scale: 10 nm) of a mixed DOPC/DPPC (1:1) bilayer was first recorded in Tris buffer before Triton X-100 addition (A). After addition of 0.48 mM TX-100, images of the same zone were acquired at different incubation times: (B) 10 min, (C) 30 min, (D) 60 min, (E) 90 min, and (F) after 120 min [24]. Reprinted with the permission from Ref. [24]. Copyright 2006 American Chemical Society.

Because artificial membranes have a definite composition, a membrane protein (e.g., the KcsA potassium channel) can be inserted into the lipid bilayer to visualize the dynamic conformation of the membrane protein (channel opening and closing) at high resolution [27]. The KcsA-reconstituted proteoliposomes were added onto the mica surface, and the liposome membrane was extended on the mica substrate, forming the flat bilayer. The cytoplasmic part of the KcsA channel protruded from the membrane surface was shown in Fig. 3.5.

Although some properties of cell membranes can be derived from artificial membranes, direct observation of natural cell membranes is more important. In 2013, AFM successfully imaged plasma membranes from the eye lens. The AFM image showed a flat bilayer membrane and its height profile with a thickness of approximately 4 nm (Fig. 3.6) [28].

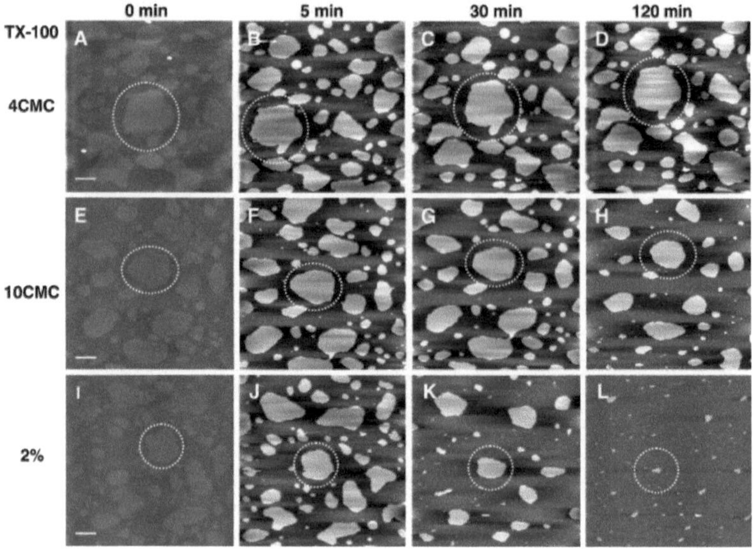

Figure 3.4 Triton X-100 interaction with supported DOPC/SM/Chol 2:1:1 (mol/mol/mol). AFM height images (20 × 20 μm; Z-scale: 10 nm) of mixed DOPC/SM/Chol 2:1:1 (mol/mol/mol) bilayers were first recorded in Tris buffer before Triton X-100 addition (A, E, and I). Different TX-100 concentrations were tested on each bilayer and images of the same area were acquired at different incubation times: 0.96 mM TX-100 (4CMC): (B) 5, (C) 30, and (D) 120 min; 2.40 mM TX-100 (10CMC): (F) 5 min, (G) 30 min, and (H) 120 min; 2% TX-100 (v/v, 143CMC, 34 mM): (J) 5 min, (K) 30 min, and (L) 120 min. Scale bars are 2.5 μm [26]. Reprinted from Ref. [26]. Copyright 2007, with permission from Elsevier.

3.3.2 Imaging Membrane Proteins

Membrane proteins are highly hydrophobic and are natively embedded in a lipid bilayer [29]. AFM is a powerful tool for imaging membrane proteins under physiological conditions. The native properties of membrane proteins can be studied since the conditions under which these proteins normally exist can be maintained during AFM imaging [30, 31]. Below we list typical examples of imaging membrane proteins by AFM at the nanometer resolution.

First, AFM revealed the packing arrangement of the light receptor rhodopsin (G protein-coupled receptor). Rhodopsin was found to be dimeric and to form higher oligomers, as shown in Fig. 3.7. The higher-order oligomerization of rhodopsin and other G protein-coupled receptors (GPCRs) has important implications for G protein recognition, binding kinetics, signal amplification, and signal termination. A white ellipse marks a rhodopsin dimer, the building block of the Para crystal. Three different surface types are evident: ① mica, ② lipids, and ③ the cytoplasmic side of the disc membrane. It is worth pointing out that AFM imaging of a native eye-disc membrane also presents the flat bilayer membrane (Fig. 3.7 ②).

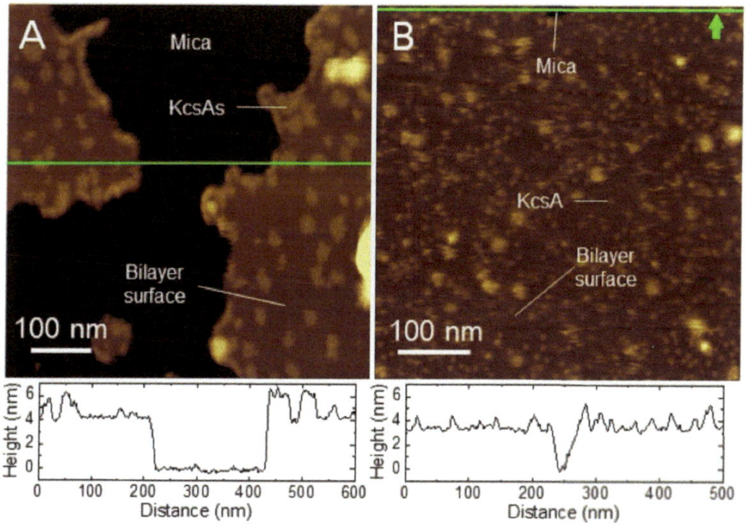

Figure 3.5 The AFM imaging of the KcsA embedded in the membrane. (A and B) AFM images of the CPD-truncated KcsA channel in lipid bilayer and the height profiles [27]. Reprinted by permission from Macmillan Publishers Ltd: Ref. [27], Copyright 2013.

A unique feature of AFM is its ability to observe single membrane proteins at work. This has been used to study the conformational changes that certain membrane proteins undergo in response to external stimuli, for example, light, pH change, and the presence of ligands. Connexin 26 (Cx26) gap junctions allow cells to communicate and respond to external signals. Upon cell death, relatively high amounts of calcium ions are released into the extracellular space. To impede influx of this second messenger into healthy cells and false initiation of signaling cascades, gap junctions close their pores. Using in situ AFM, Muller et al. studied the fine structure of the human communication channel protein Cx26 in a dense protein array. As shown in Fig. 3.8, AFM clearly displayed a hexamer structure of

Cx26. The calcium ions induced conformational change of Cx26 can be visualized at a single-molecule resolution.

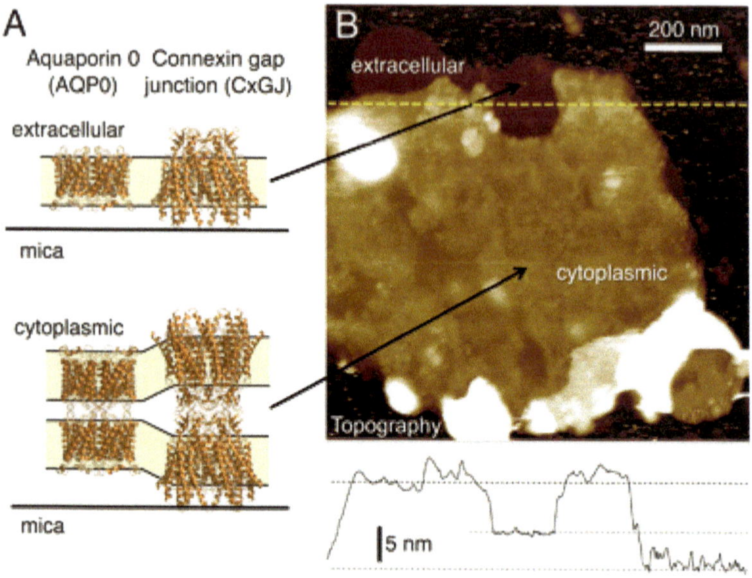

Figure 3.6 Isolated plasma membranes from the eye lens maintain the native double-layered architecture. (A) Crystal structure representation of the lens membrane proteins connexin and aquaporin 0, showing single-layer configuration (exposing the extracellular side) and the native, double-layer configuration. (B) Overview topography of an isolated lens membrane revealing a darker, single-layer region and a brighter, double-layer region. The line graph shows the cross section along the dotted line in the image and reveals the thickness of the single and double layers. The false color scale is 25 nm. Reproduced from Ref. [28] by permission of The Royal Society of Chemistry.

Usukura et al. provided direct evidence of the existence of clathrin pits on the cytoplasmic side of cell membranes using high-resolution AFM imaging (Fig. 3.9) [32]. The same type of membrane proteins was found to be randomly distributed on the cytoplasmic side of the membrane, which was verified via the use of labeled quantum dots to locate prestin (a membrane protein) in the cytoplasmic side membrane of CHO cells [33]. Furthermore, Vijayan et al. studied the changes in membrane proteins in response to acute stress that induced stress signaling pathways, and they demonstrated that the protein domains became larger after stressor exposure [34]. These results demonstrated that the pattern of membrane proteins in the cytoplasmic side of membranes is closely associated with membrane activity.

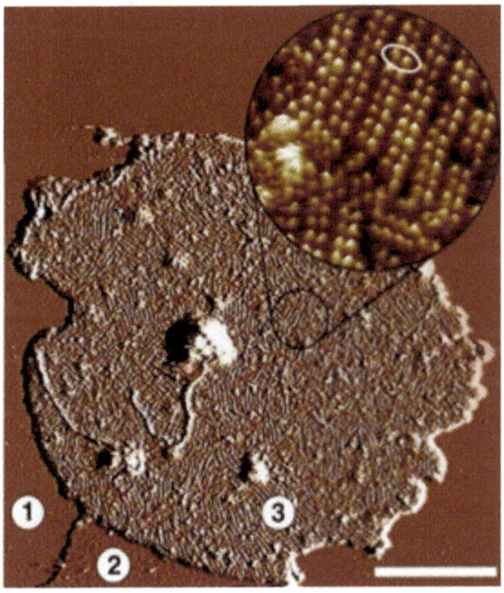

Figure 3.7 AFM imaging of a native eye-disc membrane. Contact mode AFM imaging of native disk membranes isolated from mouse retinal rods. The deflection image reveals paracrystal line arrays of the G protein-coupled receptor rhodopsin. At higher magnification (round inset), rows of rhodopsin dimers are discerned. Scale bar, 200 nm. Reprinted from Ref. [31]. Copyright 2012, with permission from Elsevier.

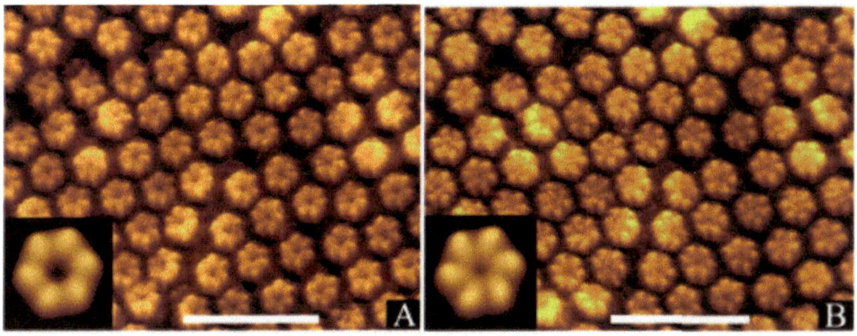

Figure 3.8 Visualization of calcium ions induced conformational changes of the extracellular connexon surface by contact mode AFM. Topography of the extracellular surfaces of a native connexin 26 gap junction plaque imaged in buffer solution in the absence (A) and presence of 0.5 mM Ca^{2+} (B). The diameter of the extracellular connexon pore is reduced significantly in the presence of calcium. Insets in (A) and (B) are correlation averages calculated from the corresponding topography with a lateral resolution of 12 Å. Scale bars represent 25 nm in (A) and (B). The frame sizes of the insets are 68 Å. Reprinted from Ref. [31]. Copyright 2012, with permission from Elsevier.

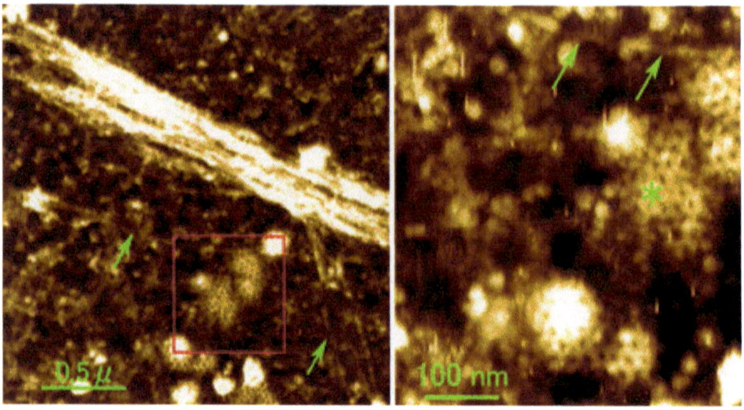

Figure 3.9 AFM images of clathrin coats and actin filaments at the cytoplasmic side of the plasma membrane. The right shows an enlarged view of the boxed area on the left. Clathrin-coated pits are clearly observed in the boxed area. Arrows indicate actin filaments. Reprinted from Ref. [32]. Copyright 2012, with permission from Oxford University Press.

3.4 Study on Erythrocyte Membrane by AFM

Erythrocytes (red blood cells, RBCs) are the most common type of blood cells in vertebrates. Erythrocytes are responsible for transporting oxygen from the lungs to various tissues in the body, while transporting the carbon dioxide produced by cell metabolism back to the lungs. Mature mammalian erythrocytes have no nucleus and endomembrane systems, so the plasma membrane of erythrocytes is the simplest and easiest biological membrane to study. When the plasma membrane of an erythrocyte is ruptured after hypoosmotic treatment, it releases hemoglobin and other intracellular soluble proteins, and the erythrocyte membrane retains its original basic shape and size. Therefore, erythrocytes provide an ideal material for studying the plasma membrane structure and the relationship between itself and the membrane skeleton [35].

3.4.1 Ectoplasmic Side of Erythrocyte Membranes

There are two reported methods for the preparation of the ectoplasmic side of the erythrocyte membrane: hypotonic lysis and the sheared open approach [36]. As shown in Fig. 3.10A,B, the ectoplasmic side of the erythrocyte membrane is smooth, with an average roughness of approximately 0.2 nm. To demonstrate that AFM could observe significant membrane protrusions on the cell membrane, type B blood antibodies were attached to the outer plasma membrane of type B human erythrocytes. Single antibody particles were observed on the membrane (green arrows in Fig. 3.10C), indicating that there was no protein projection outside the erythrocyte membrane. The fine structure of the ectoplasmic side of the membrane was imaged at high resolution with a small Z-axis scale (Fig. 3.10D); the image did not indicate the presence of any obvious protrusion or particle (Fig. 3.10G). Cross-section data were smoothened by a fast Fourier transform (FFT) filter, as shown by the red line in Fig. 3.10H. The depth of 50 curves is shown as the red bar in Fig. 3.10M, with an average depth of 0.22 ± 0.13 nm (the red bar). This result indicated that the ectoplasmic side of the membrane is rather smooth [17, 18].

To confirm the location of oligosaccharides and proteins on the cell membrane, the ectoplasmic side of the erythrocyte membrane was digested by PNGase F, which cleaves an entire saccharide from a glycoprotein. Figure 3.10E shows the image of the ectoplasmic side of the membrane following treatment by PNGase F. Interestingly, there were many indents in the membrane compared with Fig. 3.10D. From the smoothened cross-section line, the deep area of the curve was pronounced (blue bar in Fig. 3.10M). It is noteworthy that the depth of the indents was just within the range of the height of the hydrophilic lipid head (0.80 ± 0.15 nm) in the lipid bilayer despite the tip pressure on the sample, which indicated that oligosaccharides could occupy the position of hydrophilic lipid heads.

Furthermore, to reveal the location of transmembrane proteins, the ectoplasmic side of the membrane was digested by both PNGase F and trypsin, which are used to cleave the transmembrane proteins of intact erythrocytes. As shown in Fig. 3.10F, there were deeper indents in the membrane after PNGase F and trypsin treatment than those in Fig. 3.10E. The section analysis in Fig. 3.10K,L shows a typical indent from Fig. 3.10F. The average depth of the indents was 2.50 ± 0.82 nm (green bar in Fig. 3.10M), which was in the range of the thickness of the entire lipid bilayer. Although the result from PNGase F digestion was not perfect due to the low indents for the interpretation of oligosaccharide arrangement in the ectoplasmic side of the membrane, the trypsin digestion result showed low noise and deep indents; these clearly indicated the location of the transmembrane protein in the ectoplasmic side of the membrane. The width of the indents was both in the range of 20–220 nm with the peak at 50–60 nm (Fig. 3.10N,O), which indicated that these indents created by trypsin could be the same sites as those made by PNGase F. These results indicated that oligosaccharides may be spread in the middle of the lipid heads rather than protruding from the membrane surface and that the transmembrane parts of proteins may be buried in the lipid membrane under oligosaccharides.

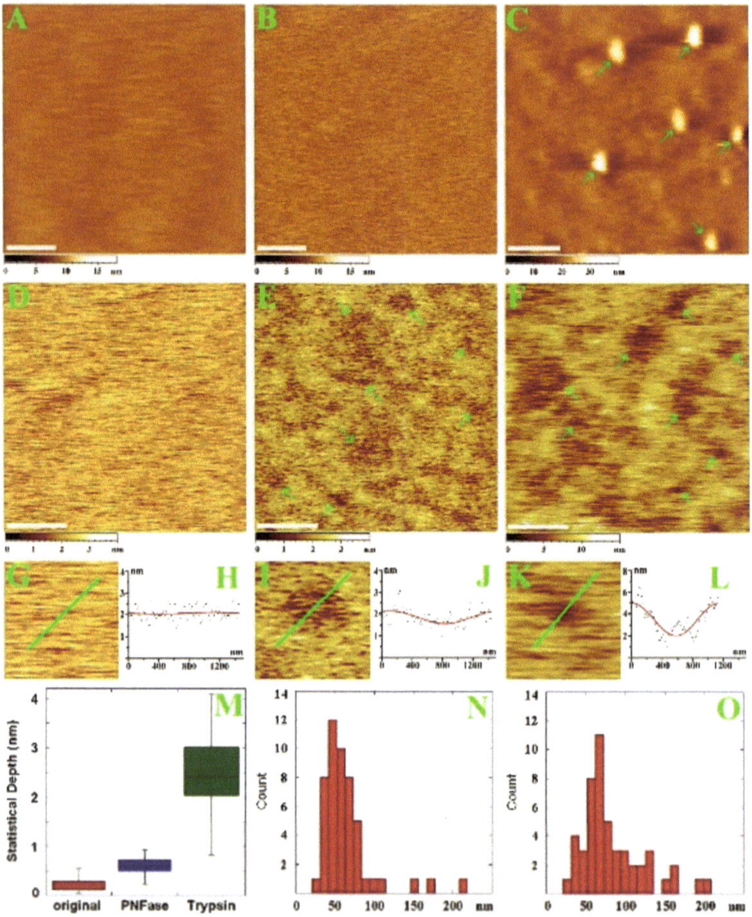

Figure 3.10 (A) The ectoplasmic side of the membrane of erythrocytes prepared by hypotonic lysis. Scale bar is 300 nm. (B) The ectoplasmic side

membrane of erythrocytes prepared by sheared open approach. Scale is 200 nm. (C) Control experiment for the smooth ectoplasmic side of the membrane. Green arrows indicate blood group B antibodies. Scale is 300 nm. (D) High-resolution image of the ectoplasmic side of the membrane. Z-axis scale is 4 nm. (E) High-resolution image of the ectoplasmic side of the membrane treated by PNGase F. Arrows point to the presumed regions of oligosaccharide aggregations in the membrane. Z-axis scale is 4 nm. (F) The ectoplasmic side of the membrane treated by PNGase F and trypsin. Arrows indicate the regions of embedded transmembrane proteins. Z-axis scale is 10 nm. Scale bar is 150 nm for Fig. 3.10D–F. (G) Magnification of the membrane in Fig. 3.10D. (H) Cross-section analysis along the line in Fig. 3.10G. (I) Magnification of the membrane from Fig. 3.10E. (J) Cross-section analysis along the line in Fig. 3.10I. (K) Magnification of the membrane in Fig. 3.10F. (L) Cross-section analysis along the line in Fig. 3.10K. (M) Statistical depths of indents in the ectoplasmic side of the membrane. Red bar corresponds to the untreated ectoplasmic side of the membrane. Blue bar corresponds to the ectoplasmic side of the membrane treated by PNGase. Green bar corresponds to the ectoplasmic side of the membrane treated by PNGase and trypsin (marked as "Trypsin" in the figure). (N) Distribution of the indent widths after PNGase F treatment. (O) Distribution of the indent widths after PNGase F and trypsin treatment. Reprinted from Ref. [38]. Copyright 2010, with permission from Elsevier.

3.4.2 Cytoplasmic Side of Erythrocyte Membranes

The cytoplasmic side of erythrocyte membranes can be prepared by the sheared open method on APTES-mica substrate and directly imaged by tapping mode AFM under quasi-native conditions. With this method, membrane proteins are found to be located in the cytoplasmic side of the membrane (Fig. 3.11A), with a height and average roughness of about 10 nm and 1.9 nm, respectively. Arrows point to free lipid bilayers on the edge of the membranes with a height of 2.94 ± 0.36 nm, and the magnified image (Fig. 3.11B) shows that the protein layer is just above the lipid bilayer. There are two clear steps shown by the cross-section analysis (Fig. 3.11C): The arrows point to the protein layer (left) and the free lipid bilayer (right). The area percentages of free lipid bilayers in cell membranes were from 0% to 10%, and not all of the patches showed free lipid bilayers. This result indicated that a fewer lipids were free in cell membranes than expected, which was consistent with the statement "membranes are more mosaic than fluid" [37].

The median resolution image of the cytoplasmic side of the cell membrane revealed dense particles that represented the proteins in the membrane (Fig. 3.11D). The proteins displayed a broad height distribution between 2.5 and 13 nm, with multiple peaks (Fig. 3.11F). Figure 3.11E shows a higher resolution image, in which the proteins were very close to each other without the free lipid bilayer being visible. The proteins showed a diameter distribution between 20 and 130 nm (Fig. 3.11G), and it was assumed that protein aggregations were present in the membrane due to the large particle size.

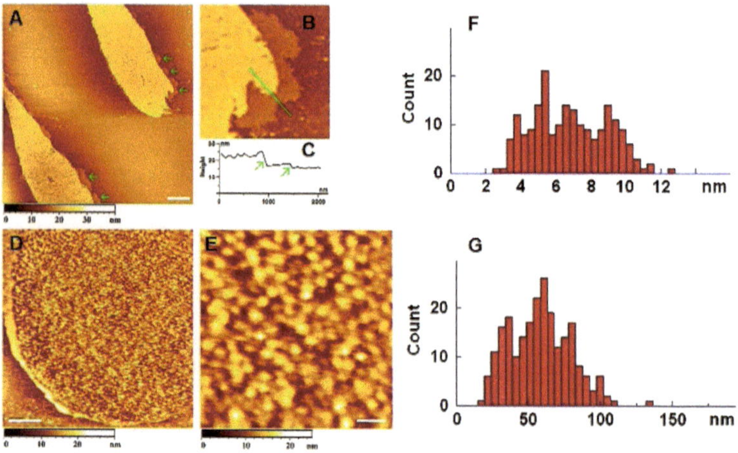

Figure 3.11 The cytoplasmic side of the erythrocyte membrane. (A) Arrows point to the free lipid bilayer under the membrane proteins. Scale bar is 3 μm. (B) Magnification of the membrane in (A). (C) Cross-section analysis along the line in (B). Arrows point to the protein layer and free lipid bilayer. (D) Cytoplasmic side of the liquid-phase membrane. Scale bar is 1 μm. (E) High-resolution image of the cytoplasmic side of the membrane. Scale bar is 200 nm. (F) Height distribution of proteins in the cytoplasmic side of the membrane. (G) Width distribution of proteins in the cytoplasmic side of the membrane. Reprinted from Ref. [38]. Copyright 2010, with permission from Elsevier.

To study the relationship between proteins and the lipid bilayer, the membrane proteins were digested by trypsin. Digestion removed most of the proteins, but some high particles of undigested proteins remained in the membrane (Fig. 3.12A). Figure 3.12B shows the magnification of the image from Fig. 3.12A. Arrows point to small particles in the membrane, which could be the transmembrane parts of membrane proteins. Figure 3.12C shows a cross section along the green line of the three small particles in Fig. 3.12B. The height distribution of these particles above the membrane was in the range of 2–10 nm with the peak at 3–4 nm, as shown in Fig. 3.12D. The average width of the small particles was 13.6 ± 2.7 nm. These results further demonstrate the ability of AFM to analyze the fine structure of cell membranes.

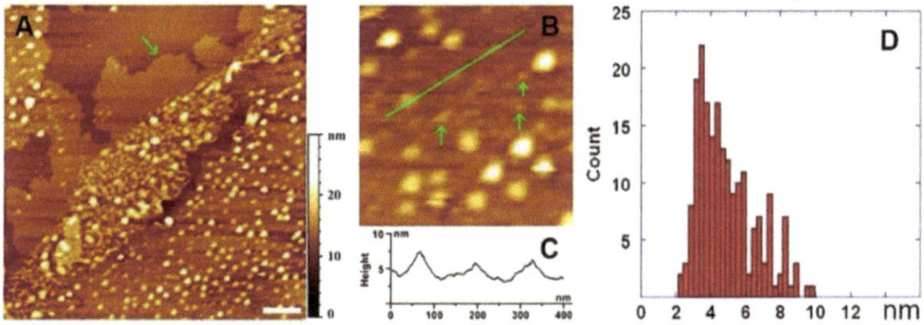

Figure 3.12 The cytoplasmic side of membranes treated with trypsin. (A) The arrow points to the free lipid bilayer. (B) Enlargement of the membrane in (A). Arrows point to possible peptides after trypsin digestion. (C) Cross-section analysis along the line in (B). (D) Height distribution on the membrane after trypsin digestion. Reprinted from Ref. [38]. Copyright 2010, with permission from Elsevier.

3.4.3 A New Model of the Erythrocyte Membrane Structure

Based on the imaging analysis of both sides of the erythrocyte membrane by AFM, Wang et al. proposed a new "semi-mosaic model" of the erythrocyte membrane, as shown in Fig. 3.13 [36]. Figure 3.13A is a top view of the semi-mosaic model, in which oligosaccharides are evenly spread in the middle of lipid heads without protein protrusion. Figure 3.13B is a bottom view of a cell membrane that shows dense proteins with less free lipids, with the bottoms of transmembrane proteins inserted in the lipid bilayer. This model suggests that the cytoplasmic side of the erythrocyte membrane is smooth and transmembrane proteins do not penetrate the membrane, with the oligosaccharides on the proteins distributed between the phospholipid heads; whereas the cytoplasmic side is covered with many proteins, has a very rough morphology, and contains fewer free lipids than expected.

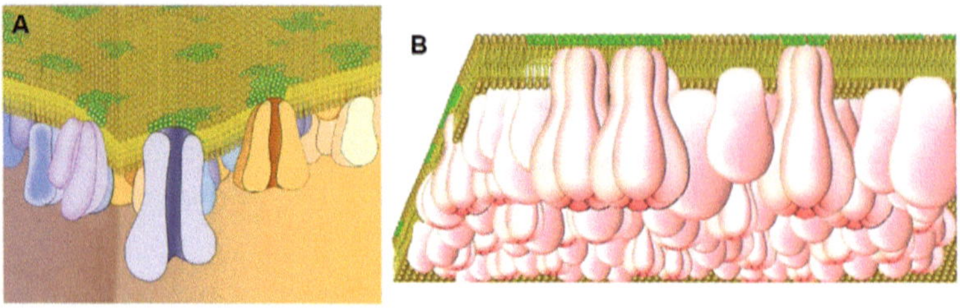

Figure 3.13 The proposed semi-mosaic model of cell membranes. (A) Top view of cell membranes. Lipids (yellow) and glycans (green) are shown on the top of the layer. Proteins are under the lipid layer. (B) Bottom view of cell membranes. Reprinted from Ref. [38]. Copyright 2010, with permission from Elsevier.

The basic structure of the erythrocyte is the conserved structure, retained within species throughout evolution. To demonstrate the universality of this structure, Wang et al. further analyzed the structural characteristics of the erythrocyte membranes of fish [38], reptiles [39], and birds [40], obtaining similar results for each. In addition, the semi-mosaic model of the erythrocyte membrane was recently verified from the 3D erythrocyte membrane structure obtained by cryogenic electron tomography. Thus, this ubiquitous asymmetry of membrane structure is considered to be the basic structural model of the erythrocyte membrane in all animals.

3.5 Nucleated Mammalian Cell Membrane Research Using AFM

Nucleated mammalian cells are more complex than erythrocytes. The various functions of nucleated mammalian cells require them to have more receptors in their membranes compared with erythrocytes.

Recent results indicated that the ectoplasmic side of the nucleated cell membrane is also smooth, with a roughness of only 1.1 ± 0.2 nm (Fig. 3.14A). In order to form such a characteristic of membrane structure, the proteins in the ectoplasmic side of the membrane must have a smooth distribution. Several pits of about 4 nm depth appeared after treating the ectoplasmic side of the membrane with protease K (Fig. 3.14B), methyl-β-cyclodextrin (M-β-CD) (Fig. 3.14C), and collagenase (Fig. 3.14H), which indicated the presence of a dense

4 nm thick protein layer on the outer membrane. In contrast to the smooth appearance of the ectoplasmic side of the membrane, the cytoplasmic side of the membrane is very rough (3.7 ± 0.2 nm) and is covered with many proteins (Fig. 3.14D). The height of proteins on the cytoplasmic side of the membrane is 11.2 ± 1.9 nm, the thickness of the ectoplasmic side protein layer is 4 nm, and the height of the phospholipid bilayer is also 4 nm; thus, the thickness of the entire membrane was calculated to be about 20 nm, which was consistent with the direct measurement of the height of the cell membrane (19.5 ± 2.8 nm, measured in Fig. 3.14F).

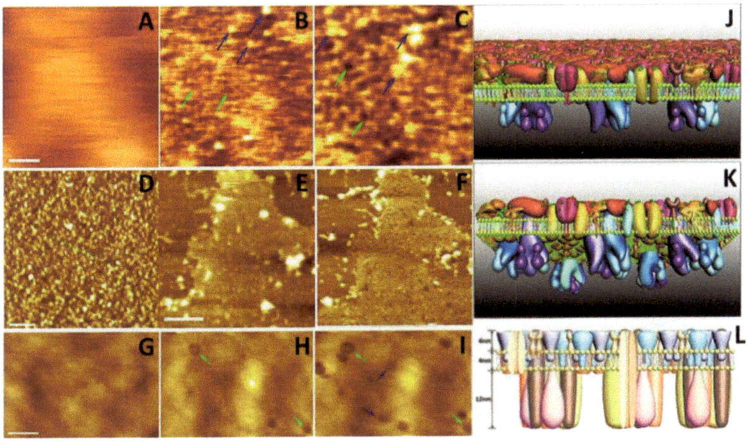

Figure 3.14 (A–C) AFM images of the ectoplasmic side of the membrane (A) and treated sequentially with proteinase K (B) and M-β-CD (C). (D–F) AFM topographic images of the cytoplasmic side of the membrane (D) and treated sequentially with proteinase K (E) and M-β-CD in (F). (G–I) AFM images of the ectoplasmic side of the membrane (G) and in situ treated sequentially with collagenase 3 (H) and M-β-CD (I). (J–L) Proposed protein layer–lipid–protein island (PLLPI) model of the cell membrane from top and bottom views, showing the size of the cell membrane. Reprinted from Ref. [43]. Copyright © 2014 Zhao et al.

As mentioned above, Wang et al. found that on the ectoplasmic side of the membrane proteins formed a dense protein layer (4 nm) on top of a lipid bilayer. On the cytoplasmic side, proteins aggregated evenly to form islands of 10–12 nm height. Lipid rafts or cholesterol-enriched domains were present in the cell membrane, with carbohydrates located in the macro-domains on the ectoplasmic side, and exposed amino groups asymmetrically distributed on both sides. Based on these observations, the authors proposed a novel structural model for nucleated mammalian cells, the protein layer–lipid–protein island (PLLPI) model (Fig. 3.14J–L) [41]. Compared with the fluid mosaic model, the dense protein layer on the ectoplasmic side forms a barrier to protect the cell membrane and maintains the cell integrity; and the conformation of aggregated proteins is conformationally less flexible, thus facilitating the recognition of membrane proteins; and the protein macro-domains are suitable for energy transfer and signal transduction.

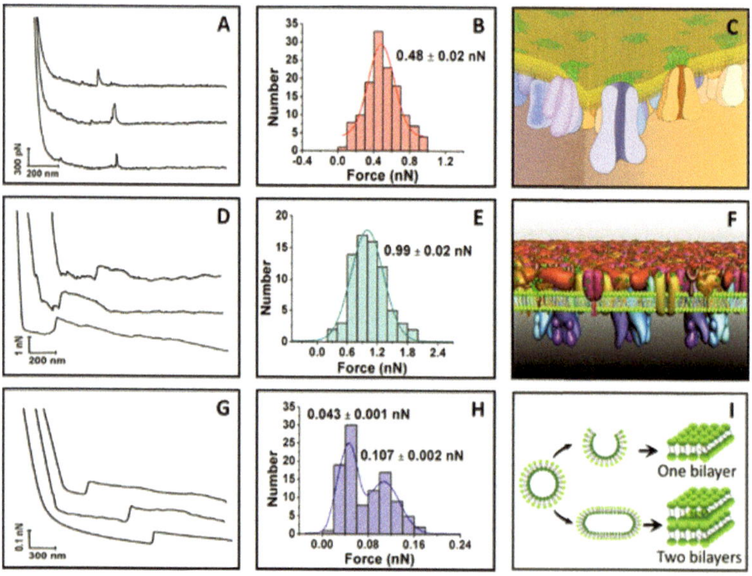

Figure 3.15 Typical force–distance curves, rupture force distribution histograms with Gauss fitting, and structural schemes of the three non-supported membranes. A–C for RBC membranes; D–F for MDCK cell membranes; G–I for phospholipid membranes. C and F show the front view of the semi-mosaic model of the red blood cell membrane and the PLLPI model of nucleated mammalian cell membranes. I shows the formation of one or two bilayers of phospholipid membranes derived from vesicles. Reproduced from Ref. [44] by permission of John Wiley & Sons Ltd.

In addition, to confirm the correction of the semi-mosaic model of the RBC membrane and the PLLPI model of the nucleated mammalian cell membrane in terms of mechanical properties, Wang's group used the AFM rupture test to obtain the mechanical properties of naked membranes. Measurement of the mechanical properties of cell membranes on whole cells is usually influenced by the support of the substrate, cytoplasm, cell condition, and stress location. Non-supported membranes on high-order pore array silicon substrates can help to reduce or eliminate these effects. The results indicate that the sequences of both rupture forces and Young's modulus of the three membranes are MDCK cell membrane > RBC membrane > phospholipid membrane (Fig. 3.15). Compared with MDCK cells, RBC membrane proteins are less abundant and not distributed on the ectoplasmic side of the cell membrane, resulting in a lower rupture force than that of MDCK cells. Additionally, the simulated cell membrane–phospholipid membranes have no intramembrane proteins or dense protein layer, so the rupture force is the lowest. The elastic character reflected by Young's modulus further verified this conclusion. Therefore, the quantity and distribution of membrane proteins have been proven to be crucial in the mechanical properties of non-supported membranes [42].

3.6 AFM Study of Lipid Rafts and Related Membrane Proteins

3.6.1 Discovery and Verification of Lipid Rafts

Studies have shown that molecules in cell membranes tend to organize into microdomains called lipid rafts, which are rich in cholesterol, sphingolipids, and special membrane proteins [43, 44]. Although lipid rafts are widely recognized to play a key role in a variety of cellular functions, there is debate about the composition, function, and even existence of lipid rafts. In 2012, the existence of lipid rafts in the cell membrane was directly demonstrated by high-resolution and time-lapse in situ AFM. Orsini et al. used sucrose-density gradient centrifugation to isolate lipid raft domains from Triton X-100-treated cells, and observed them by AFM to directly confirm that there are lipid rafts associated with specific proteins in cell membranes [45].

However, lytic cells may cause contamination of lipid rafts by membrane systems such as the Golgi apparatus and endoplasmic reticulum. To explore the real lipid raft structure, in situ treatment and observation of the cell membrane are necessary. Detergent-resistant membrane (DRM) domains were first confirmed using in situ imaging [46]. They exhibited irregular shape with a diameter of several hundred nanometers in the erythrocyte membrane (Fig. 3.16A–C). The in situ observation of the specific extraction of cholesterol by M-β-CD demonstrated that M-β-CD gradually eroded the membrane over time, and the size of the eroded irregular patches was mainly in the range of 100–300 nm (Fig. 3.16D–G), which provided convincing results of lipid rafts.

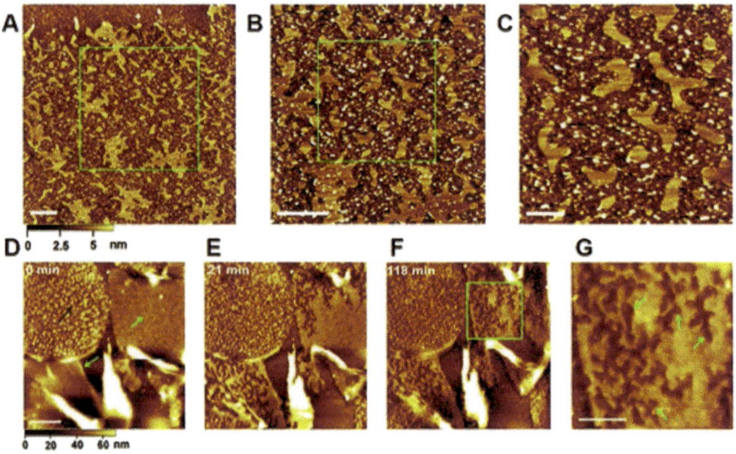

Figure 3.16 DRMs from the ectoplasmic side membrane. (A) AFM image of DRMs. (B, C) Serial magnifications of the DRMs in (A). Scale bars are 500 nm in (A) and (B) and 200 nm in (C). (D–G) In situ images of cell membranes treated with M-β-CD. (D–F) A series of images after the addition of M-β-CD for 0 min (D), 21 min (E), and 118 min (F). The green and blue arrows in (D) point to the ectoplasmic and cytoplasmic sides of the cell membranes, respectively. (G) Magnified image of the green square area in (F). The green arrows indicate the regions eroded by M-β-CD. Scale bars are 2 nm in (D) and 1 nm in (G). Reproduced from Ref. [49] by permission of John Wiley & Sons Ltd.

The analysis of the molecular structure of the lipid raft will help to elucidate the mechanism of the multiple functions of the cell membrane. Lipid rafts ought to be defined by their functions instead of the method used to isolate them. The hypothesis of lipid rafts indicates that lipid rafts execute functions through the protein complex; for example, clathrin and caveolin in lipid rafts mediate endocytosis. Moreover, the relationship between the erythrocyte anion transporter Band III and lipid rafts was studied [47]. Band III is localized in the lipid raft domains based on the molecular recognition technique (Fig. 3.17). As reported previously, Band III is not only an anion transporter but also connects with the skeleton of cell membranes, such as ankyrin and spectrin, to help maintain the mechanical properties and integrity of erythrocytes. These results by AFM imaging provide direct proof that lipid rafts are functional domains. More importantly, these results re-emphasize the important role of cholesterol in supporting the structure and function of cell membranes, especially of membrane proteins.

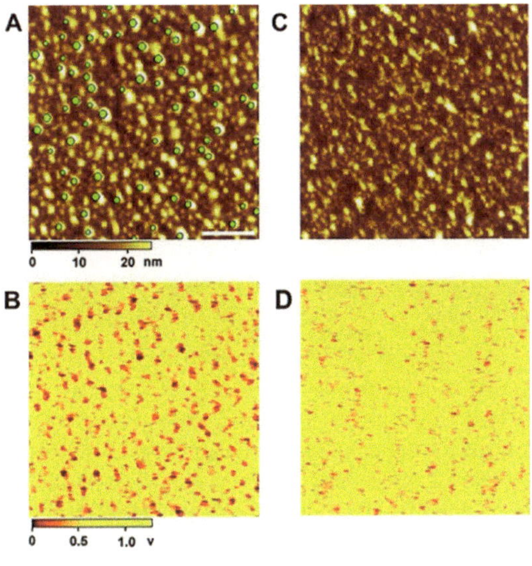

Figure 3.17 AFM recognition images of Band III in the cytoplasmic side of membrane. (A) Topographic image of the cytoplasmic side of membrane before treatment with 2 mM M-β-CD. (B) Recognition image corresponding to (A). The dark signal represents the recognition event. The recognition signal is superimposed onto the topographic image in A with green dots. (C) Topographic image of the cytoplasmic side of membrane after treatment with 2 mM M-β-CD. (D) Recognition image corresponding to (C). Scale bar is 500 nm. Reproduced from Ref. [49] by permission of John Wiley & Sons Ltd.

3.6.2 Microdomains of Membrane Proteins

Additional evidence has indicated that lipid rafts exist in nucleated mammalian cell membranes. Chen et al. [48] detected the depletion of lipid rafts from endothelial cells by M-β-CD using AFM. They found that the lipid raft is the functional unit and is associated with the underlying cytoskeletal network for the control of dynamic functions [49–51].

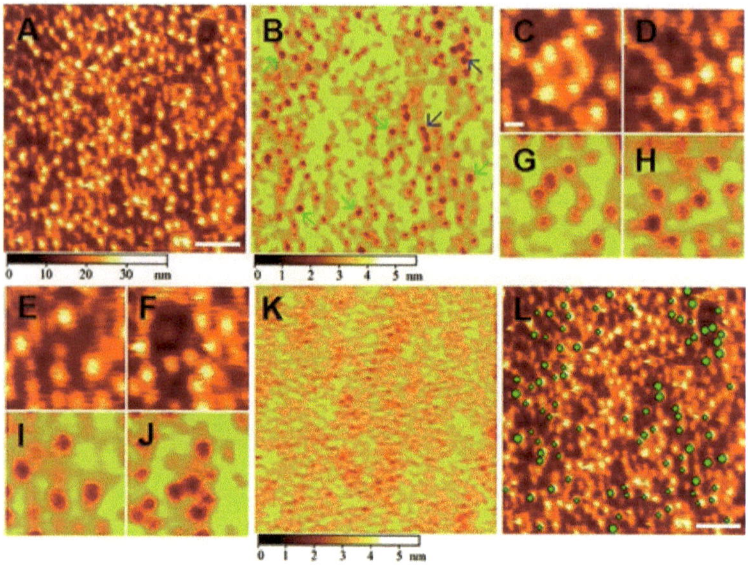

Figure 3.18 (A) Topographic image of the cytoplasmic side of erythrocyte membranes. Scale bar is 500 nm. (B) Corresponding recognition image of Na^+-K^+ ATPases (dark spots). Green and blue arrows indicate isolated Na^+-K^+ ATPases and Na^+-K^+ ATPase aggregations, respectively. (C–F) Magnified topographic images of the cytoplasmic side membrane. Scale bar is 100 nm. (G–J) Magnified recognition images corresponding to (C–F). (K) Recognition image of Na^+-K^+ ATPases after being blocked by free anti-ATPases. (L) Topographic image of the cytoplasmic side cell membrane with recognition spots superimposed by overlaying them (green dots) on the top of the topographic images. Scale bar is 500 nm. Reprinted with the permission from Ref. [55]. Copyright 2009 American Chemical Society.

Another type of membrane protein, Na^+-K^+ ATPase (NKA), has also been studied in quasi-native cell membranes by in situ AFM and

TREC [52]. Na^+-K^+ ATPase, a key transmembrane protein complex, utilizes energy from ATP hydrolysis to balance the concentration of Na^+ and K^+ across cell membranes [53]. Figure 3.18A shows a typical image of cytoplasmic side of erythrocyte membrane in which particles (supposed to be membrane proteins) are clearly observed. The proteins display a broad height distribution between 3 and 14 nm, which indicates that there are multiple types of proteins on the cytoplasmic side of cell membranes. Figures 3.18C–F show the magnification of membrane proteins with different sizes next to each other. Figure 3.18B shows the corresponding recognition image by the anti-ATPase-functionalized AFM tip. The dark spots represent the proteins recognized by the anti-ATPases. Figures 3.18G–J show the magnified recognition images corresponding to Fig. 3.18C–F. To clarify the distribution of Na^+-K^+ ATPases in the membrane, the recognition signal is marked by green dots, and the recognition signal is superimposed onto the topographic image (Fig. 3.18L). The recognition signal is mainly on proteins with heights of approximately 12–14 nm, which is consistent with the cryo-electron microscopy results.

Along with the development of single-molecule techniques, Wang's group proved that lipid rafts worked as functional domains in both erythrocyte membranes and mammalian somatic membranes. As shown in Fig. 3.14E, most of the protein disappeared after protease K digestion, thus exposing a smooth phospholipid bilayer. Next, the phospholipid bilayer was treated with M-β-CD to extract the cholesterol-rich microdomains. During the imaging process, M-β-CD was added to the cells to monitor changes in situ. Figure 3.14F shows the morphology of the membrane after 30 min of M-β-CD, and some holes appear on the membrane surface. The width of the holes formed by M-β-CD distributed at 40–200 nm, and the average hole width was 98.5 ± 25.5 nm, which was consistent with the result obtained by other optical microscopy, thus confirming the existence of the protein microdomains in cell membranes.

These results are of great significance for revealing the structure of cell membranes and lay the foundation for studying the protein–lipid interactions [54, 55] and the biomembranomics at the single molecular level.

3.6.3 Organelle Membranes

In addition to the plasma membranes, AFM also helps the study of organelle membranes, which are related to cell structure and function. The Golgi complex, also known as the Golgi apparatus or Golgi body, is one of the earliest organelles observed by light microscopy. The main function of the Golgi complex is to provide a transport system for cells. Mitochondria occupy an important part of eukaryotic cytoplasmic volume and are necessary for the evolution of higher animals. Mitochondria can efficiently convert organic matter into ATP, the energy currency of cell life activities.

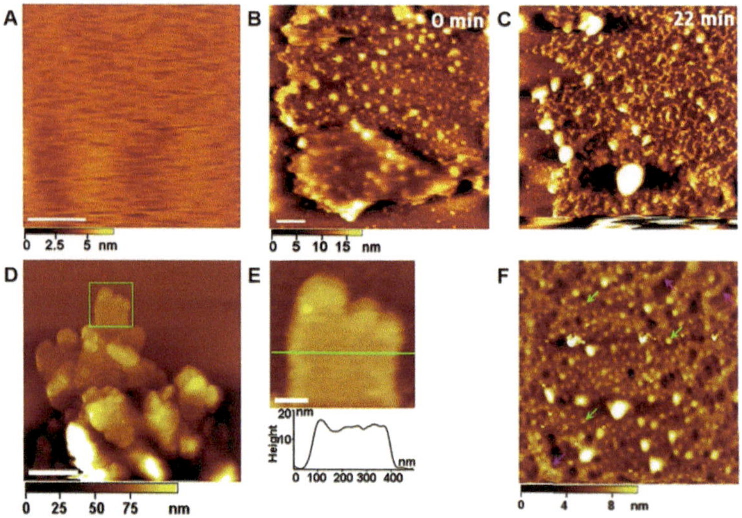

Figure 3.19 AFM imaging of individual Golgi cisternae and mitochondrial membranes. (A) The image of the ectoplasmic side of Golgi apparatus membranes. Scale bar is 100 nm. (B) The image of an opened Golgi cisternae. Protein domains are located on the cytoplasmic side of membranes. Scale bar is 500 nm. (C) Real-time image of a Golgi membrane treated with M-β-CD for 22 min. Scale bar is 200 nm. (D) AFM topographic image of flattened mitoplasts from mitochondria subjected to freeze–thaw treatment. Scale bar is 500 nm. (E) Higher magnification image (top) of the green square area in (D) with the cross-section analysis (bottom) along the green line. Scale bar is 100 nm. (F) AFM topographic image of unilamellar inner mitochondrial membranes covered by dense proteins. Scale bar is 300 nm. (A–C) are reprinted from Ref. [56], Copyright 2013, Xu et al. (D–F) are reproduced from Ref. [57] by permission of The Royal Society of Chemistry.

The membranes of Golgi apparatus and mitochondria were also imaged by in situ AFM (Fig. 3.19) [56, 57]. The results indicated that the mitochondrial and Golgi membranes exhibit features similar to erythrocyte membranes; that is, the ectoplasmic surface is flat, with protruding proteins located on the cytoplasmic side. In addition, whether there are lipid rafts in organelle membranes was not clear until Wang et al. directly visualized a Golgi vesicle. The size of the lipid rafts in Golgi membranes is very similar to that in erythrocyte membranes. The studies of organelle structure confirm that the distribution of membrane proteins is asymmetrical, which is similar to the structure of RBC membranes.

The statistical analysis of cell membrane thickness showed that the erythrocyte and Golgi membranes exhibited very similar features. The nucleated mammalian cell membranes and mitochondrial membranes are substantially thicker than the RBC membranes. Notably, the thickness of a lipid bilayer is approximately 3.0 nm, which is close to the thicknesses of erythrocyte and Golgi membranes and indicates that there is no outer protein layer for this membrane.

3.7 Protocols for AFM Experiment

3.7.1 Making APTES-Mica

Mica is an appropriate substrate for high-resolution AFM imaging because its surface is smooth at the atomic level. However, fresh mica surfaces are negatively charged, which are not suitable for the immobilization of negatively charged cell membranes. Modification of mica by APTES (3-Aminopropyltriethoxysilane, Sigma) easily creates a positively charged surface; thus, membranes can be tightly attached on the surface for imaging. The setup for producing APTES-modified mica (AP-mica) is shown in Fig. 3.20. Mica sheets with a thickness of approximately 0.5 mm and an area of 20 × 20 mm are cleaved with scotch tape several times to ensure a smooth and clean surface, followed by placement in the desiccator. After a desiccator is purged with argon for 5 min, 30 μL of APTES and 10 μL of DIPEA (N, N-diisopropylethylamine, Sigma) are each placed into small containers at the bottom of the desiccator. Purged with argon for an additional 5 min, the desiccator is sealed off. The mica is exposed to APTES vapor for 4 h. After this exposure, the APTES is carefully removed from the desiccator; the desiccator is purged with argon and sealed. AP-mica is stored in a sealed desiccator under argon for use within weeks [58, 59].

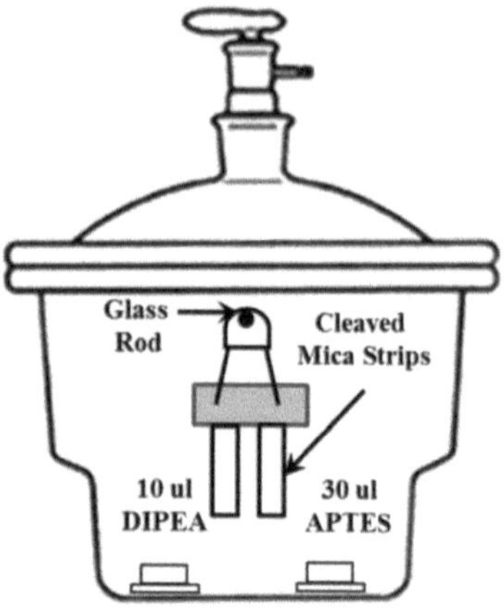

Figure 3.20 Making AP-mica. Mica strips are clamped in metal clips suspended on a glass rod of appropriate length (to nest snugly across the upper part of the desiccator). The reagents DIPEA and APTES are placed in small containers on the bottom of the desiccator. Reprinted from Ref. [58]. Copyright 2007, with permission from Elsevier.

Glass cover slips can also be used for cultured mammalian cells. Prior to cell culture, strict cleaning steps must be performed to keep cover slips clean. In general, cover slips are cleaned using a detergent (Micro 90) and sonicated in 1 mol/L potassium hydroxide for 20 min at room temperature. Then they are rinsed with sterile distilled water (18 MΩ·cm) and subsequently stored in absolute ethyl alcohol. Prior to use, the cover slips are washed three times with sterile distilled water and dried with pure argon. Clean cover slips are placed in a culture dish as the substrate for cell culture [60, 61].

3.7.2 Modification of AFM Tip

To identify specific proteins in a compositionally complex sample, an antibody against the protein of interest can be attached to the

AFM tip. During the scanning process, the AFM topographic image is generated simultaneously with a "recognition image" by exact spatial registration, which identifies the locations of antibody–antigen binding events and thus the locations of the protein of interest in the field [15]. The two images can be electronically superimposed to obtain an accurate map of specific protein locations relative to the topography.

As shown in Fig. 3.21, first, AFM tips are cleaned in a UV cleaner for 15 min to eliminate any organic contamination on the tips. The tips are placed in a dish at the bottom of a desiccator and modified with APTES, just as mentioned earlier. After the treatment, the APTES is removed, and the treated tips (AP tip) are stored in a sealed desiccator until use. One milligram of bifunctional PEG cross-linker and 5 μL of triethylamine are mixed in 1 mL $CHCl_3$. AP-modified tips are then placed into the solution and kept for 2–3 h. Next, the tips are washed with $CHCl_3$ and dried with argon. At this time, the succinimide groups attach to the amino groups on the AP tip. For the coupling of antibodies, 40 μL of biotin IgG (2.2 mg/mL) is mixed with 380 μL of buffer A (100 mM NaCl, 50 mM NaH_2PO_4, 1 mM EDTA, pH 7.5, adjusted with NaOH) and 8 μL of 1 M $NaCNBH_3$ (freshly prepared by dissolving 32 mg of $NaCNBH_3$ in 500 μL of 10 mM NaOH). Tips with aldehyde–PEG linkers are immersed in this solution for 1 h; then, the tips are washed twice with buffer A and once with PBS buffer (100 mM NaCl, 50 mM Na_2HPO_4, pH 7.5) and stored in PBS at 4°C until use (up to 3 days) [62, 63]. The functionalized tip is available for TREC imaging and single-molecule force spectroscopy (see Chapter 12).

3.7.3 AFM Recognition Imaging with Modified Tips

Silicon nitride cantilever tips for recognition imaging can be modified as described above. Recognition imaging is performed in magnetic AC mode (MAC mode) AFM 5500 with a PicoTREC recognition imaging attachment (Agilent Technologies, Chandler, AZ). Topographic and recognition images are acquired with 6–8 nm amplitude oscillation at 9 kHz, imaging at 70% set point and scan speed at <1 Hz. To test the specificity of the recognition reaction, 50 μL of a 30 μg/mL solution of a peptide antigenic is added and the sample is rescanned. If recognition is specific, the presence of the peptide will block recognition to a very high degree.

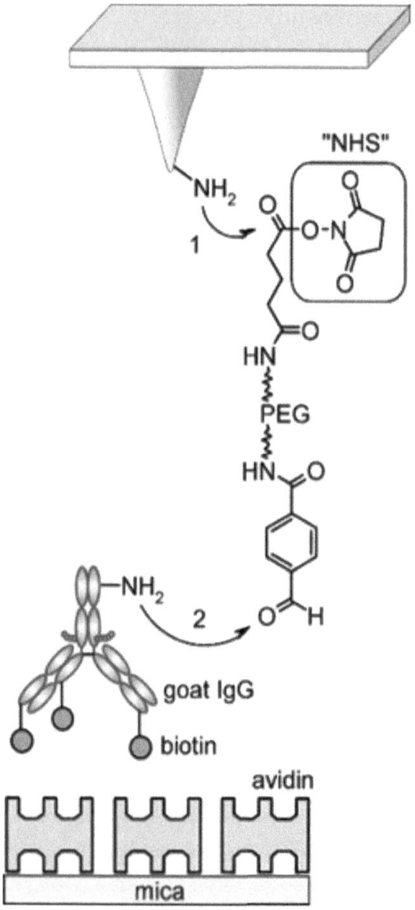

Figure 3.21 Schematic representation of the testing system for tip antibody linking via lysine–aldehyde coupling. In the first step (in organic solvent), the NHS ester of aldehyde–PEG–NHS reacts with the amino groups on the tip; in the second step (in aqueous buffer), one of the antibodies, lysine, forms a Schiff base with the aldehyde function on the free end of the cross-linker. The Schiff base is reduced by $NaCNBH_3$ to form an irreversible bond (not shown). When brought into contact with avidin-covered mica, the biotin residues on the model antibody show reversible binding to and unbinding from immobilized avidin. Reprinted with the permission from Ref. [62]. Copyright 2007 American Chemical Society.

Note: Soft cantilevers with low frequency (6–9 kHz) should be used for recognition imaging. Stiffer cantilevers (higher frequency) do not produce good recognition images. Additionally, scan speeds >1 Hz should be avoided because they increase the leakage of topography in the recognition image. It is very important to check the specificity of antibodies used for recognition imaging. We find that nonspecific reactions caused by antibodies that recognize nonantigens sometimes are quite strong in both AFM recognition imaging (including force curves) and standard ELISAs.

3.7.4 Preparation of Red Blood Cell Membranes

The sheared open method is appropriate for preparing a clean membrane with minimum damage. The steps are as follows (Fig. 3.22): (1) Two drops of blood are taken from a fingertip and centrifuged in 1 mL of PBSA buffer (136.9 mM NaCl, 2.7 mM KCl, 1.5 mM KH_2PO_4, and 8.1 mM Na_2HPO_4, pH 7.4) for five times (1000 rpm for 2 min). (2) A drop of red blood cells in PBSA buffer (100 μL) is subsequently deposited on the AP-mica surface for approximately 20 min for absorption. (3) PBSA is used to wash out the cells that are not adsorbed. (4) As shown in Fig. 3.22A, a syringe is adjusted to 20°, from the sample surface, and 10 mL of hypotonic buffer (6.850 mM NaCl, 0.135 mM KCl, 0.075 mM KH_2PO_4, 0.405 mM Na_2HPO_4, pH 7.4) is injected to flush the mica surface and obtain a flat membrane patch [64].

Using the hypotonic lysis centrifugation method, whole erythrocyte ghost membranes can be prepared as follows. Briefly, erythrocytes are collected and cleaned by centrifugation in a physiological buffer solution. The cells are lysed using low salt buffer (5 mM Na_2HPO_4, 0.2 mM EGTA) and washed three times by centrifugation at 20,000 g for 20 min at 0°C until the ghost pellet becomes white. The membrane pellet is diluted in buffer for deposition on AP-mica for AFM imaging [36].

3.7.5 Preparation of the Nucleated Mammalian Cell Membranes

Cells are cultivated overnight on glass coverslips. The cell membranes are prepared by the shearing open method [65]. Briefly, the cells are washed twice with ice-cold buffer (20 mM PIPES, 150 mM KCl, pH 6.2), incubated with cold hypotonic buffer (4 mM PIPES and 30 mM

KCl, pH 6.2) for 3 min, and then sheared open by a stream of 10 mL of hypotonic buffer through a needle at an angle of 20°. The membranes are subsequently treated with high salt buffer (2 M NaCl, 1.5 mM KH_2PO_4, 2.7 mM KCl, and 1 mM Na_2HPO_4, pH 7.2) for 30 min at room temperature to remove the cytoskeletons. The prepared membranes are immediately imaged in PBSA buffer using AFM (Fig. 3.23).

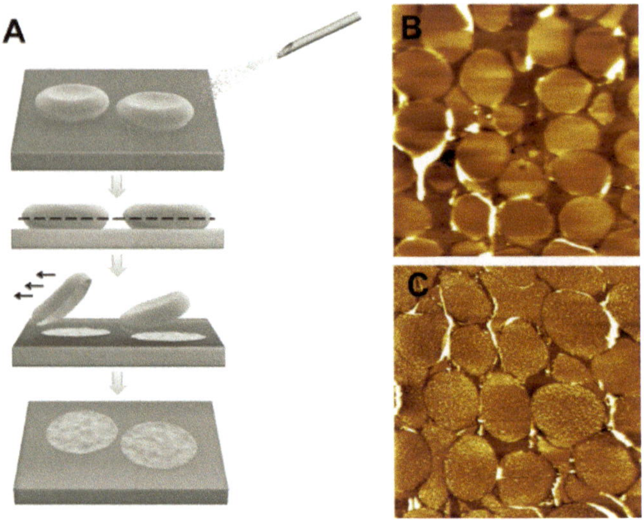

Figure 3.22 Preparation of the RBC membranes with sheared open method. (A) RBCs are exposed to fluid-flowing shear stress to open the cells. (B, C) AFM images of the ectoplasmic and cytoplasmic sides of the RBC membrane spread on the surface after shear stress.

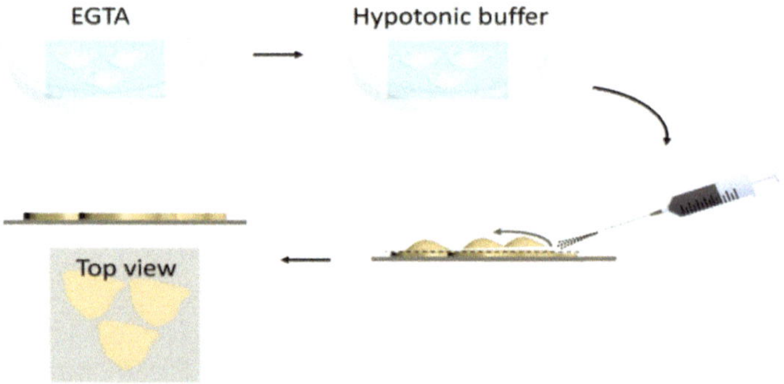

Figure 3.23 Schematic illustration of the workflow for preparing the cytoplasmic side of cell membranes.

3.8 Conclusion and Prospects

The cell membrane is the protective barrier of a cell. It separates the internal components (Golgi apparatus, mitochondria, cytoplasm, etc.) from the extracellular matrix, maintains the integrity of the cell, and determines what can enter and exit the cell. The analysis of the cell membrane structure is of great significance for studying the regulation of cell signal transduction, the mechanism of virus invasion into host cells, drug screening, cancer therapy, and many other applications.

Imaging the cell membrane in the liquid environment by tapping mode AFM minimizes the AFM tip effect on the sample. Until now, two sides of cell membranes have been directly observed by in situ AFM at molecular resolution under quasi-native conditions. With the help of high-resolution imaging and TREC technology of AFM, we believe that more novel and diverse functions of the cell membrane would be explored to improve our understanding of the functions of the cell membrane.

A new understanding of the cell membrane structure from the perspective of biomembranomics is crucial for a deep understanding of the functions of cell membrane, such as cell–cell communication, signal transduction, molecular transport, energy conversion, and cell-mediated targeting of drug delivery.

References

1. Nicolson, G. L. (2014). The fluid-mosaic model of membrane structure: Still relevant to understanding the structure, function and dynamics of biological membranes after more than 40 years, *Biochim. Biophys. Acta-Biomembr.*, **1838**, pp. 1451–1466.
2. Yoshida, M., Muneyuki, E., and Hisabori, T. (2001). ATP synthase: A marvellous rotary engine of the cell, *Nat. Rev. Mol. Cell Biol.*, **2**, pp. 669–677.
3. Binnig, G., Quate, C. F., and Gerber, C. (1986). Atomic force microscope, *Phys. Rev. Lett.*, **56**, pp. 930–933.
4. Li, M., Xi, N., Wang, Y. C., and Liu, L. Q. (2019). Advances in atomic force microscopy for single-cell analysis, *Nano Res.*, **12**, pp. 703–718.

5. Zhang, Q., Li, S., Yang, Y., Shan, Y., and Wang, H. (2021). Studying structure and functions of cell membranes by single molecule biophysical techniques, *Biophys. Rep.*, **7**, pp. 384–398.
6. Whited, A. M. and Park, P. S. H. (2014). Atomic force microscopy: A multifaceted tool to study membrane proteins and their interactions with ligands, *Biochim. Biophys. Acta-Biomembr.*, **1838**, pp. 56–68.
7. Garcia, R. and Perez, R. (2002). Dynamic atomic force microscopy methods, *Surf. Sci. Rep.*, **47**, pp. 197–301.
8. Han, W. H., Lindsay, S. M., and Jing, T. W. (1996). A magnetically driven oscillating probe microscope for operation in liquids, *Appl. Phys. Lett.*, **69**, pp. 4111–4113.
9. Ge, G. L., Han, D., Lin, D. Y., Chu, W. G., Sun, Y. X., Jiang, L., Ma, W. Y., and Wang, C. (2007). MAC mode atomic force microscopy studies of living samples, ranging from cells to fresh tissue, *Ultramicroscopy*, **107**, pp. 299–307.
10. Hansma, P. K., Cleveland, J. P., Radmacher, M., Walters, D. A., Hillner, P. E., Bezanilla, M., Fritz, M., Vie, D., and Hansma, H. G. (1994). Tapping mode atomic force microscopy in liquids, *Appl. Phys. Lett.*, **64**, pp. 1738–1740.
11. Basak, S. and Raman, A. (2007). Dynamics of tapping mode atomic force microscopy in liquids: Theory and experiments, *Appl. Phys. Lett.*, **91**, pp. 064107.
12. Allison, D. P., Mortensen, N. P., Sullivan, C. J., and Doktycz, M. J. (2010). Atomic force microscopy of biological samples, *Wiley Interdiscip. Rev.-Nanomed. Nanobiotechnol.*, **2**, pp. 618–634.
13. Muller, D. J. and Dufrene, Y. F. (2011). Atomic force microscopy: A nanoscopic window on the cell surface, *Trends Cell Biol.*, **21**, pp. 461–469.
14. Dufrene, Y. F., Ando, T., Garcia, R., Alsteens, D., Martinez-Martin, D., Engel, A., Gerber, C., and Muller, D. J. (2017). Imaging modes of atomic force microscopy for application in molecular and cell biology, *Nat. Nanotechnol.*, **12**, pp. 295–307.
15. Stroh, C., Wang, H., Bash, R., Ashcroft, B., Nelson, J., Gruber, H., Lohr, D., Lindsay, S. M., and Hinterdorfer, P. (2004). Single-molecule recognition imaging-microscopy, *Proc. Natl. Acad. Sci. U. S. A.*, **101**, pp. 12503–12507.
16. Senapati, S. and Lindsay, S. (2016). Recent progress in molecular recognition imaging using atomic force microscopy, *Accounts Chem. Res.*, **49**, pp. 503–510.

17. Shan, Y. P. and Wang, H. D. (2015). The structure and function of cell membranes examined by atomic force microscopy and single-molecule force spectroscopy, *Chem. Soc. Rev.*, **44**, pp. 3617–3638.

18. Shi, Y., Cai, M. J., Zhou, L. L., and Wang, H. D. (2018). The structure and function of cell membranes studied by atomic force microscopy, *Semin. Cell Dev. Biol.*, **73**, pp. 31–44.

19. Frederix, P., Bosshart, P. D., and Engel, A. (2009). Atomic force microscopy of biological membranes, *Biophys. J.*, **96**, pp. 329–338.

20. Goksu, E. I., Vanegas, J. M., Blanchette, C. D., Lin, W. C., and Longo, M. L. (2009). AFM for structure and dynamics of biomembranes, *Biochim. Biophys. Acta-Biomembr.*, **1788**, pp. 254–266.

21. Muller, D. J. and Engel, A. (2007). Atomic force microscopy and spectroscopy of native membrane proteins, *Nat. Protoc.*, **2**, pp. 2191–2197.

22. Lippincott-Schwartz, J. and Phair, R. D. (2010). Lipids and cholesterol as regulators of traffic in the endomembrane system, *Annu. Rev. Biophys.*, **39**, pp. 559–578.

23. van Meer, G. and de Kroon, A. (2011). Lipid map of the mammalian cell, *J. Cell Sci.*, **124**, pp. 5–8.

24. Morandat, S. and El Kirat, K. (2006). Membrane resistance to Triton X-100 explored by real-time atomic force microscopy, *Langmuir*, **22**, pp. 5786–5791.

25. El Kirat, K., Morandat, S., and Dufrene, Y. F. (2010). Nanoscale analysis of supported lipid bilayers using atomic force microscopy, *Biochim. Biophys. Acta-Biomembr.*, **1798**, pp. 750–765.

26. El Kirat, K. and Morandat, S. (2007). Cholesterol modulation of membrane resistance to Triton X-100 explored by atomic force microscopy, *Biochim. Biophys. Acta-Biomembr.*, **1768**, pp. 2300–2309.

27. Sumino, A., Sumikama, T., Iwamoto, M., Dewa, T., and Oiki, S. (2013). The open gate structure of the membrane-embedded KcsA potassium channel viewed from the cytoplasmic side, *Scientific Reports*, **3**, pp. 1063.

28. Rico, F., Picas, L., Colom, A., Buzhynskyy, N., and Scheuring, S. (2013). The mechanics of membrane proteins is a signature of biological function, *Soft Matter*, **9**, pp. 7866–7873.

29. Engel, A. and Gaub, H. E. (2008). Structure and mechanics of membrane proteins, *Annu. Rev. Biochem.*, **77**, pp. 127–148.

30. Muller, D. J. and Dufrene, Y. F. (2008). Atomic force microscopy as a multifunctional molecular toolbox in nanobiotechnology, *Nat. Nanotechnol.*, **3**, pp. 261–269.

31. Fotiadis, D. (2012). Atomic force microscopy for the study of membrane proteins, *Curr. Opin. Biotechnol.*, **23**, pp. 510–515.

32. Usukura, J., Yoshimura, A., Minakata, S., Youn, D., Ahn, J., and Cho, S. J. (2012). Use of the unroofing technique for atomic force microscopic imaging of the intra-cellular cytoskeleton under aqueous conditions, *J. Electron Microsc.*, **61**, pp. 321–326.

33. Murakoshi, M., Iida, K., Kumano, S., and Wada, H. (2009). Immune atomic force microscopy of prestin-transfected CHO cells using quantum dots, *Pflugers Arch.*, **457**, pp. 885–898.

34. Dindia, L., Faught, E., Leonenko, Z., Thomas, R., and Vijayan, M. M. (2013). Rapid cortisol signaling in response to acute stress involves changes in plasma membrane order in rainbow trout liver, *Am. J. Physiol.-Endocrinol. Metab.*, **304**, pp. E1157–E1166.

35. Bennett, V. (1985). The membrane skeleton of human erythrocytes and its implications for more complex cells, *Annu. Rev. Biochem.*, **54**, pp. 273–304.

36. Wang, H. D., Hao, X., Shan, Y. P., Jiang, J. G., Cai, M. J., and Shang, X. (2010). Preparation of cell membranes for high resolution imaging by AFM, *Ultramicroscopy*, **110**, pp. 305–312.

37. Engelman, D. M. (2005). Membranes are more mosaic than fluid, *Nature*, **438**, pp. 578–580.

38. Tian, Y. M., Cai, M. J., Zhao, W. D., Wang, S. W., Qin, Q. W., and Wang, H. D. (2014). The asymmetric membrane structure of erythrocytes from Crucian carp studied by atomic force microscopy, *Chin. Sci. Bull.*, **59**, pp. 2582–2587.

39. Tian, Y. M., Cai, M. J., Xu, H. J., Ding, B. H., Hao, X., Jiang, J. G., Sun, Y. C., and Wang, H. D. (2014). Atomic force microscopy of asymmetric membranes from turtle erythrocytes, *Mol. Cells*, **37**, pp. 592–597.

40. Tian, Y. M., Cai, M. J., Xua, H. J., and Wang, H. D. (2014). Studying the membrane structure of chicken erythrocytes by in situ atomic force microscopy, *Anal. Methods*, **6**, pp. 8115–8119.

41. Zhao, W. D., Tian, Y. M., Cai, M. J., Wang, F., Wu, J. Z., Gao, J., Liu, S. H., Jiang, J. G., Jiang, S. B., and Wang, H. D. (2014). Studying the nucleated mammalian cell membrane by single molecule approaches, *PLoS One*, **9**, pp. e91595.

42. Shi, Y., Cai, M. J., Zhou, L. L., and Wang, H. D. (2020). Measurement of mechanical properties of naked cell membranes using atomic force microscope puncture test, *Talanta*, **210**, pp. 120637.

43. Jacobson, K., Mouritsen, O. G., and Anderson, R. G. W. (2007). Lipid rafts: At a crossroad between cell biology and physics, *Nat. Cell Biol.*, **9**, pp. 7–14.

44. Lingwood, D. and Simons, K. (2010). Lipid rafts as a membrane-organizing principle, *Science*, **327**, pp. 46–50.

45. Orsini, F., Cremona, A., Arosio, P., Corsetto, P. A., Montorfano, G., Lascialfari, A., and Rizzo, A. M. (2012). Atomic force microscopy imaging of lipid rafts of human breast cancer cells, *Biochim. Biophys. Acta-Biomembr.*, **1818**, pp. 2943–2949.

46. Zidovetzki, R. and Levitan, I. (2007). Use of cyclodextrins to manipulate plasma membrane cholesterol content: Evidence, misconceptions and control strategies, *Biochim. Biophys. Acta-Biomembr.*, **1768**, pp. 1311–1324.

47. Cai, M. J., Zhao, W. D., Shang, X., Jiang, J. G., Ji, H. B., Tang, Z. Y., and Wang, H. D. (2012). Direct evidence of lipid rafts by in situ atomic force microscopy, *Small*, **8**, pp. 1243–1250.

48. Wu, L., Huang, J., Yu, X., Zhou, X., Gan, C., Li, M., and Chen, Y. (2014). AFM of the ultrastructural and mechanical properties of lipid-raft-disrupted and/or cold-treated endothelial cells, *J. Membr. Biol.*, **247**, pp. 189–200.

49. Simons, K. and Sampaio, J. L. (2011). Membrane organization and lipid rafts, *Cold Spring Harbor Perspect. Biol.*, **3**, pp. 17.

50. Head, B. P., Patel, H. H., and Insel, P. A. (2014). Interaction of membrane/lipid rafts with the cytoskeleton: Impact on signaling and function membrane/lipid rafts, mediators of cytoskeletal arrangement and cell signaling, *Biochim. Biophys. Acta-Biomembr.*, **1838**, pp. 532–545.

51. Carquin, M., Conrard, L., Pollet, H., Van Der Smissen, P., Cominelli, A., Veiga-da-Cunha, M., Courtoy, P. J., and Tyteca, D. (2015). Cholesterol segregates into submicrometric domains at the living erythrocyte membrane: Evidence and regulation, *Cell. Mol. Life Sci.*, **72**, pp. 4633–4651.

52. Jiang, J. G., Hao, X., Cai, M. J., Shan, Y. P., Shang, X., Tang, Z. Y., and Wang, H. D. (2009). Localization of Na^{+}-K^{+} ATPases in quasi-native cell membranes, *Nano Lett.*, **9**, pp. 4489–4493.

53. Morth, J. P., Pedersen, B. P., Buch-Pedersen, M. J., Andersen, J. P., Vilsen, B., Palmgren, M. G., and Nissen, P. (2011). A structural overview of the

plasma membrane Na^+, K^+-ATPase and H^+-ATPase ion pumps, *Nat. Rev. Mol. Cell Biol.*, **12**, pp. 60–70.

54. Epand, R. M. (1998). Lipid polymorphism and protein-lipid interactions, *Biochim. Biophys. Acta-Rev. Biomembr.*, **1376**, pp. 353–368.

55. Tharad, S., Promdonkoy, B., and Toca-Herrera, J. L. (2020). Protein–lipid interaction of cytolytic toxin Cyt2Aa2 on model lipid bilayers of erythrocyte cell membrane, *Toxins*, **12**, pp. 226.

56. Xu, H. J., Su, W. H., Cai, M. J., Jiang, J. G., Zeng, X. L., and Wang, H. D. (2013). The asymmetrical structure of Golgi apparatus membranes revealed by in situ atomic force microscope, *PLoS One*, **8**, pp. e61596.

57. Tian, Y. M., Li, J. H., Cai, M. J., Zhao, W. D., Xu, H. J., Liu, Y., and Wang, H. D. (2013). High resolution imaging of mitochondrial membranes by in situ atomic force microscopy, *RSC Adv.*, **3**, pp. 708–712.

58. Lohr, D., Bash, R., Wang, H., Yodh, J., and Lindsay, S. (2007). Using atomic force microscopy to study chromatin structure and nucleosome remodeling, *Methods*, **41**, pp. 333–341.

59. Shan, Y. P., Wang, Z. Y., Hao, X. A., Shang, X., Cai, M. J., Jiang, J. G., Fang, X. X., Wang, H. D., and Tang, Z. Y. (2010). Locating the Band III protein in quasi-native cell membranes, *Anal. Methods*, **2**, pp. 805–808.

60. Wu, J. Z., Gao, J., Qi, M., Wang, J. Z., Cai, M. J., Liu, S. H., Hao, X., Jiang, J. G., and Wang, H. D. (2013). High-efficiency localization of Na^+-K^+ ATPases on the cytoplasmic side by direct stochastic optical reconstruction microscopy, *Nanoscale*, **5**, pp. 11582–11586.

61. Wang, Y., Gao, J., Guo, X. D., Tong, T., Shi, X. S., Li, L. Y., Qi, M., Wang, Y. J., Cai, M. J., Jiang, J. G., Xu, C. Q., Ji, H. B., and Wang, H. D. (2014). Regulation of EGFR nanocluster formation by ionic protein-lipid interaction, *Cell Res.*, **24**, pp. 959–976.

62. Ebner, A., Wildling, L., Kamruzzahan, A. S. M., Rankl, C., Wruss, J., Hahn, C. D., Holzl, M., Zhu, R., Kienberger, F., Blaas, D., Hinterdorfer, P., and Gruber, H. J. (2007). A new, simple method for linking of antibodies to atomic force microscopy tips, *Bioconjugate Chem.*, **18**, pp. 1176–1184.

63. Wildling, L., Unterauer, B., Zhu, R., Rupprecht, A., Haselgrubler, T., Rankl, C., Ebner, A., Vater, D., Pollheimer, P., Pohl, E. E., Hinterdorfer, P., and Gruber, H. J. (2011). Linking of sensor molecules with amino groups to amino-functionalized AFM tips, *Bioconjugate Chem.*, **22**, pp. 1239–1248.

64. Schillers, H. (2008). Imaging CFTR in its native environment, *Pflugers Arch.*, **456**, pp. 163–177.

65. Ziegler, U., Vinckier, A., Kernen, P., Zeisel, D., Biber, J., Semenza, G., Murer, H., and Groscurth, P. (1998). Preparation of basal cell membranes for scanning probe microscopy, *FEBS Lett.*, **436**, pp. 179–184.

Chapter 4

Super-Resolution Imaging for Mapping Membrane Proteins and Carbohydrates

Junling Chen,[a] Jing Gao,[b] and Hongda Wang[b]
[a]*Key Laboratory of Coal Conversion and New Carbon Materials of Hubei Province, College of Chemistry and Chemical Engineering, Wuhan University of Science and Technology, Wuhan, China*
[b]*State Key Laboratory of Electroanalytical Chemistry, Research Center of Biomembranomics, Changchun Institute of Applied Chemistry, Chinese Academy of Sciences, Changchun, Jilin, China*
gaojing@ciac.ac.cn, hdwang@ciac.ac.cn

A large number of biomolecules are densely distributed on cell membranes, including proteins, lipids, and carbohydrates. Since the spatiotemporal assembly of these membrane biomolecules is increasingly thought to play an important role in biological regulation, more and more attention has been paid to revealing the organizational features and mechanisms of membrane molecules. Super-resolution fluorescence microscopy (SRM), which has nanoscale resolution and is friendly to bio-samples, has been widely applied to investigate the detailed spatial information of membrane molecules at the single-molecule level. In addition to the resolution of the microscope, the labeling characteristic of fluorescent probes

Biomembranomics: Structure, Techniques, and Applications
Edited by Hongda Wang

ISBN 978-981-4968-61-4 (Hardcover), 978-1-003-45635-3 (eBook)
www.jennystanford.com

is also the key factor to finally obtaining a high-quality super-resolution image. Here, we will introduce three typical kinds of fluorescent probes and describe the recent applications of super-resolution fluorescence imaging in revealing spatial morphology and functional mechanisms of various membrane molecules. The detailed distribution of membrane molecules can promote a better understanding of their bio-functions, as well as the systemic organization of cell membranes.

4.1 Principles of Super-Resolution Fluorescence Imaging Technology

Owing to the optical diffraction limit, conventional far-field fluorescence microscopy achieves a resolution of 200–300 nm laterally, which might be only sufficient to study tissue morphology or whole-cell dynamics [1]. Super-resolution microscopy (SRM) techniques push the resolution towards nanometer scales, thereby enabling the high-definition visualization of subcellular organization with unprecedented detail [2–5].

4.1.1 Structured Illumination Microscopy

Through regulating the excitation of fluorescent labeling samples or detection of the emitted photons, different SRM techniques have been developed based on wide field (WF), total internal reflection fluorescence (TIRF), or confocal microscope setups. One group of SRM techniques is structured illumination microscopy (SIM) [6–9], which obtains a relatively high resolution (100 nm in lateral and 300 nm in axial) by modulating interference pattern (Fig. 4.1A) [10]. Without concomitant demands and restrictions of super-high resolution, SIM methods allow routine imaging with multiple colors and conventional fluorophores. They are considered rather "gentle" and, thus, are well suitable for volumetric live-cell imaging. On the downside, SIM requires complex mathematical post-processing and carefully aligned and calibrated microscope settings. Moreover, it needs to bear an increased risk of reconstructing artifacts, which requires a lot of knowledge to detect and counteract.

4.1.2 Stimulated Emission Depletion Microscopy

The other main group of SRM is stimulated emission depletion (STED) microscopy, which is target-based inhibition of fluorescence emission by stimulated emission [11–13]. In STED, through exploiting the high-intensity emission of a "donut-shaped" depletion beam, molecules near the periphery of the excitation volume can be selectively and reversibly switched off, and spontaneous fluorescence emission is only allowed to occur at the center area of the illumination PSF (Fig. 4.1B) [14]. That is, the size of the illumination PSF appears to be effectively reduced. Commonly, expert laboratories can reach 30–80 nm lateral resolution in fixed cell and live cell with improved STED systems. Standard STED is generally considered comparably easy to use by being implemented as an add-on modality to standard confocal setups, with additional deconvolution, rather than complex computational post-processing. Additionally, by adjusting the level of laser power, STED can enhance its live-cell imaging capabilities by weighting spatial resolution against potential photo-damaging effects, particularly when combined with customized labels and optimized scanning protocols. However, based on the imaging modality, during STED imaging, the effective fluorescence observation volume is reduced, which entails a decreased scan step size and a corresponding decrease in the total signal detected. So, the final acquisition time increases. Like all point-scanning methods, imaging speed is affected by scan size. The imaging speed is, thus, comparably slow when entire cells are imaged with sufficient photon counts.

4.1.3 Single-Molecule Localization Microscopy

Compared to other super-resolution techniques, single-molecule localization microscopy (SMLM) imposes relatively few technical requirements. An inverted microscope equipped with a powerful laser source, a high NA objective, and a highly sensitive camera is most essential for single-molecule detection, which makes SMLM serve as one of the most accessible and well-established modalities [15, 16]. Collectively termed SMLM techniques mainly include (direct) stochastic optical reconstruction microscopy (dSTORM) [17, 18] and (fluorescence) photoactivation localization microscopy (fPALM) [15,

19]. Although their names sound irrelevant, the imaging principle is similar (Fig. 4.1C) [20]. The key to achieving nanoscale resolution is to realize the precise localization of individual fluorophores on the high-density labeling sample. By regulating the switching of fluorophore in on/off state by stochastic excitation and detecting fluorescent point emitters, it is ensured that only a very small subset of the population (typically <1%) is in the excited state and is precisely localized under a low probability of two emitters residing in close proximity. After bleaching or otherwise switching off the emissive fluorophores, a new subset can be activated again. On continuously repeating this process until the accumulation of sufficient localization information, the underlying structure is reconstructed from the spatial distribution of emitters. Specific to each imaging technology, they are different in achieving the on–off switching of fluorophores. dSTORM mainly exploits photoactivable fluorescent dyes [18], while fPALM utilizes specifically either photoswitchable or photoactivable fluorescent proteins [21]. As SRM is implemented at low cost on conventional, camera-based, wide-field setups, as well as relative higher resolution, SMLM approaches are very popular. Most SMLM implementations can separate individual dyes with distances reduced to 20 nm in lateral and 50 nm in axial. Therefore, SMLM can serve as an invaluable tool for the study of biological structures and functions at the nanoscale. Since the final high-definition image is reconstructed by acquiring thousands of camera frames, it leads to a longer acquisition time, which restricts the general applicability of SMLM for live-cell imaging. The other disadvantages of SMLM are caused by the complexity of the image reconstruction process, which requires careful consideration of falsely localized emitters because of high labeling densities or inappropriate photoswitching rate of fluorophores.

In general, each SRM technology has its own imaging strengths and weaknesses in biological applications. When implementing SRM into different fields of research, we need to make a balanced assessment to select an appropriate SRM technique according to the research objective and purpose. Here, we mainly focus on the application of SMLM techniques for revealing the organization of various membrane biomolecules.

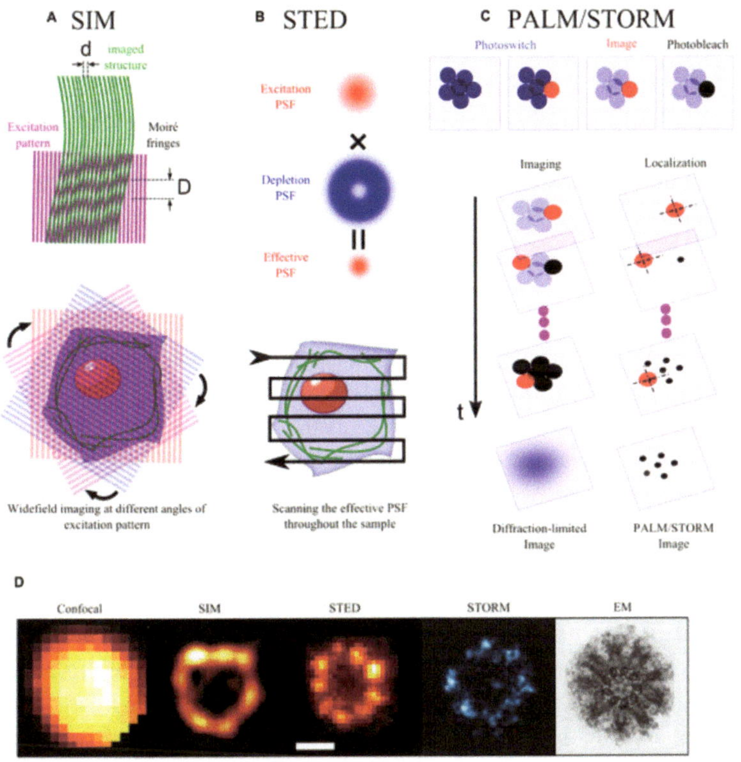

Figure 4.1 Basic principles of some super-resolution optical methods, including SIM imaging (A), STED (B), SMLM imaging (C), and the compared images of centriole from different imaging technologies: confocal, SIM, STED, STORM, and EM images (D). Reprinted with permission from Ref. [2]. Copyright 2014, Frontiers.

4.2 Key Factors in the Imaging Quality of SMLM

Based on the differences in the strategy used to achieve effective "on–off" switching of the fluorescence labels, SMLM techniques can be classified into different types as described above. To achieve a good-quality image, many factors should be considered besides the performance and accuracy of the microscope itself.

4.2.1 Types of Fluorescent Probes

PALM imaging usually uses light-excited switching of various types of fluorescent proteins, whereas STORM imaging commonly exploits inherent fluorescent dyes or light-induced blinking organic dyes that link with recognition unit, such as antibody, for labeling targets [22, 23].

In detail, generally three distinct kinds of photoswitchable fluorescent proteins are used in PALM imaging, including irreversible switching, reversible switching, and photo-convertible fluorescent proteins. For example, photoactivable green fluorescent protein (GFP) [24], the monomeric switching mEOS proteins [25], reversible switching Dronpa, and so on [21, 26, 27]. It is worth noting that genetically encoded labels demand to fuse the targets and generate new constructs, which may lose the real function or structure of the targets, as well as low/overexpression of the protein of interest [28]; besides, the photophysical properties of fluorescent proteins are sometimes not satisfactory.

dSTORM uses typical organic dyes that can reversibly switch on and off in the suitable chemical conditions to enable the precise localization of single fluorophores. These organic dyes can "blink" in the buffer systems containing the combinations of enzymatic systems such as glucose oxidase/β-D-glucose and reducing agents such as β-mercaptoethanol, β-mercaptoethylamine, or Trolox [29, 30]. Fluorescent dyes commonly used at present include cyanine derivatives such as Alexa Fluor 647, Cy5, Cy3, and rhodamines such as Alexa Fluor 488, Atto 488, Alexa Fluor 555, and Alexa Fluor 568 [31, 32]. Due to the different chemical properties of these organic dyes, the excitation power of the laser and the imaging buffer system (i.e., concentration and pH) are also different. Therefore, in the actual imaging process, it is necessary to carefully select the buffer system and illumination conditions, allowing control over the population of both the off and on states.

4.2.2 Precise Localization

To ensure the high resolution of SMLM, an accurate and precise localization of single fluorescent points is significant [33]. According to the related theoretical formula, determining the centroid position of a fluorophore with high precision mainly depends on the photon count (roughly the square root of the number of photons detected). Hence, the photophysical properties of fluorophores are critical. Generally, the fluorescence intensity of fluorescent proteins is significantly lower than that of fluorescent dyes, which leads to low signal-to-noise ratio and low localization accuracy.

4.2.3 Structural Resolution

In addition, structural resolution is an important factor that also needs careful consideration. To ensure the clarity and integrity of the target structure, the sample's labeling density and switching properties of fluorophores should meet the requirements. As a general rule of the Nyquist sampling criterion, achieving a specific structural resolution requires that the distance between neighbor localizations should be at least twofold smaller [33–35]. Moreover, in SMLM imaging, TIRF or highly inclined and laminated optical sheet (HILO) illumination is commonly used to significantly improve detection efficiency and signal-to-background ratio [36].

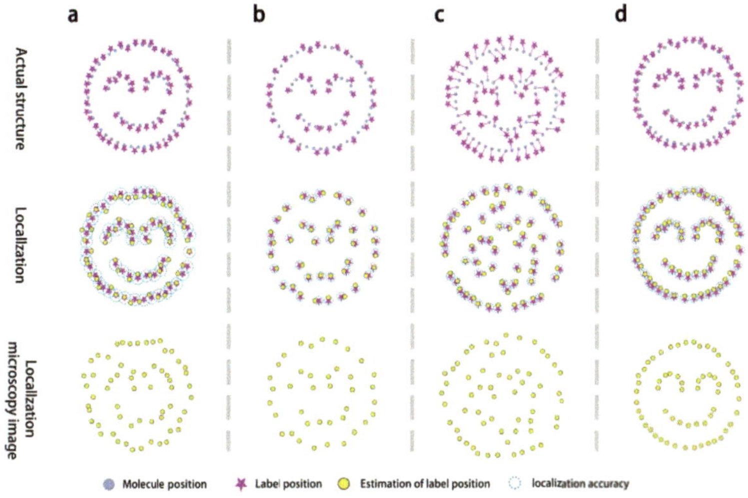

Figure 4.2 Three factors affecting the imaging resolution of SMLM. The final imaging resolution of SMLM is affected from low localization accuracy (a), low labeling density (b), and large label displacement (c). The actual structure can be faithfully represented only when these three factors are satisfied (d).

Based on the SMLM imaging principle, achieving the accurate labeling of targets and precise localization of single fluorophores is significant to the final high quality of super-resolution imaging. Therefore, the labeling probe used in the SMLM imaging is a vital factor to ensure the final imaging resolution. The desired probes need to efficiently and specifically deliver bright fluorophores into the close proximity of the targeted molecules with low background [37]. They should yield high labeling density and labeling accuracy and should have perfect photophysical characteristics to ensure high positioning accuracy. Only in this way, the final super-resolution fluorescence image can accurately display the real position and structure of the targets, as shown in Fig. 4.2.

4.2.4 Steric Effect of Labeling

In common labeling methods, whether photoactivatable fluorescent proteins directly linked to the protein of interest by transfection or fluorescent organic dyes-conjugated antibody by immunofluorescence, the location of the imaged fluorophores (fluorescent proteins or dyes) often deviates from the actual location of the targets and the deviation is random. This is due to the steric hindrance caused by labeling errors. First, as the size of the probe increases, the final linkage error increases (Fig. 4.3). Apart from the linkage error, indirect immunolabeling would also cause a variable ratio of the recognition unit and the fluorophores due to signal amplification [37], which affects the (semi-)quantitative estimation of the number of targets. The studies indicated that a hypothetical 5 nm localization uncertainty would be added to the linkage error for each probe.

For transfection technology, not only the size of the fusion fluorescent protein can affect the localization accuracy, but also there are defects in the labeling density of the target. The fused fluorescent proteins often suffer from either overexpression or low expression, which cannot represent the expression level of the target molecules. Thus, this method cannot give the real distribution of the target molecules. Additionally, the number and density of targets also affect the localization accuracy. When labeling membrane biomolecules that are densely distributed, the larger the volume of the probe, the greater the steric effect, and the more difficult it is to accurately and completely label the targets on the cell membrane (Fig. 4.4) [38]. Therefore, both fluorescent transfection labeling and immunofluorescence technique need to be improved to obtain a better labeling density and accuracy.

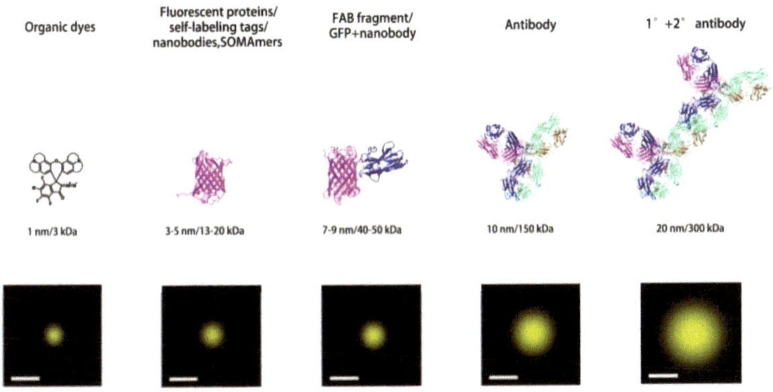

Figure 4.3 The linkage error increases as the size of probe in fluorescent imaging. Scale bars, 30 nm.

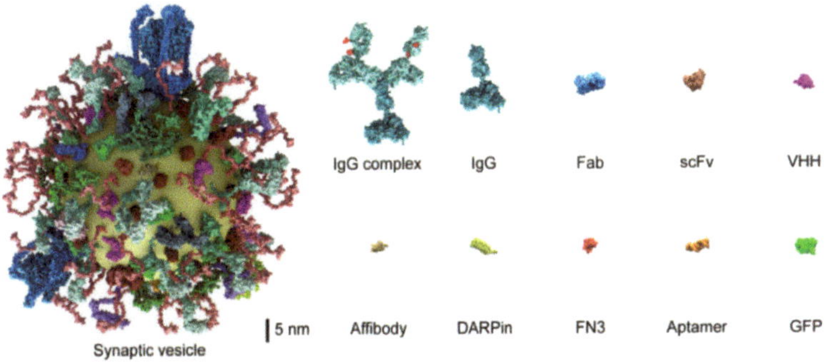

Figure 4.4 Structure and size comparison of different affinity probes. The synaptic vesicle is depicted as a size reference (~40 nm in diameter) and as an example of a subcellular structure where bulky affinity probes would have problems to bind the epitopes hidden by the high density of proteins, as in proximity of the membrane. Reproduced from Ref. [38] by permission of John Wiley & Sons Ltd.

4.3 Super-Resolution Imaging of Membrane Molecules Using Different Probes

Size, specificity, and brightness are critical characteristics of probes for localization microscopy. The present common labeling techniques usually have a trade-off between these properties. Here, we mainly focus on the labeling probes in dSTORM imaging of biomembranes. We will introduce the update progress in membrane proteins and carbohydrates using super-resolution microscopy with different types of probes.

4.3.1 Revealing the Organization of Membrane Proteins by Antibody-Probe Labeling

With a high specificity to their targets, antibodies, as the specific recognition units, are the most widely employed probes. In the development of dSTORM imaging, especially in the initial application stage, direct immunolabeling via primary antibody conjugated with organic dyes and indirect immunolabeling via primary antibody and secondary antibody conjugated organic dyes are the common labeling methods. Because the immunolabeling method has been widely used in traditional fluorescence imaging, there is a large number of commercially available antibodies that can be selected to identify different membrane molecules. In addition, due to the wide application, the immunofluorescence labeling technique is relatively mature and has a complete operation protocol that is suitable for almost all laboratories. Therefore, antibody probes have been widely used in the dSTORM imaging of the structure of various membrane molecules.

As known, membrane proteins have many important biological functions, such as material transport, signal transduction, energy conversion, cell proliferation and differentiation, which make them the focus of research. For example, Wang et al. used Alexa532-conjugated GLUT1 antibody to label GLUT1 on the cell membrane and first investigated the distribution and assembly of GLUT1 at a nanometer resolution by dSTORM imaging (Fig. 4.5) [39]. They not only characterized the clustering pattern of GLUT1, but also revealed that these protein clusters were mostly regulated by lipid rafts as well as restricted in size by both cytoskeleton and glycosylation. More importantly, they found that the activation of GLUT1 by azide or MβCD did not increase its membrane expression but induced a decrease in the large clusters, which suggested that sporadic distribution of GLUT1 may facilitate the transport of glucose, implying a potential association between the distribution and activation. Collectively, the distribution of GLUT1 and the association of the spatial structural organization to the functions can provide insights into the activation mechanism of the transporter.

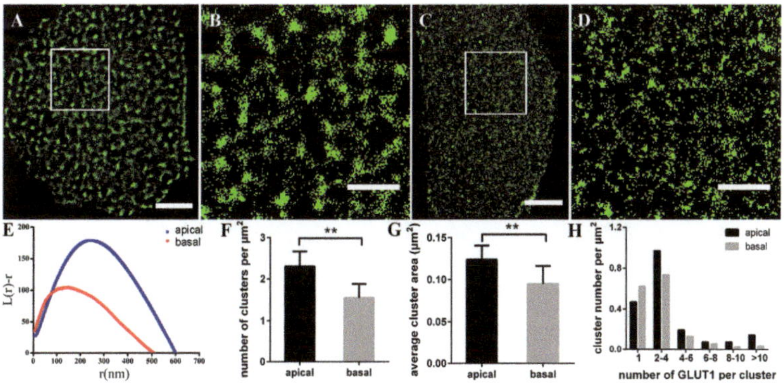

Figure 4.5 GLUT1 proteins form clusters of different sizes and amounts on HeLa apical and basal membranes. (A–D) Reconstructed dSTORM images of GLUT1 on apical (A) and basal membranes (C), and the corresponding magnified images (B and D). Scale bars: A and C, 5 μm; B and D, 2 μm. (E) Representative Ripley's K function plots of GLUT1 on different cell surfaces. (F) The average number of GLUT1 clusters per μm². (G) The average cluster area. (H) The number of sporadic GLUT1 and clusters containing different amounts of molecules per μm². Reprinted with permission from Ref. [39]. Copyright 2018, American Institute of Physics.

Besides glucose transporter, the dopamine transporter (DAT) is critical for spatiotemporal control of dopaminergic neurotransmission by mediating rapid reuptake of extracellular dopamine. Gether et al. exploited the corresponding antibody probe (primary antibody and dye-conjugated secondary antibody) to implement single color and dual-color STORM imaging of DAT distribution (Fig. 4.6) [40]. The results showed that irregular DAT nanodomains were highly sensitive to cholesterol depletion, as well as its spatial association with tyrosine hydroxylase or vesicular monoamine transporter-2. To realize live-cell imaging, they also applied PALM imaging to visualize the transfected photoswitchable Dronpa-DAT in CAD cells and found a similar dopamine transporter membrane organization in live heterologous cells. Their work demonstrates the power of SMLM imaging to dissect the distribution of an important membrane transport protein in neurons. They revealed how DAT is dynamically sequestered into irregular nanodomains in cellular structures and found that DAT function can be regulated by membrane lipids and neuronal activity at the nanoscale level.

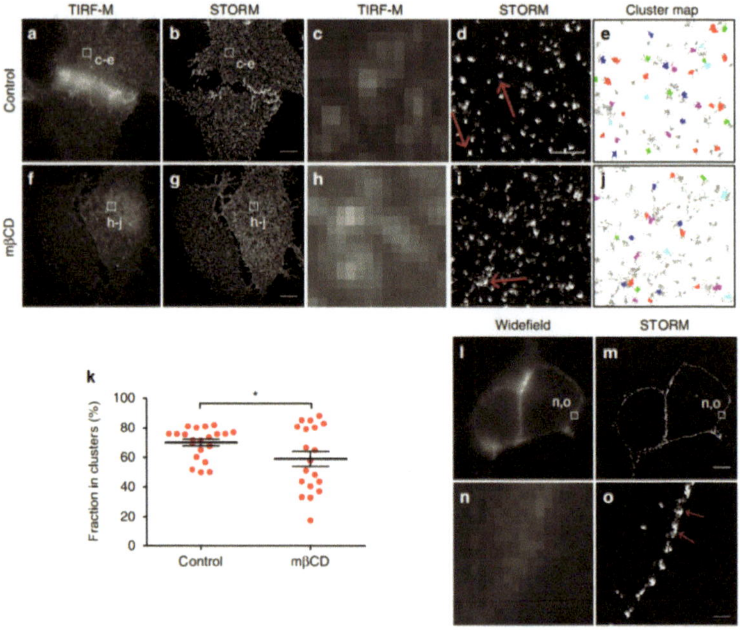

Figure 4.6 STORM imaging of DAT in CAD cells. TIRF-M (a, c, f, h), STORM (b, d, g, i, m, o), and wide-field (l, n) images of CAD cells transiently expressing DAT with or without cholesterol depletion, through immunolabeling with primary DAT antibody and Alexa405-Alexa647 conjugated-anti-rat secondary antibody. (e, j) The corresponding DBSCAN-based cluster map. Clusters are shown by color-coding. (k) Fraction of particles in nanodomains with or without cholesterol depletion. (l–o) Visualization of DAT expressing CAD cells using inclined illumination to visualize a cross section through the cell. Scale bars a, b, f, g, l, m, 5 μm; c–e, h–j, 500 nm; n, o, 400 nm. Reprinted by permission from Macmillan Publishers Ltd: Ref. [40]. Copyright 2017.

Using antibody labeling technique, researchers have revealed the assembly characteristics and mechanism of various membrane proteins via the super-resolution fluorescence imaging. These findings about the detailed spatial information of membrane molecules at the single-molecule level can promote the understanding of their biological functions and the systemic organization of cell membrane.

Nonetheless, the characteristics of the antibody itself will cause some problems in the final quality of super-resolution fluorescence images. As mentioned above, due to the large molecular mass and volume of antibodies (~150 kDa and ~15 nm in length), the effective displacement between the fluorophore and the target can reach up to 24–30 nm in indirect immunofluorescence labeling (using primary and secondary antibody); even in direct immunofluorescent labeling (only primary antibody), this linkage error is still ~10 nm. As a result, the location of the fluorescent group cannot represent the real position of the target and thus significantly deteriorates the achievable precision and accuracy of protein localization by super-resolution fluorescence microscopy. Moreover, because of the large steric effect caused by the large volume of antibodies, many target molecules densely organized on cell membranes cannot be labeled completely, resulting in a low labeling density, and the real distribution of all targets cannot be displayed. Besides, dual-valent or multi-valent binding capability of antibodies introduces significant clustering artifacts in super-resolution imaging [41, 42].

4.3.2 Small Bio-probes for Imaging Membrane Structure

With the increasing demand for the improvement in the resolution of super-resolution fluorescence microscopy, to ensure refined and accurate imaging, the labeling quality of fluorescent probes also needs to meet higher requirements. The new generation of affinity probes has been developed to have minimal size and monovalent binding to overcome some disadvantages of antibody probes [33].

4.3.2.1 Antibody fragment

The first direct approach to reduce the size of antibody probes is to generate recombinant Fabs (Fab, ~50 kDa, ~9 nm) [43–45] or single-chain variable antibody fragments (scFv, ~30 kDa, and ~6 nm) [46]. Although these antibody fragments effectively reduce the volume of the probe compared with full-length antibodies, their binding ability is often suffered from the reduction in binding sites. As a result, they cannot always perform better recognition characteristics than full-length antibodies in fluorescence imaging experiments. Additionally, laborious selection and generation of antibody fragments, as well as their low expression, always prevent them from becoming a real competitor to full-length antibodies.

4.3.2.2 Nanobody

Nanobodies [47, 48], one kind of smaller antibodies without light chains, have been developed as specific recognition units of small biomolecule fluorescent probes [3, 47, 48]. These simpler nanobodies were originally found in alpacas, llamas, and camels and can be produced in "re-programmed" bacteria [49]. Compared to normal antibodies, nanobodies are 10-fold smaller (~13 kDa, ~3 nm) because they have only one binding domain, which is of great advantage in antigen detection and labeling [47, 50].

In 2012, Ewers et al. first applied nanobody fluorescent probes to the super-resolution imaging. They developed a method to use a GFP-tagged construct in SRM imaging. The construct is composed of very small and high-affinity nanobodies to GFP5, so it can deliver bright organic fluorophores to GFP-tagged proteins. This is a simple, efficient, and versatile method that optimizes several labeling parameters and can be used to label each of the thousands of functionally tested GFP [50]. Mikhaylova et al. developed single-chain antibody fragments (nanobodies) against tubulin for the super-resolution imaging of microtubules with a smaller apparent diameter (Fig. 4.7). They created bundles of microtubes known to be 60–70 nm apart and modified the bundles with antibodies and nanobodies, respectively, to assess the labeling quality of these different probes. By showing the ability of the nanobodies to discriminate between individual microtubules, they demonstrated that nanobody-mediated labeling can outperform traditional labeling techniques in advanced single-molecule imaging applications such as single-particle tracking and single-molecule localization microscopy [51].

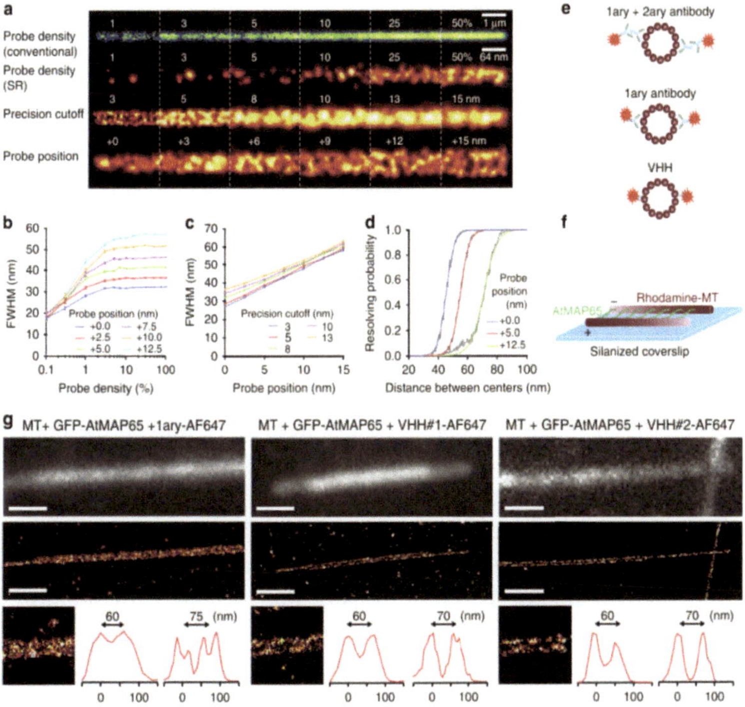

Figure 4.7 (a) Different microtubule images from conventional (top) and SMLM imaging under different probe densities, localization precision cutoffs, and probe positions. (b) FWHM of Gaussian fits to microtubule as a function of probe density and for different probe positions. (c) FWHM versus probe position for different cutoffs of the localization accuracy distribution. (d) Estimation of resolving power for the staining of microtubules with probes at increasing distance from the microtubule. (e) Compared illustration of the different labeling probes. (f) Scheme of the in vitro microtubule bundling assay to test the resolving power of different microtubule labeling probes. (g) Conventional (top) and SMLM (middle and bottom left) images and representative line scans (bottom right) of in vitro microtubule bundles stained with a primary antibody probe (1ary-AF647) or two novel tubulin nanobodies-AF647 probes. Scale bar, 1 μm. Reprinted by permission from Macmillan Publishers Ltd: Ref. [51]. Copyright 2015.

However, the method using nanobodies is not direct labeling like immunofluorescence only with primary antibody. Since it still involves transfection, the same defects could occur as labeling with fluorescent proteins by transfection. The abnormal expression of fused proteins can affect the function of target proteins. Moreover, the generation of nanobodies with high specificity is not an easy task. Nanobodies are mostly labeled at lysines by N-hydroxysuccinimide (NHS) ester fluorophores randomly and rarely quantitatively [50, 52]. Meanwhile, as shown below, it also deteriorates signal-to-noise ratios or even completely abolishes target recognition. So far, the nanobody-related labeling method has not been widely used as a popular fluorescent labeling method.

4.3.2.3 Aptamer

In parallel, Rizzoli et al. first demonstrated that aptamer, a short single-stranded oligonucleotide (~15 kDa, ~4 nm) with high specificity and small size, can serve as an alternative fluorescent probe to improve the imaging precision in STED and SMLM imaging [41].

As we know, on the cellular surface, a lot of biomolecules are aggregated to form a dense protein layer, which may promote various cell functions by better performing molecule–molecule interactions. To obtain a higher labeling density and accuracy of membrane biomolecules, the size of fluorescent probes as a key factor should be carefully considered. Chen et al. used aptamer probes and antibody probes to label the membrane protein PTK7 and obtained the compared dSTORM imaging. They found that compared to antibody probes, aptamer probes can label densely distributed targets more easily and accurately, by displaying smaller and denser protein clusters on cell membranes (Fig. 4.8) [53].

Importantly, aptamers not only have a small size compared to nanobodies, but also avoid the labeling problems of nanobodies. For example, nanobodies cannot label endogenous, nontagged proteins, whereas aptamers can conjugate organic dyes and directly label endogenous proteins on the membrane and even in the cytoplasm, which is simple and effective. Additionally, the GFP tag occasionally induces loss of function, multimerization, or mislocalization. Moreover, the structures composed of the GFP-tagged protein and the native copy (invisible to GFP nanobodies) are often discontinuously labeled. All of them affect the final imaging quality of super-resolution fluorescence microscopy. In contrast, aptamer labeling can obtain higher labeling density without all the above problems. Therefore, as a small and specific recognition unit to form small fluorescent probes, aptamers can contribute to a better super-resolution imaging of various membrane molecules.

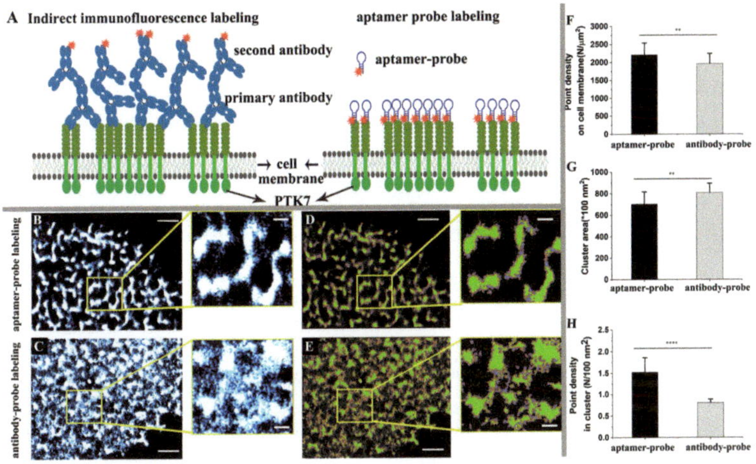

Figure 4.8 Compared dSTORM imaging of the assemble of PTK7 labeled by an aptamer probe and an antibody probe. (A) Schematic diagram showing differences between indirect immunofluorescence labeling (left) and aptamer probe (right) in labeling membrane proteins. (B, C) Reconstructed dSTORM images of PTK7 on cell membranes stained by indirect immunofluorescence labeling (B) and aptamer labeling (C). (D, E) Images of qualified clusters (in green) recognized from the distribution images by setting the appropriate thresholds. (F–H) Compared histograms of the average point density on the cell membrane (F), average cluster area (G), and average point density in cluster (H) from different labeling. Reprinted with the permission from Ref. [53]. Copyright 2020 American Chemical Society.

Wang et al. used fluorophore-conjugated aptamer to perform dSTORM imaging to observe the cancer-associated carbohydrate—Globo H (Fig. 4.9) [54]. They clarified the clustered distribution of Globo H at the nanometer scale and revealed the potential interactions between cancer-associated carbohydrates, which paved the way for further understanding of the relationship between the spatial structures and functions of carbohydrates in cancers.

Another example is the important tyrosine receptor on the cell membrane, the epidermal growth factor receptor (EGFR), which plays an important role in the physiological processes such as cell growth, proliferation, and differentiation, and is also closely related to tumor proliferation, invasion, and metastasis. Using EGFR aptamer as the small fluorescent probe, Wang et al. demonstrated that the aptamer probe can obtain a better STORM imaging than antibody probes and ligand probes [55]. The RNA aptamer labeling showed a high-quality dSTORM image of EGFR nanoscale clusters on live-cell membranes, and displayed the different organization of active EGFR and resting EGFR (Fig. 4.10), which was hidden under the antibody labeling. Their work further suggested that aptamers can be used as versatile probes in super-resolution imaging.

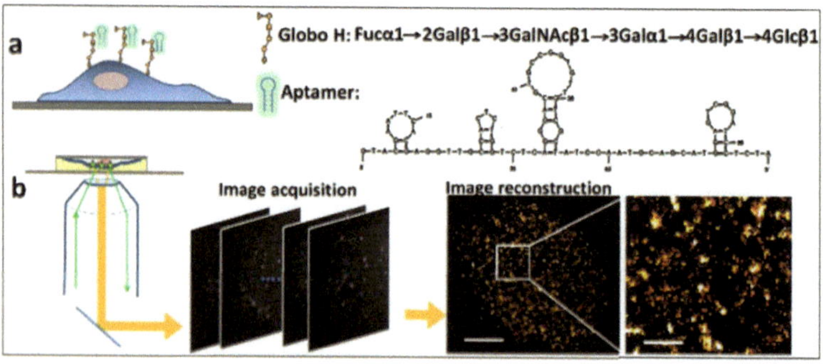

Figure 4.9 dSTORM imaging of aptamer-recognized Globo H on the cell surface. (a) Schematic of the aptamer labeling process. The composition of Globo H is shown in the right part. (b) The basic principle for dSTORM imaging. The dSTORM image shows in detail the organization of Globo H labeled with an aptamer probe. Scale bars: 5 μm in the original image, 1 μm in the magnified image. Reproduced from Ref. [54] by the permission of The Royal Society of Chemistry.

Besides, Chen et al. used the PTK7 aptamer in dSTORM imaging and found the distribution features of PTK7 on cell apical and basal membranes. They revealed the potential organization mechanism of PTK7 on different cell surfaces, that is, PTK7 clusters are associated with lipid rafts, cell skeleton, and carbohydrate chains.

Although aptamer probes can serve as a superior labeling method in super-resolution imaging, the lack of available aptamers makes it still difficult to label a majority of membrane molecules for super-resolution imaging. This is because the screening of aptamers with specific binding ability requires a complex in vitro process called systematic evolution of ligands by exponential enrichment (SELEX) [56]. This work is challenging and not suitable for ordinary laboratories without screening experience. However, compared with antibodies and nanobodies, the preparation process of aptamers is relatively simple. Therefore, to widen its application in the super-resolution imaging of membrane molecules, screening the available aptamers will be the main breakthrough.

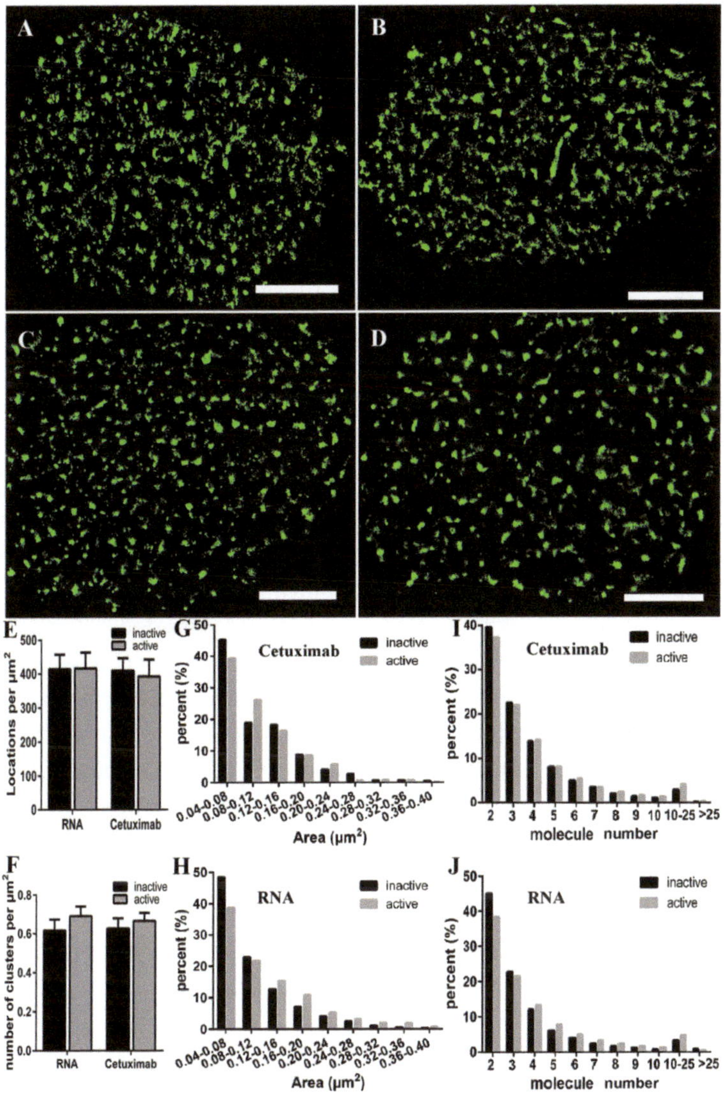

Figure 4.10 The distribution and clustering of active and resting EGFR on the serum-starved COS-7 cell membranes. (A–D) Reconstructed dSTORM images of Cy3-cetuximab-labeled inactive (A) and active (B) EGFR, and Cy3-aptamer-labeled inactive (C) and active (D) EGFR. Scale bars are 5 μm. (E) The number of localizations per μm^2 of inactive and active EGFR with aptamer or antibody labeling. (F) The number of inactive and active EGFR clusters in unit area. (G and H) Histogram of inactive and active EGFR cluster areas labeled with cetuximab (G) or the aptamer (H). (I and J) The percentage of clusters containing different molecule numbers with cetuximab (I) or aptamer (J) labeling. Reproduced from Ref. [55] by the permission of The Royal Society of Chemistry.

4.3.3 Small Chemical Probes for Revealing the Relationship between Membrane Protein Structure and Function

A high-quality imaging derives from a highly accurate labeling, which requires probes to efficiently and specifically deliver bright fluorophores into close proximity of the targeted molecules with low background [37]. Therefore, the smaller the recognition unit of the probe, the more closely the fluorophores delivered into the target, which can obtain more accurate labeling and higher labeling density. Although a small bio-recognition unit can effectively reduce the size of fluorescent probes, there is no lower limit for the volume reduction of the probes.

With interdisciplinary development, in addition to exploring new fluorescent probes in the field of biology, synthetic chemistry has gradually become an important means to generate new fluorescent probes due to its flexibility, variability, and controllability. Small organic molecules with biological activity are regarded as "chemical antibodies" because of their small volume and high specificity, so they are ideal recognition units in probes. Small and widespread natural substrates are capable of binding with a membrane transporter at high affinity and specificity, and as a result, substrate-binding sites are formed in membrane transporters [57]. Therefore, substrates preserving specific recognition ability and bioactivity can conjugate organic dyes to form small fluorescent probes for the dSTORM imaging of membrane transporters. Wu et al. selected glucose and tyrosine as substrates to synthesize probes for recognizing glucose transporters and tyrosine transporters [58]. They modified the glucose and tyrosine with organic dyes to form Glu probe (Fig. 4.11) and Tyr probe (Fig. 4.12) for the dSTORM imaging of the corresponding targets. They demonstrated that the substrate-based probe can serve as an important type of small-molecule fluorescent probes to obtain a higher labeling density and more accurate labeling without cross-linking effect when compared with antibody-probe labeling. Meanwhile, they revealed that membrane transporters are mostly organized into clusters and regulate the transport effectiveness by changing their assembled patterns.

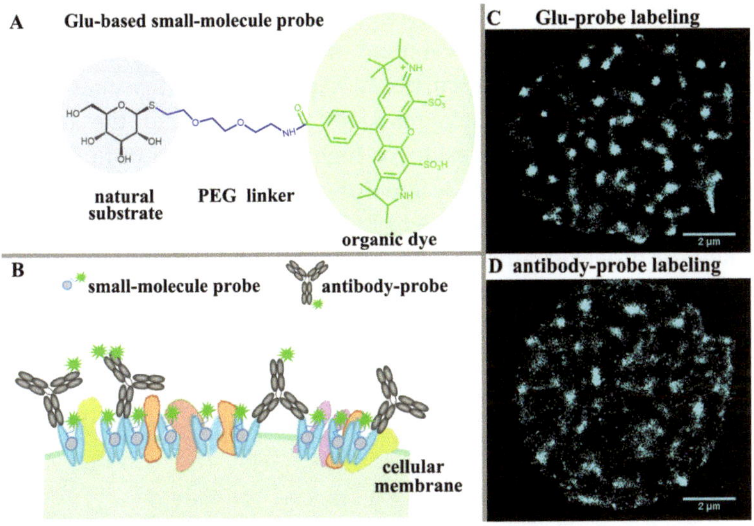

Figure 4.11 (A) Molecular structure of the Glu-Alexa532 small-molecule fluorescent probe. (B) Scheme describing the labeling superiority of Glu probe to antibody probe. (C and D) dSTORM imaging of GLUTs on the RBC membrane with Glu-Alexa532 labeling (C) and antibody-Alexa532 labeling. Scale bar is 2 μm. Reproduced from Ref. [58] by the permission of The Royal Society of Chemistry.

Moreover, with the rise of the concept of "targeted therapy," more and more small-molecule inhibitors (SMIs) with selective action sites have been developed and successfully applied to the efficient treatment of a variety of diseases, especially cancers [59, 60]. They can block the binding of antibodies or even natural ligands to membrane receptors for targeted treatment. They usually have better specificity and affinity than antibodies. So, they can be used as recognition units to form small-molecule fluorescent probes. The ideal SMI should meet the following conditions before it can be selected to synthesize fluorescent probes: (1) small size (antibody molecular weight: ~104–105 kDa, ligand molecular weight: ~103 Da, organic small molecular weight: ~102 Da), (2) strong specificity, and (3) sufficient binding sites.

Gefitinib [59, 61], as a single targeted SMI of EGFR, has been selected to form small fluorescent probes for recognizing EGFR by modification and linking with organic dyes [62]. The labeling superiority of Gefitinib probe has been exhibited through comparing with antibody and ligand probes (Fig. 4.13). The results further indicated that the whole volume of the chemical-molecule probe was quite small among currently used fluorescent probes, which greatly reduced the imaging error and improved the labeling accuracy and density. Besides, a high colocalization of fluorescent points from Gefitinib-probe and EGF-probe labeling indicated a high specificity of Gefitinib probe to EGFR. More importantly, using this kind of probes can provide more precise and refined information about spatial organization and molecular composition of membrane receptors under pathological or physiological conditions.

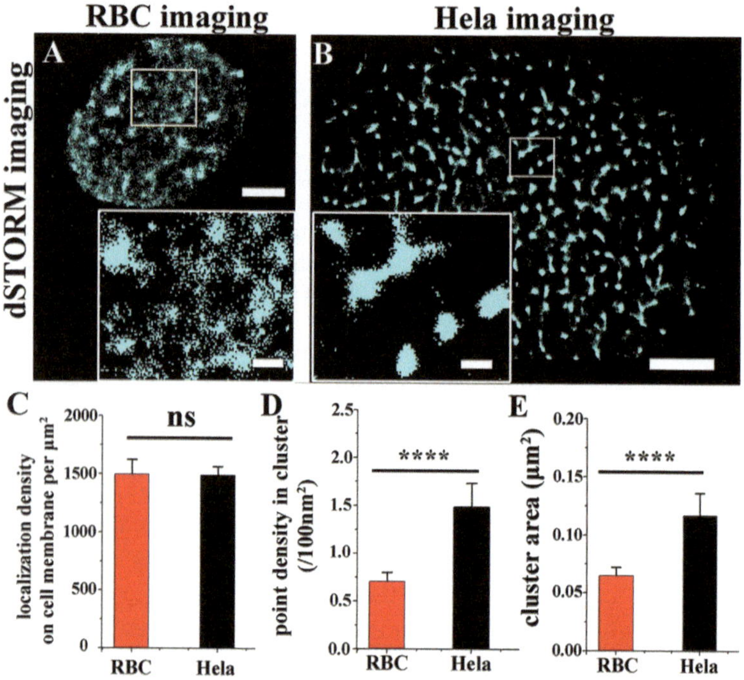

Figure 4.12 dSTORM imaging of Tyr transporters on membranes of human red blood cells and Hela cells by Tyr-probe labeling. (A and B) The representative reconstructed images of Tyr transporters on the RBC (A) and Hela membranes (B), with the corresponding enlarged images of clusters in the bottom. (C–E) The compared histograms of the localization density on cellular membrane (C), point density in cluster (D), and cluster area (E) on these two cellular membranes. Scale bars are 2 μm in A and F, 5 μm in B and G, and 500 nm in all enlarged images. Reproduced from Ref. [58] by the permission of The Royal Society of Chemistry.

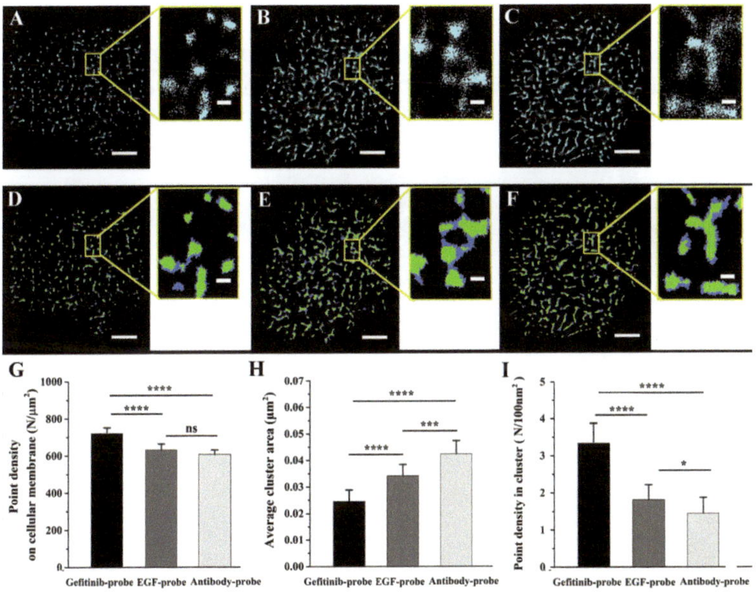

Figure 4.13 The contrast imaging of EGFR distribution on the cell membranes stained by different probes. (A–C) dSTORM images of EGFR on the fixed A549 cell membranes labeled by Gefitinib-Alexa532 (A), EGF-Alexa532 (B), and Cetuximab-Alexa532 (C), with corresponding enlarged images. (D–F) The final qualified clusters (in green) are abstracted from all fluorescent localizations. (G–I) The histograms of the average point density on cellular membrane (G), the average cluster size (I), and the average point density in cluster (I). Scale bars are 5 μm in original images, and 500 nm in enlarged images. Reproduced from Ref. [62] by the permission of The Royal Society of Chemistry.

Furthermore, unlike antibodies, natural substrates or small-molecule inhibitors can bind to the sites on the cytoplasmic side of the cell membrane without damaging the membrane and can effectively avoid steric effects of probes, which bind to extracellular sites during multicolor labeling. For example, to explore the spatial relationship between EGFR and HER3, Wu et al. used two pairs of fluorescent probes to label [62]. The first pair is HER3 antibody and Gefitinib probe, and the second is HER3 and EGFR antibody. They found a higher colocalization of EGFR and HER3 labeled by the first probe pair than by the second pair, which suggested that the Gefitinib probe with cytoplasmic binding site benefits dual-color imaging (Fig. 4.14). Their work indicates that the SMI probes are also able to serve as versatile labeling tools for high-quality super-resolution imaging. Meanwhile, this kind of probe can provide more accurate spatial colocalization and interactions between membrane molecules.

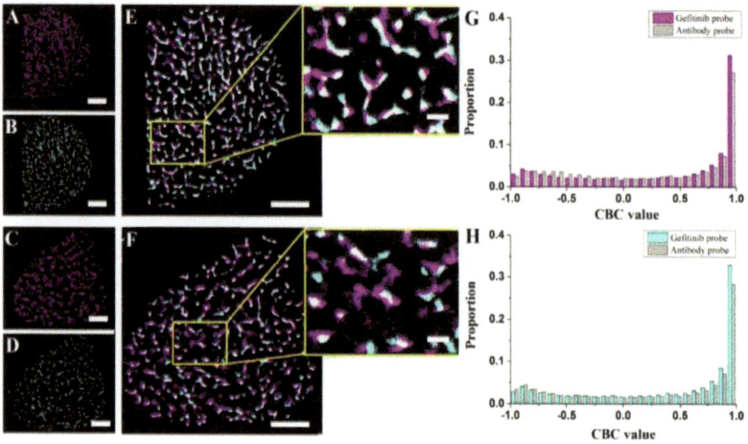

Figure 4.14 Dual-color dSTORM imaging of EGFR and HER3 labeled by different probe pairs. (A–D) The reconstructed dSTORM images of HER3 (A and C) and EGFR (B and D) on the fixed A549 cell membranes labeled by the probe pair of HER3 antibody–Alexa647 (A) and Gefitinib–Alexa532 (B) and the antibody-probe pair composite of HER3 antibody–Alexa647 (C) and Cetuximab–Alexa532 (D). Scale bars are 5 μm. (E and F) The merged images of the two channel images. Scale bars are 5 μm in A–F and 1 μm in the enlarged images. (G and H) The histograms of CBC value distribution of the red channel probe and the green channel probe of different probe pairs. Reproduced from Ref. [62] by the permission of The Royal Society of Chemistry.

Besides Gefitinib, there are many kinds of small organic molecules with biological activity, which also ensure the wide application of this kind of probe in super-resolution imaging. For example, Tariquidar-TAMRA probe, another SMI-based probe, was exploited to label P-glycoproteins (P-gp) [63]. P-gp is a common molecular pump to protect cells from external harmful molecules and has the function of an energy-dependent "drug pump" [64]. The study has found that P-gps accumulated into larger and denser clusters on cancer cells and drug-sensitive cells than on normal cells and drug-resistant cells (Fig. 4.15), indicating that P-gp can facilitate pumping efficiency by aggregating to form functional platforms. Moreover, these specific distribution patterns might serve as potential biomarkers for tumor and drug therapy screening.

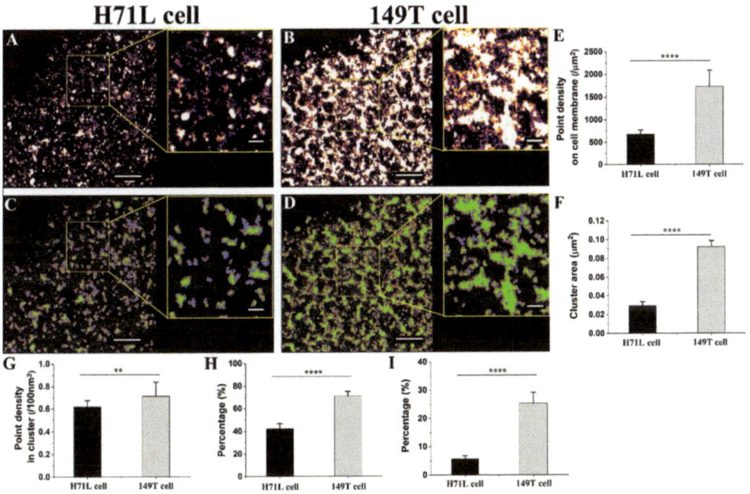

Figure 4.15 dSTORM imaging of P-gp distribution on the drug-sensitive and drug-resistant cell membranes. (A and B) The representative dSTORM images of P-gp distribution on the H71L (drug-sensitive cell) and 149T (drug-resistant cell) cell membranes by Tariquidar-TAMRA labeling. (C and D) The corresponding images of qualified clusters (green) and recognized objects (blue) analyzed by the SR-Tesseler method. (E–I) The histograms of the average point density on cell membrane €, the average cluster area (F), the average point density in cluster (G), the number percentage of total points in cluster to total points (H), and the area percentage of the total area of cluster to the total cell membrane (I). Scale bars are 2 μm in the original images and 500 nm in the enlarged images. Reproduced from Ref. [63] by the permission of The Royal Society of Chemistry.

Taken together, through linking various small chemical molecules (as the recognition unit) to different organic dyes, numerous chemical fluorescent probes can be synthesized. As the current smallest fluorescent probe, the small-molecule chemical probe can improve the accuracy of super-resolution fluorescent imaging. Moreover, the preparation process of the chemical probe is simple and standard, the quality is stable and controllable, and the reaction time is relatively fast. Additionally, as membrane proteins account for almost one-third of synthetic drug targets, many drugs targeting membrane proteins can be directly used as substrates to synthesize fluorescent probes. All the advantages will promote the wide application of the chemical fluorescent probe in the super-resolution imaging of membrane proteins.

4.4 Protocols for STORM Experiment

There are various kinds of super-resolution imaging techniques as we summarized earlier, and in addition to different instruments and equipment, they also have their own experimental methods. As the preparation of cell samples is representative in STORM imaging, here we only take it for an example to introduce how to perform STORM experiment.

4.4.1 Preparation of Fluorescent Probes

Different types of fluorescent probes should be selected for different experiments: (1) the plasmid of fluorescent protein fused target protein; (2) aptamers and chemical probes synthesized by chemical reaction; (3) antibody probes formed by antibody linked with organic dyes. Because fluorescent proteins and aptamers need to be designed and synthesized according to the targets, there is no uniform preparation method.

1. **Set up labeling reaction:** 50 μL antibodies (1 mg/mL), 6 μL $NaHCO_3$ (1 mol/L), and 1 μL Alexa Fluor 647 (1 mL/mL). Allow the reaction to proceed for about 30 min at room temperature (RT) in the dark on a rocking platform.
2. Equilibrate a Nap-5 gel filtration column (GE Healthcare) by running three column volumes of PBS through the column.
3. After the reaction, add 140 μL PBS to the reaction volume up to 200 μL (the minimum column loading volume) and gently vortex. Add the entire volume to the center of the column.
4. Allow the sample to enter the column and add sufficient PBS to wash. During this time, collect the eluent in 1.5 mL Eppendorf tubes (about 150 μL per tube).
5. Measure the absorbance at 280 and 650 nm by the UV-spectrophotometer to calculate the antibody concentration and labeling ratios of the eluent in each tube. Fractions in

which tube(s) the labeling ratio of dyes and antibodies is 0.7–1 are pooled for use.

4.4.2 Sample Preparation

For fluorescent protein probes, cells should be transfected with the constructed plasmid. For other probes, there are two situations: live cell and fixed cell. To fix cells, 4% paraformaldehyde (PFA) is commonly used. Other procedures are similar to immunofluorescence, which can be found in the protocol part of Chapter 5 or other instruction. After staining, 100 nm diameter TetraSpeck microspheres are embedded to correct the x–y drift of sample and the optical registration between different channels. The imaging buffer should be freshly prepared prior to dSTORM imaging, containing 1% (v/v) beta-mercaptoethanol (β-ME), 0.5 mg/mL of glucose oxidase, and 40 μg/mL of catalase. Of note, the buffer compositions may change according to experiment condition. For example, MEA will replace β-ME in live cells. The detailed compositions can be found in Chapter 5 of our previous book [65].

4.4.3 STORM Imaging

STORM imaging can be performed on a Nikon Ti-E microscope with a 100 × 1.49 NA TIRF lens (Nikon, Japan). The sample is illuminated in the total internal reflection fluorescence (TIRF) mode. According to the excitation spectrum of the fluorescent probe, an appropriate laser is used to excite fluorophores. Usually, a 405 nm laser is used to increase the number of on-state fluorophores by carefully controlling its irradiation intensity (<0.1 kW/cm^2). All images are captured by a cooled EMCCD. A series of images are collected for each cell with an internal time of millisecond level between frames to reconstruct a super-resolution image. TetraSpeck microspheres of 100 nm (Invitrogen) are embedded in the sample to correct the x–y drift and chromatic aberration, and a focus lock is used to correct the z drift.

4.4.4 Image Reconstruction

There are now a number of different software packages available for STORM image reconstruction, such as QuickPALM, DAOSTORM,

RainSTORM, ThunderSTORM, and other commercial software. Here we focus on ThunderSTORM [66], a free available plugin in ImageJ. The raw frames are first filtered with a "wavelet B-Spline filter" to select the PSF candidates. Next, the local maximum approach with 8-connected neighborhoods is used to give the approximate position of molecules. Then, PSFs are fitted with an integrated form of a 2D Gaussian function by a weighted least-squares method. The Gaussian function provides a very good approximation of the real PSF of a microscope. Finally, in order to obtain a high-quality image, the reconstructed super-resolution image is further filtered with a localization density filter to remove the background artifacts, and the cross-correlation method is used to determine the shift between the first image and each of the subsequent images. All candidate localizations whose parameters are not accepted by the criteria are removed.

4.4.5 Measurement of the Imaging Resolution

Single-molecule localization precision is analyzed to detect the imaging resolution of the instrument. Adequately diluted fluorescent probes are added onto the prepared cell surface, incubated for 40 min, and imaged. Five thousand images are recorded with experimental integration time per frame to generate a spot localization map. From repetitively switching fluorophores (usually 100 dyes analyzed), the localization precision can be obtained by measuring the full width at half-maximum (FWHM) of the localization distribution of individual fluorophores. Figure 4.16 shows an example of calculating the resolution using Alexa Fluor 647-conjugated anti-STAT3 antibodies [67].

Another method called Fourier ring correlation (FRC) [68] can also be used to determine the resolution of the dSTORM images. The FRC map is calculated by the NanoJ-SQUIRREL [69] ImageJ plugin. This method is based on comparing two independently acquired super-resolution images of the same field of view. First, the image sequences are separated into odd and even frames. The two new sequences generate two equal-sized blocks. FRC analysis is run locally on each block, and the FRC map provides an estimate of resolution across the whole image (Fig. 4.17) [70]. The lowest FRC value represents the minimum resolution in the image.

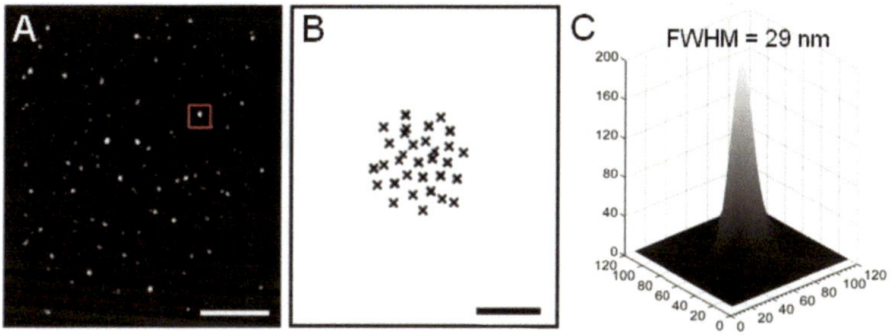

Figure 4.16 The localization precision of single Alexa Fluor 647-conjugated STAT3 antibodies on fixed and permeabilized HeLa cells. (A) The dSTORM image of adequately diluted Alexa Fluor 647-STAT3 antibodies on HeLa cells. (B) Repetitive localizations of a single Alexa Fluor 647-STAT3 antibody boxed in (A), and the number of localizations in this molecule is 31. (C) The average value of singe-molecule localization precision is 29 nm. The two-dimensional histograms of localizations were generated by aligning 100 fluorescent molecules with their center of mass and fitted to a Gaussian function to determine the full width at half-maximum (FWHM). Scale bars represent 1 μm in (A) and 20 nm in (B). Reproduced from Ref. [67] by the permission of The Royal Society of Chemistry.

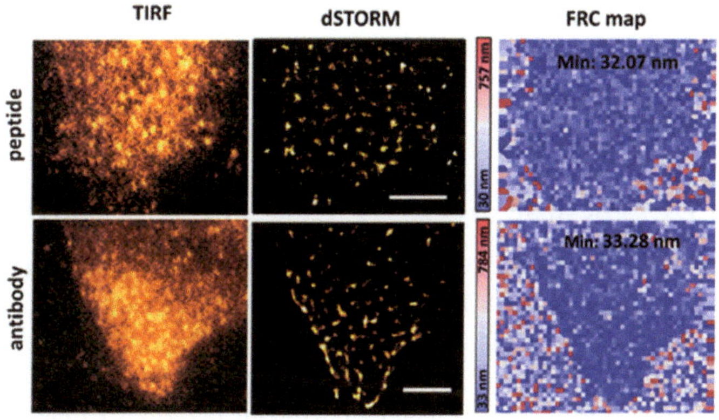

Figure 4.17 EpCAM imaging and the imaging resolution. Conventional TIRF images, corresponding dSTORM images, and FRC maps of EpCAM labeled with Cy3-conjugated peptides and antibodies, respectively. Scale bars, 5 μm. Reprinted with the permission from Ref. [70]. Copyright 2019 American Chemical Society.

4.4.6 Cluster Analysis

Various methods have been used to analyze the spatial clustered characteristic of the molecules, for example, Ripley's K function [19], DBSCAN [71], Hopkins statistic test [72], pair-correlation function [73], etc. The selection of these methods depends on the morphology and distribution of the molecules. Sometimes two or several methods need to be used to analyze and validate the data, and thus they can more comprehensively and accurately reveal the spatial localization and distribution pattern of the molecules.

Here, we mainly introduce a relatively new method, SR-Tesseler [74]. First, a file including the coordinates, intensity, and sigma of molecules is loaded in the program. The reconstructed image is shown, and the whole cell is selected as an ROI to analyze. Second, bisectors between two closest localizations are drawn, and the ROI is segmented into many polygons with different numbers of localizations. The localization density of a polygon is defined as δi1, and the average localization density of the total ROI is δ0. If δi1 > δ0, localizations in this polygon are picked up to create an "object." Similarly, the localization density of an object is set as δi2, and the average of all objects is δ1. Third, objects satisfying the requirement of δi2 > δ1 are extracted as clusters. Finally, as shown in Fig. 4.18 [75], the area, the number of localizations, coordinates, and morphological parameters of each cluster are exported.

4.4.7 Dual-Color Colocalization Analysis

4.4.7.1 CBC analysis

Two-color dSTORM data can be analyzed by the coordinate-based colocalization (CBC) [76]. This approach considers the multiple switching cycles of each set of localizations by comparing the spatial distribution of surrounding localizations from both species. The distribution function of the neighboring localizations from the same species is calculated at first; then, the function of the neighboring localizations from the other population is calculated. Afterward, the correlation coefficient for the two distribution functions (for example, A to B) is calculated, weighted by the nearest neighbor distribution from B population, and attributed to the chosen localization of A population. As a result, each molecule is assigned a CBC value ranging from −1 to 1, where −1 characterizes segregation, 0 corresponds to single species, and 1 represents high colocalization.

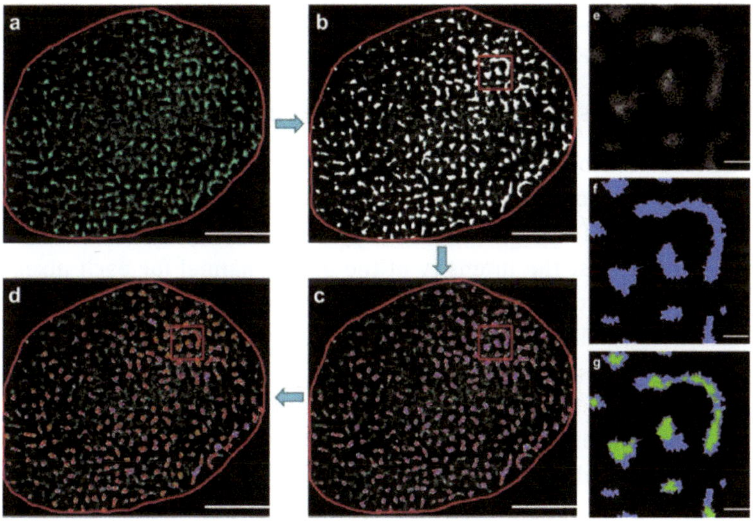

Figure 4.18 The flow diagram of analyzing clusters by the SR-Tesseler method. (a) The reconstructed dSTORM image of Trop2 on an A549 cell. The outline of the cell membrane is circled by a red line and set as an ROI. (b) The localization map is segmented into many polygons, whose edges are bisectors between the nearest localizations (white line). (c) The extracted objects (blue). (d) The extracted clusters (green regions). (e–g) The enlarged images corresponding to the red boxed regions in (b–d), respectively. Scale bars: 5 μm in (a–d), and 200 nm in (e–g). Reprinted with the permission from Ref. [75]. Copyright 2020 American Chemical Society.

4.4.7.2 Cross-correlation analysis

Cross-correlation analysis of two-color super-resolution images provides estimates of spatial scales of coclustering of two proteins at length scales down to the localization uncertainty of the experiments. Correlation functions quantify the probability of finding a second particle at a distance *r* away from a given particle. Thus, if the cross-correlation function is modeled by an exponential function, then the measured cross-correlation, $c(r)^{\text{peaks}}$, can be fitted to the following equation:

$$c(r)^{\text{peaks}} = \left(A\exp\left(\frac{-r}{\varepsilon}\right) + 1 \right) * c(r)^{\text{PSF}} \tag{4.1}$$

where $c(r)^{\text{PSF}}$ is the cross-correlation of the PSFs of the two photoactivatable probes. The two correlated species will induce a result of $c(r) > 1$ [73].

4.4.7.3 Tessellation-based colocalization analysis

The tessellation-based colocalization analysis allows us to investigate the degree of colocalization between two channels and is performed by the Coloc-Tesseler software [77]. The intensity of one channel is plotted against the intensity of the other channel for each pixel. A scatterplot is depicted as a 2D histogram, commonly used in image-based colocalization analysis where the intensity of one channel is plotted against the intensity of the other channel for each pixel. It always keeps the density of channel A in abscises and the density of channel B in ordinates. The Manders and Spearman coefficients can be derived from the scatter plot representations to quantify the spatial colocalization between the two channels.

4.5 Conclusion

In short, the spatial organization and distribution of membrane molecules are closely related to their biological functions. The changes in molecule assembly and the alteration in their biological functions can affect each other, and the mechanisms remain unclear. Therefore, characterizing the distribution of membrane biomolecules is of great significance for a comprehensive understanding of its biological function. Revealing the assembly features of various membrane biomolecules and their relationship with each other is conducive to study the whole structure and functional mechanism of cell membranes. Compared with traditional fluorescence imaging technology, super-resolution fluorescence imaging technology has nanoscale resolution and thus can reveal biological details that cannot be clearly presented before. Therefore, super-resolution fluorescence imaging technology is a vital research tool in the field of cell biology.

As different imaging techniques have their own advantages and disadvantages, appropriate super-resolution techniques should be selected according to the research purpose and observed object. For SMLM, to ensure a high-quality image of biomolecules, the localization information should be accurate and reliable. Therefore, in addition to the excellent and stable performance of the instrument, developing good fluorescent probes is also important. They should have high specificity, small volume, and high fluorescence intensity to fulfill high labeling accuracy, high labeling density, and high localization accuracy. Apart from the commonly used immunofluorescence by antibody, various small molecules have been developed and synthesized to label membrane molecules in super-resolution imaging, and their advantages of flexibility, high affinity, and high accuracy have also been confirmed.

The continuous advancement of super-resolution imaging instruments, the development of new probes, and the improvement in data analysis algorithm have brought more opportunities to study cell membrane structure and function at the single-molecule level. Many current studies indicate that membrane proteins and carbohydrates assemble into different sizes of clusters, and these functional nanoscale structures might provide platforms for their own interactions and interactions between them and other membrane components. Meanwhile, lipid rafts, cytoskeleton, glycosylation of proteins, and external stimuli like ligands, pressure, or temperature play important roles in regulating the clustered distribution of these membrane molecules. We believe that further studies using super-resolution microscopy will shed new light on revealing the nanoscale organization of membrane structure and the dynamic mechanism of membrane function.

References

1. Lichtman, J. W. and Conchello, J.-A. (2005). Fluorescence microscopy, *Nat. Methods*, **2**, pp. 910–919.
2. Truong Quang, B.-A. and Lenne, P.-F. (2014). Membrane microdomains: From seeing to understanding, *Front. Plant Sci.*, **5**, pp. 18.
3. Sahl, S. J., Hell, S. W., and Jakobs, S. (2017). Fluorescence nanoscopy in cell biology, *Nat. Rev. Mol. Cell Biol.*, **18**, pp. 685–701.

4. Baddeley, D. and Bewersdorf, J. (2018). Biological insight from super-resolution microscopy: What we can learn from localization-based images, *Annu. Rev. Biochem.*, **87**, pp. 965–989.
5. Sydor, A. M., Czymmek, K. J., Puchner, E. M., and Mennella, V. (2015). Super-resolution microscopy: From single molecules to supramolecular assemblies, *Trends Cell Biol.*, **25**, pp. 730–748.
6. Gustafsson, M. G. (2000). Surpassing the lateral resolution limit by a factor of two using structured illumination microscopy, *J. Microsc.*, **198**, pp. 82–87.
7. Heintzmann, R. and Huser, T. (2017). Super-resolution structured illumination microscopy, *Chem. Rev.*, **117**, pp. 13890–13908.
8. Wu, Y. and Shroff, H. (2018). Faster, sharper, and deeper: Structured illumination microscopy for biological imaging, *Nat. Methods*, **15**, pp. 1011–1019.
9. Gustafsson, M. G., Shao, L., Carlton, P. M., Wang, C. R., Golubovskaya, I. N., Cande, W. Z., Agard, D. A., and Sedat, J. W. (2008). Three-dimensional resolution doubling in wide-field fluorescence microscopy by structured illumination, *Biophys. J.*, **94**, pp. 4957–4970.
10. Lawo, S., Hasegan, M., Gupta, G. D., and Pelletier, L. (2012). Subdiffraction imaging of centrosomes reveals higher-order organizational features of pericentriolar material, *Nat. Cell Biol.*, **14**, pp. 1148–1158.
11. Willig, K. I., Rizzoli, S. O., Westphal, V., Jahn, R., and Hell, S. W. (2006). STED microscopy reveals that synaptotagmin remains clustered after synaptic vesicle exocytosis, *Nature*, **440**, pp. 935–939.
12. Hell, S. W. and Wichmann, J. (1994). Breaking the diffraction resolution limit by stimulated emission: Stimulated-emission-depletion fluorescence microscopy, *Opt. Lett.*, **19**, pp. 780–782.
13. Westphal, V. and Hell, S. W. (2005). Nanoscale resolution in the focal plane of an optical microscope, *Phys. Rev. Lett.*, **94**, pp. 143903.
14. Lau, L., Lee, Y. L., Sahl, S. J., Stearns, T., and Moerner, W. (2012). STED microscopy with optimized labeling density reveals 9-fold arrangement of a centriole protein, *Biophys. J.*, **102**, pp. 2926–2935.
15. Betzig, E., Patterson, G. H., Sougrat, R., Lindwasser, O. W., Olenych, S., Bonifacino, J. S., Davidson, M. W., Lippincott-Schwartz, J., and Hess, H. F. (2006). Imaging intracellular fluorescent proteins at nanometer resolution, *Science*, **313**, pp. 1642–1645.
16. Moerner, W. E. and Kador, L. (1989). Optical detection and spectroscopy of single molecules in a solid, *Phys. Rev. Lett.*, **62**, pp. 2535.

17. Rust, M. J., Bates, M., and Zhuang, X. (2006). Sub-diffraction-limit imaging by stochastic optical reconstruction microscopy (STORM), *Nat. Methods*, **3**, pp. 793–796.

18. van de Linde, S., Löschberger, A., Klein, T., Heidbreder, M., Wolter, S., Heilemann, M., and Sauer, M. (2011). Direct stochastic optical reconstruction microscopy with standard fluorescent probes, *Nat. Protoc.*, **6**, pp. 991–1009.

19. Owen, D. M., Rentero, C., Rossy, J., Magenau, A., Williamson, D., Rodriguez, M., and Gaus, K. (2010). PALM imaging and cluster analysis of protein heterogeneity at the cell surface, *J. Biophotonics*, **3**, pp. 446–454.

20. Mennella, V., Keszthelyi, B., McDonald, K., Chhun, B., Kan, F., Rogers, G. C., Huang, B., and Agard, D. (2012). Subdiffraction-resolution fluorescence microscopy reveals a domain of the centrosome critical for pericentriolar material organization, *Nat. Cell Biol.*, **14**, pp. 1159–1168.

21. Ishitsuka, Y., Nienhaus, K., and Nienhaus, G. U. (2014). *Photoswitching Proteins. Methods in Molecular Biology*, eds. Sidney Cambridge., "Photoactivatable fluorescent proteins for super-resolution microscopy" (Springer Humana Press, New York) pp. 239–260.

22. Heilemann, M., van de Linde, S., Schüttpelz, M., Kasper, R., Seefeldt, B., Mukherjee, A., Tinnefeld, P., and Sauer, M. (2008). Subdiffraction-resolution fluorescence imaging with conventional fluorescent probes, *Angew. Chem. Int. Ed.*, **47**, pp. 6172–6176.

23. Turkowyd, B., Virant, D., and Endesfelder, U. (2016). From single molecules to life: Microscopy at the nanoscale, *Anal. Bioanal. Chem.*, **408**, pp. 6885–6911.

24. Chalfie, M., Tu, Y., Euskirchen, G., Ward, W. W., and Prasher, D. C. (1994). Green fluorescent protein as a marker for gene expression, *Science*, **263**, pp. 802–805.

25. Annibale, P., Vanni, S., Scarselli, M., Rothlisberger, U., and Radenovic, A. (2011). Quantitative photo activated localization microscopy: Unraveling the effects of photoblinking, *PloS one*, **6**, pp. e22678.

26. Nienhaus, K. and Nienhaus, G. U. (2014). Fluorescent proteins for live-cell imaging with super-resolution, *Chem. Soc. Rev.*, **43**, pp. 1088–1106.

27. Zhang, L. and Wang, E. (2014). Metal nanoclusters: New fluorescent probes for sensors and bioimaging, *Nano Today*, **9**, pp. 132–157.

28. Opazo, F., Punge, A., Bückers, J., Hoopmann, P., Kastrup, L., Hell, S. W., and Rizzoli, S. O. (2010). Limited intermixing of synaptic vesicle components upon vesicle recycling, *Traffic*, **11**, pp. 800–812.
29. Aitken, C. E., Marshall, R. A., and Puglisi, J. D. (2008). An oxygen scavenging system for improvement of dye stability in single-molecule fluorescence experiments, *Biophys. J.*, **94**, pp. 1826–1835.
30. Vogelsang, J., Kasper, R., Steinhauer, C., Person, B., Heilemann, M., Sauer, M., and Tinnefeld, P. (2008). A reducing and oxidizing system minimizes photobleaching and blinking of fluorescent dyes, *Angew. Chem. Int. Ed.*, **47**, pp. 5465–5469.
31. Dempsey, G. T., Vaughan, J. C., Chen, K. H., Bates, M., and Zhuang, X. (2011). Evaluation of fluorophores for optimal performance in localization-based super-resolution imaging, *Nat. Methods*, **8**, pp. 1027.
32. Klehs, K., Spahn, C., Endesfelder, U., Lee, S. F., Fürstenberg, A., and Heilemann, M. (2014). Increasing the brightness of cyanine fluorophores for single-molecule and superresolution imaging, *ChemPhysChem*, **15**, pp. 637–641.
33. Deschout, H., Zanacchi, F. C., Mlodzianoski, M., Diaspro, A., Bewersdorf, J., Hess, S. T., and Braeckmans, K. (2014). Precisely and accurately localizing single emitters in fluorescence microscopy, *Nat. Methods*, **11**, pp. 253–266.
34. Shroff, H., Galbraith, C. G., Galbraith, J. A., and Betzig, E. (2008). Live-cell photoactivated localization microscopy of nanoscale adhesion dynamics, *Nat. Methods*, **5**, pp. 417–423.
35. Demmerle, J., Wegel, E., Schermelleh, L., and Dobbie, I. M. (2015). Assessing resolution in super-resolution imaging, *Methods*, **88**, pp. 3–10.
36. Tokunaga, M., Imamoto, N., and Sakata-Sogawa, K. (2008). Highly inclined thin illumination enables clear single-molecule imaging in cells, *Nat. Methods*, **5**, pp. 159–161.
37. Moore, R. P. and Legant, W. R. (2018). Improving probes for super-resolution, *Nat. Methods*, **15**, pp. 659–660.
38. Fornasiero, E. F. and Opazo, F. (2015). Super-resolution imaging for cell biologists: Concepts, applications, current challenges and developments, *Bioessays*, **37**, pp. 436–451.
39. Yan, Q., Lu, Y., Zhou, L., Chen, J., Xu, H., Cai, M., Shi, Y., Jiang, J., Xiong, W., Gao, J., and Wang, H. (2018). Mechanistic insights into GLUT1 activation and clustering revealed by super-resolution imaging, *Proc. Natl. Acad. Sci. U.S.A.*, **115**, pp. 7033–7038.

40. Rahbek-Clemmensen, T., Lycas, M. D., Erlendsson, S., Eriksen, J., Apuschkin, M., Vilhardt, F., Jørgensen, T. N., Hansen, F. H., and Gether, U. (2017). Super-resolution microscopy reveals functional organization of dopamine transporters into cholesterol and neuronal activity-dependent nanodomains, *Nat. Commun.*, **8**, pp. 1–14.

41. Opazo, F., Levy, M., Byrom, M., Schafer, C., Geisler, C., Groemer, T. W., Ellington, A. D., and Rizzoli, S. O. (2012). Aptamers as potential tools for super-resolution microscopy, *Nat. Methods*, **9**, pp. 938–939.

42. Li, C., Tebo, A. G., and Gautier, A. (2017). Fluorogenic labeling strategies for biological imaging, *Int. J. Mol. Sci.*, **18**, pp. 1473.

43. Porter, R. (1959). The hydrolysis of rabbit γ-globulin and antibodies with crystalline papain, *Biochem. J.*, **73**, pp. 119.

44. Grimm, J. B., Muthusamy, A. K., Liang, Y., Brown, T. A., Lemon, W. C., Patel, R., Lu, R., Macklin, J. J., Keller, P. J., and Ji, N. (2017). A general method to fine-tune fluorophores for live-cell and in vivo imaging, *Nat. Methods*, **14**, pp. 987–994.

45. Zheng, Q., Juette, M. F., Jockusch, S., Wasserman, M. R., Zhou, Z., Altman, R. B., and Blanchard, S. C. (2014). Ultra-stable organic fluorophores for single-molecule research, *Chem. Soc. Rev.*, **43**, pp. 1044–1056.

46. Winter, G., Griffiths, A. D., Hawkins, R. E., and Hoogenboom, H. R. (1994). Making antibodies by phage display technology, *Annu. Rev. Immunol.*, **12**, pp. 433–455.

47. Chakravarty, R., Goel, S., and Cai, W. (2014). Nanobody: The "magic bullet" for molecular imaging? *Theranostics*, **4**, pp. 386.

48. Muyldermans, S. (2013). Nanobodies: Natural single-domain antibodies, *Annu. Rev. Biochem.*, **82**, pp. 775–797.

49. Hamers-Casterman, C., Atarhouch, T., Muyldermans, S. A., Robinson, G., Hammers, C., Songa, E. B., Bendahman, N., and Hammers, R. (1993). Naturally occurring antibodies devoid of light chains, *Nature*, **363**, pp. 446–448.

50. Ries, J., Kaplan, C., Platonova, E., Eghlidi, H., and Ewers, H. (2012). A simple, versatile method for GFP-based super-resolution microscopy via nanobodies, *Nat. Methods*, **9**, pp. 582–584.

51. Mikhaylova, M., Cloin, B., Finan, K., Van Den Berg, R., Teeuw, J., Kijanka, M. M., and Kapitein, L. C. (2015). Resolving bundled microtubules using anti-tubulin nanobodies, *Nat. Commun.*, **6**(1), pp. 1–7.

52. Frey, S. and Görlich, D. (2014). A new set of highly efficient, tag-cleaving proteases for purifying recombinant proteins, *J. Chromatogr. A*, **1337**, pp. 95–105.

53. Chen, J., Li, H., Wu, Q., Yan, Q., Sun, J., Liang, F., and Wang, H. (2020). Organization of protein tyrosine kinase-7 on cell membranes characterized by aptamer probe-based STORM imaging, *Anal. Chem.*, **93**(2), pp. 936–945.
54. Jing, Y., Chen, J., Zhou, L., Sun, J., Cai, M., Shi, Y., Tian, Y., Gao, J., and Wang, H. (2019). Super-resolution imaging of cancer-associated carbohydrates using aptamer probes, *Nanoscale*, **11**, pp. 14879–14886.
55. Yan, Q., Cai, M., Zhou, L., Xu, H., Shi, Y., Sun, J., Jiang, J., Gao, J., and Wang, H. (2019). Using an RNA aptamer probe for super-resolution imaging of native EGFR, *Nanoscale Adv.*, **1**, pp. 291–298.
56. Ellington, A. D. and Szostak, J. W. (1990). In vitro selection of RNA molecules that bind specific ligands, *Nature*, **346**, pp. 818–822.
57. Deng, D., Sun, P., Yan, C., Ke, M., Jiang, X., Xiong, L., Ren, W., Hirata, K., Yamamoto, M., and Fan, S. (2015). Molecular basis of ligand recognition and transport by glucose transporters, *Nature*, **526**, pp. 391–396.
58. Wu, Q., Cai, M., Gao, J., Zhao, T., Xu, H., Yan, Q., Jing, Y., Shi, Y., Kang, C., Liu, Y., Liang, F., Chen, J., and Wang, H. (2020). Developing substrate-based small molecule fluorescent probes for super-resolution fluorescent imaging of various membrane transporters, *Nanoscale Horiz.*, **5**, pp. 523–529.
59. Zhang, J., Yang, P. L., and Gray, N. S. (2009). Targeting cancer with small molecule kinase inhibitors, *Nat. Rev. Cancer*, **9**, pp. 28–39.
60. Bedard, P. L., Hyman, D. M., Davids, M. S., and Siu, L. L. (2020). Small molecules, big impact: 20 years of targeted therapy in oncology, *Lancet*, **395**, pp. 1078–1088.
61. Wu, P., Nielsen, T. E., and Clausen, M. H. (2016). Small-molecule kinase inhibitors: An analysis of FDA-approved drugs, *Drug Discov. Today*, **21**, pp. 5–10.
62. Wu, Q., Jing, Y., Zhao, T., Gao, J., Cai, M., Xu, H., Liu, Y., Liang, F., Chen, J., and Wang, H. (2020). Development of small molecule inhibitor-based fluorescent probes for highly specific super-resolution imaging, *Nanoscale*, **12**, pp. 21591–21598.
63. Chen, J., Li, H., Wu, Q., Zhao, T., Xu, H., Sun, J., Liang, F., and Wang, H. (2021). A multidrug-resistant P-glycoprotein assembly revealed by tariquidar-probe's super-resolution imaging, *Nanoscale*, **13**, pp. 16995–17002.
64. Borst, P. and Schinkel, A. H. (2013). P-glycoprotein ABCB1: A major player in drug handling by mammals, *J. Clin. Invest.*, **123**, pp. 4131–4133.

65. Gao, J., Chen, J., and Wang, H. (2018). *Membrane Biophysics: New Insights and Methods*, eds. Wang, H. and Li, G., Chapter 5 "Super-resolution imaging of membrane heterogeneity" (Springer Nature Press, Singapore) pp. 117–146.

66. Ovesný, M., Křížek, P., Borkovec, J., Švindrych, Z., and Hagen, G. M. (2014). ThunderSTORM: A comprehensive ImageJ plug-in for PALM and STORM data analysis and super-resolution imaging, *Bioinformatics*, **30**, pp. 2389–2390.

67. Gao, J., Wang, F., Chen, J., Wang, J., Cai, M., Xu, H., Jiang, J., and Wang, H. (2016). Super-resolution imaging of STAT3 cellular clustering during nuclear transport, *RSC Adv.*, **6**, pp. 54597–54607.

68. Culley, S., Albrecht, D., Jacobs, C., Pereira, P. M., Leterrier, C., Mercer, J., and Henriques, R. (2018). Quantitative mapping and minimization of super-resolution optical imaging artifacts, *Nat. Methods*, **15**, pp. 263–266.

69. Laine, R. F., Tosheva, K. L., Gustafsson, N., Gray, R. D., Almada, P., Albrecht, D., Risa, G. T., Hurtig, F., Lindås, A.-C., and Baum, B. (2019). NanoJ: A high-performance open-source super-resolution microscopy toolbox, *J. Phys. D: Appl. Phys.*, **52**, pp. 163001.

70. Jing, Y., Zhou, L., Chen, J., Xu, H., Sun, J., Cai, M., Jiang, J., Gao, J., and Wang, H. (2019). Quantitatively mapping the assembly pattern of EpCAM on cell membranes with peptide probes, *Anal. Chem.*, **92**, pp. 1865–1873.

71. Ester, M., Kriegel, H.-P., Sander, J., and Xu, X. (1996). A density-based algorithm for discovering clusters in large spatial databases with noise, *KDD*, **96**, pp. 226–231.

72. Zhang, J., Leiderman, K., Pfeiffer, J. R., Wilson, B. S., Oliver, J. M., and Steinberg, S. L. (2006). Characterizing the topography of membrane receptors and signaling molecules from spatial patterns obtained using nanometer-scale electron-dense probes and electron microscopy, *Micron*, **37**, pp. 14–34.

73. Sengupta, P., Jovanovic-Talisman, T., Skoko, D., Renz, M., Veatch, S. L., and Lippincott-Schwartz, J. (2011). Probing protein heterogeneity in the plasma membrane using PALM and pair correlation analysis, *Nat. Methods*, **8**, pp. 969–975.

74. Levet, F., Hosy, E., Kechkar, A., Butler, C., Beghin, A., Choquet, D., and Sibarita, J.-B. (2015). SR-Tesseler: A method to segment and quantify localization-based super-resolution microscopy data, *Nat. Methods*, **12**, pp. 1065–1071.

75. Fu, Y., Hua, P., Lou, Y., Li, Z., Jia, M., Jing, Y., Cai, M., Wang, H., Tong, T., and Gao, J. (2020). Mechanistic insights into trop2 clustering on lung

cancer cell membranes revealed by super-resolution imaging, *ACS omega*, **5**, pp. 32456–32465.

76. Malkusch, S., Endesfelder, U., Mondry, J., Gelléri, M., Verveer, P. J., and Heilemann, M. (2012). Coordinate-based colocalization analysis of single-molecule localization microscopy data, Histochem. *Cell Biol.*, **137**, pp. 1–10.

77. Levet, F., Julien, G., Galland, R., Butler, C., Beghin, A., Chazeau, A., Hoess, P., Ries, J., Giannone, G., and Sibarita, J.-B. (2019). A tessellation-based colocalization analysis approach for single-molecule localization microscopy, *Nat. Commun.*, **10**, pp. 1–12.

Chapter 5

Fluorescence Microscopy for Studying Plasma Membrane and Intracellular Membranes

Haijiao Xu and Hongda Wang
State Key Laboratory of Electroanalytical Chemistry,
Research Center of Biomembranomics,
Changchun Institute of Applied Chemistry,
Chinese Academy of Sciences, Changchun, Jilin, China
hdwang@ciac.ac.cn

The membrane system of eukaryotic cells consists of the plasma membrane and intracellular membranes that enclose various organelles. Maintenance of physiologic cellular functions and homeostasis requires highly ordered interactions between different membranous organelles. The rapid evolution of fluorescence microscopy and the accompanying development of novel fluorescent probes have continuously improved our knowledge of organelle interactions, including the mode, function, and mechanism of organelle interactions. We believe that in the near future we will map the organelle interactions network, elucidate the establishment, maintenance, dynamic changes, and regulatory mechanisms of the

Biomembranomics: Structure, Techniques, and Applications
Edited by Hongda Wang

ISBN 978-981-4968-61-4 (Hardcover), 978-1-003-45635-3 (eBook)
www.jennystanford.com

organelle interaction network, and reveal the physiological and pathological functions of the organelle interaction network.

5.1 Introduction

The membrane system of eukaryotic cells is precisely compartmentalized into discrete membrane-bound organelles with specific morphology and function, including the plasma membrane and various intracellular organelles, such as endoplasmic reticulum (ER), Golgi apparatus, mitochondria, lysosomes, endosomes, peroxisomes, and so on, which allows for the separation of incompatible biochemical processes. However, they are not completely isolated in structure and function but coordinate with each other to perform a series of important physiological functions [1]. The fine division and proper organelle cooperation form an orderly functional network of organelle interacting essential for cell homeostasis. Thus, elucidating the mechanisms underlying the orderly organelle interactions can help deepen our knowledge of complex cellular functions and their related disease roots. Unfortunately, due to the limitations of research techniques and methods, the mode, function, and mechanism of the orderly organelle interactions have been poorly understood for a long time. The rapid development of fluorescence imaging and fluorescence labeling technology allows us to directly visualize the highly dynamic interaction process of multiple organelles at a high spatiotemporal resolution, which facilitates uncovering detailed information on organelle interactions and related mechanisms.

In the following, we will first summarize the evolution of fluorescence microscopy, which is suitable for live-cell imaging, mainly involving basic principles of various fluorescence microscopies. Then, we will introduce an overview of several fluorescent labeling techniques and analyze their respective advantages and disadvantages. Finally, we will address advances in the study of dynamically ordered organelle interactions, including models, conventional molecular mechanisms, and novel structural mechanisms of orderly organelle interactions. This chapter will provide a comprehensive understanding of ordered organelle interactions and arouse the attention of researchers to the importance of orderly organelle interplay in the field of the membrane system.

5.2 Overview of Fluorescence Imaging System

5.2.1 Fluorescence Imaging Methods

5.2.1.1 Confocal laser scanning microscopy

From its commercialization in the late 1980s, confocal laser scanning microscopy (CLSM), which can image thin optical sections in living specimens, has become one of the most prevalent fluorescence microscopy techniques in biological and medical sciences [2]. A confocal microscope is a specialized type of fluorescence microscope based on conventional fluorescence microscopy. In conventional wide-field fluorescence imaging, the specimen is uniformly and simultaneously illuminated by light from above and below the focal plane, which blurs the image [3, 4]. Progressively, confocal microscopy offers clearer images due to its ability to control the depth of field, eliminate or reduce background information, and collect serial optical sections from thick specimens.

A typical confocal microscope basically consists of pinholes, objective lenses, low-noise detectors, fast-scanning mirrors, filters for wavelength selection, and laser illumination. As shown in Fig. 5.1, the conjugate focal plane of the sample is equipped with two pinholes, namely an illumination pinhole and a detector pinhole, for point illumination and point detection, respectively. Light from the laser source passes through the illumination pinhole and is then reflected by the dichroic mirror and scans across the sample in a defined focal plane. Secondary fluorescence emitted from points on the sample (within the same focal plane) is returned through the dichroic mirror and focused on the detector pinhole aperture. The detector pinhole suppresses most of the fluorescence emission with which it is not colocalized, thus causing the detector to detect only the fluorescence emitted from that point and greatly improving image resolution.

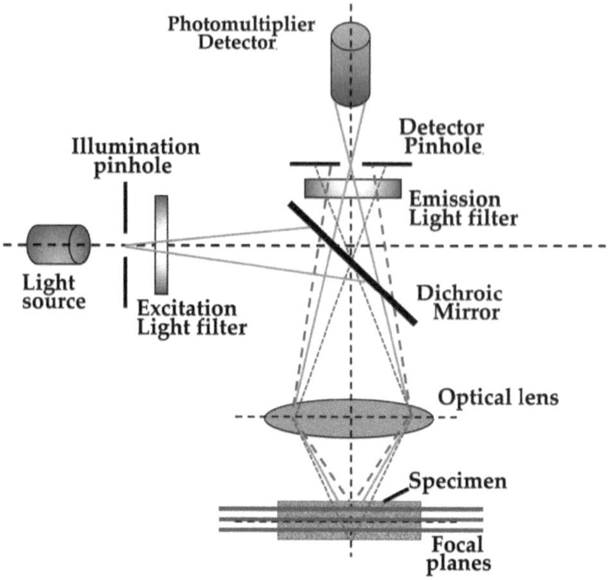

Figure 5.1 Illustration of the principle of a laser scanning confocal microscope.

With the rise of living cell technology, a new confocal imaging system, a spinning-disk confocal (SDC) microscope, has been developed. It is equipped with a rotating disk containing an array of microlenses and pinholes [5]. In this design, an excitation beam is focused onto an aperture aligned in the Nipkow disk through a disk containing 20,000 microlenses, which generates a multiplex of excitation and detection spots, thus allowing for the rapid scanning of an entire field of view [6]. As shown in Fig. 5.2, the emitted light is transmitted back through the rotating disc and reflected to a detector on a dichroic mirror, typically a CCD or sCMOS camera. Compared to the single-point scanning system of CLSM, the speed advantage of SDC stems from the way in which the imaging plane is illuminated and recorded. Moreover, the parallelized approach of multi-beam scanning is much faster. Additionally, this SDC applies a low dose of multiple excitation beams to the specimen, which reduces the fluorescent bleaching and phototoxicity of the sample. A spinning-disk confocal microscope is, therefore, most suitable for fast in vivo cell imaging.

Taken together, confocal microscopy offers significant imaging improvements compared to conventional microscopy. Confocal imaging modes include single optical sections, time-lapse imaging, 3D imaging, and 4D imaging (3D and time lapse), making it a core technique for detecting dynamic cellular processes [7].

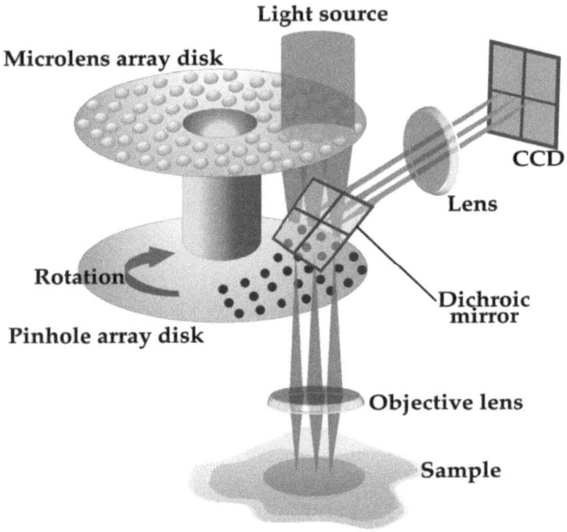

Figure 5.2 The principle schematic of spinning-disk confocal.

5.2.1.2 Total internal reflection fluorescence microscopy

Numerous key cellular processes that occur on the cell surface need to be visualized without interference from other intracellular regions. Total internal reflection fluorescence (TIRFM) microscopy is a technique for exciting fluorescence within close proximity of the coverslip surface (within ≤100 nm) [8, 9]. As shown in Fig. 5.3, the technique produces an evanescent wave by excitation at a critical angle. This totally internally reflecting laser beam intensity decays exponentially with increasing distance perpendicular to the surface. This limits the excitation depth to about 100 nm. Fluorescence excitation through this thin zone of electromagnetic energy allows imaging in very low background fluorescence without out-of-focus fluorescence. This illumination mode can be coupled with a conventional wide-field microscope to allow imaging of species close to or away from the cover glass. The advantages of TIRFM are miniaturization, enhanced depth resolution, and reduction in detection volumes, which endows it with a wide range of applications in biochemistry and cell biology. For instance, TIRFM has been employed for visualizing the topography of cell-substrate contacts, membrane protein dynamics, endocytosis, and exocytosis [10].

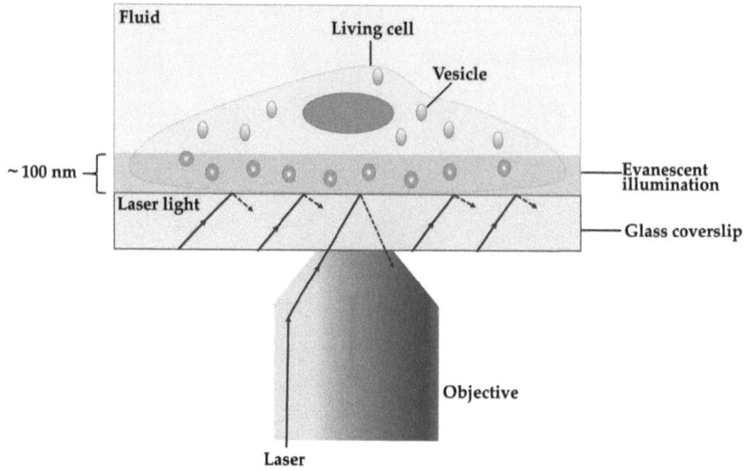

Figure 5.3 The underlying principle of TIRFM.

5.2.1.3 Structured-illumination microscopy

The capability of fluorescence microscopy, and even confocal microscopy, to observe biological processes at the molecular level is restricted by the relatively low spatial resolution due to the diffraction limit. Structured-illumination microscopy (SIM), a type of super-resolution microscopy, breaks the diffraction limit barrier by applying a patterned illumination field to the sample [11]. The possible setup of SIM is outlined in Fig. 5.4. Scrambled laser light from a multimode fiber is aligned onto a linear phase grating. Diffraction levels −1, 0, and +1 are refocused on the back focal plane of the objective. The beams recollimated by the objective intersect at the focal plane of the sample, where they interfere and produce intensity patterns with both lateral and axial structures. Emission light from the sample is observed by a camera via a dichromatic mirror (DM). The principle of SIM is that a patterned excitation light illuminates the sample through the interference of multiple light sources in the axial direction, the lateral direction, or both [12, 13]. As a result, high-resolution images can be reconstructed by acquiring multiple images

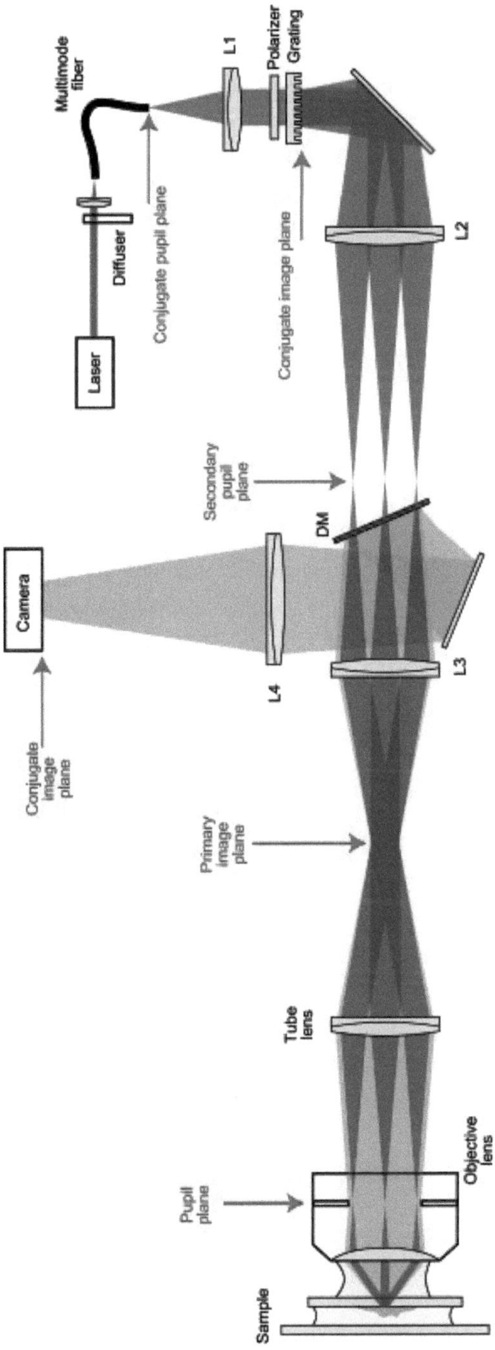

Figure 5.4 Simplified diagram of 3D structured-illumination apparatus. Reprinted from Ref. [11], Copyright © 2014 Elsevier Inc.

with illumination patterns of different phases and orientations, up to a resolution of 100 nm in the lateral direction and 300 nm in the axial direction [14]. However, because the illumination pattern itself is also limited by light diffraction, SIM can only double the spatial resolution by combining two diffraction-limited information sources [15]. Saturated structured-illumination microscopy (SSIM) has been characterized by the saturation of fluorescence emission, which is achieved by illuminating a fluorophore with very high-intensity excitation light [16]. In this approach [17], the sample is illuminated by the sinusoidal mode of strong excitation light, and the peaks of the excitation mode can be clipped by fluorescence saturation and become flat, while the fluorescence emission is still not present at the zero point in the valley. Mixing this excitation pattern with high-frequency spatial characteristics in the sample leads to the microscopic detection of subdiffraction-limited spatial features. Based on this principle, SSIM pushes the limits of diffraction and can reach a two-dimensional spatial resolution of 50 nm. SIM has the advantages of low phototoxicity, fast imaging speed, no specific dyes required and ultra-high resolution, thus enabling fast super-resolution imaging at the living cell level.

5.2.1.4 Lattice light-sheet microscopy

A revolutionary high-resolution light-sheet microscopy, lattice light-sheet microscopy (LLSM), has achieved unprecedented intracellular spatial-temporal resolution scanning at the whole-cell level [18]. The advantage of light-sheet illumination is its ability to collect accurate and reliable 3D information. Unlike frequently used point-scanning techniques, such as confocal and two-photon microscopy, light-sheet microscopy uses a thin laser beam to illuminate the specimen as a plane, while imaging this plane with a camera (Fig. 5.5a) [19]. Superior to the conventional light sheet created by several micron thick Gaussian beams (Fig. 5.5b, left), LLSM illuminates the specimen using an ultrathin light sheet of ~0.4–1 μm thickness, which is created by huge arrays of parallel mutually interfering non-diffracting light beams (Bessel beams) (Fig. 5.5b, right). This approach meets the essential requirements for live imaging: high imaging speed and signal-to-noise ratio, and low levels of photobleaching and phototoxicity; herein these features allow for long-term imaging [20–22]. The SIM based on light-sheet illumination has opened the door for real-time super-resolution imaging to precisely understand the biological processes of living cells [23].

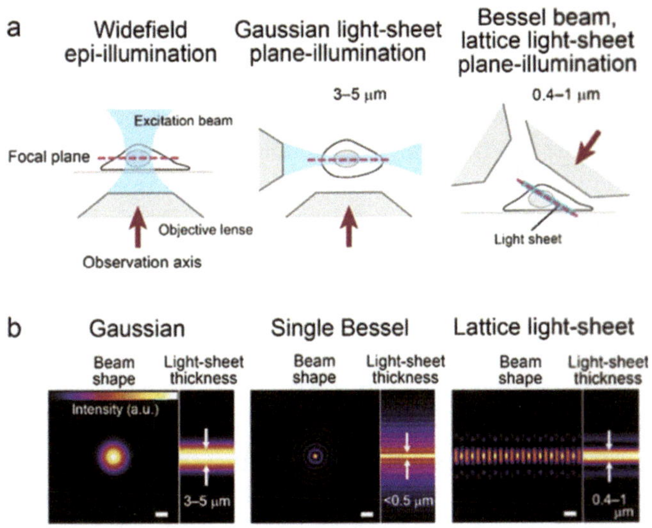

Figure 5.5 Schematic comparison of optical slice power and beam shapes used in different lighting technologies, and methods of large-volume imaging with LLSM. (a) Optical sectioning power of wide-field epi-illumination, Gaussian light-sheet, and Bessel beam/lattice light-sheet plane illumination techniques. (b) Beam shapes of three different light-sheet techniques. Light-sheet thickness is shown on the right of each panel. Reprinted by permission from Ref. [19], Copyright © 2021, Yuko Mimori-Kiyosue.

5.2.2 Fluorescent Labeling Techniques

A powerful advantage of fluorescence microscopy over electron microscopy is that it allows direct visualization of pathways, localization, and physiological events in biological systems in a minimally invasive manner depending on fluorescent labels [24]. Advances in fluorescent labeling techniques coupled with the vast array of sophisticated fluorescence microscopies make it feasible to study the dynamic processes of living cells. The fluorescent labeling techniques fall into two broad categories. One is direct fluorescent dye staining, and the other is molecular tagging by genetic engineering to introduce fluorescent proteins (FPs) or fluorochrome-specific binding motifs to targets. In the following, we will describe the most used fluorescent labeling techniques in detail.

5.2.2.1 Immunofluorescence staining

Immunofluorescence is a robust technique that utilizes fluorescent-labeled antibodies to detect specific target antigens (proteins, glycans, lipids) within cells, including direct or indirect immunofluorescence staining [25]. Direct immunofluorescence uses fluorescent-tagged antibodies to bind to the target antigen. The problem with this method lies in a relatively high staining background resulting from antibody cross-reaction. Indirect immunofluorescence involves two steps, in which an unlabeled primary antibody binds to the target and the primary antibody is then labeled with a fluorophore-labeled secondary antibody. This technique avoids antibody cross-reactivity and is more sensitive than the direct labeling method, resulting in wide applications. With the development of diverse synthetic fluorescent probes, such as FITC, Texas Red, Alexa Fluor 488, 532, 647, Cy3, Cy5, and so on, multicolor labeling can be implemented by combining secondary antibodies coupled with different fluorochromes [26]. By recognizing specific proteins, antibody labeling can be used to detect various organelles and cytoskeletons. The main advantage of the immunofluorescence method is that it can detect endogenous proteins, thus giving a true picture of the natural state of the target protein. However, this technique is not suitable for real-time imaging of live cells due to its requirement of fixation and permeabilization for cells prior to staining.

5.2.2.2 Genetically engineered labeling

Proteins can also be labeled by genetic engineering with molecular tags such as fluorescent proteins (FPs). FP-fused proteins are easily detected after cell transfection. FPs such as GFP can show not only the localization of the target molecules in the cell, but also their dynamics, thus opening the door to live-cell imaging [27]. With much attention and progress, many FPs have been identified and engineered, such as GFP variants EGFP, EBFP, ECFP, EYFP, Venus, DsRed2, mRFP1, ZsYellow, Kusabira-Orange1, etc. We can select suitable ones according to the wavelength (excitation and emission), the quantum yield that correlates to the signal intensity, and the degree of polymerization (monomer, dimer, or tetramer) [28]. This FP-fusion technique is one of the most important labeling methods for live-cell imaging, because an auto-fluorescent molecule can be genetically encoded as a fusion with the cDNA of interest without any additional components to provide fluorescence. However, the large size of FPs sometimes changes the intrinsic properties of the labeled molecules, resulting in adverse effects on their behaviors.

Luckily, small-sized peptides or single amino acid sequences, such as TC-Tag and HaloTag [29, 30], can overcome the shortcoming. Unlike the FPs, these tags themselves do not have fluorescence in any case; therefore, they need to be linked to specific cell-permeable fluorescent dyes. Even so, the introduction of exogenous genes may lead to the overexpression of target proteins, and their abnormal localization or functions. This adverse effect can be reduced by controlling the amount of exogenous gene loading. Anyway, the combination of this labeling method and conventional immunofluorescence method is still a powerful tool to detect cell functions.

5.2.2.3 Chemical fluorescent probes

Chemical reagents with fluorescent chromophores or coupled to fluorescent dyes can specifically bind to some cellular structures, such as the nucleus, plasma membrane, various organelles, and cytoskeleton, thereby allowing them to be viewed by fluorescence microscopy [31]. Chemically synthesized fluorescent probes are divided into two categories: permeable and impermeable. For the former, they can be enriched on the target organelle by simply adding to the cell culture medium and are, therefore, suitable for live-cell labeling. For the latter, the cells need to be permeabilized and fixed prior to probe addition, so they are suitable for labeling only dead cells. Taking nuclear labeling as an example, DAPI is a semi-permeant dye that can be used for nuclear staining in living cells and fixed cells. Hoechst dyes are permeable and can, therefore, be used exclusively for nuclear staining in living cells. Propidium iodide (PI) is an impermeant nuclear dye, so it can be used to evaluate cell mortality. In addition, phalloidin is a toxic substance from *Amanita phalloides* with cell impermeability, which can selectively bind F-actin in fixed cells. Various fluorescent-dyes-coupled phalloidin complexes with different wavelengths have been commercial, such as Alexa Fluor-coupled phalloidin, FITC-phalloidin, eBioscience Phalloidin eFluor 660, and ActinRed 555 ReadyProbes (Invitrogen™). Lipophilic dyes commonly used for cell membrane labeling include DiO, DiI, DiD, and their derivatives, such as CellTracker CM-DiI (Invitrogen™), which has better solubility in water. With an increase in demand for live-cell imaging, various fluorescent probes with higher fluorescence intensity have been developed. For instance, the live-cell ER dye ER-Tracker™ (green/red, Invitrogen™) is a class of cell-permeable dyes that are highly selective for the ER. For mitochondrial labeling, MitoTracker probes are cationic dyes that are selectively retained by mitochondria in living cells due to their charge. Like other mitochondrial dyes, their loading is dependent on the membrane potential. In addition, CellLight reagents (Invitrogen™) have been developed to label multiple organelles in live cells, such as ER and the Golgi complex. Fluorescein-labeled ceramides are widely used to study lipid metabolism and transport and can selectively stain Golgi complex in living cells, such as BODIPY® FL C5-Ceramide and BODIPY® TR Ceramide (Invitrogen™).

5.3 Applications of Fluorescence Imaging in Studying Dynamic Organelle Interactions

5.3.1 Ongoing Knowledge of Organelle–Organelle Interplay Network

A major way in which organelles communicate and integrate their activities is through the formation of tight contacts, often referred to as "membrane contact sites" (MCSs) [32]. A hallmark of MCSs is that membranes from two organelles are tethered in close apposition, typically within 30 nm, but not fused, which is regulated by a protein or a protein complex. A limited number of contact points between organelles ensures effective communication between organelles while maintaining their respective stable morphology. The first discoveries in cells were the MCSs between ER and the plasma membrane and between ER and the mitochondria obtained by electron microscopy [33, 34]. Over the past decade, MCS has received increasing attention, and the ER has been revealed to form close MCSs with various intracellular organelles, such as the Golgi complex, endosomes, lysosomes, and lipid droplets. The function of MCSs was initially understood to be responsible for intracellular exchange of substances. For example, Ca^{2+} exchange occurs at membrane contact sites between the ER and the plasma membrane and between the ER and the mitochondria [35, 36]. Lipid transfer is an important function of ER–Golgi MCSs [37]. As fluorescent imaging techniques progress, more insights into the function of MCSs have been gained. For example, apart from completing material exchange, in 2011, Voeltz et al. showed that the MCSs between mitochondria and ER facilitated mitochondrial fission [38]. Following this, in 2014, Voeltz et al. demonstrated that contraction and fission sites in early and late endosomes are spatiotemporally associated with ER contact sites [39]. In this study, they transfected Cos-7 cells with markers for late endosomes (mCh-Rab7) and the ER (GFP-Sec61b) and imaged them by confocal fluorescence microscopy. A representative example of a late endosome that undergoes fission is shown in Fig. 5.6. They observed that a dynamic ER tubule moved into place, crossed over, and "cups" the bud, and fission followed MCS formation.

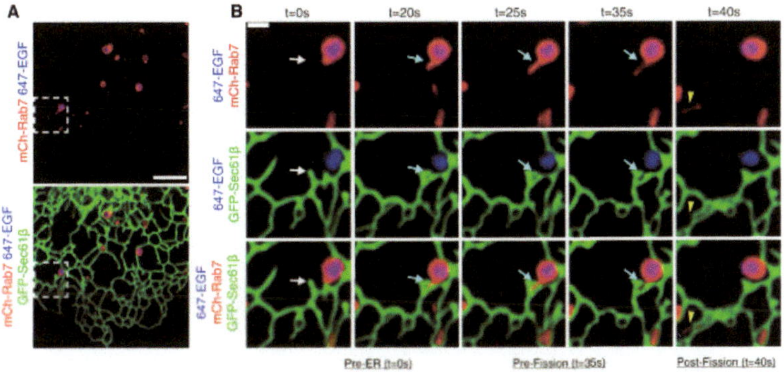

Figure 5.6 Late endosome division occurs at ER contact sites. (A) A Cos-7 cell coexpressing mCh-Rab7 (late endosome), and GFP-Sec61b was pulse-labeled with Alexa Fluor 647-conjugated EGF (cargo in blue). (B) An enlarged image of the region boxed in (A) displays an example of late endosome fission. Merged images show the relative location of Rab7, EGF, and the ER over time. Reprinted by permission from Ref. [39], Copyright © 2014 Elsevier Inc.

It has been increasingly recognized that inter-organelle contacts are critical for a variety of cellular functions, many of which require multiple paired organelle contacts working together to complete their functions. For example, lipid metabolism is carried out jointly by the ER, lipid droplets, mitochondria and peroxisomes, and lysosomes [40]. Dynamic imaging of organelle contacts with molecular specificity is possible using genetically encoded fluorescent fusion proteins. However, the spatial and temporal organization of the dynamic network of organelle interactions remains poorly characterized, as the filters are unable to distinguish between multiple fluorophores with overlapping emission spectra, limiting the simultaneous use of multiple fluorescent proteins. In 2017, Lippincott-Schwartz et al. combined live-cell confocal and lattice light-sheet instrumentation to achieve multispectral imaging that overcomes the challenge of spectral overlap in the fluorescent protein palette, and based on this technology, they systematically analyzed the dynamics of organelle contacts simultaneously among six different organelles [41]. They revealed the frequency and location of dynamic inter-organelle contacts among ER, Golgi, lysosome, peroxisome, mitochondria, and lipid droplets and how these relationships change over time [42] (Fig. 5.7). With this approach, it was found that the number of inter-organelle contacts was remarkably stable over time but was affected by microtubule (MT) perturbation.

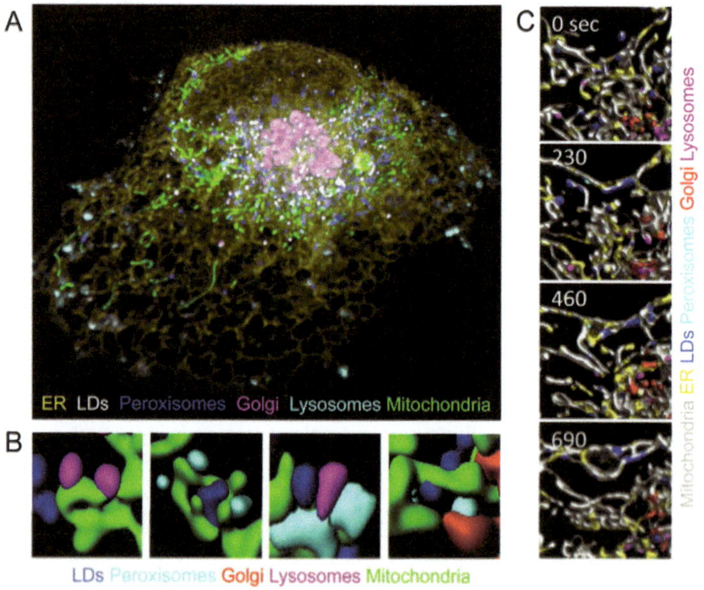

Figure 5.7 Organelle organization and contacts revealed by multispectral lattice light-sheet microscopy. (A) Maximum intensity projection of a COS-7 displays a location and distribution panorama of peroxisomes, the mitochondria, the ER and Golgi, and lysosomes and LDs. (B) Examples of complex inter-organelle contacts and organization in segmented lattice light-sheet images. (C) Dynamics of mitochondria–organelle interactions in segmented lattice light-sheet images. Reprinted from Ref. [42], Copyright © 2018 Published by Elsevier Ltd.

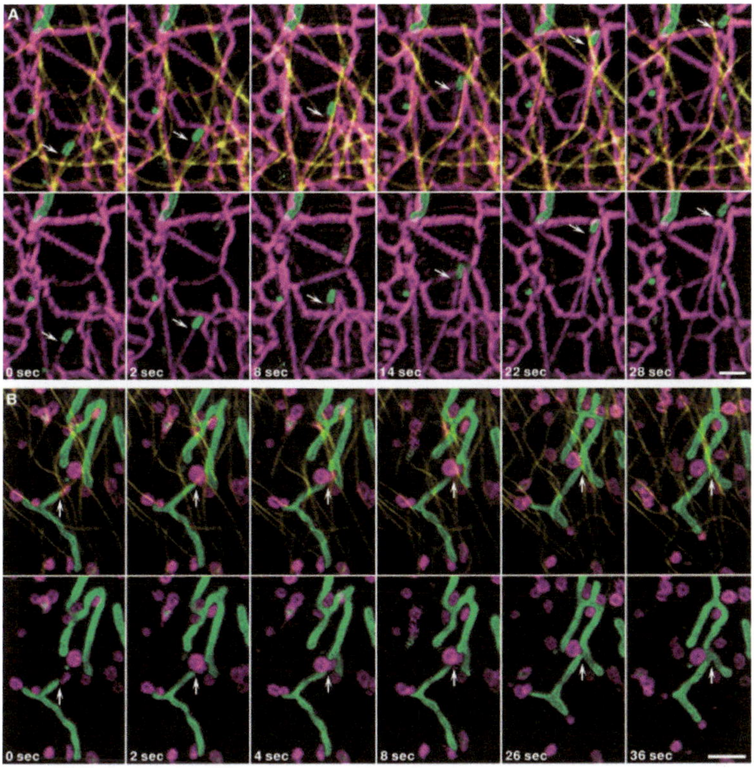

Figure 5.8 Hitchhiking interactions between intracellular organelles. (A) Generation of a new ER tubule (magenta, with tip indicated by white arrow) by hitchhiking on a mitochondrion (green) moving along an MT (yellow). (B) Mitochondrial fusion (green) is mediated by hitchhiking (white arrow) on an LE or Lyso (magenta) moving along an MT (yellow). Scale bars, 1 mm (A) and 2 mm (B). Reprinted by permission from Ref. [43], Copyright © 2018 Elsevier Inc.

Many studies have shown that the cytoskeleton is involved in organelle interactions to synergistically execute various physiological functions [32, 40]. Visualizing such interactions requires noninvasive, long-term imaging of the intracellular environment at high spatiotemporal resolution and low background. In 2018, Li et al. employed multicolor grazing incidence structure illumination microscopy (GI-SIM) to describe the rapid dynamic interactions of diverse organelles and the cytoskeleton, providing new clues to the complex behavior of these structures [43]. They uncovered new ER remodeling mechanisms, such as hitchhiking of the ER on mitochondria moving along MT (Fig. 5.8A). Meanwhile, a mitochondrial hitchhiker has been observed to anchor to mobile late endosomes or lysosomes along MT and undergo complete fusion (Fig. 5.8B).

5.3.2 Orderly Membrane Trafficking among Organelles

The other mechanism of organelle cooperation is indirect communication via membrane trafficking, which involves the transport of vesicles that bud from a donor compartment and fuse with an acceptor compartment [44]. A balance between the commonness and uniqueness of organelles requires highly directed and selective vesicle transport between them. The entire vesicle transport process includes vesicle budding, directed transport, target recognition, docking, and fusion, and many proteins involved in these reactions have been identified [45]. Rab proteins with highly selective distribution on membrane systems act as membrane identifiers to control the specificity and directionality of vesicle transport [46]. For example, Rab5 acts as an early endosome (EE) marker and mediates traffic from the plasma membrane to EE, while Rab7 as a marker for late endosomes is associated with early sorting endosome to degradative compartment [47, 48]. Following Rab-mediated accurate vesicle transport and tethering, the specific topological pairing of cognate SNARES (soluble NSF attachment protein receptor, where NSF stands for N-ethylmaleimide-sensitive fusion protein) between the transport vesicle and its target membrane ensures precision in the fusion event [49]. For instance, autophagosome-localized STX17 and SNAP29 and lysosome-localized VAMP8 or VAMP7 coordinate the fusion between autophagosomes and lysosomes [50].

Inter-organelle vesicle transport is responsible for a variety of biological processes, such as cellular metabolism, coordinated signaling, cell adhesion and motility, cell immunity, neurotransmitters, and hormone release [51–57]. The trafficking pathways mainly consist of an inward flux of endocytic vesicles from the plasma membrane and an outward flux of exocytic vesicles to the plasma membrane [58]. Endocytosis entails the internalization of nutrients, receptor–ligand complexes, fluids, lipids, extracellular proteins and viruses, and many other biomolecules, of which receptor protein endocytosis is the most intensively studied and biologically significant event. Internalization of receptor proteins is closely related to signal transduction, mainly by modulating the numbers of signal-associated receptor proteins at PM, such as receptor tyrosine kinases (RTKs) and serine/threonine kinases such as transforming growth factor beta (TGF-β) receptors, G-protein coupled receptors (GPCRs), and so on [59]. Endocytosis and intracellular sorting of transforming growth factor-β (TGF-β) receptors play an important regulatory role in TGF-β signaling [60]. In 2015, Fang et al. [61] utilized live-cell spinning-disk confocal imaging to track the intracellular dynamics of fluorescently labeled TGF-β type I receptor (TβRI). They found that the clathrin- and caveolae-mediated endocytic pathways can converge during TβRI endocytic trafficking (Fig. 5.9A). Furthermore, they demonstrated the existence of the multi-component early endosomes containing EEA1, caveolin-1, Rab5, TβRI, Smad3/SARA, Smad7/Smurf2 and Rab11, which effectively promotes TGF-β signaling and TβRI recycling and degradation (Fig. 5.9B). The endocytic sorting of signaling receptors between recycling and degradative pathways is a crucial cellular process that governs receptor surface complement.

Endocytosis occurs all the time, yet cells maintain their morphological size unchanged, thanks to exocytosis. In eukaryotic cells, transmembrane and secretory proteins are synthesized and initially modified in the ER, then transported via vesicles to the Golgi apparatus, where they are further processed, and finally as secretory vesicles flow to and fuse with the PM, a process known as the constitutive exocytosis pathway. In 2016, Scharaw et al. revealed that EGF stimulation specifically increased the efficiency of the transport of newly synthesized EGFR from the ER through the Golgi apparatus to the PM [62]. Their work established a novel regulatory mechanism that integrates the degradation and secretion of EGFR to maintain its physiological levels at the plasma membrane.

The second secretion pathway is the regulated exocytosis pathway, which is present in some specialized cells dedicated to rapid on-demand product secretion, specifically the release of soluble proteins and other substances initially stored in secretory vesicles via exocytosis under specific conditions, for example, the secretion of hormones, neurotransmitters, or digestive enzymes. Intracellular insulin is stored in a specific class of organelles called dense core vesicles (DCVs) and can be released via exocytosis under

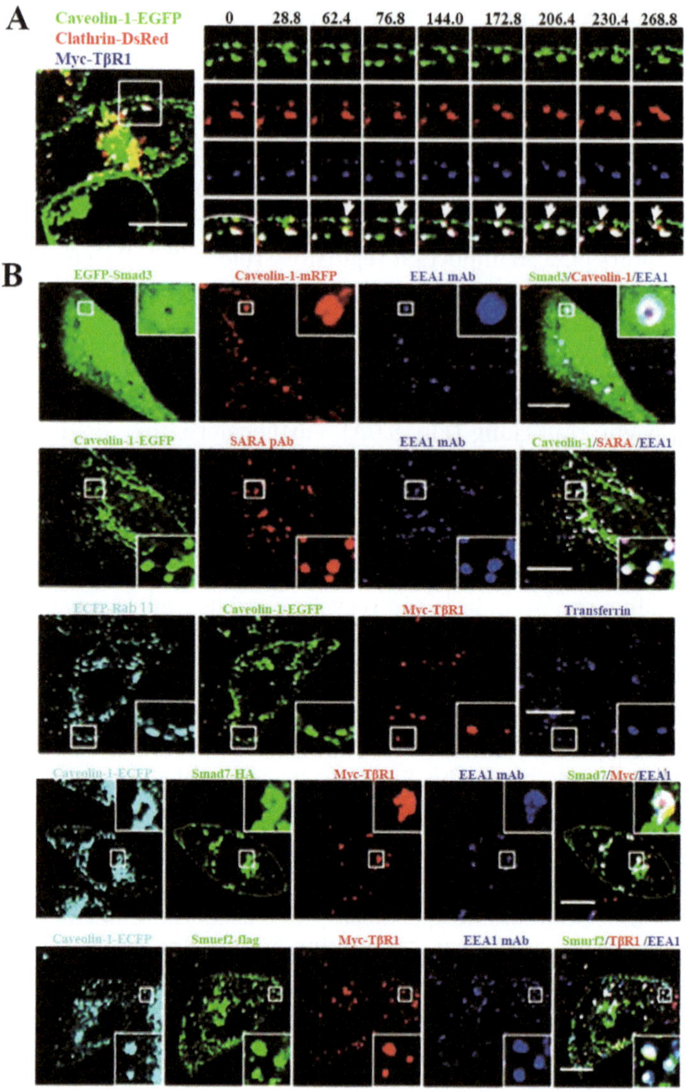

Figure 5.9 Receptor protein endocytosis and recycling pathways. (A) Triple-color live-cell confocal imaging of the cell coexpressing caveolin-1-EGFP, clathrin-DsRed, and Myc-TβRI. The boxed region was magnified (images to the right) to show the movement of a Myc-TβRI, caveolin-1-EGFP, and clathrin-DsRed triple-positive vesicle (arrows) from the lateral plasma membrane (white lines) to the cytoplasm. The corresponding kymographs of these three molecules are shown. (B) Detection of Smad3/SARA, Rab11, and Smad7/Smurf2 in caveolin-1-positive early endosomes. Scale bars, 10 μm. Reproduced from Ref. [61], Copyright © 2015 Shanghai Institutes for Biological Sciences, Chinese Academy of Sciences.

certain conditions. Visualization of insulin secretion is an urgent problem to be solved. In 2017, Schultz et al. developed a genetically encoded reporter termed RINS1 based on a fusion of proinsulin superfolder GFP (sfGFP) and mCherry for monitoring insulin secretion [63]. When insulin secretion is induced from pancreatic β-cells by glucose or drug stimulation, insulin sfGFP constructs are preferentially released, while mCherryfused C peptide remains trapped in extracellular granules. Based on this, they used TIRFM to monitor insulin secretion proportionally by detecting this physical separation (Fig. 5.10). A work applied correlative scanning ion conductance and fluorescence confocal microscopy (SICM-FCM) to realize direct, label-free, real-time display of DCV release, providing a deeper understanding of the insulin exocytosis process [64].

As our understanding of membrane transport becomes more comprehensive, in addition to endocytosis and recycling transport, as well as standard secretion (ER-Golgi-cell surface), new studies have identified several non-conventional secretion pathways. Golgi bypass secretion is a widely studied unconventional secretion pathway defined as the anterograde transport of transmembrane proteins from the ER to the plasma membrane without passing through the Golgi complex [65]. An increasing subset of proteins reach the cell surface in this way, such as cystic fibrosis transmembrane conductance regulator (CFTR), serglycin, aPS1, and so on [66–68]. Identifying more unconventionally secreted proteins and their release mechanism from the cells facilitates the understanding of how eukaryotic cells secrete cellular proteins in response to signals and cellular demands. Additionally, unconventional secretion pathways via exosomes have been uncovered, which refers to the release of different types of membrane vesicles from the endosome and plasma membrane, into the extracellular environment, called exosomes and microvesicles, respectively [69]. Exosome secretion is mainly accomplished through the fusion of multivesicular bodies (MVB) with the PM. An intuitive and dynamic insight into the mechanism of exosome biogenesis and release has been a long-needed problem in exosome research. In 2019, Martin et al. transfected cells with CD63-pHluorin and then used direct TIRF imaging to track Ca^{2+}-induced exosome secretion, and they found that this secretion process was dependent on Munc13-4 [70]. In 2020, Verweij et al. developed tetraspanin-based pH-sensitive fluorescent reporters to detect the MVB–PM fusion of single cells in real time by evaluating the simultaneous release of different cargoes from the MVB exocytosis [71].

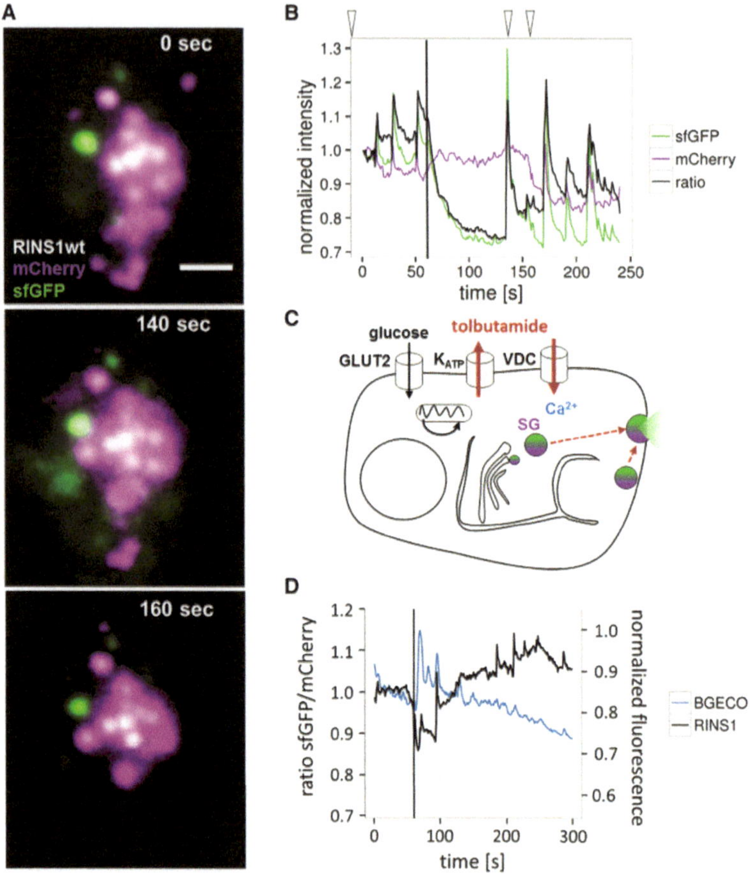

Figure 5.10 Drug-mediated modulation of insulin secretion visualized by RINS1 in MIN6 cells imaged by TIRFM. (A) Example TIRF images of RINS1-bearing granules in a single β cell at defined time points. Scale bar, 2 mm. (B) sfGFP (green), mCherry (magenta) intensity, and ratio sfGFP/mCherry (blue) changes over time. (C) Schematic drawing of sulfonylurea drugs affecting insulin secretion. (D) MIN6 cell-expressing RINS1 and B-GECO treated with tolbutamide (100 mM, 30 min) and stimulated by glucose after 60 s. Reprinted with permission from Ref. [63], Copyright 2017 Elsevier.

5.3.3 Membrane Protein Dense Distribution Determines the Orderly Organelle Interactions

Although the major proteins that regulate orderly organelle interactions have been identified, little is known about the overall membrane structure of organelles. The essence of organelle–organelle interplay is interactions between biological membranes. Given the structure-determining function, structural features of biological membranes may play an important role in regulating ordered organelle interactions. Accordingly, dissecting the relationship between the structural features of membrane itself and the modes of interaction between membranous organelles will help us to better understand the integral nature of orderly organelle interaction.

Because of complicated composition and limited research techniques, structural characteristics of the cell membranes (e.g., the distribution of the phospholipid bilayer and membrane proteins) are still hypothetical, revolving around the fluid mosaic model and lipid raft model [72, 73]. These traditional models assume that cell membranes are primarily based on phospholipid structure with mosaic proteins, resembling those of pure phospholipid vesicles. Hence, these hypotheses fail to explain the orderly organelle interactions. Importantly, Wang et al. worked on the membrane structure of various cell membranes and achieved a series of original results [74], which provided insight into the structure of cell membranes. From atomic force microscopy (AFM) observations, they proposed a model of red blood cells (semi-mosaic model), suggesting that membrane proteins are mainly located on the inner surface of the cell, and membrane proteins are partially inserted into the lipid bilayer without protruding from the outer surface of the cell (Fig. 3.13A) [75]. They further proposed a novel model of nucleated mammalian cell membranes termed the protein layer–lipid–protein island model (PLLPI), which demonstrated that proteins on the ectoplasmic side of the cell membrane formed a dense protein layer atop a lipid bilayer and that proteins aggregated as islands dispersed evenly on the cytoplasmic side of the cell membrane (Fig. 3.14J) [76]. In addition, two main organelle membranes (those of the Golgi apparatus and mitochondria) were imaged using AFM [77, 78]. The Golgi and mitochondrial membranes exhibited features similar to those of cell membranes; that is, the outer surface was flat, and the inner side was rough with protruding proteins. Overall, their systematic work showed that membrane structures in cells exhibited the characteristics of membrane asymmetry.

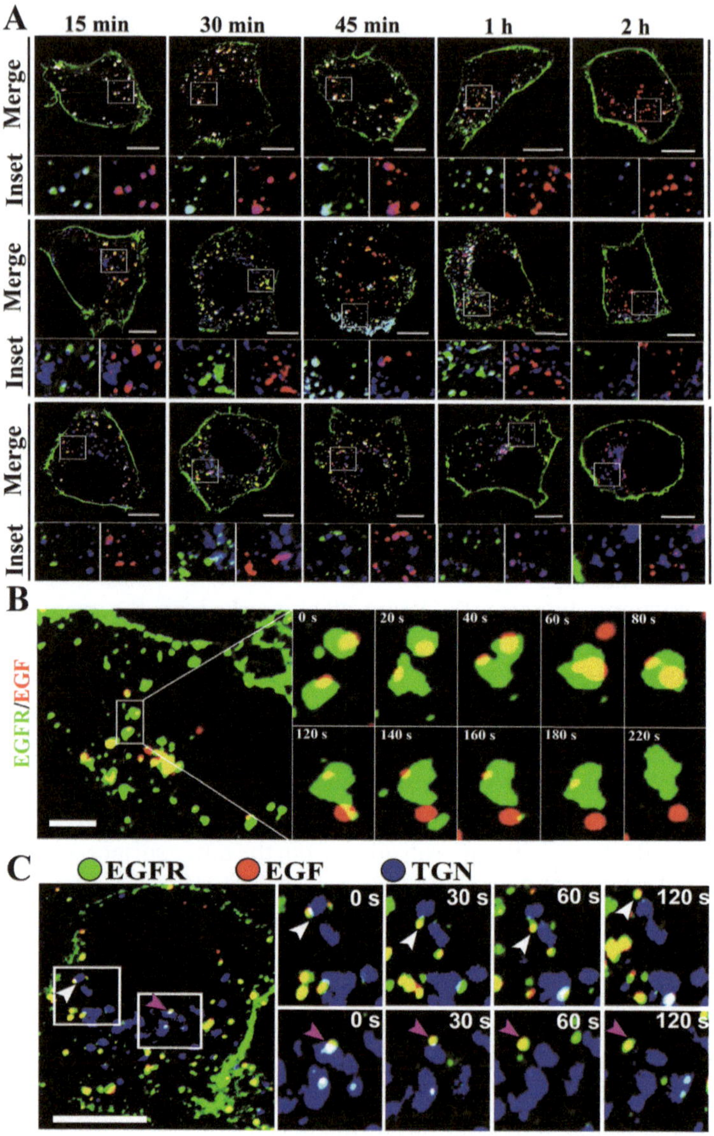

Figure 5.11 (A) Membrane sorting process via EGFR endocytosis and the recycling pathway. The treated cells were chased to the indicated time and

immuno-labeled for the markers of the early endosome (EEA1) and the recycling endosome (Rab11). For labeling the late endosome marker (Rab7), A549 cells were transiently transfected with RFP-Rab7. The A549 cells were analyzed using confocal microscopy. Scale bars = 10 μm. (B) The dynamic process of endosome fusion and fission. A549 cells were tracked by real-time imaging using a confocal microscope. Scale bar = 5 μm. (C) Endocytic vesicles dynamically interacted with the TGN through the kiss-and-run method. A549 cells stably expressing EGFP-EGFR were transfected with mDsRed-Golgi-7. Then the A549 cells were stimulated with Alexa Fluor 647-EGF and traced by real-time imaging. The arrowheads point to the endocytic vesicles, displaying the process of interaction with the TGN. Scale bar = 10 μm. Reprinted with permission from Ref. [79]. Copyright 2020, American Chemical Society.

In vesicle-transport-mediated interacting organelle network, vesicles and target organelles have the same or opposite membrane protein asymmetry that may determine the way they interact. In 2020, Wang et al. took a typical vesicle-transport process involving EGFR endocytosis and recycling pathway as the study object to reveal a correlation between the membrane structure itself and its orderly transport behavior [79]. In this study, they found that the asymmetry of the cell membrane was the same as that of the Golgi membrane, but that of the endocytic vesicles was opposite to them. On the basis of this structure characteristics analysis, fast-speed spinning-disk confocal microscopy was utilized to visualize the process of endocytic vesicle transport in details, demonstrating that endocytic vesicles were transported and sorted in an orderly manner, rather than through stochastic membrane fusion (Fig. 5.11A). Further, the dynamic process of vesicle–vesicle and vesicle–Golgi interactions was monitored using a real-time fluorescence tracing method. Time-lapse imaging results showed that two homotypic endocytic vesicles underwent complete membrane fusion with each other (Fig. 5.11B), while the endocytic vesicles interacted with the TGN via this kiss-and-run way (KAR) (Fig. 5.11C).

The above phenomena that vesicles undergo ordered transport without fusion at nonspecific sites and two different modes of fusion at specific sites, including complete membrane fusion and an effective KAR exchange mode, contribute to the orderly organelle interactions. Collectively, combining static structural observations with dynamic tracing results, they proposed a novel structural mechanistic model of orderly and efficient vesicle transport, namely, the membrane-asymmetry-determined orderly organelle transport (MADOOT) model. The model suggested that prominent proteins on the outer surface of the endocytic vesicles may lead to their orderly transport, while exposed lipids may allow complete membrane fusion between these homotypic vesicles if mediated by specific proteins. A dense protein layer on the outer surface of the TGN may account for the KAR interaction way between endocytic vesicles and the TGN (Fig. 5.12). The MADOOT model emphasized the importance of the membrane structure to orderly vesicle transport, demonstrating the correlation between morphology and function.

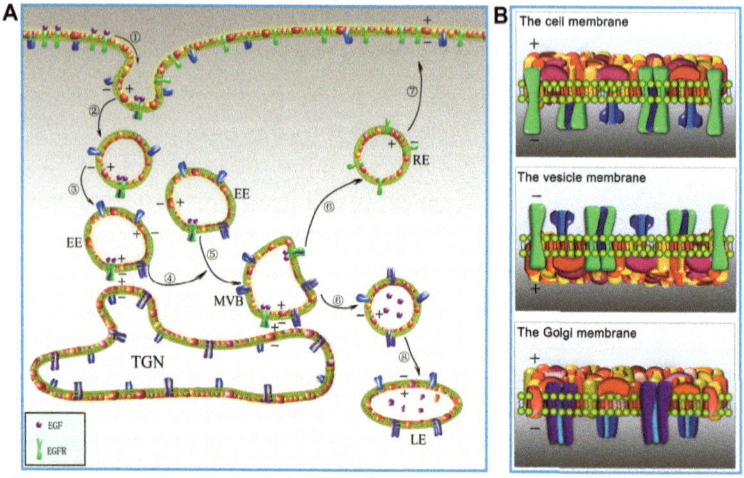

EE: early endosome; MVB: multivesicular body; RE: recycling endosome; LE: late endosome; TGN: trans Golgi network. "+" represents the smooth membrane surface covered with proteins layer; "–" represents the rough membrane surface with proteins protrusion.

Figure 5.12 Membrane trafficking via endocytosis of the receptor protein EGFR and the recycling pathway (In this schematic diagram of EGFR endocytosis and recycling, the following processes occur: ① upon ligand binding, EGFR on the surface of the plasma membrane is internalized by membrane invagination; ② the internalized vesicle forms. The orientation of membrane proteins on the plasma membrane is opposite that on the vesicle; ③ the receptor is transported to the early endosome; ④ the early endosome instantaneously contacts the TGN and then departs; ⑤ fusion of homotypic endosomes results in the multivesicular body; ⑥ fission of endosomes leads to dissociation of the receptor and ligand; ⑦ the receptor is transported to the cell surface by the recycling endosome; ⑧ the ligand is sorted to the late endosome) (A) and in-plane views of the structure characteristics of the cell membrane, the endocytic vesicle membrane, and the Golgi membrane. (B) In-plane views of the structure characteristics of the cell membrane, the endocytic vesicle membrane, and the Golgi membrane. Reprinted with permission from Ref. [79]. Copyright 2020, American Chemical Society.

5.4 Protocols for Performing Fluorescence Imaging

5.4.1 Transient Transfection of Mammalian Cells with Fluorescent Protein Expression Vectors

For cell transfection, transfection conditions vary depending on culture volume and the type of transfection reagent. Here, we will take a 35 mm glass-bottom culture dish and Lipofectamine™ 3000 Reagent (Invitrogen) as an example to illustrate the transfection process in detail.

1. Cells growing in glass-bottom dishes to be 70–90% confluent at transfection.
2. Aspirate medium from the dish and replace with fresh complete medium (2 mL).
3. Prepare master mix of plasmid by diluting 1 μg plasmid in Opti-MEM™ Medium (50 μL), then add P3000™ Reagent (2 μL). Mix well and rest for 5 min.
4. Dilute Lipofectamine™ 3000 (2 μL) Reagent in Opti-MEM™ Medium (50 μL), then rest for 5 min.
5. Add diluted plasmid to the tube of diluted Lipofectamine™ 3000 Reagent, then incubate for 15 min.
6. Add the solution containing cDNA, transfection media, and X Lipofectamine™ 3000 Transfection Reagent to the cell dish.
7. Incubate in a cell culture incubator for 24 h for fluorescence imaging analysis.

Notes:

- Perform cell passaging 24 h prior to transfection, in addition to avoiding bacterial, mycoplasma, or fungal contamination of the cells.
- Cell confluency rate, plasmid DNA concentration, and expression time all affect transfection efficiency.
- The presence or absence of serum in the culture medium at the initial stage of transfection depends on the type of transfection reagent.

- When performing transfection of different DNA or cell lines, the experimental conditions need to be optimized again according to the instructions.

5.4.2 Immuno-fluorescence

1. Aspirate medium from cells growing on 35 mm dish round coverslip.
2. Wash the cells twice with 1 mL 1 × PBS, pH 7.4.
3. Fix the cells with 1 mL 4% paraformaldehyde (PFA) in 1 × PBS, pH 7.4 for 10 min at room temperature.
4. Wash the cells thrice with 1 × PBS, for 3–5 min/time, using a shaker with gentle shaking.
5. Permeabilize the cells with 0.2–0.3% Triton X-100 (Roche) for 15 min at room temperature and wash the cells thrice with 1 × PBS.
6. Block nonspecific binding by incubating fixed cells with 1 mL 1 × PBS, pH 7.4 containing 2% bovine serum albumin (BSA) for 2 h at room temperature.
7. Wash the cells thrice with 1 × PBS, for 3–5 min/time.
8. Incubate the cells for 1 h at room temperature or overnight at 4°C with primary antibody diluted to the appropriate concentration in 0.2% BSA in 1 × PBS, pH 7.4.
9. Wash the cells thrice with 1 × PBS, for 3–5 min/time.
10. Incubate the cells for 2 h at room temperature with fluorophore-labeled secondary antibody diluted to the appropriate concentration (dilution factor is again antibody dependent) in 1 × PBS, pH 7.4 containing 0.2% BSA.
11. Wash the cells thrice with 1 × PBS, for 3–5 min/time.
12. Label the nuclei for 5–10 min at room temperature without light.
13. Add an imaging buffer containing an anti-fluorescence quencher before fluorescence imaging.

Notes:

- A thorough cleaning step before adding PFA is essential.
- Different permeation reagents target different membrane structures. Use the required permeabilization reagent as little as possible to minimize artifacts.

- Reduce the primary antibody concentration or shorten the incubation time to avoid high background staining due to antibody cross-reactivity.
- Low-temperature overnight staining has better results than room temperature for a short time.
- Stained specimens should be at best observed on the same day; otherwise, the fluorescence intensity will gradually decrease.

5.4.3 Live-Cell Labeling Probes

Preparing Stock Solutions

Dissolve the lyophilized probes targeted special organelle, such as ER-Tracker™ dyes or MitoTracker® Probes (molecular probes) in dimethylsulfoxide (DMSO) to a final concentration of 1 mM.

Cell Preparation and Staining

- **Prepare staining solution:** Dilute the 1 mM stock solution to the final working concentration in an appropriate buffer or growth medium.
- **Staining adherent cells:** Remove the medium from the culture dish, rinse with HBSS, and add prewarmed (37°C) staining solution containing probes. Incubation time varies from 15–60 min at 37°C, depending mainly on the model system and probe used. Replace the staining solution with a fresh probe-free medium and observe cells using a fluorescence microscope.
- **Staining suspension cells:** Centrifuge to obtain cell pellets and discard the supernatant. Gently resuspend the cells using a prewarmed (37°C) staining solution containing a fluorescent probe, and incubate for 15–60 min at 37°C. After staining is complete, remove the staining solution by centrifugation and resuspend the cells in a prewarm fresh medium or buffer for fluorescence imaging analysis. If immobilized cells on coverslips are needed, use poly-D-lysine to coat the slides or coverslips before mounting.

Notes:

- Before opening a vial, allow the product to warm to room temperature.

- The working concentration of the probe for optimal staining varies by application. Check the recommended concentration for specialized probe in Invitrogen.com.
- Keep the concentration of dye as low as possible to avoid artifice.
- Do not recommend using the complete media to dilute the dye.
- A dye for the cytoplasm that cannot pass the plasma membrane cannot be used for live-cell labeling.

5.4.4 Imaging System

Confocal microscope for image acquisition

Most fluorescence imaging, especially live-cell fluorescence imaging, is obtained by a confocal microscope or imaging system based on it. Andor Revolution WD (Andor, Oxford Instruments, UK) or a similar confocal microscope is generally equipped with the following devices:

- An inverted microscope with 63× or 100× oil immersion objective.
- Laser lines: 405, 488, 561, and 640 nm.
- An iXon Ultra EMCCD instrument (Andor, Oxford Instruments) or the other cameras.
- A precise motorized stage from PRIOR or other brands.
- Heated stage with CO_2 and humidity from TOKAI HIT or other brands.
- Self-contained imaging software, such as Andor iQ3 software (Andor).

Software for image analysis

- ImageJ for colocalization analysis.
- Imaris for 3D render.
- Matlab for trajectory tracking.
- Analyze Internalization GUI software tool (available from http://www.crick.ac.uk/pavel-tolar).
- Huygens software for image deconvolution, improving signal-to-noise ratio.

Notes:

- Glass-bottom dishes with bottom coverslips no thicker than 0.17 mm are suitable for high magnification observation and high-quality imaging.
- Select the excitation light that matches the fluorescent probe used and use the lowest possible laser intensity during the experiment.
- Using a microscope with faster acquisition speed, such as a spinning-disk confocal imaging system, can reduce phototoxicity and photobleaching of live-cell samples and also allows tracking intracellular rapid life processes.
- Long-term observation requires fixing the specimen to prevent movement; room temperature should also be kept stable to avoid drifting of the microscope itself.

References

1. Wu, H. X., Carvalho, P., and Voeltz, G. K. (2018). Here, there, and everywhere: The importance of ER membrane contact sites, *Science*, **361**, pp. eaan5835.
2. White, J. G. and Amos, W. B. (1987). Confocal microscopy comes of age, *Nature*, **328**, pp. 183–184.
3. Nwaneshiudu, A., Kuschal, C., Sakamoto, F. H., Anderson, R. R., Schwarzenberger, K., and Young, R. C. (2012). Introduction to confocal microscopy, *J. Invest. Dermatol.*, **132**, pp. e3.
4. Claxton, N. S., Fellers, T. J., and Davidson, M. W. (2006) *Encyclopedia of Medical Devices and Instrumentation*, eds. De Luca, C. J., "Laser scanning confocal microscopy," (John Wiley Publisher, Hoboken), pp. 1–37.
5. Nakano, A. (2002). Spinning-disk confocal microscopy: A cutting-edge tool for imaging of membrane traffic, *Cell. Struct. Funct.*, **27**, pp. 349–355.
6. Bayguinov, P. O., Oakley, D. M., Shih, C. C., Geanon, D. J., Joens, M. S., and Fitzpatrick, J. A. (2018). Modern laser scanning confocal microscopy, *Curr. Protoc. Cytom.*, **85**, pp. e39.
7. Elliott, A. D. (2020). Confocal microscopy: Principles and modern practices, *Curr. Protoc. Cytom.*, **92**, pp. e68.

8. Axelrod, D. (1981). Cell-substrate contacts illuminated by total internal reflection fluorescence, *J. Cell Biol.*, **89**, pp. 141–145.

9. Kassandra Kisler, R. H. C. and Reymundo, D. (2014). Fluorescently-labeled estradiol internalization and membrane trafficking in live N-38 neuronal cells visualized with total internal reflection fluorescence microscopy, *J Steroids Horm Sci.*, **12**, pp. 1–15.

10. Schneckenburger, H. (2005). Total internal reflection fluorescence microscopy: Technical innovations and novel applications, *Curr. Opin. Biotech.*, **16**, pp. 13–18.

11. Shao, L. and Rego, E. H. (2014). *Fluorescence Microscopy*, eds. Anda Cornea, A. and Conn, P. M., "Structured illumination microscopy" (Elsevier), pp. 213–225.

12. Bailey, B., Farkas, D. L., Taylor, D., and Lanni, F. (1993). Enhancement of axial resolution in fluorescence microscopy by standing-wave excitation, *Nature*, **366**, pp. 44–48.

13. Gustafsson, M. G. (2000). Surpassing the lateral resolution limit by a factor of two using structured illumination microscopy, *J. Microsc.*, **198**, pp. 82–87.

14. Schermelleh, L., Carlton, P. M., Haase, S., Shao, L., Winoto, L., Kner, P., Burke, B., Cardoso, M. C., Agard, D. A., and Gustafsson, M. G. (2008). Subdiffraction multicolor imaging of the nuclear periphery with 3D structured illumination microscopy, *Science*, **320**, pp. 1332–1336.

15. Gustafsson, M. G., Shao, L., Carlton, P. M., Wang, C. R., Golubovskaya, I. N., Cande, W. Z., Agard, D. A., and Sedat, J. W. (2008). Three-dimensional resolution doubling in wide-field fluorescence microscopy by structured illumination, *Biophys. J.*, **94**, pp. 4957–4970.

16. Gustafsson, M. G. (2005). Nonlinear structured-illumination microscopy: Wide-field fluorescence imaging with theoretically unlimited resolution, *Proc. Natl. Acad. Sci. USA.*, **102**, pp. 13081–13086.

17. Huang, B., Bates, M., and Zhuang, X. (2009). Super-resolution fluorescence microscopy, *Annu. Rev. Biochem.*, **78**, pp. 993–1016.

18. Gao, L., Tang, W.-C., Tsai, Y.-C., and Chen, B.-C. (2019). Lattice light-sheet microscopy using tiling lattice light sheets, *Opt. Express*, **27**, pp. 1497–1506.

19. Mimori-Kiyosue, Y. (2021). Imaging mitotic processes in three dimensions with lattice light-sheet microscopy, *Chromosome. Res.*, **29**, pp. 37–50.

20. Chen, B., Legant, W. R., Kai, W., Shao, L., Milkie, D. E., Davidson, M. W., Janetopoulos, C., Wu, X. S., Hammer, J. A., and Zhe, L. (2014). Lattice

light-sheet microscopy: Imaging molecules to embryos at high spatiotemporal resolution, *Science*, **346**, pp. 1257998.

21. Planchon, T. A., Gao, L., Milkie, D. E., Davidson, M. W., Galbraith, J. A., Galbraith, C. G., and Betzig, E. (2011). Rapid three-dimensional isotropic imaging of living cells using Bessel beam plane illumination, *Nat. Methods.*, **8**, pp. 417–423.
22. Gao, L., Shao, L., Higgins, C. D., Poulton, J. S., Peifer, M., Davidson, M. W., Wu, X., Goldstein, B., and Betzig, E. (2012). Noninvasive imaging beyond the diffraction limit of 3D dynamics in thickly fluorescent specimens, *Cell*, **151**, pp. 1370–1385.
23. Hirano, Y., Matsuda, A., and Hiraoka, Y. (2015). Recent advancements in structured-illumination microscopy toward live-cell imaging, *Microscopy*, **64**, pp. 237–249.
24. Lippincott-Schwartz, J. and Patterson, G. H. (2003). Development and use of fluorescent protein markers in living cells, *Science*, **300**, pp. 87–91.
25. Odell, I. D. and Cook, D. (2013). Immunofluorescence techniques, *J. Invest. Dermatol.*, **133**, pp. e4.
26. Suzuki, T., Matsuzaki, T., Hagiwara, H., Aoki, T., and Takata, K. (2007). Recent advances in fluorescent labeling techniques for fluorescence microscopy, *Acta. Histochem. Cytoc.*, **40**, pp. 131–137.
27. Bastiaens, P. I. and Pepperkok, R. (2000). Observing proteins in their natural habitat: The living cell, *Trends. Biochem. Sci.*, **25**, pp. 631–637.
28. Shaner, N. C., Steinbach, P. A., and Tsien, R. Y. (2005). A guide to choosing fluorescent proteins, *Nat. Methods.*, **2**, pp. 905–909.
29. Andresen, M., Schmitz-Salue, R., and Jakobs, S. (2004). Short tetracysteine tags to β-tubulin demonstrate the significance of small labels for live cell imaging, *Mol. Biol. Cell.*, **15**, pp. 5616–5622.
30. Los, G., Learish, R., Karassina, N., Zimprich, C., Mcdougall, M. G., Encell, L. P., Friedman-Ohana, R., Wood, M., Vidugiris, G., and Zimmerman, K. (2006). HalotagTM technology: Cell imaging and protein analysis, *Cell Notes*, **14**, pp. 10–14.
31. Gonçalves, M. S. T. (2009). Fluorescent labeling of biomolecules with organic probes, *Chem. Rev.*, **109**, pp. 190–212.
32. Prinz, W. A. (2014). Bridging the gap: Membrane contact sites in signaling, metabolism, and organelle dynamics, *J. Cell Biol.*, **205**(6), pp. 759–769.
33. Dan, M. (1957). Electron microscope study of mammalian cardiac muscle cells, *J. Cell Biol.*, **3**, pp. 261–268.

34. Porter, K. R. (1957). Studies on the endoplasmic reticulum. III. Its form and distribution in striated muscle cells, *J. Cell Biol.*, **3**(2), pp. 269–300.

35. Zhang, S. L., Yeromin, A. V., Zhang, X. H.-F., Yu, Y., Safrina, O., Penna, A., Roos, J., Stauderman, K. A., and Cahalan, M. D. (2006). Genome-wide RNAi screen of Ca^{2+} influx identifies genes that regulate Ca^{2+} release-activated Ca^{2+} channel activity, *Proc. Natl. Acad. Sci. USA.*, **103**, pp. 9357–9362.

36. Cárdenas, C., Miller, R. A., Smith, I., Bui, T., Molgó, J., Müller, M., Vais, H., Cheung, K.-H., Yang, J., and Parker, I. (2010). Essential regulation of cell bioenergetics by constitutive InsP3 receptor Ca^{2+} transfer to mitochondria, *Cell*, **142**, pp. 270–283.

37. Hanada, K., Kumagai, K., Yasuda, S., Miura, Y., Kawano, M., Fukasawa, M., and Nishijima, M. (2003). Molecular machinery for non-vesicular trafficking of ceramide, *Nature*, **426**, pp. 803–809.

38. Friedman, J. R., Lackner, L. L., West, M., Dibenedetto, J. R., Nunnari, J., and Voeltz, G. K. (2011). ER tubules mark sites of mitochondrial division, *Science*, **334**, pp. 358–362.

39. Rowland, A. A., Chitwood, P. J., Phillips, M. J., and Voeltz, G. K. (2014). ER contact sites define the position and timing of endosome fission, *Cell*, **159**, pp. 1027–1041.

40. Phillips, M. J. and Voeltz, G. K. (2015). Structure and function of ER membrane contact sites with other organelles, *Nat. Rev. Mol. Cell Biol.*, **17**, pp. 69–82.

41. Valm, A. M., Cohen, S., Legant, W. R., Melunis, J., Hershberg, U., Wait, E., Cohen, A. R., Davidson, M. W., Betzig, E., and Lippincottschwartz, J. (2017). Applying systems-level spectral imaging and analysis to reveal the organelle interactome, *Nature*, **546**, pp. 162–167.

42. Cohen, S., Valm, A. M., and Lippincott-Schwartz, J. (2018). Interacting organelles, *Curr. Opin. Cell. Biol.*, **53**, pp. 84–91.

43. Guo, Y., Li, D., Zhang, S., Yang, Y., Liu, J. J., Wang, X., Liu, C., Milkie, D. E., Moore, R. P., and Tulu, U. S. (2018). Visualizing intracellular organelle and cytoskeletal interactions at nanoscale resolution on millisecond timescales, *Cell*, **175**, pp. 1430–1442.

44. Jani, R. A. and Setty, S. R. G. (2014). Membrane trafficking and vesicle fusion, *Resonance*, **19**, pp. 421–445.

45. Bonifacino, J. S. and Glick, B. S. (2004). The mechanisms of vesicle budding and fusion, *Cell*, **116**, pp. 153–166.

46. Pfeffer, S. R. (2013). Rab GTPase regulation of membrane identity, *Curr. Opin. Cell. Biol.*, **25**, pp. 414–419.

47. Feng, Y., Press, B., and Wandinger-Ness, A. (1995). Rab 7: An important regulator of late endocytic membrane traffic, *J. Cell Biol.*, **131**, pp. 1435–1452.

48. Bucci, C., Parton, R. G., Mather, I. H., Stunnenberg, H., Simons, K., Hoflack, B., and Zerial, M. (1992). The small GTPase rab5 functions as a regulatory factor in the early endocytic pathway, *Cell*, **70**, pp. 715–728.

49. Jahn, R. and Scheller, R. H. (2006). SNAREs: Engines for membrane fusion, *Nat. Rev. Mol. Cell Biol.*, **7**, pp. 631–643.

50. Tian, X., Teng, J., and Chen, J. (2021). New insights regarding SNARE proteins in autophagosome-lysosome fusion, *Autophagy*, **17**, pp. 2680–2688.

51. Antonescu, C. N., Mcgraw, T. E., and Klip, A. (2014). Reciprocal regulation of endocytosis and metabolism, *Cold Spring Harb. Perspect. Biol.*, **6**, pp. a016964.

52. Villaseñor, R., Kalaidzidis, Y., and Zerial, M. (2016). Signal processing by the endosomal system, *Curr. Opin. Cell. Biol.*, **39**, pp. 53–60.

53. Sandri, C., Serini, G., Caccavari, F., Val De Mbri, D., Camillo, C., Veltel, S., Santambrogio, M., Lanzetti, L., Bussolino, F. D. N., and Ivaska, J. (2012). The R-Ras/RIN2/Rab5 complex controls endothelial cell adhesion and morphogenesis via active integrin endocytosis and Rac signaling, *Cell Res*, **22**, pp. 1479–1501.

54. Tang, B. L. (2015). Rab, Arf and Arl-regulated membrane traffic in cortical neuron migration, *J. Cell. Physiol.*, **231**, pp. 1417–1423.

55. Soria, F. N., Olatz, P., Mathieu, B., Meissner, W. G., Erwan, B., and Benjamin, D. (2017). Exosomes, an unmasked culprit in neurodegenerative diseases, *Front. Neurosci.*, **11**, pp. 1–12.

56. Singer, J. H., Lassová, L., Vardi, N., and Diamond, J. S. (2004). Coordinated multivesicular release at a mammalian ribbon synapse, *Nat. Neurosci.*, **7**, pp. 826–833.

57. Lawrence, J., Birnbaum, T. R., and Morris, J. (2003). ADP-ribosylation factor 6 regulates insulin secretion through plasma membrane phosphatidylinositol 4,5-bisphosphate, *Proc. Natl. Acad. Sci. USA.*, **100**, pp. 13320–13325.

58. Ivanov, A. I. (2014). *Exocytosis and Endocytosis*, 2nd Ed. (Humana Press, New York, NY).

59. Platta, H. W. and Stenmark, H. (2011). Endocytosis and signaling, *Curr. Opin. Cell. Biol.*, **23**, pp. 393–403.

60. Tzavlaki, K. and Moustakas, A. (2020). TGF-β Signaling, *Biomolecules*, **10**, pp. 487.

61. He, K., Yan, X., Li, N., Dang, S., Xu, L., Zhao, B., Li, Z., Lv, Z., Fang, X., Zhang, Y., and Chen, Y.-G. (2015). Internalization of the TGF-β type I receptor into caveolin-1 and EEA1 double-positive early endosomes, *Cell Research*, **25**, pp. 738–752.

62. Scharaw, S., Iskar, M., Ori, A., Boncompain, G., Laketa, V., Poser, I., Lundberg, E., Perez, F., Beck, M., and Bork, P. (2016). The endosomal transcriptional regulator RNF11 integrates degradation and transport of EGFR, *J. Cell. Biol.*, **215**, pp. 543–558.

63. Schifferer, M., Yushchenko, D. A., Stein, F., Bolbat, A., and Schultz, C. (2017). A ratiometric sensor for imaging insulin secretion in single β cells, *Cell. Chem. Biol.*, **24**, pp. 525–531.

64. Bednarska, J., Novak, P., Korchev, Y., Rorsman, P., and Shevchuk, A. (2020). Release of insulin granules by simultaneous, high-speed correlative SICM-FCM, *J. Microsc.*, **282**, pp. 21–29.

65. Grieve, A. G. and Rabouille, C. (2011). Golgi bypass: Skirting around the heart of classical secretion, *CSH. Perspect. Biol.*, **3**, pp. a005298.

66. Yoo, J.-S., Moyer, B. D., Bannykh, S., Yoo, H.-M., Riordan, J. R., and Balch, W. E. (2002). Non-conventional trafficking of the cystic fibrosis transmembrane conductance regulator through the early secretory pathway, *J. Biol. Chem.*, **277**, pp. 11401–11409.

67. Scully, O. J., Chua, P. J., Harve, K. S., Bay, B. H., and Yip, G. W. (2012). Serglycin in health and diseases, *Anat. Rec.*, **295**, pp. 1415–1420.

68. Schotman, H., Karhinen, L., and Rabouille, C. (2008). dGRASP-mediated noncanonical integrin secretion is required for Drosophila epithelial remodeling, *Dev. Cell.*, **14**, pp. 171–182.

69. Graça, R. and Willem, S. (2013). Extracellular vesicles: Exosomes, microvesicles, and friends, *J. Cell Biol.*, **200**, pp. 373–383.

70. Messenger, S. W., Woo, S. S., Sun, Z., and Martin, T. F. (2019). Correction: A Ca^{2+}-stimulated exosome release pathway in cancer cells is regulated by Munc13-4, *J. Cell. Biol.*, **218**, pp. 1423.

71. Bebelman, M. P., Bun, P., Huveneers, S., van Niel, G., Pegtel, D. M., and Verweij, F. J. (2020). Real-time imaging of multivesicular body-plasma membrane fusion to quantify exosome release from single cells, *Nat. Protoc.*, **15**, pp. 102–121.

72. Singer, S. J. and Nicolson, G. L. (1972). The fluid mosaic model of the structure of cell membranes, *Science*, **175**, pp. 720–731.

73. Simons, K. and Ikonen, E. (1997). Functional rafts in cell membranes, *Nature*, **387**, pp. 569–572.

74. Shan, Y. P. and Wang, H. D. (2015). The structure and function of cell membranes examined by atomic force microscopy and single-molecule force spectroscopy, *Chem. Soc. Rev.*, **44**, pp. 3617–3638.

75. Wang, H. D., Hao, X., Shan, Y. P., Jiang, J. G., Cai, M. J., and Shang, X. (2010). Preparation of cell membranes for high resolution imaging by AFM, *Ultramicroscopy*, **110**, pp. 305–312.

76. Zhao, W. D., Tian, Y. M., Cai, M. J., Wang, F., Wu, J. Z., Gao, J., Liu, S. H., Jiang, J. G., Jiang, S. B., and Wang, H. D. (2014). Studying the nucleated mammalian cell membrane by single molecule approaches, *PLoS One*, **9**, e91595.

77. Tian, Y. M., Li, J. H., Cai, M. J., Zhao, W. D., Xu, H. J., Liu, Y., and Wang, H. D. (2013). High resolution imaging of mitochondrial membranes by in situ atomic force microscopy, *RSC Adv.*, **3**, pp. 708–712.

78. Xu, H. J., Su, W. H., Cai, M. J., Jiang, J. G., Zeng, X. L., and Wang, H. D. (2013). The asymmetrical structure of Golgi apparatus membranes revealed by in situ atomic force microscope, *PLoS One*, **8**, e61596.

79. Xu, H. J., Gao, J., Cai, M. J., Chen, J. L., Zhang, Q. R., Li, H. R., and Wang, H. D. (2020). Structural mechanism analysis of orderly and efficient vesicle transport by high-resolution imaging and fluorescence tracking, *Anal. Chem.*, **92**, pp. 6555–6563.

Chapter 6

Single-Molecule Fluorescence for Studying the Membrane Protein Dynamics and Interactions

Hua He,[a] Qian Wang,[b] Xiaoqiang Wang,[a] and Fang Huang[a]
[a]*State Key Laboratory of Heavy Oil Processing and College of Chemistry and Chemical Engineering, China University of Petroleum (East China), Qingdao 266580, China*
[b]*College of Chemistry and Chemical Engineering, Xi'an Shiyou University, 710065, Xi'an, China*
huahe@upc.edu.cn, fhuang@upc.edu.cn

Single-molecule fluorescence (SMF) techniques have been booming since their establishment in the 1990s. The applications of SMF techniques to the dynamics and interactions of membrane proteins have revolutionized our views on many cellular processes. In this chapter, we introduce the basics of SMF techniques, including single-molecule fluorescence resonance energy transfer (smFRET), fluorescence correlation spectroscopy (FCS), total internal reflection fluorescence microscopy (TIRFM), and their combinations with single-molecule localization and deep-learning techniques. We summarize the progress of SMF techniques and their recent

Biomembranomics: Structure, Techniques, and Applications
Edited by Hongda Wang

ISBN 978-981-4968-61-4 (Hardcover), 978-1-003-45635-3 (eBook)
www.jennystanford.com

applications in the study of dynamics and interactions of membrane proteins. We finally provide the example experimental protocol and data analysis procedure for SMF imaging of membrane proteins in live cells.

6.1 Introduction

Membrane proteins constitute 20–30% of all proteins encoded by the human genome and achieve biological functions as receptors, transporters, anchors, or enzymes [1]. Membrane proteins sense extracellular stimuli and mediate signaling cascades on various cellular behaviors. Studies on the dynamics and interaction of membrane proteins allow us to gain a deep insight into many biological processes. In recent decades, the progress has been accelerated by several techniques, including fluorescence resonance energy transfer (FRET), fluorescence recovery after photobleaching (FRAP), and the SMF method [2]. Among these approaches, the SMF technique may be more informative than the other techniques, because it offers rich information on the dynamics and spatial organization of single proteins rather than on ensemble averages in heterogeneous cell environments. Historically, single-molecule observations were first made by measuring the optical absorption of single fluorescent molecules at cryogenic temperatures [3]. In 1995, a single fluorescently labeled protein was directly visualized by TIRFM combined with high-sensitive detectors [4]. Since then, SMF methods have become a powerful tool to probe the dynamics and interactions of membrane protein in vitro and in a live cell because of their superior ability to provide quantitative information about the locations, kinetics, and dynamics of individual molecules with high spatiotemporal resolution. These SMF methods include single-molecule FRET (smFRET), fluorescence correlation spectroscopy (FCS), TIRFM, and single-molecule localization microcopy (SMLM). In this chapter, we first introduce the basic principles of these SMF methods and discuss their recent advancements, particularly in combination with deep-learning algorithms. We then summarize the recent applications of SMF methods for studying the dynamics and interactions of membrane proteins.

6.2 Fundamental Principles

6.2.1 smFRET

FRET is the only tool that can measure intramolecular and intermolecular distances in real time and nanoscale precision [5, 6]. smFRET is a variant of FRET that extends its use to single-molecule level [7, 8]. Differing with X-ray crystallography and cryo-electron microscopy (cryoEM) that capture the conformational states of a protein [9], smFRET can dissect conformational change processes of the protein using a single pair of fluorophores attached to the specific sites within the protein structure [8]. FRET is a nonradiative dipole–dipole coupling process where the energy is transferred from an excited donor to an acceptor (Fig. 6.1) [10]. The fraction of energy transferred per donor excitation event is termed FRET efficiency (E), which depends on the donor–acceptor separation distance R (<10 nm) with an inverse sixth power relation:

$$E = \frac{1}{1 + \left(\frac{R}{R_0}\right)^6} \tag{6.1}$$

where R_0 is the characteristic distance.

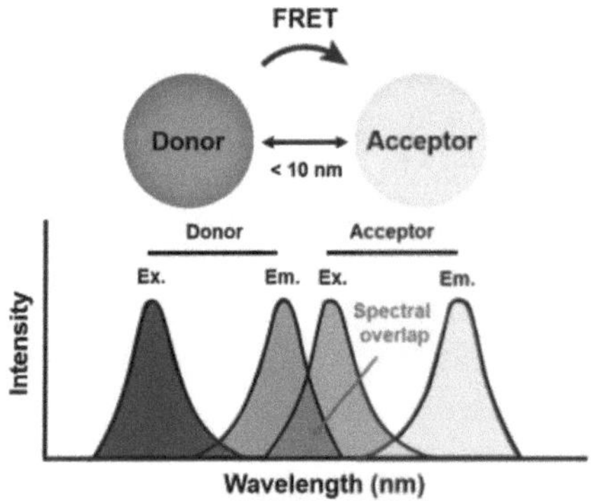

Figure 6.1 Specific requirements for a FRET measurement. The emission ($E_{m'}$) spectrum of a donor is overlapped with the excitation ($E_{x'}$) spectrum of an acceptor. FRET occurs when the distance between the donor and the acceptor falls within 10 nm. Reprinted from Ref. [10], Copyright 2017, with permission from Elsevier.

Compared with ensemble FRET, smFRET provides information on unsynchronized structural changes in a protein, enabling the detection of protein conformational dynamics [11–15]. In a typical analysis, smFRET includes the collection of photons from a single-molecule event, background subtraction, and calculation of the parameters such as FRET efficiency and photon count rates. The 1D or 2D FRET efficiency histograms are plotted to present an insight into the conformation and interaction dynamics of proteins. We refer the readers to the benchmark study for calculating these parameters and extracting the donor–acceptor distance from smFRET measurements [16]. For an smFRET experiment, a two-laser excitation scheme is recommended for addressing issues such as photophysical effect, low FRET efficiency, and requirement of accurately calculating the donor–acceptor distance [16].

6.2.2 FCS

FCS is a sensitive method that measures the intensity fluctuations as fluorescent molecules diffuse through a small detection volume (Fig. 6.2) [17]. The autocorrelation function of the intensity trace is calculated by the equation

$$G(\tau) = \frac{\delta F(t) \cdot \delta F(t+\tau)}{F(t)} \tag{6.2}$$

where G is the autocorrelation function, F is the intensity as a function of time (t), and τ is the correlation time. By fitting the autocorrelation curves to different mathematical models, one can obtain quantitative information on the average number and diffusion rate of the fluorescent molecules in a solution or a live-cell membrane. The limitation of FCS is that it fails to detect immobile or very slow molecules. An extension of FCS is fluorescence cross-correlation spectroscopy (FCCS), which measures the cross-correlation between two or more detection channels. FCCS provides unique advantages for studying protein–protein or ligand–receptor interactions. In addition, photon-counting histogram analysis can be used to analyze the size of protein complexes [18]. Interested readers can find detailed information on FCS and FCCS in the references [19].

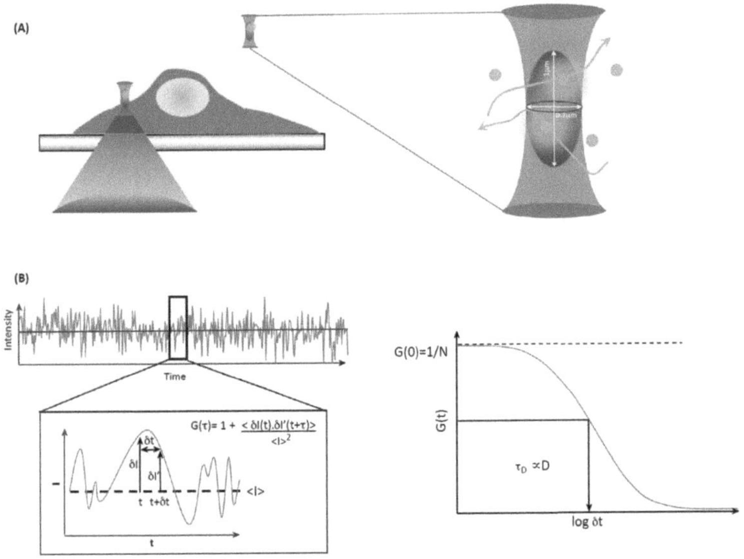

Figure 6.2 Principle of FCS. (A) Fluorescent species diffuse through the precisely positioned confocal detection volume. (B) Autocorrelation analysis of the resulting fluorescence fluctuation allows determination of dwell time (τ_D) and particle number (N). Reprinted from Ref. [17], Copyright 2018, with permission from Elsevier.

6.2.3 TIRFM

Epi-illumination is one of the simplest modes for SMF imaging but limited to a thin sample. For a thick sample such as a cell, total internal reflection (TIR) illumination is preferred as it uses the evanescent field to excite the sample (Fig. 6.3). Because the intensity of the evanescent field decreases exponentially with the distance from the interface, this method allows illuminating a thin layer of about 100–200 nm with little or no out-of-focus fluorescence, resulting in a high signal-to-noise ratio and thus facilitating single-molecule visualization. TIRFM is divided into two setups: prism-type and objective-type. The former uses a prism to direct a light beam on the interface, producing a "cleaner" evanescent-field fluorescence than the latter. However, the prism restricts the space above the objective, which is inaccessible to a thick sample. The objective-type setup uses an objective with high numerical aperture (>1.45) to achieve evanescent-field illumination and collects the fluorescence as well. This setup provides a free space above the objective, which enables real-time imaging of single molecules in a cell during culture as well as permits the combination of other techniques such as micromanipulators or scanning probe microscopes.

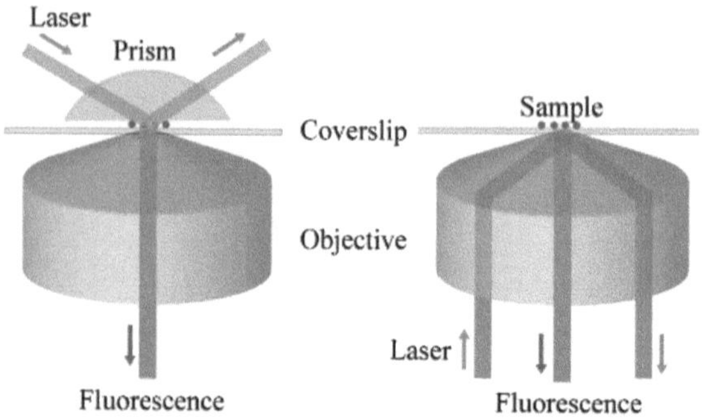

Figure 6.3 Prism- and objective-type TIRFM. Reprinted from Ref. [20], with permission from Springer Science + Business Media.

In addition, TIRFM has the advantage of visualizing hundreds of single molecules simultaneously and tracking their movements in two- and three-dimensional spaces for long durations [21]. By analyzing the movement trajectories, one can obtain the information on the dynamics and interactions of proteins. Because of the use of an electron-multiplying charge coupled device (EMCCD) as the detector, however, TIRFM suffers from low time-resolution down to 10–30 ms, which limits the observation to slow diffusion processes or conformational changes of proteins. For the detection of rapid dynamic processes, a confocal-type setup combined with FCS or smFRET is recommended. In this setup, a tightly focused laser beam defines a tiny excitation volume. When the concentration of fluorophore-labeled proteins is below ~1 nM, the volume is empty most of the time, and each molecule diffusing through it will be detected as a fluorescent "burst" [22]. The detection configuration can use one or more detectors, typically avalanche photodiodes (APD), that afford single-photon sensitivity and the temporal resolution down to the picoseconds scale.

6.2.4 SMLM

Conventional fluorescence microscopy works with low densities of target molecules because of the resolution of >~200 nm and >~500 nm for lateral and axial direction, respectively. The recent advent of SMLM such as photoactivatable localization microscopy (PALM), stochastic optical reconstruction microscopy (STORM), or direct STORM (dSTORM) has overcome this limitation by sequential activation and localization of individual fluorophores, enabling the observation of single molecules at high density [23]. The principle of STORM/PALM relies on the use of photoactivatable fluorescent proteins or photoswitchable dyes [24]. By modulating the activation and excitation lasers, the fluorophores are randomly switched between fluorescent ("on") and dark ("off") states. Single fluorophores are sequentially localized and reconstructed for a super-resolved image where the resolution is defined by the localization precision ranging from 5 to 50 nm. For dSTORM, only single laser is needed by using a fluorophore with spontaneous blinking behaviors [25, 26]. SMLM has become a popular tool for imaging single proteins with nanoscale spatial precision, especially for revealing protein complexes on a cell membrane [24, 27–29]. However, the current issues to be considered for SMLM are undercounting or overcounting of molecules [30]. In STORM/PALM, overcounting can be caused by the spontaneous blinking of fluorophores [31], while undercounting may occur due to incomplete labeling or prior photobleaching [32]. In recent years, various strategies have been developed to correct the counting errors for performing quantitative analysis of proteins or protein complexes on cell membranes [31, 33–37]. For example, Platzer et al. offered a methodology to determine, optimize, and quantitatively account for the blinking behavior of STORM/PALM-compatible fluorophores such as fluorescent proteins and organic fluorophores, which enables robust evaluation of molecular clustering based on the localization maps [38]. In addition, a localization-based imaging method called quantitative points accumulation in nanoscale topography (qPAINT) has been recently developed by transient binding of dye-labeled DNA to target DNA labeled with antibodies [39]. Instead of the stochastic switching, qPAINT can avoid undercounting and overcounting errors due to the photobleaching or blinking of a fluorophore. However, the accuracy of qPAINT relies on stoichiometric labeling of proteins, labeling efficiency, and the accessibility of DNA probe to the target DNA.

6.2.5 Deep-Learning-Assisted SMF Techniques

Deep learning (DL) is a subset of machine-learning algorithms, which has gained increasing attention in recent years as it can be used to improve the accuracy and speed of fluorescence imaging [40–42]. Saguy et al. developed a quantitative algorithm for fluorescent kinetics analysis (QAFKA) [43], which includes an automated localization algorithm that can extract emission features per localization cluster, and a deep neural-network-based estimator that reports the ratios of cluster types within the population. This method allows an accurate analysis of the monomer/dimer equilibrium of membrane receptors in single cells. Xu et al. designed a DL model containing convolutional and long-short-term memory deep-learning neural network (CLDNN) for counting photobleaching events, which was used to measure the stoichiometry of single-dye-labeled molecules in vitro and epidermal growth factor receptors (EGFR) in a cell [44]. Williamson et al. developed a supervised machine-learning approach for fast and accurate analysis of clusters from the SMLM data set, enabling further measurement of cluster area, shape, and point density. This approach was demonstrated by the kinase Csk and the adaptor PAG in primary human T-cell immunological synapses [45]. In addition, Li et al. developed a rapid and automatic SMF trace selector for improving the sensitivity and specificity of an assay for a DNA point mutation. The DL-based selector was used for the automated screening of smFRET data to identify traces for further analysis and achieved ~90% concordance with manual selection while requiring less processing time [46]. Yuan et al. proposed an unsupervised DL model to analyze SMF traces, which could be used to extract membrane protein interaction dynamics [47]. Recently, our group has proposed two DL-assisted SMF methods for unveiling the organization of membrane receptors [48, 49]. One is the DL convolutional neutral network (DLCNN) model, which can extract the features of fluorescent spots produced from single or multiple quantum dots (QDs). This model has been trained to reach an accuracy of >98% for precisely distinguishing the monomer and complex as well as for real-time tracking of protein interactions on live-cell membranes [48]. Another is named the deep-blinking fingerprint recognition (deep-BFR) model, which can recognize fluorescent blinking of carbon dots (CDs) (Fig. 6.4). This DL model integrates convolutional layers, long-short-term memory, and fully connected layers to extract time-dependent blinking features of CDs and has been trained to a high accuracy (~95%) for automatic classification of CD-labeled receptor organizations on the cell membrane [49].

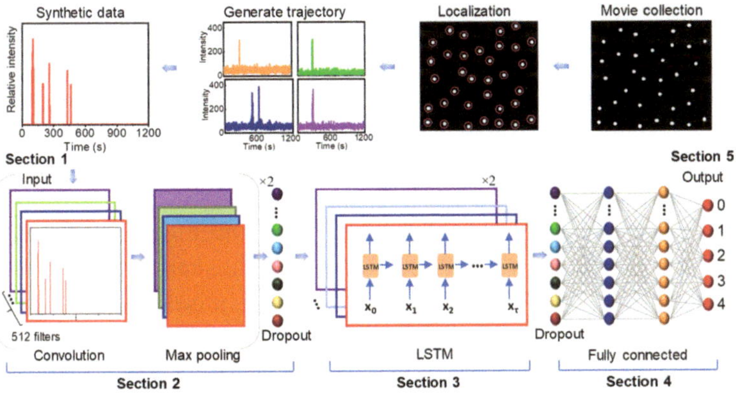

Figure 6.4 Architecture of the deep-BFR model for automatic analysis of protein organizations by classifying the blinking fingerprints of CDs from single-molecule imaging. Reprinted from Ref. [49], Copyright 2022, with permission from American Chemical Society.

6.3 Fluorophore Selection and Labeling

6.3.1 Fluorescent Proteins

Fluorescent proteins (FPs) can label a target protein with one-to-one ratio through gene fusion [50–52]. In the past decades, the FPs such as EGFP and mCherry and TagRFP have been widely used for SMF studies in a live cell [53–56]. The demerits of FPs are that they show low brightness, poor photostability, and large size, all of which are unfavorable for SMF tracking. To overcome these limits, many new FPs have been developed. For example, the engineered monomeric FPs, including mNeonGreen and mRuby2, improve the brightness of green–yellow or red FPs by 2~3 times [57, 58], while the red fluorescent TagRFP-T exhibits ~9-fold more photostability than TagRFP [59]. Particularly, the advent of SMLM such as PALM has prompted the development of FPs with specific photophysics such as photoactivatable, photoswitchable, and reversibly photoswitchable properties [60]. The photoactivatable FP remains non-fluorescent until activated by UV light to a fluorescent state. The photoswitchable FPs can switch fluorescent state from one to another after excitation by UV or blue–green light [61]. The reversibly photoswitchable FP can quickly photoswitch between nonfluorescent and fluorescent states in response to activation light [62]. These photo-controlled FPs provide a variety of choices for SMLM.

6.3.2 Organic Dyes

Organic dyes are preferred for the imaging of single proteins on cell membranes. The most common dyes include commercial series of Alexa Fluor, ATTO, and Cy families [63]. Dye molecules have smaller size and higher brightness than FPs, such that a single-dye-labeled molecule can be accurately localized down to a 1.5 nm precision by fitting the point spread function of the bright emission with a 2D Gaussian function [64]. Moreover, some organic dyes exhibit photoswitchable properties in aqueous solution containing thiols [65, 66]. The photoswitchable ability, combined with the high brightness and small size, makes them favorable for SMLM imaging. The disadvantage of organic dyes is that it is difficult to count single target proteins by using dye-labeled antibodies because each antibody molecule may bind to more than one dye [67, 68]. To guarantee that a single antibody has a single dye, nanobodies can be used [69], but this strategy is costly and time consuming. In addition, fluorescent antibodies may be used to label a target protein, but care must be taken to avoid overcounting and undercounting errors caused by nonspecific binding or incomplete labeling.

6.3.3 Nanoparticles

Fluorescent nanoparticles such as semiconductor QDs, CDs, and polymer dots (PDs) have been developed for SMF imaging in the recent decades because they have many more excellent properties than organic dyes and FPs [70–72]. Among them, QDs are one of the most famous fluorescent probes owing to their high brightness and strong resistance to photobleaching [73], which makes them an attractive candidate for SMF imaging and tracking [74, 75]. A limitation of QDs is the blinking behavior, that is, the alternation between bright and dark states [21, 76], which may disrupt the tracking processes. This drawback can be overcome by the development of non-blinking QDs [77]. Another potential limit of QDs is the large size (10–50 nm), which may affect the diffusion of QD-labeled membrane proteins. Although recent reports support the independence of lateral diffusion on the particle size, some reports show that large-sized QDs may sterically hinder the mobility of the labeled membrane receptors [78]. Whether QDs are small enough to probe membrane protein dynamics remains a debate. Except for QDs, we and other groups have found that CDs, a class of carbon-based nanoparticles with a smaller size than QDs, have burst-like blinking properties, which can be explored for dSTORM and super-resolution optical fluctuation imaging (SOFI) [70, 79]. We demonstrated the more superior performance of CDs for dSTORM than other fluorophores such as Cy3, Cy5, Alexa Fluor 647, and CdSe/ZnS QDs, enabling visualization of protein clusters on the cell membrane [80]. Using blinking CDs, we further developed a quantitative SMLM to count single proteins on the cell membrane at 10 nm localization precision and to discriminate the protein oligomers and clusters [71].

6.4 In Vitro Studies of Membrane Proteins

Membrane proteins can be removed from the natural lipid bilayer for in vitro analysis of protein conformation and interaction because of its complex and dynamic surrounding. Morrison et al. used TIRFM-based smFRET and NMR spectroscopy to measure in vitro conformational exchanges of EmrE, a multidrug transporter protein, between two states. They confirmed the antiparallel arrangement of monomers within an EmrE dimer [81]. smFRET has also been used to study the conformational changes in the prokaryotic neurotransmitter: Na^+ symporters (NSS) homologue LeuT on the substate. The authors revealed molecular details of the modulation of intracellular gating of LeuT by substrates and inhibitors, as well as by mutations that alter binding, transport, or both [11]. An approach that combines cell-derived nanovesicles, microfluidics, and SMF imaging has been developed to track single binding events at a cyclic nucleotide-gated TAX-4 ion channel. The results revealed the dynamics of nucleotide binding and a subsequent conformational change likely preceding pore opening [82]. Vafabakhsh et al. used smFRET to probe the activation mechanism of full-length mammalian group II metabotropic glutamate receptors (mGluRs) [15]. They demonstrated the interconversion of ligand-binding domains between three conformations, including the resting, activated, and a short-lived intermediate state. The transitions between these states could be induced by orthosteric agonists with efficacy determined by the occupancy of the active conformation. In addition, smFRET has been used to investigate conformational dynamics of β_2 adrenergic receptor (β_2AR) [83] and adenosine A2A receptor (A2AR) [84]. Gregorio et al. probed TM6 movements in β_2AR upon exposure to its ligands with different efficacies in the absence and presence of the G_s heterotrimer. They found that partial and full agonists differentially affect TM6 motions to regulate the rate at which GDP-bound β_2AR-G_s complexes are formed and the efficiency of nucleotide exchange leading to Gs activation [83]. Asher et al. also performed TIRF-based smFRET imaging of β-arrestin1 (βarr1) tethered to a surface interacting with GPCR kinase 2-phosphorylated receptor (pβ_2V_2R) free in solution and bound to ligands of varying efficacy (Fig. 6.5) [85]. They investigated the activation mechanism of βarr1 by examining the release of its C-terminal tail region upon the binding of phosphorylated receptors.

6.5 In Situ Studies of Membrane Proteins

6.5.1 Diffusion Dynamics

SMF has been widely used to characterize the diffusion dynamics of membrane proteins for investigating their functional mechanisms. Notelaers et al. elaborated on the diffusion pattern of α3-containing glycine receptor (GlyR) in the membrane of HEK 293 cells [86]. Different receptor variants were found to display distinct confinement probability level and residence time, which suggested changes in receptor diffusion patterns because of RNA splicing. Lippert et al. compared the diffusion coefficients of Wnt3A in Drosophila S2 cells and S2 receptor-plus (S2R+) cells influenced by Fz proteins (Fig. 6.6) [87]. They found that the diffusion coefficient of Wnt3A in the S2R+ cells expressing Frizzled proteins was significantly decreased. The results suggested the interaction of Wnt3A with live-cell membranes. Using QDs as a fluorescent label, Chung et al. performed the SMF tracking of epidermal growth factor receptors (EGFRs) on live-cell membranes. The analysis of dimerization dynamics of EGFRs revealed that the dimers after the addition of epidermal growth factor (EGF) had very slow diffusivity state that correlated with activation [88]. Calebiro et al. used the SMF method to monitor the diffusion of β_1ARs, β_2ARs, and $GABA_B$ receptors on live-cell membranes. They found that β_1ARs and β_2ARs diffused differently with $GABA_B$. β_1-/β_2ARs diffused freely, while $GABA_B$ was organized into arrays on the cell membrane by interactions with the actin cytoskeleton [89]. Moreover, SMF can be used to characterize the monomer–dimer equilibrium of G-protein-coupled receptors (GPCRs) on cell membranes [90]. Kasai et al. determined the dynamic equilibrium for the N-formyl peptide receptor (FPR) at 37°C [91]. Hern et al. directly determined the mobility, clustering, and dimerization kinetics of M1 muscarinic acetylcholine receptor with a resolution of ~30 ms and ~20 nm [92]. A class-A GPCR, dopamine D2 receptor (D2R), was also demonstrated to form transient dimers with a lifetime of 68 ms in its resting state [93]. In addition, Decker's group developed fluorescent ligands for the SMF tracking of opioid receptors (ORs) on live-cell membranes. They found short-life dimers between μ-ORs [94], but not between κ-ORs [95]. Recently, Briddon et al. developed a machine-learning method that classifies the mobility of membrane-bound receptor tyrosine kinase (MET) to three diffusion modes (immobile, slow, and fast). This method was used to discriminate between internalin B-treated and -untreated cells with an accuracy of >99% [96].

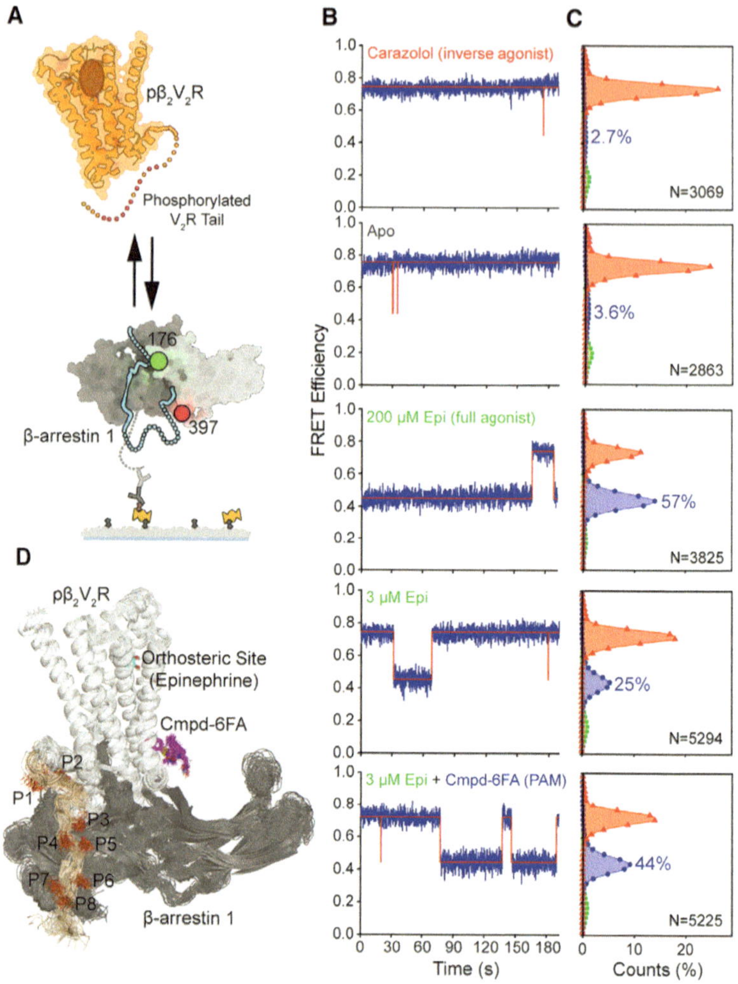

Figure 6.5 Activation of β-arrestin1 by phosphorylated, agonist-bound receptor. (A) Schematic of the experiment showing βarr1 tethered to an imaging surface interacting with $p\beta_2V_2R$ free in solution and bound to ligands of varying efficacy. (B and C) (B) Representative FRET (blue) and state assignment (red) traces and (C) population FRET histograms from smFRET imaging of βarr1 tail sensor conducted in the presence of 1 mM $p\beta_2V_2R$ and 10 mM carazolol, no ligands, 200 mM epi, 3 mM epi, or 3 mM epi with 10 mM Cmpd-6FA, respectively. (D) Representative MD frames of the simulated $p\beta_2V_2R$-βarr complex model bound to epinephrine (cyan) and Cmpd-6FA (magenta), showing the positions and dynamics of P1 to P8 in V_2Rpp (wheat color). Reprinted from Ref. [85], under a CC BY 4.0 license.

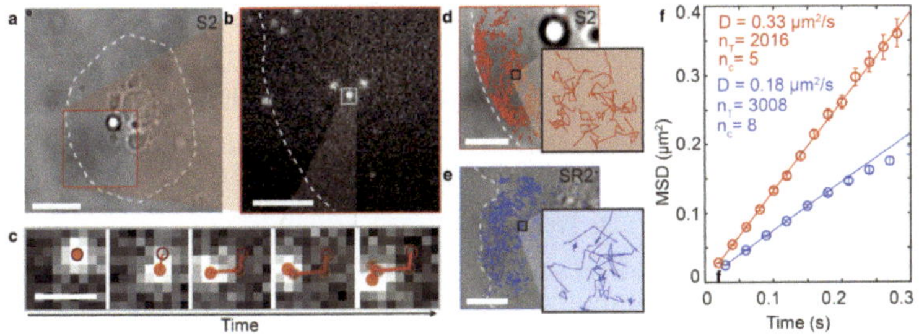

Figure 6.6 Wnt diffusion is slowed by the presence of the Wnt receptor in S2R+ cells. The fluorophore-labeled Wnt protein was imaged on S2 (red) and S2Rþ (blue) cells using highly inclined and laminated optical sheet microscopy ((a) shows a white-light image and (b) shows an SMF image). (c) A tracking algorithm was used to link fluorescent puncta (red circles, starting (dark red) and current (filled) positions in consecutive frames (scale bars, 1 μm), resulting in tracks with a localization precision of ~23 nm). (d and e) Overlay of tracks with white-light images (scale bars, 5 μm). (f) The ensemble diffusion coefficient was determined by fitting a linear function, which considers static and dynamic errors, to the mean-square displacement (MSD) values versus time. Shown are average MSD values for S2 (red) and S2R+ (blue) cells with mean ± SE, as well as the linear corresponding fit. Reprinted from Ref. [87], under a CC BY 4.0 license.

Furthermore, the diffusion dynamics of receptor and ligand/receptor–receptor complexes on live-cell membranes has been investigated by FCS [17, 97], smFRET [98], and SMLM [99–101]. Asher et al. used smFRET to track transmembrane proteins on live-cell membranes. They revealed agonist-induced structural dynamics within individual metabotropic glutamate receptor dimers. This method can provide evidence for receptor monomers, density-dependent dimers, and constitutive dimers in class A, B, and C receptors [98]. Gormal et al. used single-particle tracking PALM (spPALM) to unveil the diffusion dynamics of activated and inactivated endogenous conformers by the agonist treatment. The results showed that activated β_2AR (Nb80) was highly immobile and organized in nanoclusters, while the Gαs-GPCR complex has higher mobility with similar nanoclustering dynamics to that of Nb80 [101].

6.5.2 Conformational Dynamics

The conformational processes of a protein in a live cell can be dissected by SMF techniques [102, 103]. König et al. demonstrated the feasibility of confocal-type smFRET for monitoring in situ protein dynamics from milliseconds down to the nanosecond scale [104]. To date, smFRET has been applied to the conformational states and dynamics of the antibacterial peptide exporter McjD [105] and GPCRs such as β_2AR and metabotropic glutamate receptor 2 (mGluR2) [106, 107]. These results showed that the conformation of β_2AR in a live cell changed more rapidly than that of purified β_2AR in vitro. mGluR2 was found to interconvert between four conformational states and undergo activation via a conformational selection mechanism [107]. The intermediate states of mGluR2 serve as conformational checkpoints for activation and regulate allosteric modulation of signaling. Terry et al. also used smFRET to unveil β-arrestin dynamics. They found that β-arrestin was autoinhibited in basal state, and upon activation, the released β-arrestin tail showed at least two distinct conformations [108]. In addition, FCS and FCCS have been used to investigate the allosteric effects and binding kinetics of membrane receptors by using ligand residence time [109]. The ligand-dependent conformational dynamics and activation of receptors could be directly visualized. Wei et al. performed SMF studies on the conformational dynamics of human A2A adenosine receptor (A2AAR) [110, 111]. They observed ligand-dependent slow and reversible conformational exchange of A2AAR in the native-like environment. They further demonstrated the transition of agonist-bound A2AAR among three distinct conformational states and their occurrence in a specific order.

6.5.3 Signaling Transduction

SMF has been applied to investigate and reveal the signaling mechanism of membrane receptors because of the available information such as receptor localization, mobility patterns, and organization [112]. Sako et al. first used the SMF method to visualize the dimerization of EGFR and revealed their links to signaling transduction [113]. Since then, a great number of membrane proteins such as N-formyl peptide receptor (FPR), γ-aminobutyric acid receptor, and μ-OR have been investigated by SMF for exploring the role of protein organization in the signaling regulation [89–91, 114, 115]. SMF also permits direct visualization of the interactions between GPCRs and G proteins on a live-cell membrane. The results demonstrated that the receptor and G protein formed activity-dependent complexes with a lifetime of ~1 s in the basal state. The hot spots were also found on cell membrane and partially defined by the cytoskeleton and clathrin-coated pit. The results further showed the occurrence of GPCR signaling at these hot spots [116]. Luo et al. used SMF tracking, combined with the photobleaching step counting method, to probe the formation and change in GABAB dimers and tetramers, which offered new evidence for understanding the molecular basis for receptor aggregation and activation (Fig. 6.7) [117]. In recent years, our group used SMF imaging and the chemotaxis test to study the dimerization of EGFP-CXCR4 on cell membranes and demonstrated the correlation between the oligomeric status of CXCR4 and their signal transduction [118]. We also used the DL-assisted SMF method to automatically distinguish the monomer and complex of CXCR4 on live-cell membranes. The

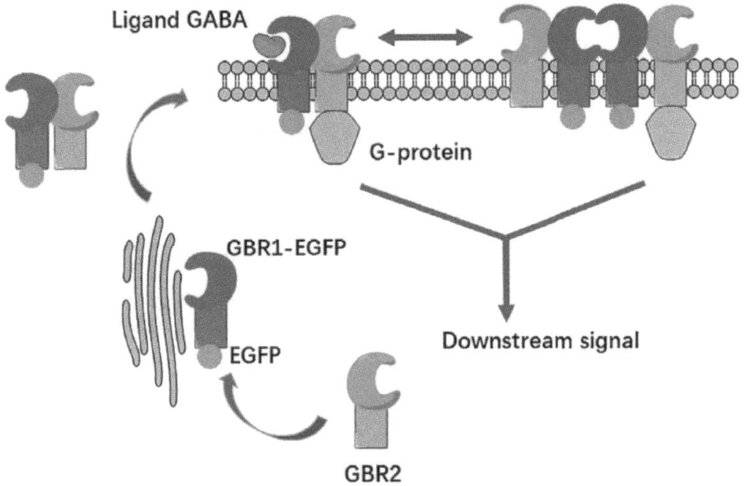

Figure 6.7 Schematic illustration of GABAB receptor activation and its aggregation at cell membrane. The GABAB receptor consists of two subunits: GBR1 and GBR2. GBR1 carries the endoplasmic reticulum (ER) retention signal (RSRR); thus, it is retained in the endoplasmic reticulum. When GBR1 binds with GBR2 to form a heterodimer, GBR2 blocks the RRSR of GBR1, and both could dock to the membrane. On the cell membrane, the heterodimer can also be associated to form a tetramer. After GABA ligand stimulation, heterodimers and tetramers bind to Gi/o proteins to activate the downstream signaling pathway. Either receptor dimer or tetramer only binds with 1 G protein. Reprinted from Ref. [117], under a CC BY 4.0 license.

results showed that the CXCR4 underwent a dynamic process to form a complex upon SDF-1α stimulation. We further found that the complex tends to be internalized at 2.5-fold higher rate into the cell interior than the monomer via the clathrin-dependent pathway [48].

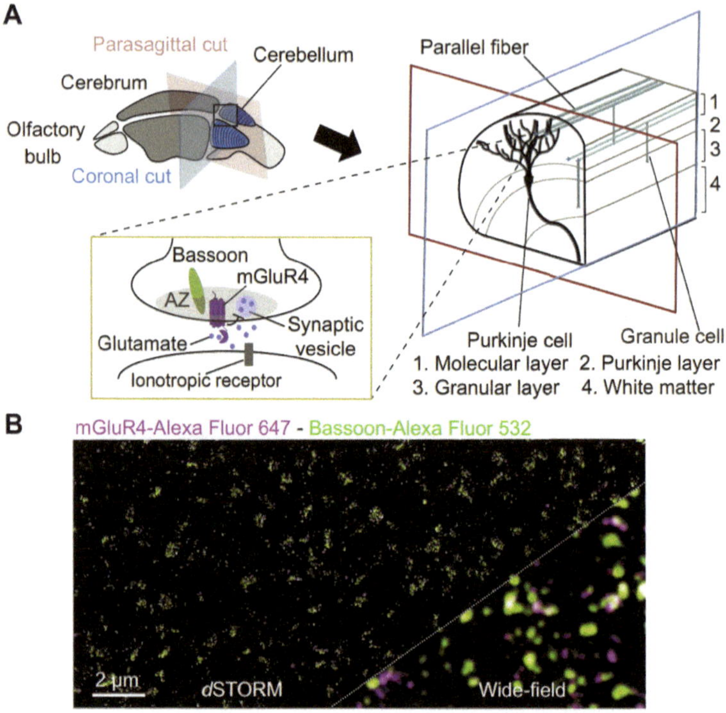

Figure 6.8 dSTORM imaging reveals the nanoscale organization of mGluR4 at parallel fiber AZs. (A) Schematic view of the organization of the mouse cerebellum, showing the ordered arrangement of the parallel fibers, which originate from granule cells and form dense synapses with the dendritic spines of Purkinje cells. mGluR4s located on the presynaptic membrane of parallel fiber synapses regulate synaptic transmission by inhibiting the release of glutamate (yellow box). Two different planes were used for cutting the cerebellum: coronal (blue; parallel to parallel fibers) and parasagittal (red; perpendicular to parallel fibers). (B) Two-color dSTORM imaging of mGluR4 (magenta) and bassoon (green). An image of a coronal section acquired in a region corresponding to the molecular layer of the cerebellum is shown. The corresponding wide-field fluorescence image is given for comparison. Reprinted from Ref. [119], under a CC BY 4.0 license.

In addition, SMLM is a powerful tool to investigate the signaling mechanism associated with the receptor organizations. Möller et al. used dSTORM to investigate the dynamic monomer–dimer equilibrium of μ-OR. They found that the formation of dimer was induced by the agonist DAMGO but not morphine in the process that correlates with β-arrestin2 binding to the receptors, which suggested a new level of GPCR regulation for downstream signals by an agonist-induced dimer [114]. Furthermore, dSTORM, combined with two-color imaging, has been used to uncover the nanoscale organization of mGluR4 at parallel fiber AZs in the mouse cerebellum (Fig. 6.8) [119], which implies the close association between mGluR4 and the secretory machinery in modulating synaptic transmission. Karathanasis et al. have developed quantitative SMLM to study the assembly of tumor necrosis factor receptor 1 (TNFR1) in native cellular settings [120]. They found that TNFR1 assembled into monomers and dimers in the basal states, but clustered into trimers and higher-order oligomers after the ligand (TNF) binding, which suggested ligand-independent TNFR1 dimerization in the κ-light-chain-enhancer of activated B cells signaling.

6.5.4 Membrane Organization

The membrane organization mediates protein–protein interactions for regulating specific receptor-mediated responses [121–126]. The two models, lipid-dependent microdomain and cytoskeleton-based meshwork, are commonly recognized to account for membrane organizations [127, 128]. The development of SMF techniques enables the direct observation of membrane organizations as well as depicting a more dynamic picture than previously thought. Fresnel et al. performed FCS measurements on various spatial scales for identifying two processes in a live cell [129]. The data showed that putative raft markers were dynamically compartmented into small microdomains that were sensitive to the levels of cholesterol and sphingomyelin, whereas actin-based cytoskeleton barriers accounted for the confinement of the transferrin receptor protein. Komura et al. used the SMF imaging method to demonstrate that gangliosides are dynamically exchanged between raft domains and the bulk domain on the cell membrane, suggesting that raft domains act as dynamic entities [130]. The recent combination of FCS and stimulated emission depletion (STED) microscopy has been developed to precisely monitor the diffusion of sphingolipids and GPI-anchored proteins in nanosized areas in live-cell membranes [131, 132]. The results support the formation of small, transient cholesterol-assisted lipid–protein complexes or nanodomains rather than stable liquid-ordered domains. In addition, SMF tracking and SMLM imaging of α-amino-3-hydroxy-5-methyl-4-isoxazolepropionic acid (AMPA) receptors in neurons have demonstrated that their localization and lateral mobility are governed by interactions with the membrane skeleton [133]. Similar findings have been observed for GPCRs. For example, metabotropic GABAB receptors have been shown to be largely immobile and align along actin fibers on the cell membrane [89]. This organization results from GABAB receptor interactions with the actin cytoskeleton. Moreover, SMF was also used to investigate the interactions between the somatostatin receptor type 2 (SSTR2), filamin A, and the actin cytoskeleton [134]. The results indicated the transient interaction between SSTR2 and filamin A, which occurred preferentially along actin fibers and restrained SSTR2 diffusion.

In recent years, we demonstrated the clustering of chemokine receptors and their links to lipid raft [71, 80]. By using CDs as a dSTORM probe, we have detected the clustering and distribution of CCR3 and CXCR4 on cell membranes. Quantitative SMLM suggested the dependence of CXCR4 clusters on lipid raft. Furthermore, we have used the DL-assisted blinking fingerprint recognition method to automatically classify the oligomeric status for CD-labeled CXCR4 on cancer cell membranes. We found that CXCR4 mainly existed as monomers and dimers under native conditions, which could be regulated by different agonists. Our results also verified that the disruption of actin cytoskeleton could promote the formation of CXCR4 oligomers (Fig. 6.9) [49].

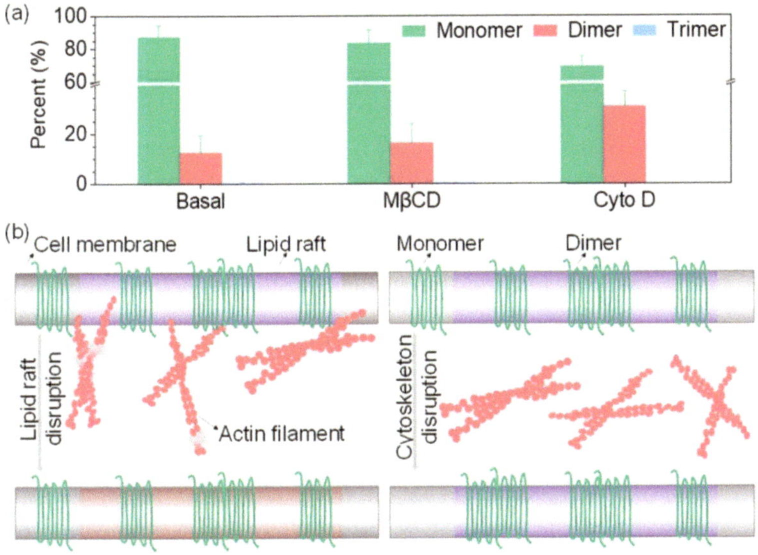

Figure 6.9 (a) Influence of cholesterol sequestration and actin microfilament disruption on CXCR4 organization on a HeLa cell membrane. (b) Schematic diagram illustrating the effect of cell membrane disruption on the receptor organization. Reprinted from Ref. [49], Copyright 2022, with permission from American Chemical Society.

6.6 Protocols for SMF Analysis of Membrane Proteins

6.6.1 Cell Culture

The CXCR4 serves as an example of how to perform SMF imaging and data analysis for membrane proteins. The CXCR4 or CXCR4-EGFP plasmid need to be first constructed. The cells were seeded into 8-well plate overnight with high Dulbecco's modified eagle medium (DMEM) containing fetal bovine serum (FBS, 10%), penicillin (100 μg/mL), streptomycin (100 μg/mL) at 37°C under 5% CO_2. The plasmid was transfected into cells using lipofectamine 2000 according to the manufacturer's instructions. After transfection, the cells were cultured for different times to control the protein expression level.

6.6.2 Fluorescence Labeling

For EGFP-labeled CXCR4, the cells expressing EGFP-CXCR4 are washed with PBS (pH 7.4), and then imaged in DMEM without phenol red using TIRFM. For QD-labeled CXCR4 tracking, the cells were washed with DMEM (1% FBS), and blocked by DMEM containing 20% FBS at 37°C for 0.5 h. Thereafter, 120 μL (2.5 μg/mL) anti-CXCR4 primary antibody was added for 1 h incubation at 37°C. After removing the antibody solution, 120 μL secondary antibody IgG-labeled QDs (emission peaked at 605 nm) were added for 0.5 h incubation followed by PBS washes. The ligand or inhibitor treatment experiments were performed by the addition of 120 μL SDF-1α (100 nM) or chlorpromazine hydrochloride (2.5 μM) for 20 min incubation. After removing the incubation solution, the phenol red- and serum-free medium was added for cell imaging.

For the SMLM imaging of CD-labeled CXCR4, the conjugates of CDs to antibody Fab fragments need to be first prepared via the 1-ethyl-3(3-dimethyl aminopropyl) carbodiimide (EDC)/N-hydroxysuccinimide (NHS) method. Briefly, 20 μL of anti-CXCR4 antibody (1 mg/mL), 1 mg of papain, 1 μL of cysteine (0.2 mol/L), and 1 μL of ethylene diamine tetraacetic acid solution (0.1 mol/L) were mixed into 100 μL of PBS solution and reacted for 2 h at 37°C. Subsequently, iodoacetamide (1 μL, 0.2 mol/L) was mixed and ice-bathed for 0.5 h to stop the enzymatic reaction. The resultant Fab fragments were isolated from the Fc fragments by using a protein L column. The purified Fab fragments were used to label the CDs. Shortly, the CDs (50 mg/mL) were activated by EDC (100 mg/mL) and NHS (100 mg/mL) in an MES buffer of pH 5.5 (10 mM) for half an hour at 25°C, followed by ultrafiltration (3 kDa MW cut off) to remove the MES buffer. The CDs were then added to the PBS buffer of pH 7.4 (10 mM) containing Fab fragments for the reaction of greater than 4 h. The solution was ultrafiltered with a 50 kDa MW cut-off membrane for obtaining Fab-conjugated CDs. Before use, the Fab-CDs were characterized by FCS to demonstrate a 1:1 labeling ratio of Fab to CDs. For cell labeling, the cells were rinsed with PBS and then fixed with paraformaldehyde (4%) for 15 min, followed by a 0.5 h incubation in the blocking solution containing 20% FBS for removing nonspecific binding sites. Thereafter, the cells were reacted with 5 μg/mL Fab-CDs at 4°C overnight. The CD-labeled cells were rinsed thoroughly to remove unbound CDs for imaging.

6.6.3 Fluorescence Microscopy Setup

SMF imaging was conducted with a Nikon objective-type TIRFM. EGFP and QDs were excited by 488 nm laser in conjunction with 505 nm dichroic mirror (DM) and long-pass (LP) filter (520 nm for EGFP and 590 nm for QDs). CDs were excited by 532 nm laser in conjunction with 575 nm DM and 590 nm LP filter. The image was taken by an EMCCD camera (Andor iXon 897, 16 μm/pixel) using a 100× objective combined with 1.5× magnification changer lens, and the effective dimension for each pixel was 106 nm. The electron multiplying (EM) gain of EMCCD was set to 300. During imaging, the TIR angle was adjusted to accurately locate the cell membrane. In addition, different excitation intensity and exposure time should be optimized for achieving a high signal-to-noise ratio as well as avoiding rapid photobleaching.

6.6.4 Data Processing and Analysis

For DL-assisted QD-labeled CXCR4 tracking, the image stack (50 ms/frame) was recorded with TIRFM. Fluorescent spots in the images were localized by an ImageJ plugin, ThunderSTROM, to obtain information, including spot coordinates, intensity, and goodness of fit (χ^2) [135]. A series of regions (9 × 9 pixels^2) centered at each spot were cropped and inputted into the trained DL model for identification as a single molecule or complex. For the diffusion trajectories of CXCR4, the mean squared displacement (MSD) was computed by the equation

$$\mathrm{MSD}(t) = \mathrm{MSD}(n \cdot fr_{\mathrm{time}}) = \frac{1}{N}\sum_{i=1}^{N}\left[(x_{i+n} - x_i)^2 + (y_{i+n} - y_i)^2\right] \quad (6.3)$$

where N is the number of steps, n is the step size in frames, fr_{time} is the time between two consecutive frames, and x and y describe the particle position at the frame. The diffusion coefficients (D) were calculated by fitting data with the equation [89]

$$\mathrm{MSD} = 4Dt \quad (6.4)$$

For the SMLM analysis of CD-labeled CXCR4, the image stack (500 ms/frame) was recorded in 20 min. The coordinate list of single particles was obtained with ThunderSTORM, at default parameter settings except for using 3 × standard deviations as threshold. The raw coordinates were grouped into the average position of single particles using the algorithm developed in our previous work [71]. The distribution pattern of CXCR4 was analyzed by DBSCAN and Ripley's K function. DBSCAN was performed with the DBSCAN function in MATLAB2020a. A minimum number of molecules (minPts) per cluster and search radius (ε) were specified. Ripley's K-function was calculated as

$$K(r) = \frac{A}{N^2} \sum_{i=1}^{N} \sum_{j=1, j \neq i}^{N} \delta_{ij} \quad (6.5)$$

where A is the region area, N is the molecule number, r is the spatial radius for the K-function calculation, and δ_{ij} is the distance between the ith and jth molecules. Whenever δ_{ij} is less than r, the value will be 1; otherwise, $\delta_{ij} = 0$. The linear transformation of $K(r)$ was used to interpret the spatial randomness

$$L(r) - r = \sqrt{\frac{K(r)}{\pi}} - r \quad (6.6)$$

The amplitude of $L(r) - r$ would be 0 for particles with random distribution, and positive for clustering particles.

Acknowledgments

This work was supported by the National Natural Science Foundation of China (No. 21874154 and 22177133).

References

1. Cournia, Z., Allen, T. W., Andricioaei, I., Antonny, B., Baum, D., Brannigan, G., Buchete, N. V., Deckman, J. T., Delemotte, L., Del Val, C., Friedman, R., Gkeka, P., Hege, H. C., Henin, J., Kasimova, M. A., Kolocouris, A., Klein, M. L., Khalid, S., Lemieux, M. J., Lindow, N., Roy, M., Selent, J., Tarek, M., Tofoleanu, F., Vanni, S., Urban, S., Wales, D. J., Smith, J. C., and Bondar, A. N. (2015). Membrane protein structure, function, and dynamics: A perspective from experiments and theory, *J. Membr. Biol.*, **248**, pp. 611–640.

2. Harriss, L. M. and Wallace, M. I. (2009) *Single Molecule Biology,* eds. Knight, A. E., Chapter 9 "Single molecule fluorescence in membrane biology," (Academic Press, San Diego) pp. 253–288.

3. Moerner, W. E. and Kador, L. (1989). Optical detection and spectroscopy of single molecules in a solid, *Phys. Rev. Lett.*, **62**, pp. 2535–2538.

4. Funatsu, T., Harada, Y., Tokunaga, M., Saito, K., and Yanagida, T. (1995). Imaging of single fluorescent molecules and individual ATP turnovers by single myosin molecules in aqueous solution, *Nature*, **374**, pp. 555–559.

5. Lerner, E., Cordes, T., Ingargiola, A., Alhadid, Y., Chung, S., Michalet, X., and Weiss, S. (2018). Toward dynamic structural biology: Two decades of single-molecule forster resonance energy transfer, *Science*, **359**, pp. eaan1133.

6. Clegg, R. M. (2006) *Reviews in Fluorescence 2006,* eds. Geddes, C. D. and Lakowicz, J. R., Chapter 1 "The history of FRET," (Springer, Boston) pp. 1–45.

7. Weiss, S. (2000). Measuring conformational dynamics of biomolecules by single molecule fluorescence spectroscopy, *Nat. Struct. Biol.*, **7**, pp. 724–729.

8. Roy, R., Hohng, S., and Ha, T. (2008). A practical guide to single-molecule FRET, *Nat. Methods*, **5**, pp. 507–516.

9. Hite, R. K. and MacKinnon, R. (2017). Structural titration of Slo2.2, a Na^+-dependent K^+ channel, *Cell*, **168**, pp. 390–399.

10. Broussard, J. A. and Green, K. J. (2017). Research techniques made simple: Methodology and applications of forster resonance energy transfer (FRET) microscopy, *J. Invest. Dermatol.*, **137**, pp. e185–e191.

11. Zhao, Y., Terry, D., Shi, L., Weinstein, H., Blanchard, S.C., and Javitch, J. A. (2010). Single-molecule dynamics of gating in a neurotransmitter transporter homologue, *Nature*, **465**, pp. 188–193.

12. Akyuz, N., Altman, R. B., Blanchard, S. C., and Boudker, O. (2013). Transport dynamics in a glutamate transporter homologue, *Nature*, **502**, pp. 114–118.

13. Erkens, G. B., Hanelt, I., Goudsmits, J. M., Slotboom, D. J., and van Oijen, A. M. (2013). Unsynchronised subunit motion in single trimeric sodium-coupled aspartate transporters, *Nature*, **502**, pp. 119–123.

14. Wang, Y., Liu, Y., Deberg, H. A., Nomura, T., Hoffman, M. T., Rohde, P. R., Schulten, K., Martinac, B., and Selvin, P. R. (2014). Single molecule FRET reveals pore size and opening mechanism of a mechano-sensitive ion channel, *Elife*, **3**, pp. e01834.

15. Vafabakhsh, R., Levitz, J., and Isacoff, E. Y. (2015). Conformational dynamics of a class C G-protein-coupled receptor, *Nature*, **524**, pp. 497–501.

16. Hellenkamp, B., Schmid, S., Doroshenko, O., Opanasyuk, O., Kuhnemuth, R., Rezaei Adariani, S., Ambrose, B., Aznauryan, M., Barth, A., Birkedal, V., Bowen, M. E., Chen, H., Cordes, T., Eilert, T., Fijen, C., Gebhardt, C., Gotz, M., Gouridis, G., Gratton, E., Ha, T., Hao, P., Hanke, C. A., Hartmann, A., Hendrix, J., Hildebrandt, L. L., Hirschfeld, V., Hohlbein, J., Hua, B., Hubner, C. G., Kallis, E., Kapanidis, A. N., Kim, J. Y., Krainer, G., Lamb, D. C., Lee, N. K., Lemke, E. A., Levesque, B., Levitus, M., McCann, J. J., Naredi-Rainer, N., Nettels, D., Ngo, T., Qiu, R., Robb, N. C., Rocker, C., Sanabria, H., Schlierf, M., Schroder, T., Schuler, B., Seidel, H., Streit, L., Thurn, J., Tinnefeld, P., Tyagi, S., Vandenberk, N., Vera, A. M., Weninger, K. R., Wunsch, B., Yanez-Orozco, I. S., Michaelis, J., Seidel, C. A. M., Craggs, T. D., and Hugel, T. (2018). Precision and accuracy of single-molecule FRET measurements-a multi-laboratory benchmark study, *Nat. Methods*, **15**, pp. 669–676.

17. Briddon, S. J., Kilpatrick, L. E., and Hill, S. J. (2018). Studying GPCR pharmacology in membrane microdomains: Fluorescence correlation spectroscopy comes of age, *Trends Pharmacol. Sci.*, **39**, pp. 158–174.

18. Kitamura, A. and Kinjo, M. (2018). State-of-the-art fluorescence fluctuation-based spectroscopic techniques for the study of protein aggregation, *Int. J. Mol. Sci.*, **19**, pp. 964.

19. Krieger, J. W., Singh, A. P., Bag, N., Garbe, C. S., Saunders, T. E., Langowski, J., and Wohland, T. (2015). Imaging fluorescence (cross-) correlation spectroscopy in live cells and organisms, *Nat. Protoc.*, **10**, pp. 1948–1974.

20. He, H., Wang, X., and Huang, F. (2018) *Membrane Biophysics: New Insights and Methods*, eds. Wang, H. and Li, G., Chapter 6 "Analysis and applications of single-molecule fluorescence in live cell membranes," (Springer, Singapore) pp. 147–173.

21. Manzo, C. and Garcia-Parajo, M. F. (2015). A review of progress in single particle tracking: From methods to biophysical insights, *Rep. Prog. Phys*, **78**, pp. 124601.

22. Nie, S., Chiu, D. T., and Zare, R. N. (1994). Probing individual molecules with confocal fluorescence microscopy, *Science*, **266**, pp. 1018–1021.

23. Schermelleh, L., Ferrand, A., Huser, T., Eggeling, C., Sauer, M., Biehlmaier, O., and Drummen, G. P. C. (2019). Super-resolution microscopy demystified, *Nat. Cell Biol.*, **21**, pp. 72–84.

24. Betzig, E., Patterson, G. H., Sougrat, R., Lindwasser, O. W., Olenych, S., Bonifacino, J. S., Davidson, M. W., Lippincott-Schwartz, J., and Hess, H. F. (2006). Imaging intracellular fluorescent proteins at nanometer resolution, *Science*, **313**, pp. 1642–1645.

25. Klein, T., Loschberger, A., Proppert, S., Wolter, S., van de Linde, S., and Sauer, M. (2011). Live-cell dSTORM with SNAP-tag fusion proteins, *Nat. Methods*, **8**, pp. 7–9.

26. Sauer, M. and Heilemann, M. (2017). Single-molecule localization microscopy in eukaryotes, *Chem. Rev.*, **117**, pp. 7478–7509.

27. Bates, M., Huang, B., Dempsey, G. T., and Zhuang, X. (2007). Multicolor super-resolution imaging with photo-switchable fluorescent probes, *Science*, **317**, pp. 1749–1753.

28. Heilemann, M., van de Linde, S., Schuttpelz, M., Kasper, R., Seefeldt, B., Mukherjee, A., Tinnefeld, P., and Sauer, M. (2008). Subdiffraction-resolution fluorescence imaging with conventional fluorescent probes, *Angew. Chem. Int. Ed.*, **47**, pp. 6172–6176.

29. Furstenberg, A. and Heilemann, M. (2013). Single-molecule localization microscopy-near-molecular spatial resolution in light microscopy with photoswitchable fluorophores, *Phys. Chem. Chem. Phys.*, **15**, pp. 14919–14930.

30. Annibale, P., Vanni, S., Scarselli, M., Rothlisberger, U., Radenovic, A. (2011). Identification of clustering artifacts in photoactivated localization microscopy, *Nat. Methods*, **8**, pp. 527–528.

31. Annibale, P., Vanni, S., Scarselli, M., Rothlisberger, U., and Radenovic, A. (2011). Quantitative photo activated localization microscopy: Unraveling the effects of photoblinking, *PLoS ONE*, **6**, pp. e22678.

32. Sugiyama, Y., Kawabata, I., Sobue, K., and Okabe, S. (2005). Determination of absolute protein numbers in single synapses by a GFP-based calibration technique, *Nat. Methods*, **2**, pp. 677–684.

33. Lee, S. H., Shin, J. Y., Lee, A., and Bustamante, C. (2012). Counting single photoactivatable fluorescent molecules by photoactivated localization microscopy (PALM), *Proc. Natl. Acad. Sci. U.S.A.*, **109**, pp. 17436–17441.

34. Hummer, G., Fricke, F., and Heilemann, M. (2016). Model-independent counting of molecules in single-molecule localization microscopy, *Mol. Biol. Cell*, **27**, pp. 3637–3644.

35. Puchner, E. M., Walter, J. M., Kasper, R., Huang, B., and Lim, W. A. (2013). Counting molecules in single organelles with superresolution microscopy allows tracking of the endosome maturation trajectory, *Proc. Natl. Acad. Sci. U.S.A.*, **110**, pp. 16015–16020.

36. Nicovich, P. R., Owen, D. M., and Gaus, K. (2017). Turning single-molecule localization microscopy into a quantitative bioanalytical tool, *Nat. Protoc.*, **12**, pp. 453–460.

37. Sengupta, P., Jovanovic-Talisman, T., Skoko, D., Renz, M., Veatch, S. L., and Lippincott-Schwartz, J. (2011). Probing protein heterogeneity in the plasma membrane using PALM and pair correlation analysis, *Nat. Methods*, **8**, pp. 969–975.

38. Platzer, R., Rossboth, B. K., Schneider, M. C., Sevcsik, E., Baumgart, F., Stockinger, H., Schutz, G. J., Huppa, J. B., and Brameshuber, M. (2020). Unscrambling fluorophore blinking for comprehensive cluster detection via photoactivated localization microscopy, *Nat. Commun.*, **11**, pp. 4993.

39. Jungmann, R., Avendaño, M. S., Dai, M., Woehrstein, J. B., Agasti, S. S., Feiger, Z., Rodal, A., and Yin, P. (2016). Quantitative super-resolution imaging with qPAINT, *Nat. Methods*, **13**, pp. 439–442.

40. Möckl, L., Roy, A. R., and Moerner, W. (2020). Deep learning in single-molecule microscopy: Fundamentals, caveats, and recent developments, *Biomed. Opt. Express*, **11**, pp. 1633–1661.

41. Nehme, E., Weiss, L. E., Michaeli, T., and Shechtman, Y. (2018). Deep-STORM: Super-resolution single-molecule microscopy by deep learning, *Optica*, **5**, pp. 458–464.

42. Thomsen, J., Sletfjerding, M. B., Stella, S., Paul, B., Jensen, S. B., Malle, M. G., Montoya, G., Petersen, T. C., and Hatzakis, N. S. (2020). DeepFRET: Rapid and automated single molecule FRET data classification using deep learning, *bioRxiv*.

43. Saguy, A., Baldering, T. N., Weiss, L. E., Nehme, E., Karathanasis, C., Dietz, M. S., Heilemann, M., and Shechtman, Y. (2021). Automated analysis of fluorescence kinetics in single-molecule localization microscopy data reveals protein stoichiometry, *J. Phys. Chem. B*, **125**, pp. 5716–5721.

44. Xu, J., Qin, G., Luo, F., Wang, L., Zhao, R., Li, N., Yuan, J., and Fang, X. (2019). Automated stoichiometry analysis of single-molecule fluorescence imaging traces via deep learning, *J. Am. Chem. Soc*, **141**, pp. 6976–6985.

45. Williamson, D. J., Burn, G. L., Simoncelli, S., Griffié, J., Peters, R., Davis, D. M., and Owen, D. M. (2020). Machine learning for cluster analysis of localization microscopy data, *Nat. Commun.*, **11**, pp. 1–10.

46. Li, J., Zhang, L., Johnson-Buck, A., and Walter, N. G. (2020). Automatic classification and segmentation of single-molecule fluorescence time traces with deep learning, *Nat. Commun.*, **11**, pp. 1–11.

47. Yuan, J., Zhao, R., Xu, J., Cheng, M., Qin, Z., Kou, X., and Fang, X. (2020). Analyzing protein dynamics from fluorescence intensity traces using unsupervised deep learning network, *Commun. Biol.*, **3**, pp. 1–10.

48. Wang, Q., He, H., Zhang, Q., Feng, Z., Li, J., Chen, X., Liu, L., Wang, X., Ge, B., and Yu, D. (2021). Deep-learning-assisted single-molecule tracking on a live cell membrane, *Anal. Chem.*, **93**, pp. 8810–8816.

49. Wang, Q., Zhang, Q., He, H., Feng, Z., Mao, J., Hu, X., Wei, X., Bi, S., Qin, G., and Wang, X. (2022). Carbon dot blinking fingerprint uncovers native membrane receptor organizations via deep learning, *Anal. Chem.*, **94**, pp. 3914–3921.

50. Rollins, G. C., Shin, J. Y., Bustamante, C., and Presse, S. (2015). Stochastic approach to the molecular counting problem in superresolution microscopy, *Proc. Natl. Acad. Sci. U.S.A.*, **112**, pp. E110–E118.

51. Harms, G. S., Cognet, L., Lommerse, P. H., Blab, G. A., and Schmidt, T. (2001). Autofluorescent proteins in single-molecule research: Applications to live cell imaging microscopy, *Biophys. J.*, **80**, pp. 2396–2408.

52. Iino, R., Koyama, I., and Kusumi, A. (2001). Single molecule imaging of green fluorescent proteins in living cells: E-cadherin forms oligomers on the free cell surface, *Biophys. J.*, **80**, pp. 2667–2677.

53. Murakoshi, H., Iino, R., Kobayashi, T., Fujiwara, T., Ohshima, C., Yoshimura, A., and Kusumi, A. (2004). Single-molecule imaging analysis of Ras activation in living cells, *Proc. Natl. Acad. Sci. U.S.A.*, **101**, pp. 7317–7322.

54. Ulbrich, M. H. and Isacoff, E. Y. (2007). Subunit counting in membrane-bound proteins, *Nat. Methods*, **4**, pp. 319–321.

55. Elf, J., Li, G.W., and Xie, X. S. (2007). Probing transcription factor dynamics at the single-molecule level in a living cell, *Science*, **316**, pp. 1191–1194.

56. Zhang, W., Jiang, Y., Wang, Q., Ma, X., Xiao, Z., Zuo, W., Fang, X., and Chen, Y. G. (2009). Single-molecule imaging reveals transforming growth factor-beta-induced type II receptor dimerization, *Proc. Natl. Acad. Sci. U.S.A.*, **106**, pp. 15679–15683.

57. Shaner, N. C., Lambert, G. G., Chammas, A., Ni, Y., Cranfill, P. J., Baird, M. A., Sell, B. R., Allen, J. R., Day, R. N., Israelsson, M., Davidson, M. W., and Wang, J. (2013). A bright monomeric green fluorescent protein derived from branchiostoma lanceolatum, *Nat. Methods*, **10**, pp. 407–409.

58. Lam, A. J., St-Pierre, F., Gong, Y., Marshall, J. D., Cranfill, P. J., Baird, M. A., McKeown, M. R., Wiedenmann, J., Davidson, M. W., Schnitzer, M. J.,

Tsien, R. Y., and Lin, M. Z. (2012). Improving FRET dynamic range with bright green and red fluorescent proteins, *Nat. Methods*, **9**, pp. 1005–1012.

59. Shaner, N. C., Lin, M. Z., McKeown, M. R., Steinbach, P. A., Hazelwood, K. L., Davidson, M. W., and Tsien, R. Y. (2008). Improving the photostability of bright monomeric orange and red fluorescent proteins, *Nat. Methods*, **5**, pp. 545–551.

60. Shcherbakova, D. M., Sengupta, P., Lippincott-Schwartz, J., and Verkhusha, V. V. (2014). Photocontrollable fluorescent proteins for superresolution imaging, *Annu Rev Biophys*, **43**, pp. 303–329.

61. Zhang, M., Chang, H., Zhang, Y., Yu, J., Wu, L., Ji, W., Chen, J., Liu, B., Lu, J., Liu, Y., Zhang, J., Xu, P., and Xu, T. (2012). Rational design of true monomeric and bright photoactivatable fluorescent proteins, *Nat. Methods*, **9**, pp. 727–729.

62. Li, N., Zhao, R., Sun, Y., Ye, Z., He, K., and Fang, X. (2017). Single-molecule imaging and tracking of molecular dynamics in living cells, *Natl. Sci. Rev.*, **4**, pp. 739–760.

63. Liu, Z., Lavis, L.D., and Betzig, E. (2015). Imaging live-cell dynamics and structure at the single-molecule level, *Mol. Cell*, **58**, pp. 644–659.

64. Yildiz, A., Forkey, J. N., McKinney, S. A., Ha, T., Goldman, Y. E., and Selvin, P. R. (2003). Myosin V walks hand-over-hand: Single fluorophore imaging with 1.5-nm localization, *Science*, **300**, pp. 2061–2065.

65. van de Linde, S., Loschberger, A., Klein, T., Heidbreder, M., Wolter, S., Heilemann, M., and Sauer, M. (2011). Direct stochastic optical reconstruction microscopy with standard fluorescent probes, *Nat. Protoc.*, **6**, pp. 991–1009.

66. Dempsey, G. T., Vaughan, J. C., Chen, K. H., Bates, M., and Zhuang, X. W. (2011). Evaluation of fluorophores for optimal performance in localization-based super-resolution imaging, *Nat. Methods*, **8**, pp. 1027–1036.

67. Panchuk-Voloshina, N., Haugland, R. P., Bishop-Stewart, J., Bhalgat, M. K., Millard, P. J., Mao, F., Leung, W.-Y., and Haugland, R. P. (1999). Alexa dyes, a series of new fluorescent dyes that yield exceptionally bright, photostable conjugates, *J. Histochem. Cytochem.*, **47**, pp. 1179–1188.

68. Ehmann, N., van de Linde, S., Alon, A., Ljaschenko, D., Keung, X. Z., Holm, T., Rings, A., DiAntonio, A., Hallermann, S., Ashery, U., Heckmann, M., Sauer, M., and Kittel, R. J. (2014). Quantitative super-resolution imaging of bruchpilot distinguishes active zone states, *Nat. Commun.*, **5**, pp. 4650.

69. Ries, J., Kaplan, C., Platonova, E., Eghlidi, H., and Ewers, H. (2012). A simple, versatile method for GFP-based super-resolution microscopy via nanobodies, *Nat. Methods*, **9**, pp. 582–584.

70. Jin, D., Xi, P., Wang, B., Zhang, L., Enderlein, J., and van Oijen, A. M. (2018). Nanoparticles for super-resolution microscopy and single-molecule tracking, *Nat. Methods*, **15**, pp. 415–423.

71. He, H., Liu, L., Chen, X., Wang, Q., Wang, X., Nau, W. M., and Huang, F. (2021). Carbon dot blinking enables accurate molecular counting at nanoscale resolution, *Anal. Chem.*, **93**, pp. 3968–3975.

72. Leménager, G., De Luca, E., Sun, Y.-P., and Pompa, P. P. (2014). Super-resolution fluorescence imaging of biocompatible carbon dots, *Nanoscale*, **6**, pp. 8617–8623.

73. Michalet, X., Pinaud, F. F., Bentolila, L. A., Tsay, J. M., Doose, S., Li, J. J., Sundaresan, G., Wu, A. M., Gambhir, S. S., and Weiss, S. (2005). Quantum dots for live cells, in vivo imaging, and diagnostics, *Science*, **307**, pp. 538–544.

74. Bruchez, M., Jr., Moronne, M., Gin, P., Weiss, S., and Alivisatos, A. P. (1998). Semiconductor nanocrystals as fluorescent biological labels, *Science*, **281**, pp. 2013–2016.

75. Chan, W. C. and Nie, S. (1998). Quantum dot bioconjugates for ultrasensitive nonisotopic detection, *Science*, **281**, pp. 2016–2018.

76. Nirmal, M., Dabbousi, B. O., Bawendi, M. G., Macklin, J., Trautman, J., Harris, T., and Brus, L. E. (1996). Fluorescence intermittency in single cadmium selenide nanocrystals, *Nature*, **383**, pp. 802–804.

77. He, H., Qian, H., Dong, C., Wang, K., and Ren, J. (2006). Single nonblinking cdte quantum dots synthesized in aqueous thiopropionic acid, *Angew. Chem., Int. Ed.*, **45**, pp. 7588–7591.

78. Abraham, L., Lu, H. Y., Falcão, R. C., Scurll, J., Jou, T., Irwin, B., Tafteh, R., Gold, M. R., and Coombs, D. (2017). Limitations of Qdot labelling compared to directly-conjugated probes for single particle tracking of B cell receptor mobility, *Sci. Rep.*, **7**, pp. 1–13.

79. He, H., Chen, X., Feng, Z., Liu, L., Wang, Q., and Bi, S. (2021). Nanoscopic imaging of nucleolar stress enabled by protein-mimicking carbon dots, *Nano Lett.*, **21**, pp. 5689–5696.

80. He, H., Liu, X., Li, S., Wang, X., Wang, Q., Li, J., Wang, J., Ren, H., Ge, B., Wang, S., Zhang, X., and Huang, F. (2017). High-density super-resolution localization imaging with blinking carbon dots, *Anal. Chem.*, **89**, pp. 11831–11838.

81. Morrison, E. A., DeKoster, G. T., Dutta, S., Vafabakhsh, R., Clarkson, M. W., Bahl, A., Kern, D., Ha, T., and Henzler-Wildman, K. A. (2012). Antiparallel EmrE exports drugs by exchanging between asymmetric structures, *Nature*, **481**, pp. 45–50.

82. Patel, V. R., Salinas, A. M., Qi, D., Gupta, S., Sidote, D. J., and Goldschen-Ohm, M. P. (2021). Single-molecule imaging with cell-derived nanovesicles reveals early binding dynamics at a cyclic nucleotide-gated ion channel, *Nat. Commun.*, **12**, pp. 1–13.

83. Gregorio, G. G., Masureel, M., Hilger, D., Terry, D. S., Juette, M., Zhao, H., Zhou, Z., Perez-Aguilar, J. M., Hauge, M., Mathiasen, S., Javitch, J. A., Weinstein, H., Kobilka, B. K., and Blanchard, S. C. (2017). Single-molecule analysis of ligand efficacy in β2AR–G-protein activation, *Nature*, **547**, pp. 68–73.

84. Fernandes, D. D., Neale, C., Gomes, G.-N. W., Li, Y., Malik, A., Pandey, A., Orazietti, A. P., Wang, X., Ye, L., Scott Prosser, R., and Gradinaru, C. C. (2021). Ligand modulation of the conformational dynamics of the A2A adenosine receptor revealed by single-molecule fluorescence, *Sci. Rep.*, **11**, pp. 5910.

85. Asher, W. B., Terry, D. S., Gregorio, G. G. A., Kahsai, A. W., Borgia, A., Xie, B., Modak, A., Zhu, Y., Jang, W., Govindaraju, A., Huang, L.-Y., Inoue, A., Lambert, N. A., Gurevich, V. V., Shi, L., Lefkowitz, R. J., Blanchard, S. C., and Javitch, J. A. (2022). GPCR-mediated β-arrestin activation deconvoluted with single-molecule precision, *Cell*, **185**, pp. 1661–1675.e1616.

86. Notelaers, K., Rocha, S., Paesen, R., Smisdom, N., De Clercq, B., Meier, J. C., Rigo, J.M., Hofkens, J., and Ameloot, M. (2014). Analysis of α3 GlyR single particle tracking in the cell membrane, *Biochim. Biophys. Acta - Mol. Cell Res.*, **1843**, pp. 544–553.

87. Lippert, A., Janeczek, A. A., Fürstenberg, A., Ponjavic, A., Moerner, W., Nusse, R., Helms, J. A., Evans, N. D., and Lee, S. F. (2017). Single-molecule imaging of Wnt3A protein diffusion on living cell membranes, *Biophys. J.*, **113**, pp. 2762–2767.

88. Chung, I., Akita, R., Vandlen, R., Toomre, D., Schlessinger, J., and Mellman, I. (2010). Spatial control of EGF receptor activation by reversible dimerization on living cells, *Nature*, **464**, pp. 783–787.

89. Calebiro, D., Rieken, F., Wagner, J., Sungkaworn, T., Zabel, U., Borzi, A., Cocucci, E., Zürn, A., and Lohse, M. J. (2013). Single-molecule analysis of fluorescently labeled g-protein–coupled receptors reveals complexes with distinct dynamics and organization, *Proc. Natl. Acad. Sci. U.S.A.*, **110**, pp. 743-748.

90. Kasai, R. S. and Kusumi, A. (2014). Single-molecule imaging revealed dynamic GPCR dimerization, *Curr. Opin. Cell Biol.*, **27**, pp. 78–86.

91. Kasai, R. S., Suzuki, K. G., Prossnitz, E. R., Koyama-Honda, I., Nakada, C., Fujiwara, T. K., and Kusumi, A. (2011). Full characterization of GPCR monomer-dimer dynamic equilibrium by single molecule imaging, *J. Cell Biol.*, **192**, pp. 463–480.

92. Hern, J. A., Baig, A. H., Mashanov, G. I., Birdsall, B., Corrie, J. E., Lazareno, S., Molloy, J. E., and Birdsall, N. J. (2010). Formation and dissociation of M1 muscarinic receptor dimers seen by total internal reflection fluorescence imaging of single molecules, *Proc. Natl. Acad. Sci. U.S.A.*, **107**, pp. 2693–2698.

93. Kasai, R. S., Ito, S. V., Awane, R. M., Fujiwara, T. K., and Kusumi, A. (2018). The Class-A GPCR dopamine D2 receptor forms transient dimers stabilized by agonists: Detection by single-molecule tracking, *Cell Biochem. Biophys.*, **76**, pp. 29–37.

94. Gentzsch, C., Seier, K., Drakopoulos, A., Jobin, M. L., Lanoiselee, Y., Koszegi, Z., Maurel, D., Sounier, R., Hubner, H., Gmeiner, P., Granier, S., Calebiro, D., and Decker, M. (2020). Selective and wash-resistant fluorescent dihydrocodeinone derivatives allow single-molecule imaging of μ-opioid receptor dimerization, *Angew. Chem. Int. Ed.*, **59**, pp. 5958–5964.

95. Drakopoulos, A., Koszegi, Z., Lanoiselee, Y., Hubner, H., Gmeiner, P., Calebiro, D., and Decker, M. (2020). Investigation of inactive-state κ opioid receptor homodimerization via single-molecule microscopy using new antagonistic fluorescent probes, *J. Med. Chem.*, **63**, pp. 3596–3609.

96. Malkusch, S., Rahm, J. V., Dietz, M. S., Heilemann, M., Sibarita, J.-B., and Lötsch, J. (2022). Receptor tyrosine kinase MET ligand-interaction classified via machine learning from single-particle tracking data, *Mol. Biol. Cell*, **33**, pp. ar60.

97. Briddon, S. J. and Hill, S. J. (2007). Pharmacology under the microscope: The use of fluorescence correlation spectroscopy to determine the properties of ligand–receptor complexes, *Trends Pharmacol. Sci.*, **28**, pp. 637–645.

98. Asher, W. B., Geggier, P., Holsey, M. D., Gilmore, G. T., Pati, A. K., Meszaros, J., Terry, D. S., Mathiasen, S., Kaliszewski, M. J., McCauley, M. D., Govindaraju, A., Zhou, Z., Harikumar, K. G., Jaqaman, K., Miller, L. J., Smith, A. W., Blanchard, S. C., and Javitch, J. A. (2021). Single-molecule FRET imaging of GPCR dimers in living cells, *Nat. Methods*, **18**, pp. 397–405.

99. Bayle, V., Fiche, J.-B., Burny, C., Platre, M. P., Nollmann, M., Martinière, A., and Jaillais, Y. (2021). Single-particle tracking photoactivated localization microscopy of membrane proteins in living plant tissues, *Nat. Protoc.*, **16**, pp. 1600–1628.

100. Wäldchen, F., Schlegel, J., Götz, R., Luciano, M., Schnermann, M., Doose, S., and Sauer, M. (2020). Whole-cell imaging of plasma membrane receptors by 3D lattice light-sheet dSTORM, *Nat. Commun.*, **11**, pp. 1–6.

101. Gormal, R. S., Padmanabhan, P., Kasula, R., Bademosi, A. T., Coakley, S., Giacomotto, J., Blum, A., Joensuu, M., Wallis, T. P., and Lo, H. P. (2020). Modular transient nanoclustering of activated β2-adrenergic receptors revealed by single-molecule tracking of conformation-specific nanobodies, *Proc. Natl. Acad. Sci. U.S.A.*, **117**, pp. 30476–30487.

102. Krainer, G., Keller, S., and Schlierf, M. (2019). Structural dynamics of membrane-protein folding from single-molecule FRET, *Curr. Opin. Struct. Biol.*, **58**, pp. 124–137.

103. Kedrov, A., Sustarsic, M., de Keyzer, J., Caumanns, J. J., Wu, Z. C., and Driessen, A. J. (2013). Elucidating the native architecture of the YidC: Ribosome complex, *J. Mol. Biol.*, **425**, pp. 4112–4124.

104. König, I., Zarrine-Afsar, A., Aznauryan, M., Soranno, A., Wunderlich, B., Dingfelder, F., Stüber, J.C., Plückthun, A., Nettels, D., and Schuler, B. (2015). Single-molecule spectroscopy of protein conformational dynamics in live eukaryotic cells, *Nat. Methods*, **12**, pp. 773–779.

105. Husada, F., Bountra, K., Tassis, K., de Boer, M., Romano, M., Rebuffat, S., Beis, K., and Cordes, T. (2018). Conformational dynamics of the ABC transporter McjD seen by single-molecule FRET, *EMBO J.*, **37**, pp. e100056.

106. Nakanishi, J., Takarada, T., Yunoki, S., Kikuchi, Y., and Maeda, M. (2006). FRET-based monitoring of conformational change of the β2 adrenergic receptor in living cells, *Biochem. Biophys. Res. Commun.*, **343**, pp. 1191–1196.

107. Liauw, B. W.-H., Afsari, H. S., and Vafabakhsh, R. (2021). Conformational rearrangement during activation of a metabotropic glutamate receptor, *Nat. Chem. Biol.*, **17**, pp. 291–297.

108. Terry, D. (2022). Single-molecule analysis of GPCR-mediated β-arrestin activation, *Biophys. J.*, **121**, pp. 85a.

109. Christie, S., Shi, X., and Smith, A. W. (2020). Resolving membrane protein–protein interactions in live cells with pulsed interleaved excitation fluorescence cross-correlation spectroscopy, *Acc. Chem. Res.*, **53**, pp. 792–799.

110. Wei, S., Thakur, N., Ray, A.P., Jin, B., Obeng, S., McCurdy, C. R., McMahon, L. R., Gutiérrez-de-Terán, H., Eddy, M. T., and Lamichhane, R. (2022). Single-molecule view of slow and sequential conformational transitions of A2A adenosine receptor, *Biophys. J.*, **121**, pp. 429a.

111. Wei, S., Thakur, N., Ray, A. P., Jin, B., Obeng, S., McCurdy, C. R., McMahon, L. R., Gutiérrez-de-Terán, H., Eddy, M. T., and Lamichhane, R. (2022). A2A adenosine receptor conformational dynamics revealed by single-molecule fluorescence, *Biophys. J.*, **121**, pp. 193a.

112. Snaar-Jagalska, B., Cambi, A., Schmidt, T., and de Keijzer, S. (2013). Single-molecule imaging technique to study the dynamic regulation of GPCR function at the plasma membrane, *Methods Enzymol.*, **521**, pp. 47–67.

113. Sako, Y., Minoghchi, S., and Yanagida, T. (2000). Single-molecule imaging of EGFR signalling on the surface of living cells, *Nat. Cell Biol.*, **2**, pp. 168–172.

114. Möller, J., Isbilir, A., Sungkaworn, T., Osberg, B., Karathanasis, C., Sunkara, V., Grushevskyi, E. O., Bock, A., Annibale, P., and Heilemann, M. (2020). Single-molecule analysis reveals agonist-specific dimer formation of μ-opioid receptors, *Nat. Chem. Biol.*, **16**, pp. 946–954.

115. Zhao, R., Li, N., Xu, J., Li, W., and Fang, X. (2018). Quantitative single-molecule study of TGF-β/Smad signaling, *Acta Biochim. Biophys. Sin.*, **50**, pp. 51–59.

116. Sungkaworn, T., Jobin, M.-L., Burnecki, K., Weron, A., Lohse, M. J., and Calebiro, D. (2017). Single-molecule imaging reveals receptor–G protein interactions at cell surface hot spots, *Nature*, **550**, pp. 543–547.

117. Luo, F., Qin, G., Wang, L., and Fang, X. (2021). Single-molecule fluorescence imaging reveals GABAB receptor aggregation state changes, *Front. Chem.*, **9**, pp. 779940.

118. Lao, J., He, H., Wang, X., Wang, Z., Song, Y., Yang, B., Ullahkhan, N., Ge, B., and Huang, F. (2017). Single-molecule imaging demonstrates ligand regulation of the oligomeric status of CXCR4 in living cells, *J. Phys. Chem. B*, **121**, pp. 1466–1474.

119. Siddig, S., Aufmkolk, S., Doose, S., Jobin, M. L., Werner, C., Sauer, M., and Calebiro, D. (2020). Super-resolution imaging reveals the nanoscale organization of metabotropic glutamate receptors at presynaptic active zones, *Sci Adv*, **6**, pp. eaay7193.

120. Karathanasis, C., Medler, J., Fricke, F., Smith, S., Malkusch, S., Widera, D., Fulda, S., Wajant, H., van Wijk, S. J., and Dikic, I. (2020). Single-molecule imaging reveals the oligomeric state of functional TNFα-

induced plasma membrane TNFR1 clusters in cells, *Sci. Signal.*, **13**, pp. eaax5647.

121. Insel, P., Head, B., Patel, H., Roth, D., Bundey, R., and Swaney, J. (2005). Compartmentation of G-protein-coupled receptors and their signalling components in lipid rafts and caveolae, *Biochem Soc Trans.*, **33**, pp. 1131–1134.

122. Rybin, V. O., Xu, X., Lisanti, M. P., and Steinberg, S. F. (2000). Differential targeting of β-adrenergic receptor subtypes and adenylyl cyclase to cardiomyocyte caveolae: A mechanism to functionally regulate the cAMP signaling pathway, *J. Biol. Chem.*, **275**, pp. 41447–41457.

123. Xiang, Y., Rybin, V. O., Steinberg, S. F., and Kobilka, B. (2002). Caveolar localization dictates physiologic signaling of β2-adrenoceptors in neonatal cardiac myocytes, *J. Biol. Chem.*, **277**, pp. 34280–34286.

124. Aprigliano, O., Rybin, V., Pak, E., Robinson, R., and Steinberg, S. (1997). Beta 1-and beta 2-adrenergic receptors exhibit differing susceptibility to muscarinic accentuated antagonism, *Am. J. Physiol. - Heart Circ. Physiol.*, **272**, pp. H2726–H2735.

125. Wright, P. T., Nikolaev, V. O., O'Hara, T., Diakonov, I., Bhargava, A., Tokar, S., Schobesberger, S., Shevchuk, A. I., Sikkel, M. B., and Wilkinson, R. (2014). Caveolin-3 regulates compartmentation of cardiomyocyte beta2-adrenergic receptor-mediated cAMP signaling, *J. Mol. Cell. Cardiol.*, **67**, pp. 38–48.

126. Nikolaev, V. O., Moshkov, A., Lyon, A. R., Miragoli, M., Novak, P., Paur, H., Lohse, M. J., Korchev, Y. E., Harding, S. E., and Gorelik, J. (2010). β2-adrenergic receptor redistribution in heart failure changes cAMP compartmentation, *Science*, **327**, pp. 1653–1657.

127. Jacobson, K., Sheets, E. D., and Simson, R. (1995). Revisiting the fluid mosaic model of membranes, *Science*, **268**, pp. 1441–1442.

128. Morone, N., Fujiwara, T., Murase, K., Kasai, R. S., Ike, H., Yuasa, S., Usukura, J., and Kusumi, A. (2006). Three-dimensional reconstruction of the membrane skeleton at the plasma membrane interface by electron tomography, *J. Cell Biol.*, **174**, pp. 851–862.

129. Lenne, P. F., Wawrezinieck, L., Conchonaud, F., Wurtz, O., Boned, A., Guo, X. J., Rigneault, H., He, H. T., and Marguet, D. (2006). Dynamic molecular confinement in the plasma membrane by microdomains and the cytoskeleton meshwork, *EMBO J.*, **25**, pp. 3245–3256.

130. Komura, N., Suzuki, K. G., Ando, H., Konishi, M., Koikeda, M., Imamura, A., Chadda, R., Fujiwara, T. K., Tsuboi, H., Sheng, R., Cho, W., Furukawa, K., Furukawa, K., Yamauchi, Y., Ishida, H., Kusumi, A., and Kiso, M.

(2016). Raft-based interactions of gangliosides with a GPI-anchored receptor, *Nat. Chem. Biol.*, **12**, pp. 402–410.

131. Eggeling, C., Ringemann, C., Medda, R., Schwarzmann, G., Sandhoff, K., Polyakova, S., Belov, V. N., Hein, B., von Middendorff, C., Schonle, A., and Hell, S. W. (2009). Direct observation of the nanoscale dynamics of membrane lipids in a living cell, *Nature*, **457**, pp. 1159–1162.
132. Honigmann, A., Mueller, V., Ta, H., Schoenle, A., Sezgin, E., Hell, S. W., and Eggeling, C. (2014). Scanning STED-FCS reveals spatiotemporal heterogeneity of lipid interaction in the plasma membrane of living cells, *Nat. Commun.*, **5**, pp. 5412.
133. Compans, B., Choquet, D., and Hosy, E. (2016). Review on the role of AMPA receptor nano-organization and dynamic in the properties of synaptic transmission, *Neurophotonics*, **3**, pp. 041811.
134. Treppiedi, D., Jobin, M. L., Peverelli, E., Giardino, E., Sungkaworn, T., Zabel, U., Arosio, M., Spada, A., Mantovani, G., and Calebiro, D. (2018). Single-molecule microscopy reveals dynamic FLNA interactions governing SSTR2 clustering and internalization, *Endocrinology*, **159**, pp. 2953–2965.
135. Ovesny, M., Krizek, P., Borkovec, J., Svindrych, Z., and Hagen, G. M. (2014). ThunderSTORM: A comprehensive imagej plug-in for PALM and STORM data analysis and super-resolution imaging, *Bioinformatics*, **30**, pp. 2389–2390.

Chapter 7

Cryo-Transmission Electron Microscopy for Studying Cell Membranes

Guanfang Zhao, Jing Gao, and Hongda Wang
State Key Laboratory of Electroanalytical Chemistry,
Research Center of Biomembranomics,
Changchun Institute of Applied Chemistry,
Chinese Academy of Sciences, Changchun, Jilin, China
hdwang@ciac.ac.cn

The cell membrane separates the cell from the external environment and maintains a stable internal environment for cell activities. It is a complicated complex composed of lipids, proteins, and some carbohydrates, in which many biological functions are accomplished through the interaction of these components. Although many membrane models have been proposed based on their functions, a high-resolution and three-dimensional (3D) structure of the cell membrane in situ and the nano-organization of the membrane proteins have not been studied clearly. Cryo-electron microscopy (cryo-EM), with its unique imaging modality, can be used to explore the native 3D structures of the cell membranes and the membrane proteins at an ultra-high resolution. Here we introduce the principles

Biomembranomics: Structure, Techniques, and Applications
Edited by Hongda Wang

ISBN 978-981-4968-61-4 (Hardcover), 978-1-003-45635-3 (eBook)
www.jennystanford.com

and different imaging modes of cryo-EM and its application in the field of cell-membrane and membrane-protein studies. We hope it will provide up-to-date insights into membrane structures and functions and help people to have a deeper understanding of cell membranes.

7.1 Introduction

The cell membrane is the outermost layer of the cell, separating the cell from the external environment and performing a variety of biological functions, such as material transport, energy conversion, and signal transduction. Although its importance has been recognized, the structure of the cell membrane at the molecular level is poorly understood and is still in the stage of model hypotheses. Initially, Gorter et al. discovered that the cell membranes are composed of phospholipid bilayers through the extraction technique [1]. With the advent of electron microscopy (EM), Singer et al. found that membrane proteins are mosaic in phospholipid bilayers and proposed the classical flow mosaic model (FMM) [2]. With the development of ultracentrifuges, Simons et al. discovered that some membrane proteins are present in the cholesterol-rich region and proposed the lipid raft model [3]. Although there is an increasing number of techniques and tools to study cell membranes, each of them has its own shortcomings and limitations. For example, single-molecule imaging techniques such as atomic force microscopy (AFM) and single-molecule fluorescence microscopy (SMFM) are widely used to study cell-membrane structures at the molecular level. They can provide 3D topography of the cell membranes and the distribution of the membrane proteins. However, the resolution of these two techniques is still low (~10 nm for AFM and ~20 nm for SMFM) and could not reveal the information of single molecules. Moreover, the in situ 3D structure of the cell membranes and membrane proteins cannot be directly observed.

With the development of freeze-etching, EM has been used to image cell membranes and membrane proteins with sufficient resolution (~3 nm) and reveal various mechanisms underlying cell-membrane functions. It is worth noting that membrane functions depend on the interactions between various membrane biomolecules. Thus, there is a great need for high-resolution imaging of undisturbed or in situ cell-membrane structures and membrane proteins. However, the traditionally prolonged invasive sample preparation method of EM leads to sample deformation and artifacts, which prevents us from observing the structural features of cell membranes and membrane proteins in situ. Fortunately, over the past few decades, the gradually maturing cryo-EM technique has become the mainstream technique for studying the structure of biological samples due to its higher resolution of less than 1 nm. Furthermore, cryo-EM preserves the structural integrity of biological samples by preserving them in vitrified ice through plunge freezing. Therefore, more and more structures of membrane components have been resolved by cryo-EM, facilitating our understanding of the membrane structures and functions.

In the following sections, we will review the development of cryo-EM and its contribution to the study of membrane structures and functions. First, we introduce the imaging modes and imaging principles of cryo-EM. Then we focus on the processing and analysis of cryo-EM data. And finally, we review some applications and advances of cryo-EM imaging in cell-membrane studies. We expect it will help to fully understand the membrane structures and functions and open a new avenue for the study of cell membranes.

7.2 Cryo-Electron Microscopy

Cryo-electron microscopy [4] is a microscopic technique developed in the 1980s to observe samples with transmission electron microscopy (TEM) at a low temperature. Together with X-ray crystallography [5] and nuclear magnetic resonance (NMR) spectroscopy [6, 7], they are the three significant tools in structural biology (Fig. 7.1A). In recent years, due to the rapid development of EM hardware and image-processing software, the resolution of 3D structures resolved by cryo-EM has advanced from the initial nanometer level to the atomic level, and a large number of previously unresolvable biological macromolecule structures have become solvable (Fig. 7.1B). As a result, cryo-EM technology has gained extensive attention from biologists and has gradually become a crucial and irreplaceable research tool in structural biology.

Three scientists, Jacques Dubochet, Joachim Frank, and Richard Henderson, who made pioneering contributions to the development of this technology, were awarded the Nobel Prize in Chemistry in 2017 (Fig. 7.1C) [8].

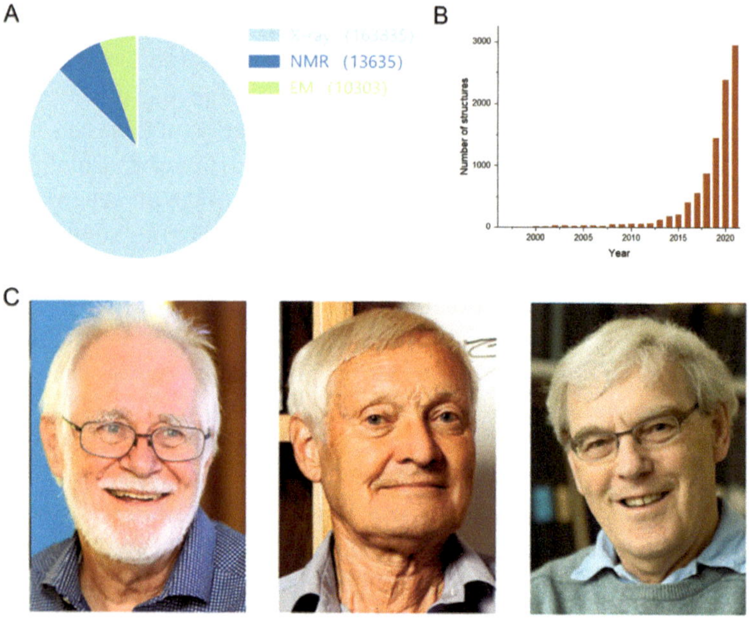

Figure 7.1 Statistics of atomic model data in the International Protein Structure Database and 2017 Nobel Prize winners. (A) The number of structures resolved by the three structural biology methods (statistics from Protein Data Bank). (B) The number of atomic models resolved by cryo-electron microscopy in recent years (statistics from EM DataBank). (C) Three Nobel Prize winners in Chemistry for the year 2017. From left to right: Jacques Dubochet, Joachim Frank, and Richard Henderson.

Cryo-EM is based on TEM and includes the basic steps of sample preparation, imaging, image processing, and structural analysis. In the TEM, electrons can be used as "light waves" for microscopic imaging after being accelerated to near the speed of light by an electric field. The precisely designed magnetic field with a specific shape can be used as a lens to focus the electron beam on the sample, and the transmitted electrons are captured by the camera as an image, and then the 3D structure of the sample is resolved by a computer program based on the information contained in the image. However, there are several difficulties in observing biological samples through cryo-EM [9]. First, the electron beams would cause radiation damage to biological samples [10]. It was found that the biological samples were quickly damaged by the bombardment of high-energy electron beams when people first observed them with EM, so the fine atomic structure of the biological samples could not be obtained. Second, the interaction between biological samples and electrons is weak, leading to a relatively low signal-to-noise ratio (SNR) of the collected images. Third, the vacuum environment of EM can cause the water contained in the biological sample to evaporate, resulting in the dehydration of the biological sample and the destruction of the biological sample structures. Collectively, the above problems about radiation damage and imaging conditions of biological samples have become the biggest challenges to overcome for EM in biology.

To reduce the radiation damage of the biological sample and improve the SNR of the collected images, two approaches are commonly employed. The first way is to image the biological samples that are plunged to freezing at the temperature of liquid nitrogen or liquid helium. It was found that the degree of radiation damage could be reduced by a factor of six by imaging biological samples at liquid nitrogen temperature by comparing the diffraction intensity decay at room temperature with that at low temperature [11]. This means that EM imaging can be performed at a lower temperature with higher electron doses, which not only greatly reduces the radiation damage to the sample, but also improves the SNR of the image. Prof. Jacques Dubochet's outstanding contribution is the breakthrough in the preparation of vitreous ice, improving the technique of preparing frozen biological samples [12]. The second approach is to image biological samples with low electron dose to reduce radiation damage, and then average a large number of identical structures to improve the SNR. The signal of structures in the images that we want to reconstruct is stable, so the high-resolution structures can be obtained by superimposing the same signal in multiple images. In addition, noise cannot be enhanced by superposition because it is random information. So, when averaging multiple identical images, their signals are steadily enhanced, whereas the noise stays the same, ultimately improving the SNR of the image [13].

Since cryo-EM overcomes the above drawbacks, it has gradually become a crucial technical tool in structural biology with its unique imaging advantages. First, compared to other structural biology research methods, cryo-EM imaging needs only small quantities of samples. For example, for samples with good biochemical properties and conformational homogeneity, it usually needs tens to hundreds of thousands of molecules to obtain a near-atomic resolution 3D structure [14]. In contrast, for large molecular complexes such as viruses with a high degree of symmetry, only a few thousand molecules are needed to obtain a 3D structure with atomic resolution [15]. In operation, only 3–5 μL of 0.1–1 μmol of protein solution is required for a frozen EM grid, in contrast to the large sample volume required for X-ray crystallography and NMR spectroscopy. Second, the cryo-EM plunge-freezing techniques can rapidly freeze samples that are suspended in aqueous solutions to a deep-freezing temperature. As a result, the biological samples are preserved in the vitreous ice and retain their state in solution, which is closer to their natural physiological state [16]. Third, cryo-EM imaging has lower requirements for sample homogeneity. Unlike NMR spectroscopy and X-ray crystallography, which analyze the average information of samples, cryo-EM acquires electron projection images of samples at magnifications of tens to hundreds of thousands of times and obtains structural information by the statistical analysis of multiple molecular images. This process of statistical analysis allows the classification of multiple molecular structures that may be present in a sample, thereby distinguishing molecules with different compositions and conformations. A variety of algorithms for 2D and 3D classification of molecular structures have been used in practice, making it possible to resolve the structure of poorly homogeneous samples at high resolution [17]. Finally, cryo-EM imaging techniques are applicable to a wide range of research objects, from protein macromolecules at the molecular level, to organelles at the subcellular level, and even tissue structures at the cellular level.

7.2.1 Cryo-EM Imaging Modes

In the process of imaging by cryo-EM, different imaging modes are adopted depending on the properties and characteristics of different biological samples. The main modes of cryo-EM for imaging biological samples are single-particle analysis (SPA) and cryo-electron tomography (cryo-ET). In the following sections, we introduce these two different cryo-EM imaging modes, respectively.

7.2.1.1 Single-particle cryo-electron microscopy

Single-particle analysis [18], which originated in the 1980s [19, 20], involves freezing biological macromolecules and then uses cryo-EM to image structurally homogeneous and dispersed particles of the whole sample at low temperatures, followed by image processing to obtain the 3D structure of the sample (Fig. 7.2A). Ideally, the frozen sample will contain homogeneous macromolecules or complexes with different orientations. Then, the random projections and photographs of these macromolecules or complexes are taken, and the large number (usually including hundreds of thousands or even millions of macromolecular particles) of identical or similarly oriented macromolecules or complex structures in the taken projection images are extracted for alignment and averaging to obtain the 2D image with a high SNR. Next, according to the central slice theorem [21], the 2D Fourier transform of the 2D projection image is exactly the central section normal to the direction of electron incidence of the 3D Fourier transform of that 3D object (Fig. 7.2B). Therefore, we can reconstruct the 3D structures of biomolecules from 2D images with high SNR from all angles. SPA allows the study of structural and conformational changes of biomolecules in solution. It does not require crystallization, but only relatively small amounts of samples, and can analyze a wide range of molecules. Moreover, the larger the molecular weight and the higher the sample symmetry, the easier it is to reconstruct a high-resolution structure. In December 2013, Yifan Cheng and D. Julius obtained the 3.4 Å resolution structure of the membrane protein TRPV1 for the first time using the SPA method and successfully built a model of this macromolecule [14, 22]. This caused great repercussions in the field of structural biology and marked a new era of SPA.

To date, developments in cryo-electron microscopy hardware (e.g., electron direct detection device (DDD), phase plates, etc.) [25, 26] and software (e.g., Relion, EMAN2, etc.) [27, 28] have greatly expanded the application of SPA techniques in structural biology, especially for large molecular weight membrane proteins and protein complexes that were previously difficult to obtain crystals (Fig. 7.3A). At the same time, the structural resolution resolved by the SPA method has been improving (Fig. 7.3B). For example, in 2015, the resolution of the 20S proteasome resolved by SPA methods reached 2.8 Å [29]. In 2020, the resolution of SPA methods improved to 1.15 Å by resolving the structure of human Apo protein [30]. Besides, the development of some hardware has led to great progress in the resolving of structures with small molecular weights by SPA methods. In general, the smaller the molecular weight of a biomolecule, the more difficult it is to resolve the high-resolution structure. For example, the 1.8 Å resolution glutamate dehydrogenase has a molecular weight of 334 kD, while the resolutions of lactate dehydrogenase with a molecular weight of 145 kD and isocitrate dehydrogenase with a molecular weight of 93 kD are only 2.8 Å and 3.8 Å, respectively [31]. All these advances have made SPA a rapid and efficient method comparable to X-ray crystallography for the structural analysis of biological macromolecules. Moreover, with years of development, SPA has developed a mature and complete process, which includes sample extraction and purification, frozen sample preparation, data collection, and data processing.

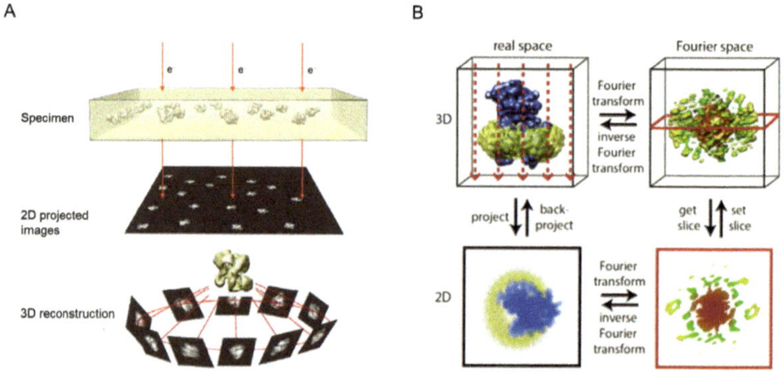

Figure 7.2 Principles of structural analysis by SPA. (A) The principle of the SPA technique. Reprinted from Ref. [23], with permission from Springer Nature. (B) The principle of the reconstruction of 3D structure by Fourier inversion. Reprinted from Ref. [24], Copyright 2015, with permission from Elsevier.

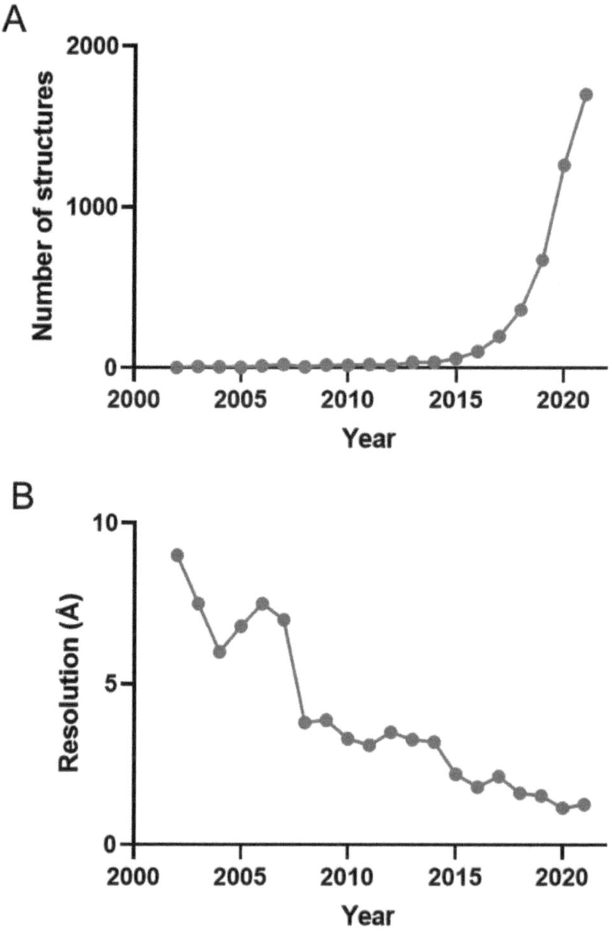

Figure 7.3 Application and resolution of single-particle analysis (statistics from EM DataBank)). (A) Number of structures resolved by SPA in recent years. (B) The highest resolution that can be achieved using SPA in recent years.

7.2.1.2 Cryo-electron tomography

The development of SPA methods has greatly increased the application of cryo-EM in the field of structural biology. However, for biological systems, most physiological functions are the results of the interactions between various biomolecules [32]. Moreover, the structural morphology of biomolecules and the execution of their physiological functions are strictly dependent on the physiological environment in which they are located, and it is impossible to isolate and purify biomolecules from their physiological environment without destroying their structural integrity. Therefore, characterizing the 3D structure of biomolecules in their native environment at high resolution is of great importance for understanding their functions [33]. Based on this, cryo-electron tomography [34], proposed in the 1970s, becomes another important imaging modality for cryo-EM after the SPA method. Unlike the SPA method, it is capable of directly visualizing the 3D structure of biomolecules with multiple conformations at high resolution in the native environment. Cryo-ET is performed by tilting the sample to a certain angle (typically −60 to +60°) with a fixed angular interval and acquiring the projection images of the sample at each angle (Fig. 7.4A,B). The obtained tilt series are reconstructed by specific image-processing software to generate a 3D tomogram (Fig. 7.4C). We can then identify and segment the target sample of interest in this tomogram or obtain a medium- to high-resolution structure of that sample by extracting and averaging multiple homogeneous target samples (Fig. 7.4D) [35].

Cryo-ET imaging is not yet able to obtain the high-resolution structures available with NMR spectroscopy, X-ray crystallography, and SPA methods at the current state, but it has an irreplaceable and important role in studying the 3D structure and function of amorphous, asymmetric, and non-holomorphic biological samples. To date, many biological samples such as isolated organelles [37, 38], viruses [39], bacteria [40], eukaryotic cells [41], and tissue sections [42] have been successfully studied by cryo-ET. The development of cryo-EM hardware and improvements in data-processing algorithms, such as template matching [43], contrast transfer function (CTF) correction [44], sub-tomogram averaging and classification [45, 46], have increased the resolution of the target samples to the sub-nanometer level. Together with its advantages of accurately reflecting the physiological conformation of target samples in the native environment, cryo-ET has gained growing attention in the field of cryo-EM and was named one of the top 10 scientific and technological breakthroughs in 2002 by the journal *Science* [47].

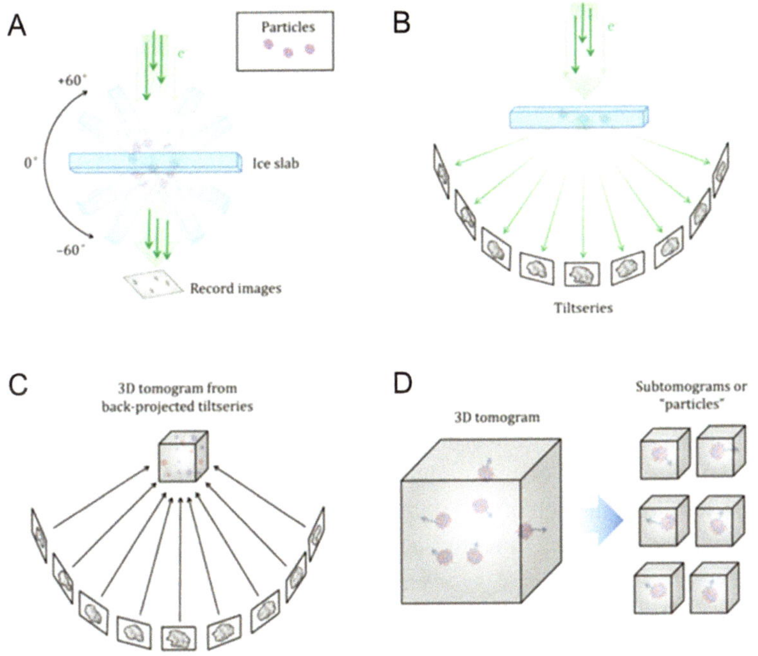

Figure 7.4 Principles of data collection and processing in cryo-ET. (A) Samples embedded in ice are tilted by a certain range of angles, and the projected images are recorded at each angle. (B) The tilt series of projected images collected around a common axis. (C) The projected images in the tilt series are aligned with their common axis through calculation and reconstructed into a 3D tomogram by weighted projection or other methods. (D) Sub-tomograms representing 3D view of single macromolecules can be extracted from the reconstructed tomogram, and then aligned and averaged. Reprinted from Ref. [36], with permission from Springer Nature.

7.3 Applications of Cryo-EM in Cell Membranes and Membrane Proteins

In the last few years, major technological breakthroughs in cryo-EM have led to a much higher resolution of this technology, which undoubtedly has had a major impact on structural biology. An impressive array of biological questions are beginning to be addressed. The determination of membrane-protein structure is one of the few areas of structural biology that have been greatly impacted by the revolutionary technological breakthrough in cryo-EM. Nowadays, cryo-EM has become a mature method and is applied to the characterization of cell membranes and the resolving of the 3D structures of many membrane proteins.

7.3.1 Applications of SPA in Cell Membranes and Membrane Proteins

Until now, the structures of most membrane proteins have been determined by X-ray crystallography or electron crystallography, which requires the membrane proteins to form 3D or 2D crystals in solution or in lipid bilayers. Thus, obtaining good diffraction crystals is a major factor to successfully obtain high-resolution structures. For many mammalian membrane proteins, however, crystallization remains a major bottleneck. SPA requires neither sample crystallization nor absolute sample homogeneity and is, therefore, often used to study large integral membrane proteins because they are difficult to crystallize. SPA has led to a resolution revolution in cryo-EM, because it introduces DDD and various novel image-processing algorithms. This resolution revolution transformed SPA from a complementary of X-ray crystallography to atomic structure determination in many areas of structural biology, especially for determining membrane-protein structures, which is one of the important breakthroughs in the field of structural biology.

Membrane proteins play an important role in cell membranes, influencing cellular functions and biological behaviors. Structural information is essential for understanding the biological mechanisms of these protein functions. However, the structural characterization of membrane proteins is a major challenge, mainly because it is difficult to extract them from their natural lipid environment while maintaining structural and functional integrity. For many years, detergents and detergent-like amphiphilic polymers have been used to solubilize membrane proteins or protein complexes so that they can be stabilized in solution (Fig. 7.5A) [48–50]. By this approach, an increasing number of membrane-protein structures have been successfully resolved by SPA, including ryanodine receptors [51, 52], Trp channels [53, 54], γ-secretase [55], glutamate receptors

[56], calcium channel Cav1.1 [57], and mitochondrial complexes [58–60]. These successful results demonstrate that it is possible to achieve high resolution in detergents for certain kinds of proteins, but protein stability and the presence of excess detergent remain a stumbling block in obtaining high resolution. A recently reported gradient-based method, GraDeR [61], removed excess boundary detergent and provided a new method for enhancing membrane-protein stability in the detergent-soluble state (Fig. 7.5B). The GraDeR method achieves detergent reduction by slowly removing excess amphiphilic detergents. The structure of the innexin-6 channel was recently obtained by this method [62].

Although membrane proteins solubilized with these reagents have remarkable stability and solubility, they are still very different from those in natural lipid membranes. Extensive studies have shown that the functions of membrane proteins usually depend on direct interactions with lipids [64–66]. In recent years, the most used lipid bilayer environment is the nanodisc (Fig. 7.6) [67]. Nanodiscs are disk-shaped lipid bilayers stabilized by two circular, amphiphilic bands of helical proteins (called membrane scaffold proteins) that stabilize proteins in a better way and have been widely used for the functional and structural analysis of membrane proteins. Many structures of membrane proteins have been solved in the lipid bilayer by SPA using nanodisc systems, including bacterial ribosomes–SecYE complex [68], ribosome–YidC complex [69], the muscle isoform of the ryanodine receptor (RyR1) [70], Tc toxin from *Photorhabdus luminescens* [71], and the transient receptor potential cation channel sub-family V member 1 (TRPV1) [72] and polycystin-2 (PKD2) [73]. In particular, in the case of TRPV1, the resolution of TRPV1 in complex with double knot toxin (DkTx) and resiniferatoxin (RTX) was 2.9 Å in the form of nanodisc embedding, which represented a substantial improvement from the 3.8 Å resolution obtained from aminophenol-substituted specimens (Fig. 7.6B,C) [14]. Significant qualitative improvements were observed in the side chain density within the transmembrane region or in the linking loops facing the lipids, and even further improvements were observed in the cytoplasmic domain. In this structure, the interaction between tightly bound lipids and amino acid side chains from TRPV1 and DkTx is clearly demonstrated.

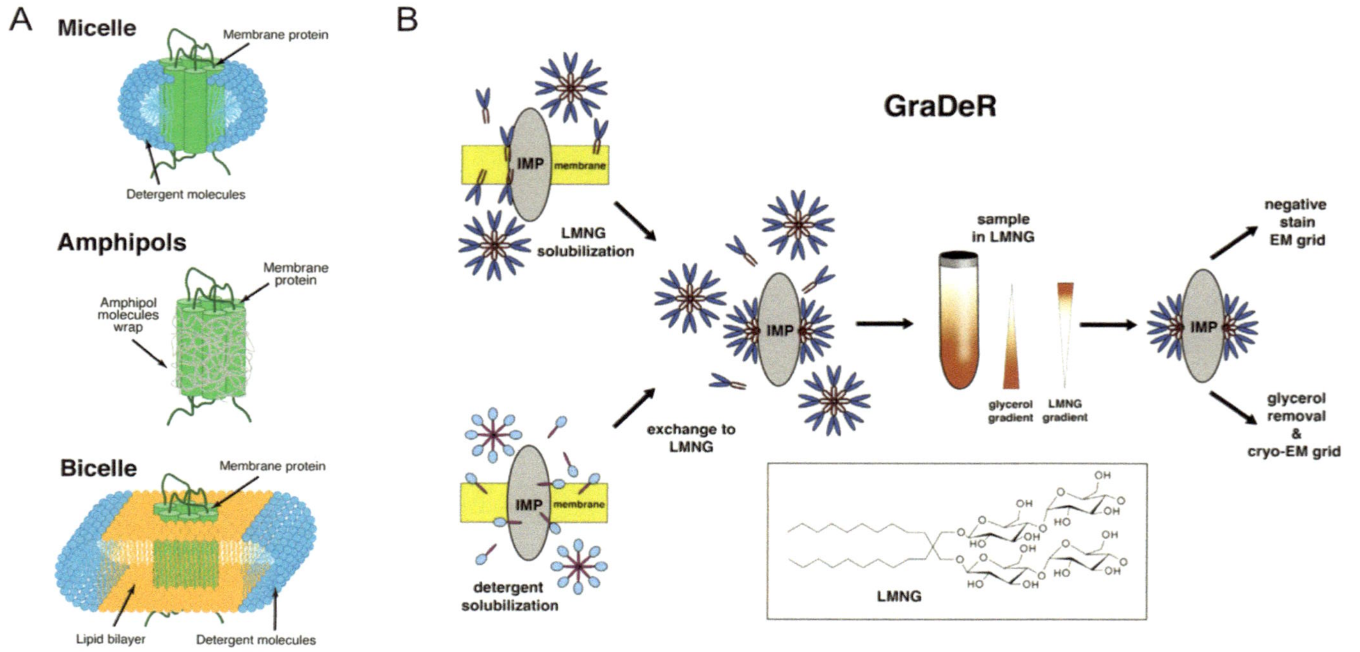

Figure 7.5 Structures of various artificial membranes. (A) Structures of artificial membranes. Micelles are spherical vesicles in which the detergent hydrophobic chains face inward and the hydrophilic polar heads face outward. Bicelles are obtained by a mixture of lipids and short chain detergents. The lipids interact with the protein to form a lipid bilayer, and the detergent forms the rim of the bicelle. Amphipol polymers wrap around the hydrophobic patches of the membrane protein to form a stable complex in solution. Reprinted from Ref. [63], Copyright 2018 Sgro and Costa. (B) GraDeR workflow, IMP, integral membrane proteins, LMNG, lauryl maltose-neopentyl glycol. Reprinted from Ref. [61], Copyright 2012, with permission from Elsevier.

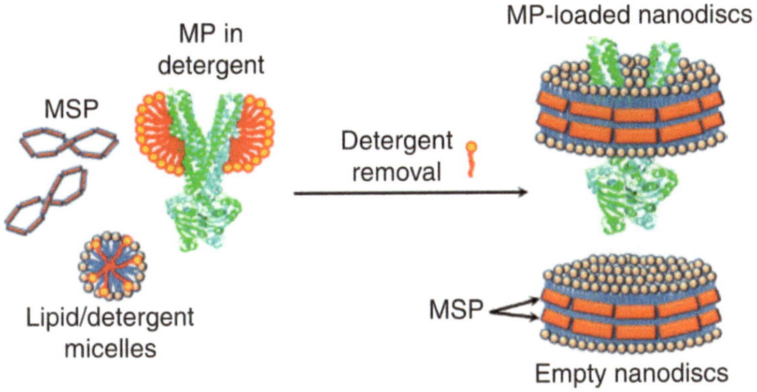

Figure 7.6 Assembly of nanodiscs. Schematic representation of the self-assembly of nanodiscs from a mixture containing an MSP (membrane scaffold protein), phospholipid/detergent mixed micelles (yellow circles: hydrophilic head; blue: phospholipid hydrophobic chains; orange circles: hydrophilic detergent moiety; red: hydrophobic detergent moiety) and detergent-solubilized MP (membrane protein) upon detergent removal. Reprinted from Ref. [74], Copyright 2017 Walter de Gruyter GmbH, Berlin/Boston.

Despite these advances, all the above approaches for the isolation of membrane proteins eliminate electrochemical gradients and membrane curvature, which in turn disrupts the membrane topology. To preserve these important properties, membrane-protein structures can be embedded in liposomes, and this method has been widely used for the functional analysis of membrane proteins (Fig. 7.7) [75–78]. A convenient workflow was proposed by Nieng Yan's group. The structure of AcrB, a membrane protein embedded in liposomes, was resolved by SPA. The 3D structure of AcrB embedded in liposomes was obtained at a resolution of 3.9 Å [79]. The conformation of AcrB remained unchanged when the surrounding membrane showed different curvatures. This method can be widely applied to the cryo-EM analysis of membrane proteins with different soluble domains, laying the foundation for the cryo-EM analysis of integral or peripheral membrane proteins whose transmembrane electrochemical gradient or/and membrane curvature affects their function.

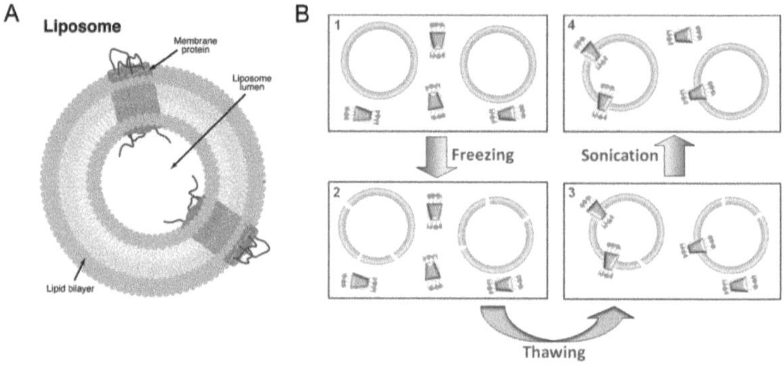

Figure 7.7 Membrane proteins embedded in liposomes resolved by SPA. (A) Schematic diagram of artificial spherical lipid membranes in which membrane proteins can be assembled. Reprinted from Ref. [63], Copyright 2018 Sgro and Costa. (B) The four steps of freeze–thaw sonication: (1) mixture of preformed liposome with protein and detergent; (2) freezing of the mixture in liquid N_2 with consequent rupture of vesicle membranes; (3) slow thawing at 0°C causing the insertion of the protein, solubilized in detergent, into the membrane; (4) mild sonication step (low energy in pulse mode), which facilitates the sealing of proteoliposomes. Reprinted from Ref. [80], Copyright 2013 by the authors.

Technological advances have given SPA the ability to determine the structure of intact membrane proteins in multiple conformational states. Indeed, SPA has great potential to analyze membrane proteins, especially those of mammalian origin, with multiple functions and even transient conformations, which could be used for ligand screening or drug design, pointing the way to the development of the field of drug structure screening. For example, the first G protein-coupled receptor, which is one of the most abundant receptor proteins on the cell surface and is targeted by 30% of the drugs on the market, was resolved by single-particle analysis at 4.1 Å resolution in 2016 [81]. The heterotrimeric structures of salmon calcitonin, G protein, and human calcitonin receptor (class B GPCR) in the activated state were determined. These side chains are clearly identifiable and may help resolve higher-resolution GPCR protein complexes and also aid in the design of new drugs for the treatment of type II diabetes and obesity. Overall, the structural analysis of membrane proteins is increasingly becoming an interesting and challenging area of research. SPA is definitely becoming a powerful technique.

7.3.2 Applications of Cryo-ET in Cell Membranes and Membrane Proteins

Although the structures of many important membrane proteins have been successfully obtained at the atomic level by SPA, this approach has the limitation of not being able to study large intracellular molecules and biomolecules that are difficult to purify and show continuous conformational flexibility. In general, it is difficult to reveal the biological functions of molecules that have been isolated and purified in their native environment. Fortunately, cryo-ET can image biomolecules in the native environment. In other words, it bridges the gap between light microscopy and in vitro structure determination techniques. In recent years, with the development of direct electron detector devices, energy filters, and phase plates, the resolution of cryo-ET has been greatly improved. Therefore, cryo-ET is widely used to study the 3D structures of cell membranes, organelle membranes, and membrane proteins.

A unique advantage of cryo-ET is the ability to characterize intact cells and intracellular macromolecules, obtaining the structural blueprint of undisturbed macromolecular organization within the cell at a resolution of a few nanometers. Thus, a realistic view of the cellular landscape in the physiological environment is obtained. For cryo-ET, the thickness of the sample is the main limitation for imaging, but some eukaryotic cells with thin edges can be directly imaged by cryo-ET. For example, Lindsay A. Baker and other researchers at the University of Oxford, UK, have labeled target proteins on cell membranes with DNA origami signposts that allow precise identification of individual protein complexes in tomograms (Fig. 7.8A) [82]. They labeled sfGFP-gB (super folder GFP-glycoprotein B) transfected on BHK-21 cells by signposted origami tags (SPOTs) and imaged the thin periphery of the cells by cryo-ET. They observed many regions of SPOTs binding to the cell surface at low magnification (Fig. 7.8B). By observing the reconstructed 3D tomograms of these regions, the binding of SPOTs to the plasma membrane can be easily observed. And the distribution characteristics of the tagged membrane proteins on the cell membrane were studied by SPOTs (Fig. 7.8C).

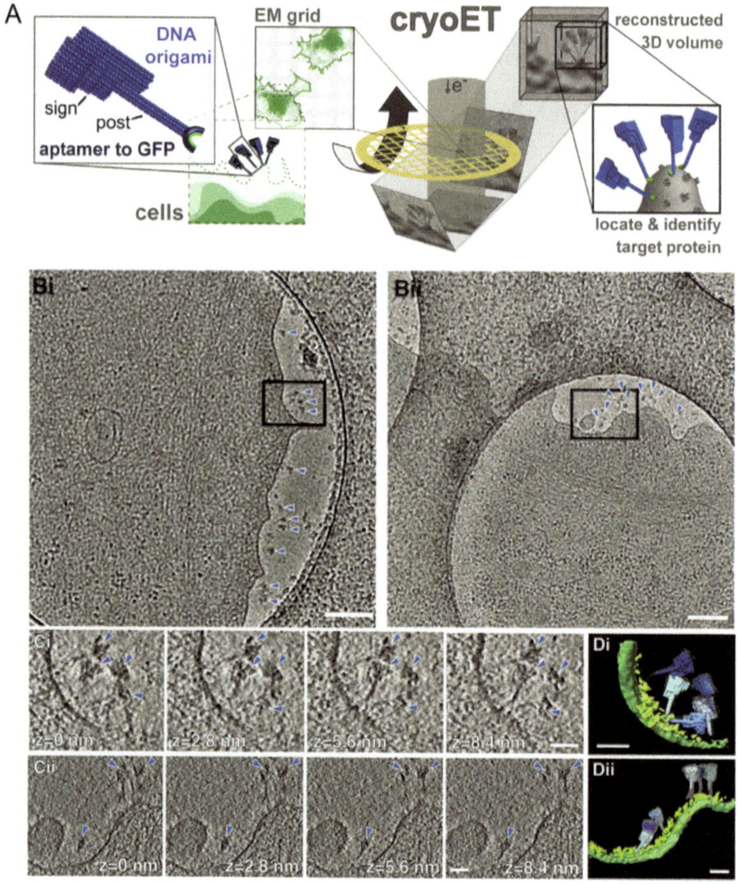

Figure 7.8 Signposted origami tags (SPOTs) binding to cells expressing sfGFP-gB. (A) Signpost origami tagging. (B) Two low-magnification cryo-EM projection images of SPOTS bound to the surface of BHK-21 cells expressing sfGFP-gB (i and ii). SPOTs are indicated with blue arrowheads. Scale bar, 250 nm. (C) Sequential computed tomographic slices of the areas indicated in (Bi) and (Bii), respectively. Although the SPOTs (blue arrowheads) often extend across multiple slices, it is still possible to trace the post structure to the membrane in many cases. Slice thickness is ~2.8 nm. Scale bars, 50 nm. (D) Manual segmentation of the tomograms in (C). Scale bars, 50 nm. Reprinted from Ref. [82], Copyright 1969, with permission from Elsevier.

For some thicker cells, the samples can be thinned by cryo-sectioning or focused ion beam (FIB) milling before being imaged by cryo-ET (Fig. 7.9A) [83]. For example, the researchers prepared

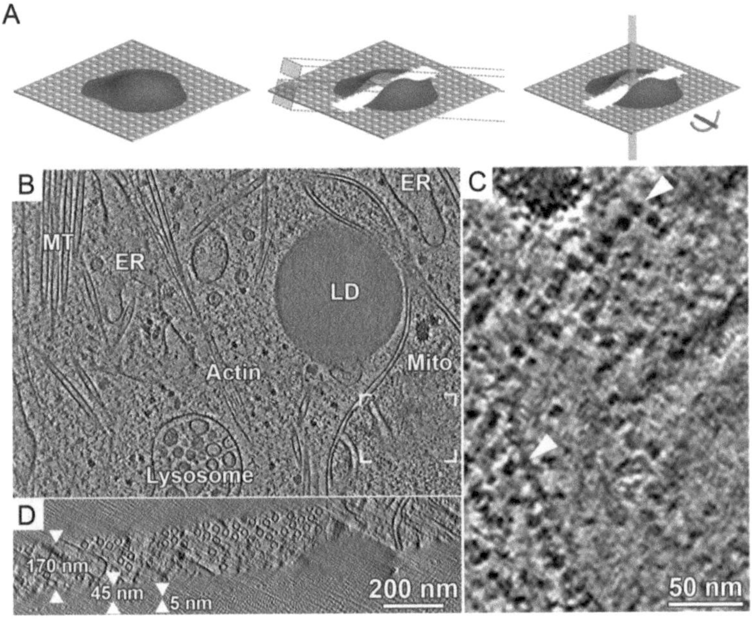

Figure 7.9 Imaging the FIB-thinned HeLa cells by cryo-ET. (A) The workflow for cellular tomography via cryo-FIB milling. Reprinted from Ref. [83], Copyright 2013, with permission from Elsevier. (B) xy slice from a tomographic volume showing a variety of organelles and cytoskeletal structures within the cytoplasm. MT: microtubules, ER: endoplasmic reticulum, LD: lipid droplet, mito: mitochondrion. (C) An enlarged area within the mitochondrion (the framed region in B rotated by 90°). A row of ATP synthase complexes is visible along the cristae membranes in top view (top arrowhead) and side view (bottom arrowhead). (D) Corresponding xz slice of the tomographic volume in B. Reprinted from Ref. [84], Copyright 2016, with permission from Elsevier.

thin layers spanning the entire Hela cells by FIB milling [84]. Imaging them by Cryo-ET revealed the molecular details of various organelles, including the ER membrane, lysosomal compartments, lipid droplets, mitochondria, and cytoskeleton (Fig. 7.9B). The rows of ATP synthase complexes can be seen within the mitochondria decorating the native cristae membrane. Features of ATP synthase are clearly visible, including the F1 head structural domain of about 10 nm and the stalk connecting F1 and the membrane-embedded F0 structural domain (Fig. 7.9C). The tomogram has a lamella thickness of 170 nm, covered by an additional 45 nm of condensed water vapor followed by a 5 nm conductive layer of sputtered Pt

(Fig. 7.9D). Thus, imaging of FIB-milled thin cell sections by cryo-ET allows visualization of many macromolecular complexes without computational averaging methods. With the help of synergistic applications of thinning techniques such as FIB milling, cryo-ET holds promise for revealing the molecular organization of cellular functions in an undisturbed environment.

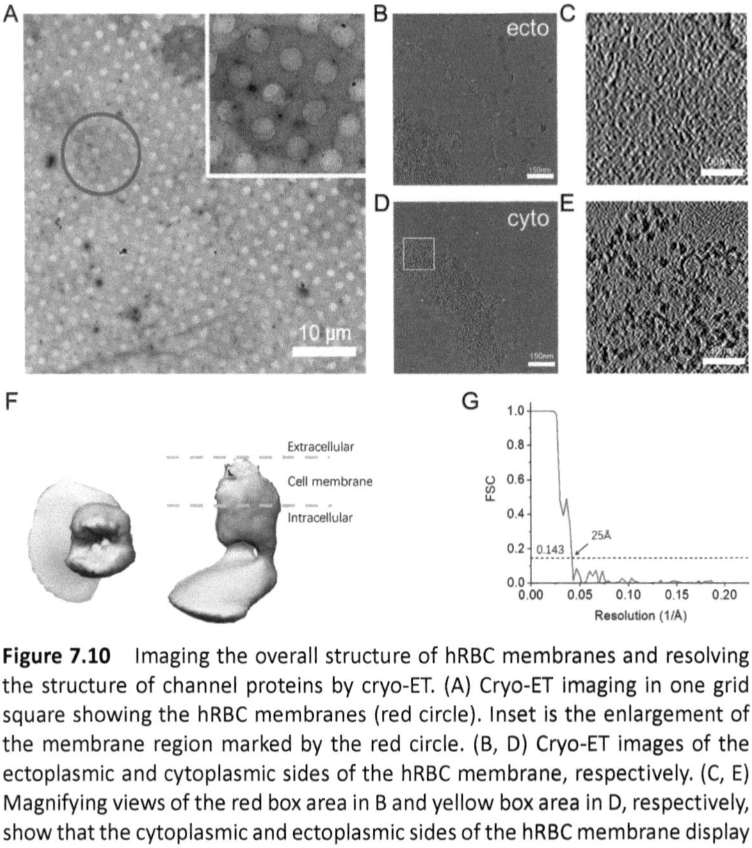

Figure 7.10 Imaging the overall structure of hRBC membranes and resolving the structure of channel proteins by cryo-ET. (A) Cryo-ET imaging in one grid square showing the hRBC membranes (red circle). Inset is the enlargement of the membrane region marked by the red circle. (B, D) Cryo-ET images of the ectoplasmic and cytoplasmic sides of the hRBC membrane, respectively. (C, E) Magnifying views of the red box area in B and yellow box area in D, respectively, show that the cytoplasmic and ectoplasmic sides of the hRBC membrane display completely different characteristics, and many proteins are visible on the cytoplasmic side. (F) Sub-tomogram average of the transmembrane proteins. (G) Fourier shell correlation of the sub-tomogram average of the transmembrane proteins. FSC, Fourier shell correlation.

In addition, cryo-ET has been applied to characterize the in situ 3D structure of cell membranes. Wang's group prepared human red blood cell (hRBC) membranes by the developed method of extracting cell membranes and observed the high-resolution 3D structure of hRBC membranes and the characteristics of membrane proteins by cryo-ET in situ for the first time (Fig. 7.10A) [85]. By analyzing the tomograms of the hRBC membranes, an asymmetric distribution of membrane proteins on both sides of the cell membrane was found: membrane proteins were mainly located on the cytoplasmic side of the hRBC membranes, with protein sizes ranging from 6 nm to 8 nm, in contrast to the ectoplasmic side with basically no protein (Fig. 7.10B–E). In addition, 3D classification and sub-tomogram averaging of the selected membrane proteins yielded the channel protein-like structure with a resolution of 25 Å (Fig. 7.10F,G). These results represent the first in situ characterization of cell membranes and membrane proteins by cryo-ET and open the door for studying cell membranes in situ by cryo-ET.

Besides the study of intact cells and cell membranes, cryo-ET has also characterized the 3D structures of some organelles and organelle membranes. Abraham J. Koster's team studied the ultrastructure of intact mitochondria in primary human umbilical vein endothelial cells by cryo-ET and found that the mitochondrial cristae are connected to the intermembrane space through large slits, challenging the then-held view that such connections are established exclusively through small circular pores (Fig. 7.11) [86]. Werner Kühlbrandt investigated the structure and organization of the dimerization of ATP synthase on the inner mitochondrial membrane cristae of wild-type and mutant *Saccharomyces cerevisiae* by cryo-ET [87]. They found that the composition of the dimeric row of ATP synthase is driven by a reduction in membrane elastic energy rather than by direct protein contact and that the dimeric row drives the formation of highly curved cristae in the mitochondrial cristae (Fig. 7.12). Moreover, they also revealed profound age-dependent changes in mitochondrial membrane structure by cryo-ET imaging (Fig. 7.13) [88]. Besides mitochondria, Wolfgang Baumeister's team also characterized the structure of the endoplasmic reticulum (ER) by cryo-ET and investigated the structure of local translocations in evolutionarily distinct biological and disease-associated TRAP mutant fibroblasts (Fig. 7.14) [89]. The team assigned positions to the four TRAP subunits in the complex to understand their respective functions. Revealing the molecular structure of a central translocation subcomponent has advanced our understanding of membrane protein biogenesis and elucidated the role of TRAP in human congenital glycosylation diseases. Besides, the ultrastructure of the Golgi at molecular resolution in its native environment was also obtained by cryo-ET imaging [90].

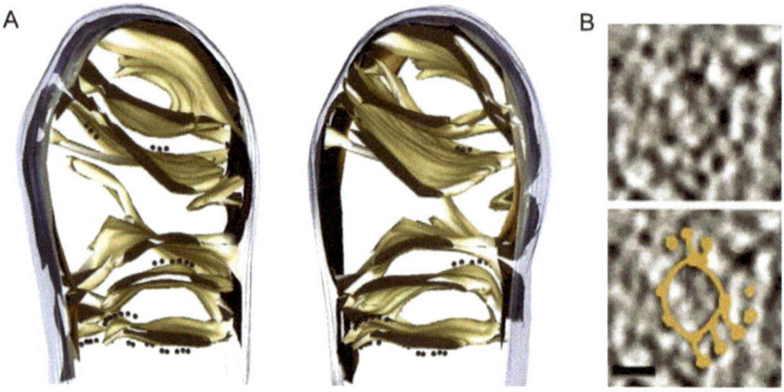

Figure 7.11 Investigating the ultrastructure of mitochondria by combining cryo-CLEM and ET. (A) A three-dimensional model of the mitochondria is shown from two different angles. The outer membrane is colored in blue; the inner boundary membrane is shown in dark brown; the cristae are shown in light brown; and several ATP synthases are shown in black. Note that the cristae are disc-shaped, not tubular, and that the junction between the cristae and the inner membrane is often elongated. (B) As shown above, the ATP synthase plotted in (A) forms a repeating density pattern, sometimes connected to a dense region on the membrane by a stalk. The same image is superimposed with a schematic representation of the putative ATPase in the bottom panel. Reprinted from Ref. [86], Copyright 2009, with permission from Elsevier.

Cryo-ET is being more and more widely used due to its unique imaging advantages and has become an increasingly important technical tool in structural biology research. As many researchers say, tomography is the king of the future.

7.4 Protocols for Cryo-EM Experiment

7.4.1 Protocols for SPA Experiment

7.4.1.1 Extraction and purification of samples

For proteins with small molecular weight and low endogenous abundance, extraction and purification can be performed by exogenous recombinant overexpression. For some large molecular complexes with larger molecular weight, such as ribosomes, extraction can be done from tissues. The purity of the samples should be as high as possible, generally more than 90–95%.

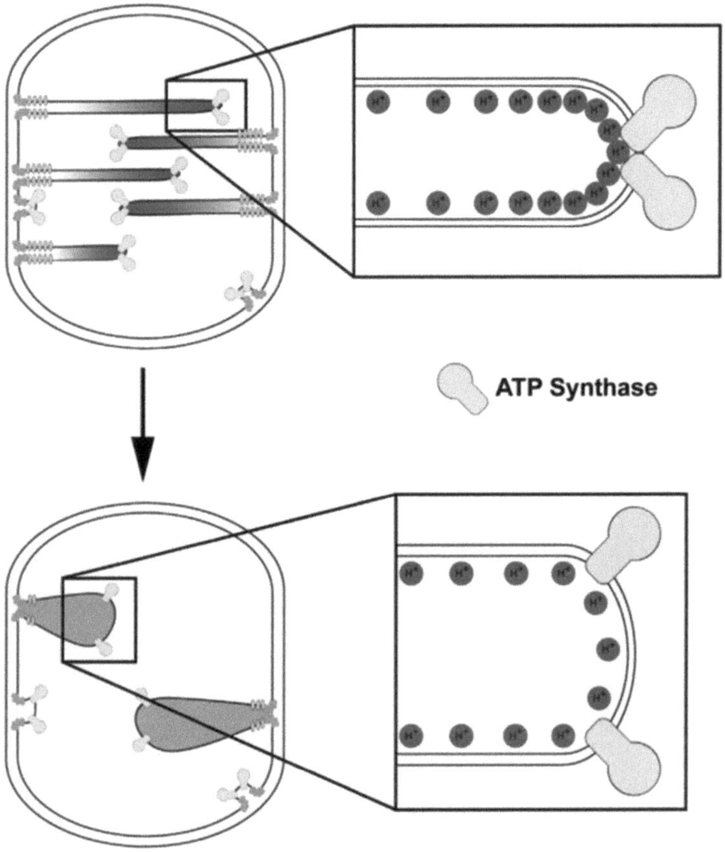

Figure 7.12 The role of ATP synthase dimers in shaping the mitochondrial cristae. Reprinted from Ref. [38], Copyright 1969, with permission from Elsevier.

7.4.1.2 Sample preparation and data acquisition

The extracted and purified samples are plunge frozen to obtain a well-dispersed and homogeneous sample for data acquisition. Plunge freezing involves (1) keeping the sample in solution and maintaining its natural state; (2) adding a drop of the sample onto an EM grid covered with holey carbon film; (3) blotting off the excess sample with filter paper; and (4) immersing the sample rapidly in liquid ethane refrigerant at liquid nitrogen temperature [11]. Since the sample layer is very thin and can be frozen very fast (up to tens of thousands of degrees per second), the water in the sample forms an amorphous ice, and the molecules are suspended in this layer of vitreous ice, preserving the original structural information and conformational state of the sample. The prepared frozen-hydrated samples are subsequently imaged by cryo-EM to obtain a large number of 2D electron projections of the target samples. Generally, a large number of particles need to be collected, and the higher the target resolution, the more particles are needed, especially for samples without symmetry and those with poor homogeneity. The amount of data collected depends on many factors such as sample symmetry, homogeneity, imaging quality, and target resolution.

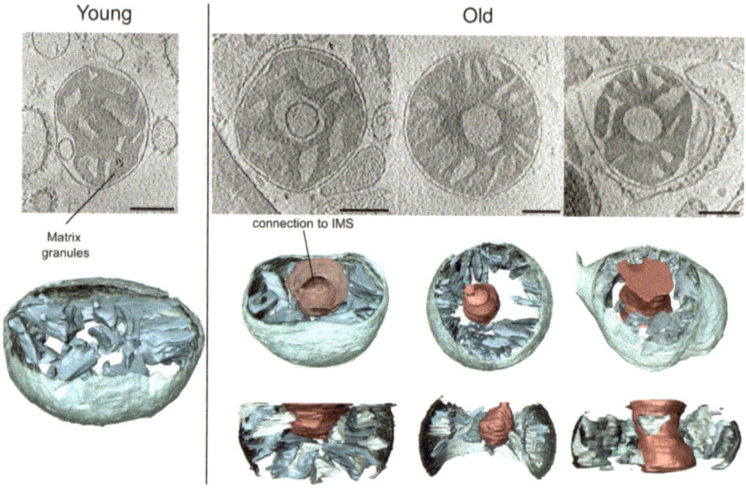

Figure 7.13 Cryo-ET of liver mitochondria from young (left, 20 weeks old) and old (right, 80 weeks old) mice. Upper panel: slices through tomographic volumes (scale bars, 250 nm). Lower panels: segmented 3D volumes. About 25% of the mitochondria from old animals have large central voids (red). The voids were connected to the intermembrane space (IMS) by openings of variable sizes. Reprinted from Ref. [88], Copyright 2017, Brandt et al.

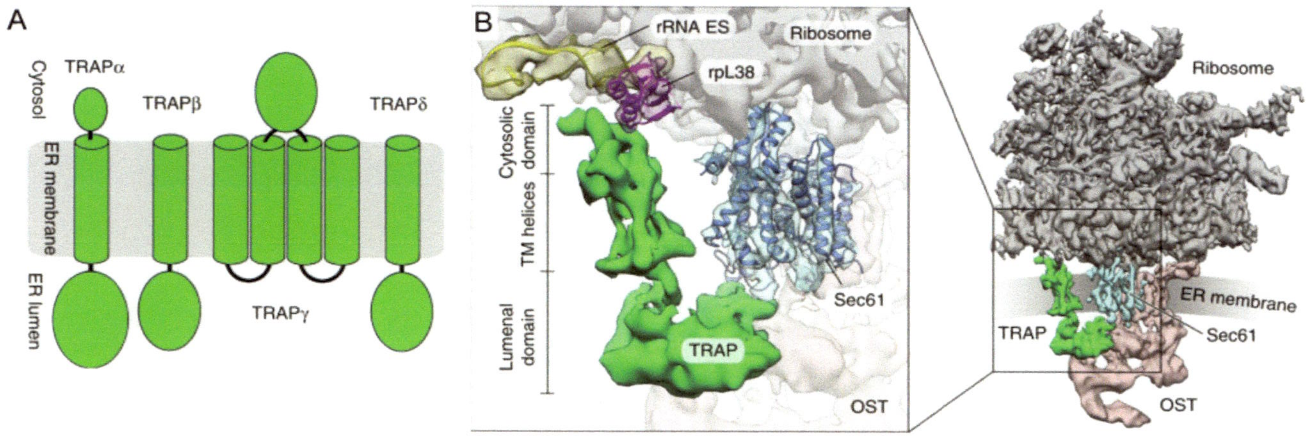

Figure 7.14 Overall structure and subunit composition of the mammalian TRAP complex. (A) TRAP subunit composition and membrane topology as predicted by bioinformatic analysis. (B) Isolated densities for the mammalian ribosome (grey), the Sec61 protein-conducting channel (blue), TRAP (green), and OST (red) extracted from the high-resolution tomography density (EMD-3068) of the ER membrane-associated ribosome. In the magnified insert (left), atomic models for Sec61 (blue), rpL38 (magenta), and an rRNA ES (yellow) are superposed with the EM density to visualize interactions of the TRAP complex with the ribosome and the protein-conducting channel. Reprinted from Ref. [89], with permission from Springer Nature.

7.4.1.3 Data processing

To obtain high-resolution structures, we first need to pick the target particles from the obtained 2D electron projections, also known as "particle picking," which can be divided into manual picking, semi-automatic picking, and fully automatic picking. The semi-automatic and fully automatic particle picking are mainly based on template matching with templates and deep learning without templates to identify the target structures in the raw data. Then, the selected particle images are matched and classified in 2D by the maximum likelihood method. The errors of rotation and translation of the particles are eliminated by image matching, and the image classification is performed using the principle of tight within-class and discrete between-class, which can eventually average the particle images within classes and improve the SNR, thus realizing the construction of high-resolution 3D structures. The basis of 3D construction is the central section theorem, and the key problem in the construction process is how to determine the spatial angle (orientation determination) of each particle image. Most of the 3D construction and optimization algorithms are based on projection-matching iterative methods. In brief, a rough 3D structure model is used to project the reference image, compare it with the experimental particle image, update the spatial orientation parameters according to the results, and then construct a new 3D structure, correct the spatial orientation of the experimental image, and form an iterative process until convergence to obtain the final 3D structure. There are many integrated software packages to complete the whole analysis process (e.g., SPIDER [91], EMAN2 [27], FREALIGN [92], SPARX [93], RELION [28], etc.). Finally, the overall resolution of the obtained 3D structure is evaluated by the Fourier shell correlation (FSC) curve, and the local resolution of the 3D structure is evaluated by the ResMap software [94].

7.4.2 Protocols for Cryo-ET Experiment

7.4.2.1 Sample preparation

Smaller or thinner samples that can be transmitted by the electron beam, such as viruses, bacteria, and some small cells, can be prepared for cryo-ET studies by the plunge-freezing method. For larger or thicker samples, the depth and volume are limited by plunge freezing. High-pressure freezing can be used to lower the freezing point of water to make plunge freezing reach deeper levels [95]. In addition, thick samples can be thinned by techniques such as cryo-sectioning [96] or FIB milling [83] to meet the requirements of cryo-ET.

7.4.2.2 Cryo-ET data collection

To obtain cryo-ET data, the prepared frozen-hydrated sample is rotated to a certain angle at a fixed angular interval and the electron projection is collected at the corresponding angle. Data can be collected either by starting at an angle of 0 degrees and tilting toward the ends, or by starting from the maximum negative tilt angle to the maximum positive tilt angle. The most common tilt series for collection range from ±60 to ±70 degrees, using linear steps of 1- to 3-degree intervals. The data collection process is as follows: (1) search for a suitable area, usually at low magnification; (2) adjust the sample stage tilt center height (eucentricity) so that the sample stays in the imaging center range when the stage is tilted; (3) focus; (4) locate the exact tracking of the same imaging area (tracking); (5) take a picture (exposure); (6) find the center of the sample position again after the sample is tilted to a higher angle, and then repeat the above four modes, i.e., search, focus, tracking, and exposure alternately. Through the above data collection process and method, the tilt series consisting of electronic projections of the sample at different angles were obtained.

7.4.2.3 Cryo-ET data processing and analysis

The drift correction is applied to each angle of the tilt series by MotionCorr [97], and each 2D projected image is projected into 3D space by weighted back projection in IMOD with the backward direction of its tilt angle at the time of recording, and all back-projected images are superimposed in 3D space to form the 3D structure of the sample [98]. In addition, there are other reconstruction methods, including iterative reconstruction methods published in the early 1970s, such as algebraic reconstruction technique (ART) [99] and simultaneous iterative reconstruction technique (SIRT) [100], etc. To further analyze the target structures in tomograms, the target structures in tomograms can be identified and selected by manual or semi-automatic methods (e.g., template matching [43], convolutional neural networks [101], Pyseg [102], etc.). Subsequently, the annotated sub-tomograms containing the target structures are aligned and averaged by some software (e.g., EMAN2 [27], Relion [28], emClarity [103], etc.) to improve the SNR of the target structures and obtain high-resolution target structures.

References

1. Gorter, E. and Grendel, F. (1925). On biomolecular layers of lipoids on the chromocytes of the blood, *J. Exp. Med.*, **41**, pp. 439–443.
2. Singer, S. J. and Nicolson Garth, L. (1972). The fluid mosaic model of the structure of cell membranes, *Science*, **175**, pp. 720–731.
3. Simons, K. and Ikonen, E. (1997). Functional rafts in cell membranes, *Nature*, **387**, pp. 569–572.
4. Taylor Kenneth, A. and Glaeser Robert, M. (1974). Electron diffraction of frozen, hydrated protein crystals, *Science*, **186**, pp. 1036–1037.
5. Dessau, M. A. and Modis, Y. (2011). Protein crystallization for X-ray crystallography, *J Vis Exp*, pp. 2285.
6. Foster, M. P., McElroy, C. A., and Amero, C. D. (2007). Solution NMR of large molecules and assemblies, *Biochemistry*, **46**, pp. 331–340.
7. Sprangers, R., Velyvis, A., and Kay, L. E. (2007). Solution NMR of supramolecular complexes: Providing new insights into function, *Nat. Methods*, **4**, pp. 697–703.
8. Cressey, D. and Callaway, E. (2017). Cryo-electron microscopy wins chemistry Nobel, *Nature*, **550**, pp. 167–167.
9. Taylor, K. A. and Glaeser, R. M. (2008). Retrospective on the early development of cryoelectron microscopy of macromolecules and a prospective on opportunities for the future, *J. Struct. Biol.*, **163**, pp. 214–223.
10. Egerton, R. F. (2013). Control of radiation damage in the TEM, *Ultramicroscopy*, **127**, pp. 100–108.
11. Dubochet, J., Adrian, M., Chang, J.-J., Homo, J.-C., Lepault, J., McDowall, A. W., and Schultz, P. (1988). Cryo-electron microscopy of vitrified specimens, *Q. Rev. Biophys.*, **21**, pp. 129–228.
12. Dubochet, J. (2012). Cryo-EM—the first thirty years, *J. Microsc.*, **245**, pp. 221–224.

13. Frank, J., Goldfarb, W., Eisenberg, D., and Baker, T. S. (1978). Reconstruction of glutamine synthetase using computer averaging, *Ultramicroscopy*, **3**, pp. 283–290.

14. Liao, M., Cao, E., Julius, D., and Cheng, Y. (2013). Structure of the TRPV1 ion channel determined by electron cryo-microscopy, *Nature*, **504**, pp. 107–112.

15. Zhang, X., Ge, P., Yu, X., Brannan, J. M., Bi, G., Zhang, Q., Schein, S., and Zhou, Z. H. (2013). Cryo-EM structure of the mature dengue virus at 3.5-Å resolution, *Nat. Struct. Mol. Biol.*, **20**, pp. 105–110.

16. Adrian, M., Dubochet, J., Lepault, J., and McDowall, A. W. (1984). Cryo-electron microscopy of viruses, *Nature*, **308**, pp. 32–36.

17. Fernández Israel, S., Bai, X.-C., Hussain, T., Kelley Ann, C., Lorsch Jon, R., Ramakrishnan, V., and Scheres Sjors, H. W. (2013). Molecular architecture of a eukaryotic translational initiation complex, *Science*, **342**, pp. 1240585.

18. Bai, X.-C. (2021). Seeing atoms by single-particle cryo-EM, *Trends Biochem. Sci.*, **46**, pp. 253–254.

19. Taylor, K. A. and Glaeser, R. M. (1976). Electron microscopy of frozen hydrated biological specimens, *J. Ultrastruct. Res.*, **55**, pp. 448–456.

20. Lepault, J., Booy, F. P., and Dubochet, J. (1983). Electron microscopy of frozen biological suspensions, *J. Microsc.*, **129**, pp. 89–102.

21. De Rosier, D. J. and Klug, A. (1968). Reconstruction of three-dimensional structures from electron micrographs, *Nature*, **217**, pp. 130–134.

22. Cao, E., Liao, M., Cheng, Y., and Julius, D. (2013). TRPV1 structures in distinct conformations reveal activation mechanisms, *Nature*, **504**, pp. 113–118.

23. Wang, H. (2015). Cryo-electron microscopy for structural biology: Current status and future perspectives, *Science China Life Sciences*, **58**, pp. 750–756.

24. Nogales, E. and Scheres, S. H. W. (2015). Cryo-EM: A unique tool for the visualization of macromolecular complexity, *Mol. Cell*, **58**, pp. 677–689.

25. Danev, R. and Baumeister, W. (2016). Cryo-EM single particle analysis with the Volta phase plate, *Elife*, **5**, pp. e13046.

26. Wu, S., Armache, J.-P., and Cheng, Y. (2016). Single-particle cryo-EM data acquisition by using direct electron detection camera, *Microscopy*, **65**, pp. 35–41.

27. Tang, G., Peng, L., Baldwin, P. R., Mann, D. S., Jiang, W., Rees, I., and Ludtke, S. J. (2007). EMAN2: An extensible image processing suite for electron microscopy, *J. Struct. Biol.*, **157**, pp. 38–46.

28. Scheres, S. H. W. (2012). RELION: Implementation of a Bayesian approach to cryo-EM structure determination, *J. Struct. Biol.*, **180**, pp. 519–530.

29. Campbell, M. G., Veesler, D., Cheng, A., Potter, C. S., and Carragher, B. (2015). 2.8 Å resolution reconstruction of the *Thermoplasma acidophilum* 20S proteasome using cryo-electron microscopy, *Elife*, **4**, pp. e06380.

30. Yip, K. M., Fischer, N., Paknia, E., Chari, A., and Stark, H. (2020). Atomic-resolution protein structure determination by cryo-EM, *Nature*, **587**, pp. 157–161.

31. Merk, A., Bartesaghi, A., Banerjee, S., Falconieri, V., Rao, P., Davis, M. I., Pragani, R., Boxer, M. B., Earl, L. A., Milne, J. L. S., and Subramaniam, S. (2016). Breaking cryo-EM resolution barriers to facilitate drug discovery, *Cell*, **165**, pp. 1698–1707.

32. Lučić, V., Rigort, A., and Baumeister, W. (2013). Cryo-electron tomography: The challenge of doing structural biology in situ, *J. Cell Biol.*, **202**, pp. 407–419.

33. Robinson, C. V., Sali, A., and Baumeister, W. (2007). The molecular sociology of the cell, *Nature*, **450**, pp. 973–982.

34. Doerr, A. (2017). Cryo-electron tomography, *Nat. Methods*, **14**, pp. 34–34.

35. Harapin, J., Eibauer, M., and Medalia, O. (2013). Structural analysis of supramolecular assemblies by cryo-electron tomography, *Structure*, **21**, pp. 1522–1530.

36. Galaz-Montoya, J. G. and Ludtke, S. J. (2017). The advent of structural biology in situ by single particle cryo-electron tomography, *Biophysics Reports*, **3**, pp. 17–35.

37. von Appen, A. and Beck, M. (2016). Structure determination of the nuclear pore complex with three-dimensional cryo electron microscopy, *J. Mol. Biol.*, **428**, pp. 2001–2010.

38. Siegmund, S. E., Grassucci, R., Carter, S. D., Barca, E., Farino, Z. J., Juanola-Falgarona, M., Zhang, P., Tanji, K., Hirano, M., Schon, E. A., Frank, J., and Freyberg, Z. (2018). Three-dimensional analysis of mitochondrial crista ultrastructure in a patient with leigh syndrome by in situ cryoelectron tomography, *iScience*, **6**, pp. 83–91.

39. Vijayakrishnan, S., McElwee, M., Loney, C., Rixon, F., and Bhella, D. (2020). Structure determination of virus capsids imaged within cell nuclei by correlative light and cryo-electron tomography, *bioRxiv*, https://doi.org/10.1101/2020.02.06.936948.

40. Melia, C. E. and Bharat, T. A. M. (2018). Locating macromolecules and determining structures inside bacterial cells using electron cryotomography, *Biochim. Biophys. Acta*, **1866**, pp. 973–981.

41. Mahamid, J., Pfeffer, S., Schaffer, M., Villa, E., Danev, R., Kuhn Cuellar, L., Förster, F., Hyman Anthony, A., Plitzko Jürgen, M., and Baumeister, W. (2016). Visualizing the molecular sociology at the HeLa cell nuclear periphery, *Science*, **351**, pp. 969–972.

42. Lewis, A. J., Genoud, C., Pont, M., van de Berg, W. D. J., Frank, S., Stahlberg, H., Shahmoradian, S. H., and Al-Amoudi, A. (2019). Imaging of post-mortem human brain tissue using electron and X-ray microscopy, *Curr. Opin. Struct. Biol.*, **58**, pp. 138–148.

43. Böhm, J., Frangakis Achilleas, S., Hegerl, R., Nickell, S., Typke, D., and Baumeister, W. (2000). Toward detecting and identifying macromolecules in a cellular context: Template matching applied to electron tomograms, *Proc. Natl. Acad. Sci. U.S.A.*, **97**, pp. 14245–14250.

44. Turoňová, B., Schur, F. K. M., Wan, W., and Briggs, J. A. G. (2017). Efficient 3D-CTF correction for cryo-electron tomography using NovaCTF improves subtomogram averaging resolution to 3.4 Å, *J. Struct. Biol.*, **199**, pp. 187–195.

45. Briggs, J. A. G. (2013). Structural biology in situ: The potential of subtomogram averaging, *Curr. Opin. Struct. Biol.*, **23**, pp. 261–267.

46. Chen, Y., Pfeffer, S., Fernández, José J., Sorzano, Carlos Oscar S., and Förster, F. (2014). Autofocused 3D classification of cryoelectron subtomograms, *Structure*, **22**, pp. 1528–1537.

47. Couzin, J. and Enserink, M. J. S. (2002). Breakthrough of the year: Small RNAs make big splash, *Science*, **298**, pp. 2296–2297.

48. Seddon, A. M., Curnow, P., and Booth, P. J. (2004). Membrane proteins, lipids and detergents: Not just a soap opera, *Biochim. Biophys. Acta*, **1666**, pp. 105–117.

49. Privé, G. G. (2007). Detergents for the stabilization and crystallization of membrane proteins, *Methods*, **41**, pp. 388–397.

50. Paulsen, C. E., Armache, J.-P., Gao, Y., Cheng, Y., and Julius, D. (2015). Structure of the TRPA1 ion channel suggests regulatory mechanisms, *Nature*, **520**, pp. 511–517.

51. des Georges, A., Clarke, O. B., Zalk, R., Yuan, Q., Condon, K. J., Grassucci, R. A., Hendrickson, W. A., Marks, A. R., and Frank, J. (2016). Structural basis for gating and activation of RyR1, *Cell*, **167**, pp. 145–157.
52. Peng, W., Shen, H., Wu, J., Guo, W., Pan, X., Wang, R., Chen, S. R. W., and Yan, N. (2016). Structural basis for the gating mechanism of the type 2 ryanodine receptor RyR2, *Science*, **354**, pp. aah5324.
53. Huynh, K. W., Cohen, M. R., Jiang, J., Samanta, A., Lodowski, D. T., Zhou, Z. H., and Moiseenkova-Bell, V. Y. (2016). Structure of the full-length TRPV2 channel by cryo-EM, *Nat. Commun.*, **7**, pp. 11130.
54. Grieben, M., Pike, A. C. W., Shintre, C. A., Venturi, E., El-Ajouz, S., Tessitore, A., Shrestha, L., Mukhopadhyay, S., Mahajan, P., Chalk, R., Burgess-Brown, N. A., Sitsapesan, R., Huiskonen, J. T., and Carpenter, E. P. (2017). Structure of the polycystic kidney disease TRP channel Polycystin-2 (PC2), *Nat. Struct. Mol. Biol.*, **24**, pp. 114–122.
55. Sun, L., Zhao, L., Yang, G., Yan, C., Zhou, R., Zhou, X., Xie, T., Zhao, Y., Wu, S., Li, X., and Shi, Y. (2015). Structural basis of human γ-secretase assembly, *Proc. Natl. Acad. Sci. U.S.A.*, **112**, pp. 6003–6008.
56. Meyerson, J. R., Chittori, S., Merk, A., Rao, P., Han, T. H., Serpe, M., Mayer, M. L., and Subramaniam, S. (2016). Structural basis of kainate subtype glutamate receptor desensitization, *Nature*, **537**, pp. 567–571.
57. Wu, J., Yan, Z., Li, Z., Qian, X., Lu, S., Dong, M., Zhou, Q., and Yan, N. (2016). Structure of the voltage-gated calcium channel Cav1.1 at 3.6 Å resolution, *Nature*, **537**, pp. 191–196.
58. Vinothkumar, K. R., Zhu, J., and Hirst, J. (2014). Architecture of mammalian respiratory complex I, *Nature*, **515**, pp. 80–84.
59. Fiedorczuk, K., Letts, J. A., Degliesposti, G., Kaszuba, K., Skehel, M., and Sazanov, L. A. (2016). Atomic structure of the entire mammalian mitochondrial complex I, *Nature*, **538**, pp. 406–410.
60. Zhu, J., Vinothkumar, K. R., and Hirst, J. (2016). Structure of mammalian respiratory complex I, *Nature*, **536**, pp. 354–358.
61. Hauer, F., Gerle, C., Fischer, N., Oshima, A., Shinzawa-Itoh, K., Shimada, S., Yokoyama, K., Fujiyoshi, Y., and Stark, H. (2015). GraDeR: Membrane protein complex preparation for single-particle cryo-EM, *Structure*, **23**, pp. 1769–1775.
62. Oshima, A., Tani, K., and Fujiyoshi, Y. (2016). Atomic structure of the innexin-6 gap junction channel determined by cryo-EM, *Nat. Commun.*, **7**, pp. 13681.
63. Sgro, G. G. and Costa, T. R. D. (2018). Cryo-EM grid preparation of membrane protein samples for single particle analysis, *Front. Mol. Biosci.*, **5**, 74.

64. Phillips, R., Ursell, T., Wiggins, P., and Sens, P. (2009). Emerging roles for lipids in shaping membrane-protein function, *Nature*, **459**, pp. 379–385.

65. Zhou, H.-X. and Cross, T. A. (2013). Influences of membrane mimetic environments on membrane protein structures, *Annu Rev Biophys*, **42**, pp. 361–392.

66. Saliba, A.-E., Vonkova, I., and Gavin, A.-C. (2015). The systematic analysis of protein–lipid interactions comes of age, *Nat. Rev. Mol. Cell Biol.*, **16**, pp. 753–761.

67. Denisov, I. G. and Sligar, S. G. (2016). Nanodiscs for structural and functional studies of membrane proteins, *Nat. Struct. Mol. Biol.*, **23**, pp. 481–486.

68. Frauenfeld, J., Gumbart, J., Sluis, E. O. V. D., Funes, S., Gartmann, M., Beatrix, B., Mielke, T., Berninghausen, O., Becker, T., Schulten, K., and Beckmann, R. (2011). Cryo-EM structure of the ribosome–SecYE complex in the membrane environment, *Nat. Struct. Mol. Biol.*, **18**, pp. 614–621.

69. Kedrov, A., Wickles, S., Crevenna, A. H., van der Sluis, E. O., Buschauer, R., Berninghausen, O., Lamb, D. C., and Beckmann, R. (2016). Structural dynamics of the YidC:Ribosome complex during membrane protein biogenesis, *Cell Rep.*, **17**, pp. 2943–2954.

70. Efremov, R. G., Leitner, A., Aebersold, R., and Raunser, S. (2015). Architecture and conformational switch mechanism of the ryanodine receptor, *Nature*, **517**, pp. 39–43.

71. Gatsogiannis, C., Merino, F., Prumbaum, D., Roderer, D., Leidreiter, F., Meusch, D., and Raunser, S. (2016). Membrane insertion of a Tc toxin in near-atomic detail, *Nat. Struct. Mol. Biol.*, **23**, pp. 884–890.

72. Gao, Y., Cao, E., Julius, D., and Cheng, Y. (2016). TRPV1 structures in nanodiscs reveal mechanisms of ligand and lipid action, *Nature*, **534**, pp. 347–351.

73. Shen, P. S., Yang, X., DeCaen, P. G., Liu, X., Bulkley, D., Clapham, D. E., and Cao, E. (2016). The structure of the polycystic kidney disease channel PKD2 in lipid nanodiscs, *Cell*, **167**, pp. 763–773.

74. Zoghbi, M. E. and Altenberg, G. A. (2017). Membrane protein reconstitution in nanodiscs for luminescence spectroscopy studies, *Nanotechnol. Rev.*, **6**, pp. 33–46.

75. Racker, E. and Stoeckenius, W. (1974). Reconstitution of purple membrane vesicles catalyzing light-driven proton uptake and adenosine triphosphate formation, *J. Biol. Chem.*, **249**, pp. 662–663.

76. Vitanen, P., Newman, M., Foster, D., Wilson, T., and Kaback, H. J. M. E. (1986). Purification, reconstitution and characterization of the lac permease of *E. coli, Meth. Enzymol.*, **125**, pp. 429–452.

77. Rigaud, J.-L. and Research, B. (2002). Membrane proteins: Functional and structural studies using reconstituted proteoliposomes and 2-D crystals, *Braz. J. Med. Biol. Res.*, **35**, pp. 753–766.

78. Chan, Y.-H. M. and Boxer, S. G. (2007). Model membrane systems and their applications, *Curr. Opin. Chem. Biol.*, **11**, pp. 581–587.

79. Yao, X., Fan, X., and Yan, N. (2020). Cryo-EM analysis of a membrane protein embedded in the liposome, *Proc. Natl. Acad. Sci. U.S.A.*, **117**, pp. 18497–18503.

80. Scalise, M., Pochini, L., Giangregorio, N., Tonazzi, A., and Indiveri, C. (2013). Proteoliposomes as tool for assaying membrane transporter functions and interactions with xenobiotics. *Pharmaceutics*, **5**, pp. 472–497.

81. Liang, Y.-L., Khoshouei, M., Radjainia, M., Zhang, Y., Glukhova, A., Tarrasch, J., Thal, D. M., Furness, S. G. B., Christopoulos, G., Coudrat, T., Danev, R., Baumeister, W., Miller, L. J., Christopoulos, A., Kobilka, B. K., Wootten, D., Skiniotis, G., and Sexton, P. M. (2017). Phase-plate cryo-EM structure of a class B GPCR–G-protein complex, *Nature*, **546**, pp. 118–123.

82. Silvester, E., Vollmer, B., Pražák, V., Vasishtan, D., Machala, E. A., Whittle, C., Black, S., Bath, J., Turberfield, A. J., Grünewald, K., and Baker, L. A. (2021). DNA origami signposts for identifying proteins on cell membranes by electron cryotomography, *Cell*, **184**, pp. 1110–1121.

83. Villa, E., Schaffer, M., Plitzko, J. M., and Baumeister, W. (2013). Opening windows into the cell: Focused-ion-beam milling for cryo-electron tomography, *Curr. Opin. Struct. Biol.*, **23**, pp. 771–777.

84. Schaffer, M., Mahamid, J., Engel, B. D., Laugks, T., Baumeister, W., and Plitzko, J. M. (2017). Optimized cryo-focused ion beam sample preparation aimed at in situ structural studies of membrane proteins, *J. Struct. Biol.*, **197**, pp. 73–82.

85. Zhao, G., Cheng, S., Yu, Y., Zou, T., Wang, H., Tao, C., Bi, G., Zhou, Z. H., and Wang, H. (2021). The high-resolution structure of cell membranes revealed by in situ cryo-electron tomography, *bioRxiv*, https://doi.org/10.1101/2021.12.03.471052.

86. van Driel, L. F., Valentijn, J. A., Valentijn, K. M., Koning, R. I., and Koster, A. J. (2009). Tools for correlative cryo-fluorescence microscopy and cryo-electron tomography applied to whole mitochondria in human endothelial cells, *Eur. J. Cell Biol.*, **88**, pp. 669–684.

87. Davies Karen, M., Anselmi, C., Wittig, I., Faraldo-Gómez José, D., and Kühlbrandt, W. (2012). Structure of the yeast F1Fo-ATP synthase dimer and its role in shaping the mitochondrial cristae, *Proc. Natl. Acad. Sci. U.S.A.*, **109**, pp. 13602–13607.

88. Brandt, T., Mourier, A., Tain, L. S., Partridge, L., Larsson, N.-G., and Kühlbrandt, W. (2017). Changes of mitochondrial ultrastructure and function during ageing in mice and Drosophila. *Elife*, **6**, e24662.

89. Pfeffer, S., Dudek, J., Schaffer, M., Ng, B. G., Albert, S., Plitzko, J. M., Baumeister, W., Zimmermann, R., Freeze, H. H., Engel, B. D., and Förster, F. (2017). Dissecting the molecular organization of the translocon-associated protein complex, *Nat. Commun.*, **8**, pp. 14516.

90. Engel Benjamin, D., Schaffer, M., Albert, S., Asano, S., Plitzko Jürgen, M., and Baumeister, W. (2015). In situ structural analysis of Golgi intracisternal protein arrays, *Proc. Natl. Acad. Sci. U.S.A.*, **112**, pp. 11264–11269.

91. Shaikh, T. R., Gao, H., Baxter, W. T., Asturias, F. J., Boisset, N., Leith, A., and Frank, J. (2008). SPIDER image processing for single-particle reconstruction of biological macromolecules from electron micrographs, *Nat. Protoc.*, **3**, pp. 1941–1974.

92. Grigorieff, N. (2007). FREALIGN: High-resolution refinement of single particle structures, *J. Struct. Biol.*, **157**, pp. 117–125.

93. Hohn, M., Tang, G., Goodyear, G., Baldwin, P. R., Huang, Z., Penczek, P. A., Yang, C., Glaeser, R. M., Adams, P. D., and Ludtke, S. J. (2007). SPARX, a new environment for Cryo-EM image processing, *J. Struct. Biol.*, **157**, pp. 47–55.

94. Swint-Kruse, L. and Brown, C. S. (2005). Resmap: Automated representation of macromolecular interfaces as two-dimensional networks, *Bioinformatics*, **21**, pp. 3327–3328.

95. Studer, D., Humbel, B. M., and Chiquet, M. (2008). Electron microscopy of high-pressure frozen samples: Bridging the gap between cellular ultrastructure and atomic resolution, *Histochem. Cell Biol.*, **130**, pp. 877–889.

96. Pierson, J., Fernández, J. J., Bos, E., Amini, S., Gnaegi, H., Vos, M., Bel, B., Adolfsen, F., Carrascosa, J. L., and Peters, P. J. (2010). Improving the technique of vitreous cryo-sectioning for cryo-electron tomography: Electrostatic charging for section attachment and implementation of an anti-contamination glove box, *J. Struct. Biol.*, **169**, pp. 219–225.

97. Zheng, S. Q., Palovcak, E., Armache, J.-P., Verba, K. A., Cheng, Y., and Agard, D. A. (2017). MotionCor2: Anisotropic correction of beam-

induced motion for improved cryo-electron microscopy, *Nat. Methods*, **14**, pp. 331–332.

98. Mastronarde, D. N. and Held, S. R. (2017). Automated tilt series alignment and tomographic reconstruction in IMOD, *J. Struct. Biol.*, **197**, pp. 102–113.

99. Gordon, R., Bender, R., and Herman, G. T. (1970). Algebraic reconstruction techniques (ART) for three-dimensional electron microscopy and X-ray photography, *J. Theor. Biol.*, **29**, pp. 471–481.

100. Wolf, D., Lubk, A., and Lichte, H. (2014). Weighted simultaneous iterative reconstruction technique for single-axis tomography, *Ultramicroscopy*, **136**, pp. 15–25.

101. Chen, M., Dai, W., Sun, S. Y., Jonasch, D., He, C. Y., Schmid, M. F., Chiu, W., and Ludtke, S. J. (2017). Convolutional neural networks for automated annotation of cellular cryo-electron tomograms, *Nat. Methods*, **14**, pp. 983–985.

102. Martinez-Sanchez, A., Kochovski, Z., Laugks, U., Meyer zum Alten Borgloh, J., Chakraborty, S., Pfeffer, S., Baumeister, W., and Lučić, V. (2020). Template-free detection and classification of membrane-bound complexes in cryo-electron tomograms, *Nat. Methods*, **17**, pp. 209–216.

103. Himes, B. A. and Zhang, P. (2018). emClarity: Software for high-resolution cryo-electron tomography and subtomogram averaging, *Nat. Methods*, **15**, pp. 955–961.

Chapter 8

Cryo-Scanning Electron Microscope for Studying Plasma Membranes

Sihang Cheng, Jing Gao, and Hongda Wang
State Key Laboratory of Electroanalytical Chemistry,
Research Center of Biomembranomics,
Changchun Institute of Applied Chemistry,
Chinese Academy of Sciences, Changchun, Jilin, China
hdwang@ciac.ac.cn

Scanning electron microscopy (SEM) is an intermediate observation method between transmission electron microscopy (TEM) and optical microscopy. It characterizes the surface microscopic morphology of the sample through the interaction between high-energy electrons and substance. Cryo-scanning electron microscopy (cryo-SEM) integrates ultra-low-temperature frozen sample preparation and transmission technique into the SEM, enabling direct observation of liquid, semi-liquid, and electron-beam-sensitive samples such as biological and high polymer materials. With the advantages of simple operation and fast imaging, SEM has become an effective tool for biological membrane observation. This chapter will briefly introduce the SEM technique, including its imaging principle, sample

Biomembranomics: Structure, Techniques, and Applications
Edited by Hongda Wang

ISBN 978-981-4968-61-4 (Hardcover), 978-1-003-45635-3 (eBook)
www.jennystanford.com

preparation, and basic parameter setting, especially the advantages of cryo-SEM and its application in studying plasma membranes.

8.1 Introduction

The cell membrane is one of the essential structures of cells. Its important physiological function is to prevent extracellular materials from freely entering the barrier of the cell and to maintain the relative stability of the intracellular environment [1–3]. The structure of cell membranes has been observed and studied using different techniques and tools, for example, the commonly used optical microscope. Given that the size of substances such as phospholipids and proteins that constitute the cell membrane is much smaller than the optical resolution limit of traditional optical microscopy, indirect imaging must rely on techniques that break through the limitation of low resolution, such as atomic force microscopy (AFM), super-resolution fluorescence microscopy, cryo-transmission electron microscopy (cryo-TEM), and scanning electron microscopy (SEM) [4–6]. Other techniques will be introduced and discussed in detail in other chapters. Here we focus on SEM. Compared with the above techniques, SEM has imaging principles similar to light microscopy, which can directly observe the sample surface morphology, with a large depth of field of view and high resolution. The sample can be rotated in three dimensions in the sample chamber and viewed from various angles. Moreover, the sample preparation process of SEM is very simple, facilitating quick experiments. Moreover, by combining ultra-low temperature freezing sample preparation and freezing transfer technology, freezing SEM can directly observe the biological samples containing water. Due to this advantage, cryo-scanning electron microscopy (cryo-SEM) is increasingly used in the field of cell biology.

In this section, we first introduce the imaging principle and basic parameters of SEM as well as the relevant contents applicable to cell membrane imaging. Then, we focus on the basic structure and operating process of cryo-SEM and summarize its application in cell biology. Finally, we discuss the operating steps and preliminary results when using cryo-SEM in the observation of cytoplasmic membranes. Through the combination of different techniques, we expect that cryo-SEM will play its unique role in cell membrane studies and help to deeply understand the characteristics of cell membranes.

8.2 Scanning Electron Microscopy

SEM is an instrument that uses a finely focused electron beam to bombard the surface of a sample to observe and analyze the surface or fracture morphology of the sample through secondary electrons and backscattered electrons generated by the interaction between electrons and the sample, which is currently widely used in the fields of materials, metallurgy, minerals, and biology (Fig. 8.1). SEM has the advantages of easy sample preparation, large and thick sample observation, a wide range of magnification, dynamic observation, etc. [5, 7].

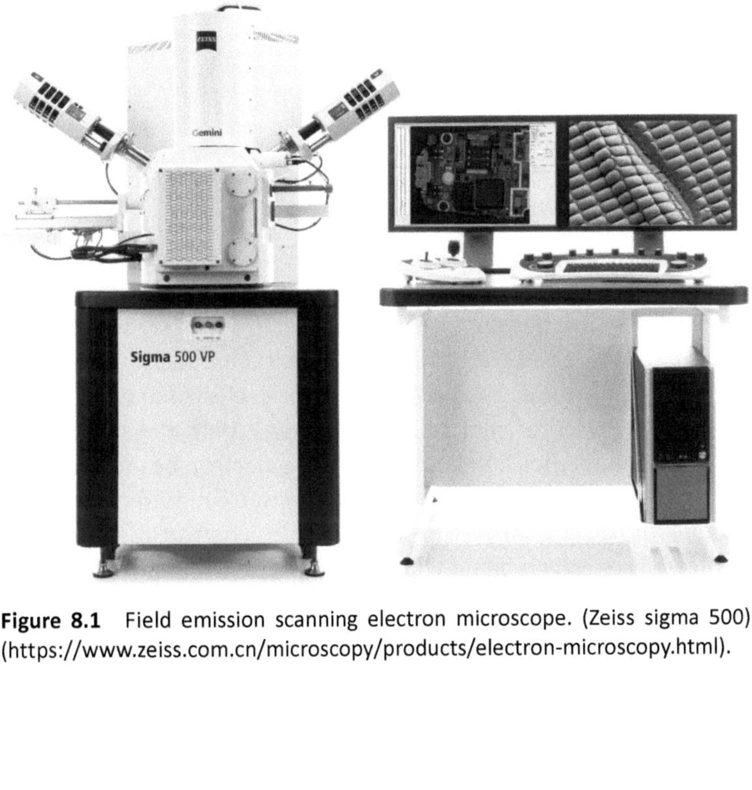

Figure 8.1 Field emission scanning electron microscope. (Zeiss sigma 500) (https://www.zeiss.com.cn/microscopy/products/electron-microscopy.html).

8.2.1 Imaging Principle of SEM

The imaging principle of SEM is different from that of optical microscopy and TEM. The electron optical system of SEM consists of an electron gun, electromagnetic lens, diaphragm, sample chamber, scanning coil, etc. (Fig. 8.2) [8].

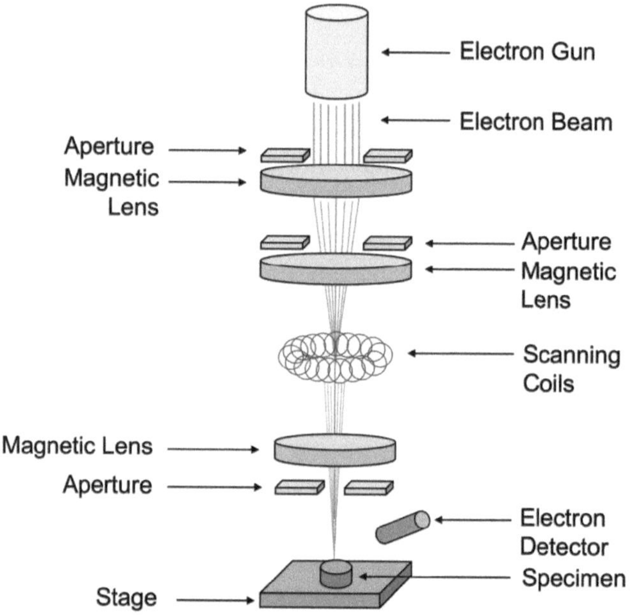

Figure 8.2 The basic structure of a scanning electron microscope.

The electron beam emitted by the electron gun is focused by the gate to become a fine electron beam focused on the sample surface after passing downward through several electromagnetic lenses under the action of an accelerating voltage. With the scanning coil above the upper lens, the direction of the electron beam turns when the electron beam enters the upper deflection coil, and then it enters the lower deflection coil to make the direction of the beam turn twice. The electron beam with the second deflection is shot on the sample surface through the optical center of the upper lens, and a line scan action occurs while the beam is deflected.

The sample generates various physical signals under the action of the high-energy electron beam, such as secondary electrons, backscattered electrons, characteristic X-rays, cathodoluminescence, and transmitted electrons. Different physical signals require corresponding signal detection and amplification systems, among which secondary electrons, backscattered electrons, and transmission electrons are usually measured by a scintillation counter, which is the most important signal detection and amplification system in SEM.

The signal is received by the detector and amplified and transmitted to the gate of the tube to adjust the brightness of the tube. It means that the electron beam scans the sample, line by line, under the action of the scanning coil and that the signals of different intensities generated by the interaction between each point on the sample and the electron beam are converted into video signals by a specific signal detection and amplification system in the same position and in equal proportion, so that we can observe a scanned image of the sample surface with certain features on the fluorescent screen [9–11].

8.2.2 Surface Topography Contrast of SEM

Various signal detection and amplification systems can be used to obtain different image contrasts of the sample. Atomic number contrast is an image contrast that indicates the difference in the chemical composition of the microregion through physical signals that are sensitive to changes in the atomic number or chemical composition of the sample as modulation signals. Signals such as backscattered electrons, absorbed electrons, and characteristic X-rays are sensitive to changes in the atomic number or chemical composition of the microregion, so they can be used to show differences in atomic number or chemical composition. Considering that the atomic number of different membrane components is not much different and that the resolution of atomic number contrast is not high, the atomic number contrast is rarely applied to imaging the plasma membrane, which is, therefore, not discussed here [12–14].

The surface topography contrast is a kind of image contrast obtained through the secondary electron signal as the modulating signal. Since the secondary electron signal mainly comes from the sample surface in the depth range of 5~10 nm, its intensity does not have a quite clear relationship with the atomic number, and only the micro-area faceted phase is very sensitive to the bit direction of the incident electron beam. The secondary electron image resolution is relatively high, so it is especially suitable for displaying the topography contrast.

Secondary electrons are emitted by the inelastic scattering of incident electrons and weakly bound valence electrons in the sample generally with energy less than 50 eV, which are generated within 10 nm of the surface of the sample. When the electrons are incident perpendicular to the surface of the sample, the amount of secondary electrons generated is the smallest, and the amount of secondary electrons gradually increases with the tilt of the sample increasing. For samples with a certain morphology, the morphology can be regarded as the composition of many details such as convex tips, steps, and pits with different inclinations, and the number of secondary electrons emitted from different parts of these details is different, thus leading to the morphological contrast [15, 16].

The secondary electron yield actually has a certain relationship with the composition, especially at a low atomic number ($Z < 20$). The secondary electrons can also clearly reflect the difference between the compositions, so the secondary electron image can also reflect the difference in composition for samples with low atomic numbers or large differences in atomic numbers.

Secondary electrons can be divided into four categories: SE1, SE2, SE3, and SE4. SE1 is excited by the incident electrons in the sample; SE2 is excited by the backscattered electrons in the sample; SE3 is generated by the backscattered electrons of the sample at a point far from the incident point of the electron beam; and SE4 is excited by the incident electron beam in the mirror barrel. SE4 is not discussed here in that it does not produce any effect on imaging.

Different types of secondary electrons have different performance in the contrast and depth of action, which makes the secondary electron images collected by different secondary electron detectors to have relatively large differences. SE1 is excited by the original electron beam, which has the shallowest depth of action and the smallest volume, so the SE1 signal comes from the polar surface of the sample with a high resolution. However, due to the high angle of the SE1 shot, its output receives less influence from the angle of the sample surface, so its stereo sense is relatively weak. SE2 and SE3 are generated by the excitation of backscattered electrons because the backscattered electrons themselves act on a larger area, so the range of secondary electrons when they return to the sample surface is relatively small. SE2 and SE3 are generated by the excitation of backscattered electrons. Considering that the backscattered electrons themselves act on a larger area, the range of secondary electrons generated when they return to the sample surface is also much larger than that of SE1, and the resolutions of SE2 and SE3 are lower than that of SE1. SE2 and SE3 are generated by the excitation of backscattered electrons in different directions, so their emission angles are very wide and they are distributed from high to low [17, 18].

However, no matter what type of secondary electrons are classified as low-energy electrons, the detectors do not distinguish between them when they are collected. Therefore, detectors in different positions are often used in SEM to collect different types of secondary electrons. At present, most of the secondary electron detectors are divided into side electron detectors (Everhart–Thornley detector, ETD) and detectors in the mirror barrel. Almost all SEMs have a side electron detector that is generally set in the side of the pole shoe. It depends on the characteristic that the relatively small secondary electron energy is easily influenced by other electric fields to deflect. With a voltage of 250~350 V in the front of the detector metal network, the secondary electrons scattered in all directions are attracted by the electric field and change their original trajectory, so that most of the secondary electrons can be received by the detector. But the ETD receives a mixed signal, which can also receive the backscattered electrons scattered by the sample in the direction of the detector in addition to the three secondary electron signals. Hence, the ETD actually receives a mixture of SE and BSE electrons, while its contrast is the combination of surface contrast and component contrast with lower resolution. In order to improve the resolution of the image, the SEM usually sets up an in-barrel detector, namely that the in-barrel detector is located inside the barrel and perpendicular to the horizontal direction of the sample surface. Only SE1 can enter the barrel and be collected by the in-barrel detector due to the high angle of SE1 emission. Meanwhile, different objective techniques contribute to the fact that the in-barrel electron detector has a larger improvement in resolution when compared to ETD [19, 20].

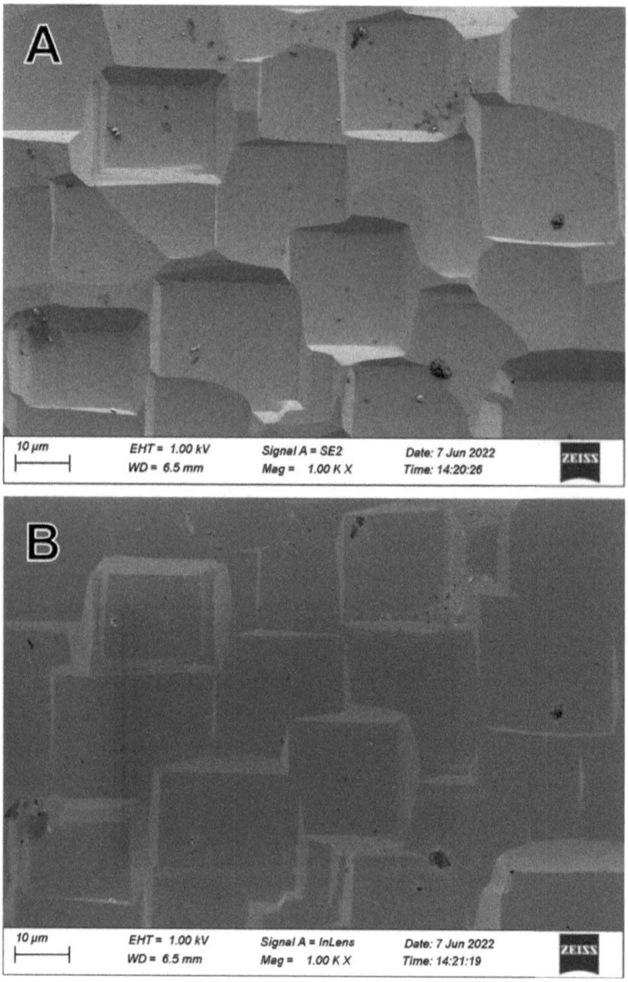

Figure 8.3 Imaging the same sample (wafer back surface) with different secondary electron detectors: (A) side electron detector and (B) in-barrel detector.

8.2.3 Sample Requirements for SEM

Taking the imaging principle of SEM into consideration, the sample preparation for SEM generally has the following requirements [21–26].

8.2.3.1 Anhydrous

Since the SEM sample chamber is in the high-vacuum environment, if the sample contains water or a volatile solvent, it will be volatilized in the vacuum environment, and the water vapor or volatile solvent will accelerate the volatilization of the filament cathode material, which will greatly reduce the filament life and even cause filament failure. A large number of gas molecules in the vacuum chamber will also lead to the vacuum error of the electron microscope and even cause molecular pump failure. Meanwhile, the gas molecules will scatter electrons, which will increase the electron beam energy dispersion and reduce the resolution and signal-to-noise ratio of the instrument. Therefore, samples containing water or volatile solvents need to be dried before observation, and critical point-drying and freeze-drying can be used if necessary to ensure that the surface morphology is not destroyed during the drying process. For biological samples containing water, it is necessary to maintain intact tissue and cellular morphology but also to ensure that the samples are sufficiently dried, so they need to be fixed, dehydrated, and dried by chemical or physical methods before observation.

8.2.3.2 Dust-free

Since SEM imaging can display surface feature of the sample, it is important to keep the sample clean during preparation, pickup, and storage. The main contaminants include dust, handprints, human secretions (saliva and sweat), and traces of grease. Therefore, clean powder-free gloves are required for sample pickup and preparation, and scissors, forceps, and other sample preparation tools should be kept clean. When the powder samples are prepared, it is supposed to ensure that the thickness is uniform, the surface is flat, and the amount is not too much, which must be firmly adhered to the conductive adhesive and needs to be purged with a larger airflow before putting it into the sample cavity in order to blow off the unglued sample particles.

8.2.3.3 Conductive

SEM uses the interaction between the electron beam and the sample for imaging. With the electron beam continuously bombarded onto the sample, the sample does not increase or decrease electrons to achieve charge balance only when the number of incident electrons and secondary electrons is equal, and the sample will have excess electrons when the incident electron beam does not meet this condition, which calls for the formation of an absorption current to meet the charge balance. In doing so, the sample should have good electrical conductivity. If the sample is not conductive or poorly conductive, the excess electrons on the sample cannot flow away and will form a charge accumulation on the sample surface, generating an electrostatic field that interferes with the incident electron beam and the emission of secondary electrons to result in the distortion and displacement of the image. For non-conductive samples, carbon or metal spraying can be used to improve the electrical conductivity of the material and also enhance the amount of sample signal, while for some samples that cannot be sprayed, the sample can be observed in low vacuum mode to mitigate the charge effect.

8.2.3.4 Thermally stable

In the normal observation by SEM, the beam current of the electron probe is usually at nA level. Therefore, heat is not a problem for most samples, but some heat-sensitive materials are prone to damage, blistering, cracking, holes, and other thermal damage phenomena in the observation site, and even some samples decompose at high temperatures and release gas or other substances. When they enter the inside of the electron microscope, it will inevitably affect the performance of the electron microscope, or even cause failure. This kind of samples can be coated on the surface to improve the thermal stability of the sample, and it is supposed to choose a smaller acceleration voltage and beam intensity when observing to reduce the damage of the electron beam on the sample.

8.2.3.5 Demagnetized

The magnetism of the sample can affect the trajectory of the electron beam movement, resulting in abnormal images and the image-scattering phenomena. If small magnetic particles are not firmly attached, they can easily be sucked inside the vacuum chamber of the electron microscope, causing contamination and imaging malfunction of the electron microscope. Therefore, extra care is needed for observing magnetic materials, while strict purging through compressed gas is also required. Some stronger magnetic materials must be demagnetized.

8.2.4 Basic Parameters of SEM

The composition, morphology, and observation focus of different samples are diversified when SEM is used to observe samples, so the requirements for resolution, depth of field, contrast, morphological fidelity, and other analysis needs are also different. Different observation focus requires various electron microscope parameters for shooting, which sometimes even contradict each other. Therefore, it is important to select the appropriate electron microscope parameters for each sample [5, 11, 27].

8.2.4.1 Acceleration voltage

The acceleration voltage of SEM is usually adjustable between 0.02 kV and 30 kV. Generally speaking, the higher the acceleration voltage of any electron microscope, the higher the resolution. But it does not mean that the higher acceleration voltage is suitable for any sample. The following factors must be considered to obtain the ideal image:

(i) Sample radiation damage and electric charge effect

The acceleration voltage should ensure that it does not cause significant radiation damage and charge effect on the sample; otherwise, the image obtained cannot truly reflect the sample. For samples with good electrical and thermal conductivity, such as metal materials, a higher acceleration voltage of 10 kV or more can be chosen for observation. For samples with poor electrical conductivity but strong thermal stability, an acceleration voltage of about 5 kV can be used. For samples that easily receive radiation damage, such as biological tissues, polymer materials, etc., an acceleration voltage of 2 kV or less should be used for observation.

(ii) Electronic production

Elements have various electron yields at different voltages. For some samples of mixed materials, it is necessary to choose an accelerating voltage range with a large difference in electron yields for imaging in order to have the morphological contrast with an excellent composition contrast. For some metallic materials, there are large differences in electron yields at higher accelerating voltages, while for low-atomic-number samples, lower voltages tend to cause larger differences in electron yields.

(iii) Penetration depth

The size of acceleration voltage is also related to the imaging depth. The higher the acceleration voltage, the higher the energy, and the larger the scattering area of electrons. There will be more secondary electrons and backscattered electrons from the depth of the sample, so the larger acceleration voltage will lead to the missing part of the sample surface details even with better horizontal resolution. However, the lower acceleration voltage with relatively low resolution can reflect the sample pole surface. When the acceleration voltage is as low as 0.2 kV or less, the image will not produce edge effect and tip effect because the electron beam area is fairly small. Therefore, for different sample requirements, low-voltage imaging is preferred if more surface detail is needed, while higher acceleration voltage is required if the interior of the sample is to be observed.

(iv) Other factors

In addition to the above factors, the selection of acceleration voltage also needs to consider the contrast balance, effective magnification, signal-to-noise ratio, and other factors. For example, the high-magnification images focusing on the composition of the contrast and signal-to-noise ratio call for a relatively high-acceleration voltage, which requires the operation of the SEM to adjust according to the needs of different samples and focus.

8.2.4.2 Working distance

The working distance is the distance from the lower surface of the SEM objective to the surface of the sample. Generally speaking, the smaller working distance leads to the higher resolution regardless of any voltage beam conditions. However, the working distance is very close when the sample is easy to collide with the polar shoe, especially for samples with large surface height differences, which need to be operated carefully. Moreover, the closer working distance requires a greater limitation of the tilting angle, which is allowed for the sample.

In addition to the resolution, the depth of field is usually an indispensable factor in electron microscope pictures. The depth of field will be significantly reduced especially when the magnification is large, so that the sample cannot be focused clearly in the case of light undulation. The depth of field under the long working distance is better than the short working distance, but a very long working distance will lead to a decline in resolution. Therefore, it is necessary to balance according to the focus of the sample.

8.2.4.3 Beam spot and beam current

The beam current and beam spot are not completely independent. An increase in the beam current will inevitably lead to an expansion of the beam spot due to the Boersch effect [28], so a larger beam current leads to a lower resolution but a better signal-to-noise ratio. In the case of high-resolution images, a smaller beam current is required to obtain a small beam spot, while conventional magnification can be used to meet the signal-to-noise ratio by increasing the beam current. However, a much larger beam current is often required for elemental analysis than the image captured.

Although a smaller beam spot contributes to a higher theoretical resolution, a too-small beam spot may also lead to a loss of image information. The electron beam is controlled by the pulse signal from the scanning coil, and it does not scan the sample surface continuously but in a point-by-point jumping pattern. Then, the electron beam has a certain coverage area beam. When the beam spot is larger, its coverage area forms a continuous trajectory. However, when the beam spot decreases, the coverage area of the beam spot also gets smaller, and some characteristic morphology passes through the middle of the coverage area of the beam spot without being scanned. Therefore, the corresponding morphological characteristics are not reflected on the image, which leads to a loss of information. Therefore, it is necessary to reduce the beam spot to match a single pixel in order to get a better resolution while obtaining the true shape of the sample [29].

8.2.4.4 Selection of detectors

The difference between these two types of detectors lies in that the side-mounted electronic detector receives SE1, SE2, SE3, and part of the BSE signal, which has a strong stereo sense but relatively low resolution, while the in-barrel detector receives most of the SE1 signal, which can respond to the information of the sample pole surface with the higher resolution. In addition, most SEMs are equipped with a backscattered electron detector. In terms of the image performance, the image of the secondary electron detector is mainly shaped contrast with a strong sense of stereo and a small amount of composition contrast. The image of the backscattered electron detector is mainly composition contrast and a certain amount of shape contrast. Different detectors can be selected for imaging various samples, and many electron microscopes have independent channel amplifiers, allowing the simultaneous imaging of multiple detectors to meet different imaging needs for the same sample location [30, 31].

8.3 Cryo-Scanning Electron Microscopy

As mentioned above, the sample for SEM imaging must be dry and volatile-free. However, for biological and polymer material samples with high water content, the process of drying may change their native structure, for example, vesicles and other self-assembled structures, gel materials, etc. Hence, the cryo-SEM technique is a swift and effective way to overcome the problem of water content in samples, which is also widely used to observe unstable samples that are sensitive to electron beams [32–35].

8.3.1 Basic Construction of Cryo-SEM

Cryo-SEM usually consists of a pre-evacuation chamber, a specimen preparation chamber, and an SEM column equipped with cold stages. The sample is fixed on the base, frozen and evacuated in the pre-evacuation chamber, and transferred to the specimen preparation chamber. The specimen preparation chamber is equipped with not only a cold stage but also a cold knife for fracturing frozen samples and a metal-coating system for coating freeze-fracture faces to facilitate the radiation of secondary electrons as well as to inhibit electric charging. The specimen preparation chamber is directly connected to the SEM column with another cold stage on which frozen samples are observed with cover by a cold trap for the decontamination of samples. The cold stage is cooled by a connected copper braid cooled by liquid nitrogen (LN_2), by a piped system for circulating LN_2, or by the Joule–Thomson refrigeration principle. Metal coating is done by resistance heating, sputter coating, or electron-beam guns.

8.3.2 Sample Preparation for Cryo-SEM

In cryo-SEM, the sample preparation is divided into freeze fixation, freeze fracture, sublimation, and conductive spraying. First, the sample is freeze fixed to maintain the real structure of the sample. The water is not crystallized but becomes glassy water in the process of curing, so that the volume of water does not expand and, thus, destroy the original structure of the sample. The method of high-pressure freezing or fast freezing with a liquid nitrogen slurry is usually used. The high-pressure freezing method involves freezing the sample with liquid nitrogen at a pressure of 2100 atm, under which the water remains in a glassy state. In freezing with a liquid nitrogen slurry, the liquid nitrogen is placed in a vacuum (about 0.1 Pa), where it is in the form of a "slurry" and does not boil, thus freezing the sample quickly and reducing the possibility of water crystallization in the sample. After freeze-fixing the liquid sample, the sample is broken down by a special device under vacuum and at a low temperature (−100°C) to reveal a fresh section, and then the water wrapped around the sample is sublimated under vacuum at −90°C, followed by the conductive coating (usually Pt). The sample is placed on a cold stage of the SEM (down to −160°C) via a cryo-transfer system for observation [36–39].

8.3.3 Applications of Cryo-SEM in Cell Biology

A cell is the basic structural and functional unit of an organism, and all organisms, except viruses, are known to be composed of cells. But viral life activities must also be expressed in cells. Water is the most abundant compound in the cell, generally accounting for 60–95% of the cell. The application of cryo-SEM for observing the cell structure can restore the original shape of the cell in its physiological state to the greatest extent.

One of the structural features of early somatic embryos (ESEs) is the presence of an extracellular matrix (ECM) network on the surface of embryogenic cells [40]. The ECM network has been specifically related to the induction and early stages of somatic embryogenesis [41]. Cryo-SEM allows the life-like appearance of the sample to be preserved and recorded in a fully hydrated and chemically unmodified state. The cryo-method has a great advantage. The material is rapidly frozen, so volatile and sensitive biological structures are well preserved. Nedela et al. [42] used cryo-SEM to observe the surface structure of Norway spruce ESEs. The presence of a smooth and compact surface of the ECM was proven by results obtained using the cryo-SEM (see Fig. 8.4A,B). The early somatic embryo fully covered by the ECM is apparent from Fig. 8.4C,D, where Fig. 8.4D shows its elongated state. Figure 8.4E shows the forming head of the early somatic embryo. Its more developed phase is apparent from Fig. 8.4F. The fracture of frozen somatic embryo tissue shows the inner structure of suspensor cells, recorded by the BSE-YAG detector in the SEM with cryo-attachment (Fig. 8.4G,H).

In addition, the ECM of *Actinidia deliciosa* was also observed using the ESEM and cryo-SEM on the surface of cell clusters as a membranous layer or reticulated network, shrunken or wrinkled, depending on the procedure. Smoother surface layers without visible fibrils and showing porosity were published by Ref. [43].

Given that cryo-SEM can directly observe liquids, we can also use it to observe some microscopic changes during cell freezing, such as the dormant compound buds of grapevines.

In regions where the air temperature drops to subfreezing, grapevines can be damaged by freezing during fall, winter, and spring, and dormant compound buds are the most susceptible to freezing injury during winter [44–46]. The use of cryo-SEM has enabled direct visualization of the freezing behavior of plant cells. Cryo-SEM allows detailed observation of frozen cell morphology and ice crystal distribution in frozen plant tissues with high resolution, thereby resulting in the clarification of the complex freezing behavior of plant cells [47–50].

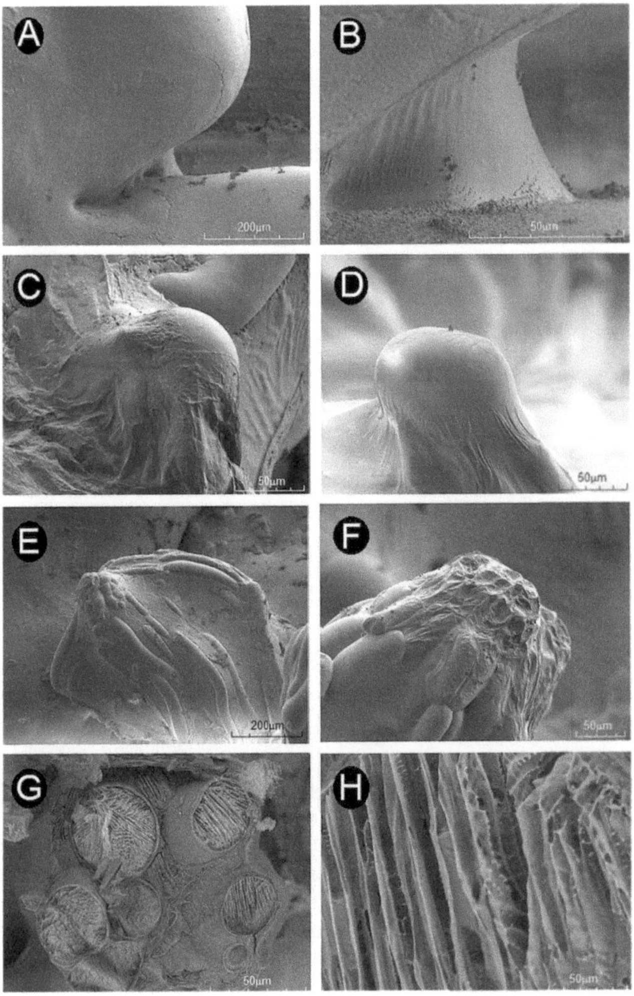

Figure 8.4 Spruce embryogenic culture (*Picea abies*) micrographs by FESEM Jeol JSM 7401F with cryo-attachment. Figures (A–F): the detector of secondary electrons in low magnification mode and figures; (G, H): the BSE-YAG detector. Figures (A, B): early somatic embryo with microstructures of the ECM; (C, D): early somatic embryo covered by the ECM; (E, F): early somatic embryo; (G, H): fracture morphology of the inner surface of suspensor cells. Reprinted from Ref. [40], Copyright 2016, with permission from Elsevier.

Kasuga et al. [51] used cryo-SEM to investigate the freezing behavior of dormant compound buds of freeze-resistant interspecific hybrid grape "Yamasachi"; the level of freezing dehydration may be important for consideration about the freezing resistance of the dormant buds of Vitis species. To confirm the occurrence of freeze dehydration, they performed cryo-SEM observation of leaf primordial cells in primary buds (Fig. 8.5).

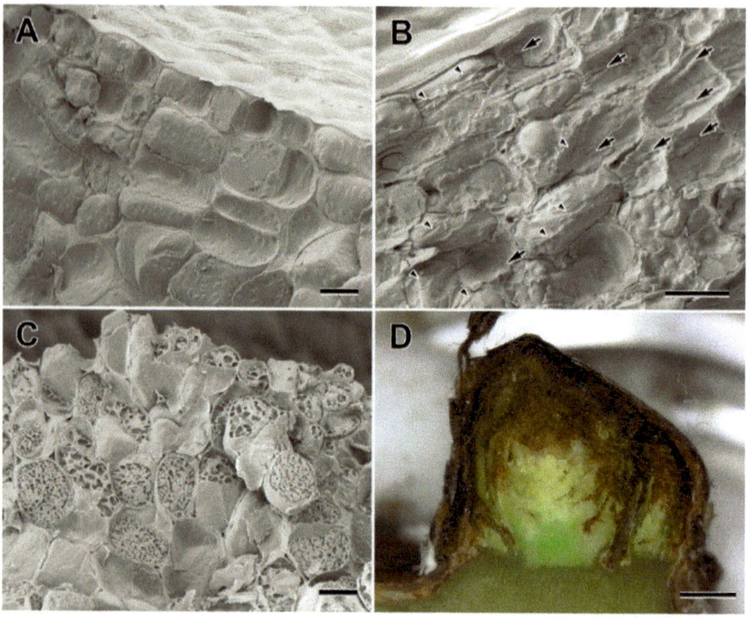

Figure 8.5 Microscopic observation of the freezing behavior of Yamasachi buds. (A) A cryo-scanning electron microscopic image of leaf primordium cells in a bud cryofixed from room temperature. (B) Cryo-scanning electron microscopic image of leaf primordium cells in a bud cryofixed after freezing at −15°C for 12 h. Arrows and arrowheads indicate wrinkles on the inner surfaces of cell walls and those on the outer surfaces of plasma membranes, respectively. (C) A cryo-scanning electron microscopic image of leaf primordium cells in a bud cryofixed after stepwise freezing to −40°C. Holes in the cells, which are evidence of intracellular freezing, were generated by the sublimation of intracellular ice crystals. (D) Stereomicroscopic image of a compound bud of Yamasachi frozen at −15°C for 12 h. Horizontal bars, 10 μm (A–C) and 1 mm (D). Reprinted from Ref. [49], Copyright 2020, with permission from Elsevier.

Cells were densely packed in leaf primordia (Fig. 8.5A–C), and no extracellular ice crystals were observed in the intercellular spaces of the frozen tissues (Fig. 8.5B,C). Before freezing, the inner surfaces of cell walls and the outer surfaces of plasma membranes were smooth, except for primary pit fields containing many plasmodesmata (Fig. 8.5A). In contrast, many wrinkles were observed on the different surfaces of leaf primordial cells in buds frozen to −15°C, which suggests partial dehydration (Fig. 8.5B).

Although no visible intracellular ice crystals were observed in leaf primordia cooled to −15°C, leaf primordial cells frozen to −40°C had numerous large intracellular ice crystals (more than 0.5 μm in diameter; Fig. 8.5C). Because such large ice crystals are generated by the freezing of supercooled cellular water [52], this observation indicates that intracellular freezing occurs between −15°C and −40°C.

In this study, cryo-SEM observation enabled the detection of such small levels of dehydration. These results suggest the existence of partial dehydration in dormant-bud primordial cells under subfreezing temperatures. The apparent absence of extracellular ice crystals in bud primordial tissues under subfreezing temperatures suggests that Yamasachi dormant buds adapt to subfreezing temperatures by extra-organ freezing.

Another example is the freezing and thawing of boar semen. Frozen-thawed (FT) boar semen has low cryo-survival rates, and the most critical injuries for cell survival occur at sub-zero temperatures. During re-warming, the characteristics of a container are vital for maximizing the number of viable cells after thawing. Cryo-SEM can directly observe a series of changes in cells and containers during the freezing process in real time to determine the state of cells [53–55].

Ekwall et al. [56] employed cryo-SEM to examine the ultrastructure of samples and determine whether the amounts of solid-state water in the extracellular, outer-extender areas of frozen straws differed between two test packages—a 0.5 mL volume plastic medium straw (MS) or MiniFlatPacks (MFP, 0.7 mL volume). The degree of hydration was monitored in relation to the areas of ice crystals formed outside the extended semen (free water, lakes)—the areas of frozen, concentrated extender (veins) where spermatozoa were presumably located and a degree of compartmentalization (size and number of lakes) was present.

Figure 8.6 depicts the surfaces of a fractured MFP (Fig. 8.6A) and an MS (Fig. 8.6B) as seen by cryo-SEM. The MPFs showed apparently

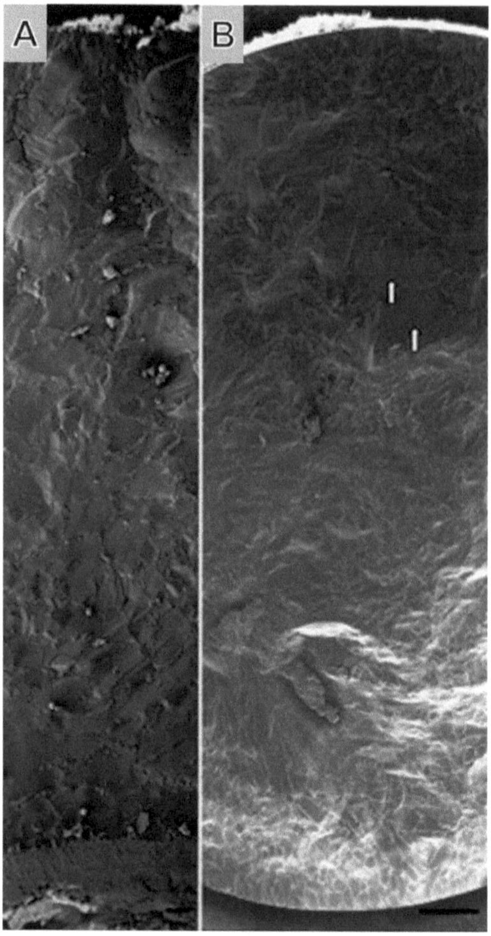

Figure 8.6 (A) Low-power magnification of a cross-sectioned MiniFlatPack (MFP) as seen after fracture. The lakes are more or less similar in size and shape throughout the bag profile. Bar = 100 mm. (B) Low-power magnification of a cross-sectioned medium straw (MS) as seen after cryo-fracture. Note the presence of larger lakes in some intermediate areas of the MS (arrows), which indicate unequal freezing. Bar = 100 mm. Reprinted from Ref. [54], Copyright 2007, with permission from Elsevier.

larger lakes than the MSs, which, in turn, had a highly asymmetrical organization of lakes and veins depending on the position of the section, with apparently smaller lakes in the periphery and the center of the straw with intermediate areas of larger lakes (arrows in

Fig. 8.6B). Random surfaces used for measurements are depicted in Fig. 8.7A for MFP and in Fig. 8.7B for MS. Figure 8.7C shows a higher magnification of veins depicting sperm heads and tails embedded in the frozen extender. The degree of dehydration was apparently higher in the MFPs, since the area of individual lakes appeared larger, thus leading to thinner/smaller veins. After image analysis of the fractured material, it was found that the relative surfaces of lakes and veins in each fracture region differed between packages ($P < 0.05$), indicating that the amount of free water was larger in MFPs and, therefore, the veins were thinner than in MSs. It was then concluded that the utilization of a plastic hexahedron-flat-shaped bag (MFP) instead of the usual cylindrical straw package (MS) for freezing boar semen resulted in a more homogenous dehydration of the spermatozoa/frozen extender allowing for a somewhat better sperm quality post-thaw.

During this slow freezing of boar semen, large areas of globular ice are formed extracellularly, leaving dehydrated spermatozoa surrounded by veins of concentrated extender. Both the modification of the ionic cell environment and the formation of intracellular ice crystals in both heads and tails that occur affect the post-thaw survival [57].

8.4 Observation of Human Red Blood Cell Membrane through Cryo-SEM

The cell membrane is a barrier that prevents extracellular substances from freely entering the cell, which ensures the relative stability of the intracellular environment so that various biochemical reactions can operate in an orderly manner. The cell membrane consisting of lipids and proteins is located on the cell surface with a thickness of usually 7 to 8 nm. Erythrocytes are the most abundant type of cells in the blood. A normal mature erythrocyte has no nucleus, no organelles such as Golgi apparatus and mitochondria, and no other membrane structures except the cell membrane. The erythrocyte skeleton is relatively thin and easy to remove, so it is often studied as a model for biological membranes. In the previous chapters of this book, the human red blood cell (hRBC) membrane is observed by AFM and cryo-transmission microscopy, respectively, while SEM with its advantages of simple operation and rapid imaging will also become an effective tool for biological membrane observation [58, 59].

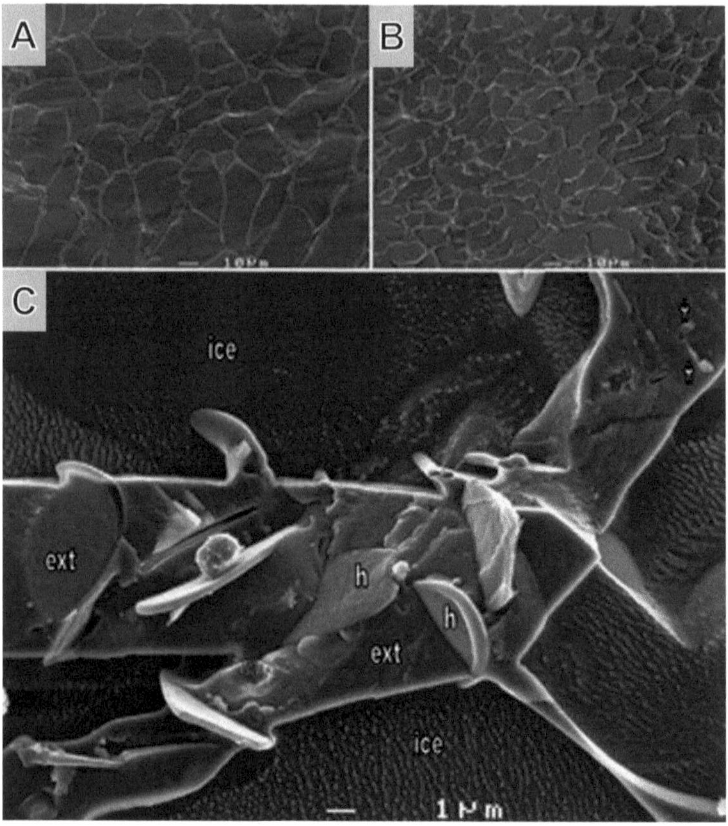

Figure 8.7 (A) Detail of MiniFlatPack (MFP) fractured and used for the calculation of lakes and veins. Bar = 10 mm. (B) Detail of medium straw (MS) fractured and used for the calculation of lakes and veins. Bar = 10 mm. (C) Details from a vein depicting several individual sperm heads (h) as well as tail profiles (*) embedded in the extender (ext). Bar = 1 mm. Reprinted from Ref. [54], Copyright 2007, with permission from Elsevier.

8.4.1 Cryo-SEM Imaging of Erythrocyte Membrane

The sample is prepared and imaged according to the procedures described in Section 8.5. Figure 8.8 shows the cryo-SEM image of the erythrocyte membrane. Figure 8.8A shows the closely arranged erythrocyte membrane on the silicon wafer, while the individual cell membrane (Fig. 8.8B) is sub-circular with the inner membrane leaflet facing upward and a higher buildup at the edge of the membrane, which is the portion of the unattached cell membrane remaining after the removal of the apical membrane. The magnified image (Fig. 8.8C) shows the distribution of membrane proteins on the cell membrane with the darker part of the contrast being proteins.

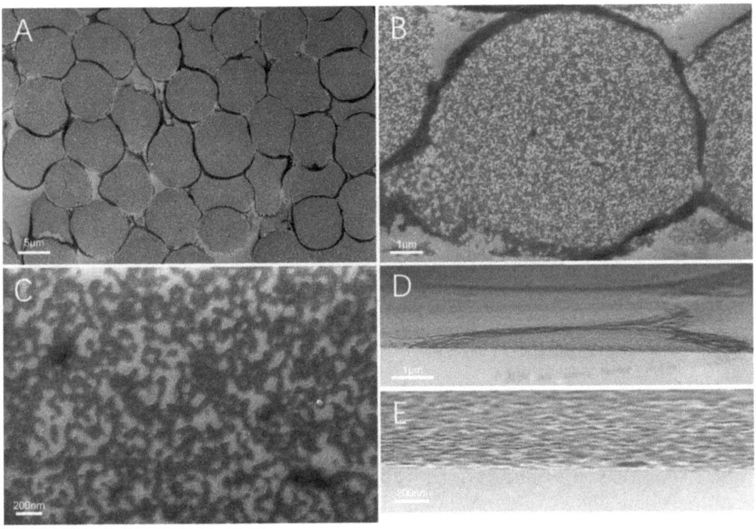

Figure 8.8 Cryo-SEM image of red blood cell membrane. (A) Red blood cell membrane at low magnification (1.5k ×). Diameter distribution in 7–10 microns. (B) Image of a single red blood cell membrane. (C) High magnification (40k ×) observation of red blood cell membrane surface; many proteins and protein aggregates are distributed (dark area). (D) A single erythrocyte membrane was observed at a vertical tilt of 5°, and the edge was obviously raised. (E) Observing the red blood cell membrane surface at a vertical tilt of 10°, the darker contrast area is indeed a higher area, confirming that it is a membrane surface protein.

Conventional SEM secondary electron images usually present the morphological contrast of the sample. However, since the cell membrane is mainly composed of proteins and phospholipids, these non-conductive light elements (C, N, O) present a darker contrast in the SEM image, while the conductive substrate (silicon wafer) presents a brighter contrast [15]. Meanwhile, considering that the thickness of the cell membrane is small and the in-lens mode imaging has a poor sense of stereo, which result in the poor stereoscopic sense of the image, the morphological difference of the sample could not be observed from the imaging angle of the vertical sample. For this reason, the silicon wafer of the settled film sample is broken from the middle, pasted vertically on the sample stage, and tilted 5°/10° from the direction parallel to the sample to observe the sample (Fig. 8.8D,E), which clearly shows that the darker contrast area is the higher part of the morphology. Based on the structure of cell membranes, it is assumed that these bumps are proteins on the membranes.

8.4.2 Cryo-SEM Coupled with AFM for Imaging

Wang et al. [60] applied AFM to directly observe the overall structure of human erythrocyte membranes in situ at the single-molecule level and under near-physiological conditions and proposed the semi-mosaic model accordingly. In order to confirm the accuracy of cryo-SEM for imaging the human erythrocyte inner membrane, AFM and cryo-SEM are successively used to image and observe the samples at the same location.

The silicon wafer is first marked by scribing a number of square marks with a glass knife, and then the sample is prepared according to the procedures described in Section 8.5. The markings are searched for in the bright field through an atomic force microscope and the position is recorded, and then the sample near the markings is prepared and imaged according to the procedures described in Chapter 3, after which the same wafer is put into the SEM for observation (Fig. 8.9A,B). It can be found that the darker region in the SEM image shows a higher morphology in the AFM image (Fig. 8.9C,D), confirming that the darker region should be the membrane protein region. The atomic force images obtained are also similar to the results of Wang et al. These results suggest that cryo-SEM can image cell membrane with high resolution.

8.4.3 Tagging of Specific Proteins on Erythrocyte Membrane through Gold Nanoparticle

In the earlier section, we obtained SEM images of the erythrocyte membrane and could see many different kinds of proteins in the inner membrane of the erythrocyte. In order to find the specific protein, gold nanoparticle is used to conjugate the antibody to label the specific protein on the membrane.

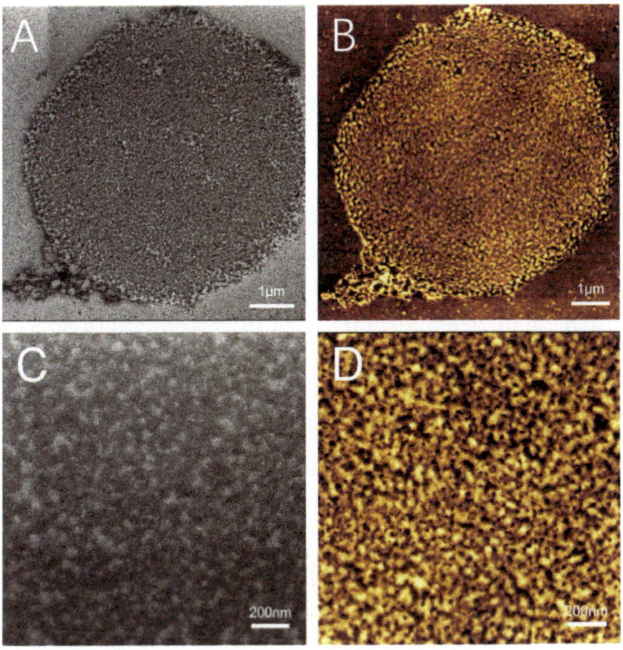

Figure 8.9 (A–B) The same red cell membrane was imaged with SEM and AFM, and the brighter areas in the AFM image corresponded to the darker areas in the SEM image. (C–D) AFM and SEM images of red blood cell membranes at high magnification. Imaging of the same location is not possible due to difficulty in marking and focusing.

Gold nanoparticles have high electron density and excellent electrical conductivity and can be clearly imaged under the electron microscope without affecting the activity of biological macromolecules, making them ideal imaging markers for an electron microscope [61–63]. The antibody can specifically recognize the corresponding antigen and bind to it, so the gold nanoparticle connected with the antibody can be labeled under the electron microscope for specific proteins.

Gold nanoparticles and proteins themselves feature the non-specific physical adsorption [64–66]. In order to reduce the nonspecific adsorption of gold nanoparticles and connect them with the antibody, they are modified with PEG1000-COOH. PEG can coat the gold nanoparticles so that they cannot contact the proteins and thus produce non-specific adsorption [67, 68]. The carboxyl group at the end of the PEG chain is activated and condensed with the amino group in the antibody molecule to connect the antibody with the gold nanoparticles [69, 70].

The samples are frozen and imaged according to the steps and parameters described in Section 8.5. The antibodies to Glut1 protein, which are more distributed on the erythrocyte membrane, are selected for this experiment to image clearly without obscuring the protein. Gold nanoparticles with 5 nm particle size are selected for this experiment. The morphology of the cell membrane is the same as that of the unlabeled sample, and obvious gold nanoparticles are visible at high magnification (Fig. 8.10A,B). The amount of rice particles is distributed in the darker lined sample region, which also confirms that the darker lined region is exactly the membrane protein region.

In order to demonstrate that high salt solution inversion washing can remove the non-specific adsorbed gold nanoparticles on the cell membrane, two control groups are set up in the experiment in addition to the experimental group. The first control group is washed by PBS buffer without high salt solution inversion washing. The results show that gold nanoparticles are still present on the cell membrane even after washing, and there are no antibodies attached to the gold nanoparticles, indicating that the gold nanoparticles are attached to the cell membrane by non-specific physical adsorption. The second control group is incubated with the Au-PEG1000-COOH molecule without an antibody and is washed in the same way as the experiment group. The images show that there are no gold nanoparticles on the cell membrane. The comparison between the two control groups indicates that the gold nanoparticle on the cell membrane in the experimental group is labeled by antigen–antibody interaction, and that the non-specific adsorbed gold nanoparticle could be removed by the inversion washing step of high salt solution.

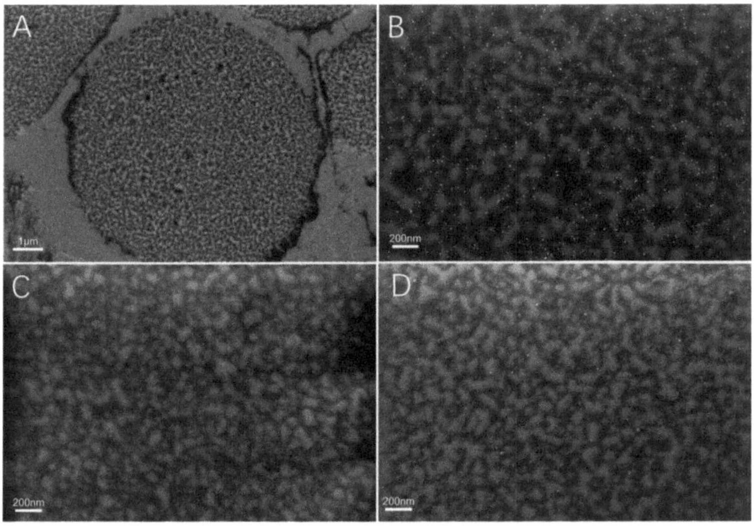

Figure 8.10 (A) Gold-nanoparticles-antibody-labeled erythrocyte membrane did not change morphologically. (B) Gold particles on the cell membrane can be clearly seen at high magnification, and they are all distributed in the protein area (darker area). (C) The control group without the conjugated antibody was washed in the same way as the experimental group (high-salt inversion). No gold particles were present on the erythrocyte membrane. (D) For the control group without the antibody attached, the conventional washing method was used. Non-specifically linked gold particles are present on the cell membrane.

8.5 Protocols for Cryo-SEM Imaging of hRBC Membrane

8.5.1 Preparation of hRBC Membrane on Silicon Wafer

Since SEM imaging requires the sample to be electrically conductive, a silicon wafer is selected as the substrate for cell membrane adhesion. The silanization modification of the silicon wafer is required to enhance the adsorption force between the sample and the substrate (silicon wafer). The wafer is cleaned using detergent and ethanol and dried with an alcohol lamp. Then the wafer and two clean Petri dishes are placed in a desiccator with argon gas for 5 min. After that 15 μl of DIPEA and 50 μl of APTES are added to the two Petri dishes in the argon environment with argon gas passed continuously for 5 min after the addition of reagents, and the desiccator is kept airtight at room temperature for more than 4 h.

Human red blood cell (hRBC) membrane samples are prepared with reference to literature methods [60]. Twenty microliters of fresh human fingertip blood is washed five times by centrifugation (1000 r/min, 2 min) in PBS buffer (150 mmol/L, pH 7.5), and the collected red blood cells (RBCs) are diluted 500 times with PBS buffer to obtain a suitable concentration of the sample. The diluted hRBC solution is added dropwise to silanized silica wafer for 30 min to settle, followed by washing off the unadhered RBCs with PBS buffer. The erythrocytes are sheared with a rapid hypotonic PBS buffer (7.5 mmol/L, pH 7.5) injection to remove the apical membrane to obtain a monolayer hRBC membrane, and the prepared hRBC membranes are subsequently stored in PBS buffer along with the silicon wafer.

8.5.2 Au-PEG1000-Antibody Conjugated with Erythrocyte Membrane Protein

First, we added EDC and NHS to 1 ml synthesized Au-PEG1000-COOH (50 μg/ml) at the ratio of 1.5 : 1.5 : 1. After mixing well, the antibody (200 μg/ml) was added to the mixture and the ligation reaction was carried out for 4 h at room temperature.

The ligation product from the previous step is diluted thrice with 1/20 PBS, and the erythrocyte membrane adhered to the silicon wafer is prepared according to the procedures described in Section 8.4.1. It is placed in a small dish with the diluted ligation product and incubated overnight at 4°C.

After incubation, the solution is aspirated from the Petri dish, and the cell membrane is washed thrice for 10 min each with 1/20 PBS buffer, after which the silicon wafer is taken out and the back of the wafer is dried and taped to a piece of tape while keeping the upper surface with the cell membrane wet. The ends of the tape are taped to the outer wall of the Petri dish so that the side with the cell membrane adhered to the silicon wafer is washed down and suspended in the center of the Petri dish. An appropriate amount of NaCl solution (0.3 M) is added to the Petri dish so that it could be immersed. The purpose of washing with high salt solution in inversion is to remove the non-specific adsorbed gold nanoparticles from the cell membrane. After the washing is completed, the samples are washed thrice with PBS buffer, and finally the prepared samples are stored in PBS buffer.

The first control group is incubated with the synthesized Au-PEG1000-COOH molecule without antibodies attached and directly with the cell membrane overnight at the same dilution. The cell membranes are washed thrice with 1/20 PBS buffer for 10 min each time, followed by 4 h of shaking in PBS buffer and three more washes before freezing and imaging. The results show that gold nanoparticles are still present on the cell membrane even after washing, and there are no antibodies attached to the gold nanoparticles, indicating that the gold nanoparticles are attached to the cell membrane by non-specific physical adsorption.

The second control group is also incubated overnight with the synthesized Au-PEG1000-COOH molecule without antibodies and directly with the cell membrane at the same dilution level. After incubation, the gold pellet solution is aspirated and the cell membrane is washed thrice with 1/20 PBS buffer for 10 min each, followed by washing with 0.3 M NaCl solution in inversion for 4 h as in the experimental group. After washing with PBS buffer three more times, the samples are frozen and imaged.

8.5.3 Frozen Sample Preparation

Since the erythrocyte membrane is relatively thin (6–10 nm), rapid freezing of liquid nitrogen sludge is chosen to prepare the frozen samples. First and foremost, SEM is required to run under high vacuum, cool a cold stage and a cold trap in the SEM column as well as a cold stage and a cold knife in the preparation chamber. The liquid nitrogen is poured into the container in the pre-evacuation chamber until complete cooling (below −160°C) and covered with a lid to evacuate the liquid nitrogen while waiting for it to become the liquid nitrogen slurry. Meanwhile, dry the bottom of the silicon wafer with RBC membrane adhered to it, stick it on the cryogenic sample carrier with special glue, suck the moisture from the side of the sample surface as much as possible, and then put it into the liquid nitrogen slurry quickly. At this time the liquid nitrogen warms up and dissolves to the liquid state with the evacuation started. When the liquid nitrogen stops bubbling and is about to reach the slurry state, pull out the sample stage from the liquid nitrogen to the upper end of the rod, close the lid to make an airtight vacuum environment, and transfer to the freezing transfer device where the temperature is raised to −95°C for 10 min, allowing the ice on the surface to sublimate and expose the cell membrane, which is transferred to the SEM for observation.

8.5.4 Selection of Electron Microscope Parameters

The hRBC membrane is a thin film of low-atomic-number elements with a thickness of less than 10 nm. The Zeiss Sigma 300 equipped with the Quorum freezing system (PP3010T) is used in the experiment. According to the properties of the cell membrane and imaging needs, the in-barrel secondary electron detector (called in-lens detector in the Zeiss Sigma 300 used in this experiment) is used for imaging [71]. Considering that the surface of the RBC membrane is a fine protrusion formed by smaller proteins and protein clusters and that the spraying conductive layer may hide its true shape, it is directly transferred to the cryo console for observation after sublimation. Since organic matter is not resistant to electron irradiation, it is imaged with low voltage (0.5–3.0 kV), which is beneficial to obtaining information only in the depth range of the sample. The sample surface is flat enough; hence, the smallest possible working distance (<2 mm) can be used to obtain high-resolution images.

8.6 Conclusion

This chapter introduces the basic structure and imaging principle of SEM, which uses various electron detectors to collect and image the signals generated through the interaction between the electron beam and the sample, and briefly explores the sample preparation conditions and parameters of SEM. For plasma membrane samples, the liquid nitrogen slurry freezing method and low-temperature conditions can protect the structure of the plasma membrane to the maximum extent and observe its morphology in the near-physiological state. Meanwhile, considering that the thickness of plasma membrane samples is small and the microstructure is fine, the application of the in-barrel secondary electron detector in the microscope and no conductive spraying of the samples can better observe the membrane samples and ensure the authenticity of their morphology.

We use cryo-SEM to observe the erythrocyte membrane adsorbed on silicon wafer, while the glut1 protein on the erythrocyte membranes can be labeled through the gold-nanoparticle–antibody linker. Electron microscopy images obtained show the erythrocyte membrane to be round with a diameter mostly distributed in the range of 7–9 μm, which is consistent with the size of normal hRBCs. The protruding protein on the surface of the inner membrane is observed at high magnification, which shows a darker contrast in the electron microscopy images due to the absence of conductive coating. Through the verification of lateral observation, it is also the same as the results obtained by AFM, while the molecular weight and the thickness of phospholipid molecules are relatively small. According to the imaging depth of secondary electrons within 10 microns, the phospholipid layer of the secondary electron image shows almost all the contrast of the substrate (silicon wafer). With the samples expanding to 40,000 times, the 5 nm gold-labeled nanoparticles could be clearly observed. After the control group comparison, it can be concluded that the experimental group of gold nanoparticles is labeled on the cell membrane by the interaction between antigen and antibody. However, due to the presence of PEG chains between gold nanoparticles and antibodies, the relationship between the position of gold nanoparticles and their labeled proteins is not clear. In summary, cryo-SEM can be used to obtain the images of the plasma membrane in the near-physiological state.

The cryo-SEM technique does not require drying of the samples and can directly observe samples containing water, which is quite suitable for life science research, especially for imaging soft materials such as tissues, cells, biological materials, and organic systems. Cryo-SEM features a series of advantages such as fast sample preparation process, easy imaging operation, high resolution, and wide magnification range. It can quickly and easily image the plasma membrane, with broad development prospects in the study and auxiliary imaging of plasma membranes.

References

1. Singer, S. J. and Nicolson, G. L. (1972). The fluid mosaic model of the structure of cell membranes. *Science*, **175**, pp. 720–731.
2. Simons, K. and Ikonen, E. (1997). Functional rafts in cell membranes. *Nature*, **387**, pp. 569–572.
3. Nichols, B. (2005). Cell biology: Without a raft. *Nature*, **436**, pp. 638–639.
4. Schutz, G. J., Kada, G., Pastushenko, V. P., and Schindler, H. (2000). Properties of lipid microdomains in a muscle cell membrane visualized by single molecule microscopy. *EMBO J.*, **19**, pp. 892–901.
5. Lewis, D. B. (2017). Scanning electron microscopy and X-ray microanalysis. *Trans. Inst. Met. Finish.*, **70**, pp. 198–202.
6. Jena, B. P. (2006). Cell secretion machinery: Studies using the AFM. *Ultramicroscopy*, **106**, pp. 663–669.
7. Everhart, T. E. and Hayes, T. L. (1972). The scanning electron microscope. *Sci. Am.*, **226**, pp. 55–69.
8. Zhou, W. and Wang, Z. L. (2007). *Scanning Microscopy for Nanotechnology* (Springer New York, NY).
9. Pease, R. F. W. (1967). The scanning electron microscope. *IEEE Spectrum*, **4**, pp. 96–102.
10. Oatley, C. W. (1985). The detective quantum efficiency of the scintillator/photomultiplier in the scanning electron microscope. *J. Microsc. (Oxford, U. K.)*, **139**, pp. 153–166.
11. Krinsley, D. (1988). Advanced scanning electron microscopy and X-ray microanalysis. *Trans., Am. Geophys. Union*, **69**, pp. 365.
12. Lyman, C. E., Newbury, D. E., Goldstein, J. I., Williams, D., Romig, A. D., Armstrong, J. T., Echlin, P., Fiori, C. E., Joy, D. C., and Lifshin, E. J. (2013). Scanning electron microscopy, X-ray microanalysis, and analytical electron microscopy. *J. Electron Microsc. Tech.*, **69**, pp. 821–821.
13. Lyman, C. E., Goldstein, J. I., Romig, A. D., Echlin, P., and Peters, K. R. (1990). *Scanning Electron Microscopy, X-Ray Microanalysis, and Analytical Electron Microscopy* (Springer New York, NY).
14. Oatley, C. W. (1968). Scanning electron microscopy. *Electron. Power*, **14**, pp. 54–56.
15. Seiler, H. (1983). Secondary-electron emission in the scanning electron-microscope. *Br. J. Appl. Phys.*, **54**, pp. R1–R18.

16. Sim, K. S., Thong, J. T. L., and Phang, J. C. H. (2004). Effect of shot noise and secondary emission noise in scanning electron microscope images. *Scanning*, **26**, pp. 36–40.

17. Castell, M. R., Perovic, D. D., and Lafontaine, H. (1997). Electronic contribution to secondary electron compositional contrast in the scanning electron microscope. *Ultramicroscopy*, **69**, pp. 279–287.

18. Reimer, L. (2000). Scanning electron microscopy: Physics of image formation and microanalysis, Second Edition. *Meas. Sci. Technol.*, **11**, pp. 1826–1826.

19. Zhang, J. Z. and Yang, Z. M. (2005). Observation of the microstructure of preferred orientation Srtio3 thin films by scanning electron microscope with an in-lens detector. *Heat Treat. Met.*, **30**, pp. 3.

20. Cazaux, J. (2004). About the role of the various types of secondary electrons (Se; Se; Se) on the performance of Lvsem. *J. Microsc. (Oxford, U. K.)*, **214**, pp. 341–347.

21. Vos, K., Vandenberghe, N., and Elsen, J. (2014). Surface textural analysis of quartz grains by scanning electron microscopy (SEM): From sample preparation to environmental interpretation. *Earth-Sci. Rev.*, **128**, pp. 93–104.

22. Lee, J. T. and Chow, K. L. (2012). SEM sample preparation for cells on 3D scaffolds by freeze-drying and Hmds. *Scanning*, **34**, pp. 12–25.

23. Kofronova, O., Nguyen, L. D., Weiser, J., and Benada, O. (2002). Streptomycetes cultured on glass beads: Sample preparation for SEM. *Microsc. Res. Tech.*, **58**, pp. 111–113.

24. Wang, L., Wei, Y., and Niu, Z. (2008). Spatial and temporal variations of vegetation in Qinghai Province based on satellite data. *J. Geogr. Sci.*, **18**, pp. 73–84.

25. Echlin, P. (2009). *Handbook of Sample Preparation for Scanning Electron Microscopy and X-Ray Microanalysis* (Springer Science & Business Media, USA).

26. Newbury, D. E. and Yakowitz, H. (1975). *Specimen Preparation, Special Techniques and Applications to Scanning Electron Microscope* (Springer, Boston, MA).

27. Fischer, E. R., Hansen, B. T., Nair, V., Hoyt, F. H., and Dorward, D. W. (2012). Scanning electron microscopy. *Curr. Protoc. Microbiol.*, **2**, pp. 2.

28. Hawkes, P. W. and Kasper, E. (1996). *Principles of Electron Optics* (Academic Press, USA).

29. Phifer, D., Tuma, L., Vystavel, T., Wandrol, P., and Young, R. J. (2009). Improving SEM imaging performance using beam deceleration. *Microsc. Today*, **17**, pp. 40–49.

30. Reimer, L. (2000). *Scanning Electron Microscopy Physics of Image Formation and Microanalysis*, 2nd Ed. (IOP Publishing, UK).

31. Jirak, J., Cudek, P., and Nedela, V. (2012). Scintillation secondary electron detector for ESEM and SEM. *Microsc. Microanal.*, **18**, pp. 1266–1267.

32. Hassan, A. N., Frank, J. F., and Elsoda, M. (2003). Observation of bacterial exopolysaccharide in dairy products using cryo-scanning electron microscopy. *Int. Dairy J.*, **13**, pp. 755–762.

33. Achor, D. S., Ochoa, R., Erbe, E. F., Aguilar, H., Wergin, W. P., and Childers, C. C. (2001). Relative advantages of low temperature versus ambient temperature scanning electron microscopy in the study of mite morphology. *Int. J. Acarology*, **27**, pp. 3–12.

34. Johnson, D. M., Meinzer, F. C., Woodruff, D. R., and McCulloh, K. A. (2009). Leaf xylem embolism, detected acoustically and by cryo-SEM, corresponds to decreases in leaf hydraulic conductance in four evergreen species. *Plant, Cell Environ.*, **32**, pp. 828–836.

35. Sriamornsak, P., Thirawong, N., Cheewatanakornkool, K., Burapapadh, K., and Sae-Ngow, W. (2008). Cryo-scanning electron microscopy (cryo-SEM) as a tool for studying the ultrastructure during bead formation by ionotropic gelation of calcium pectinate. *Int. J. Pharm.*, **352**, pp. 115–122.

36. Craig, S. and Beaton, C. D. (1996). A simple cryo-SEM method for delicate plant tissues. *J. Microsc. (Oxford, U. K.)*, **182**, pp. 102–105.

37. Walther, P. and Muller, M. (1999). Biological ultrastructure as revealed by high resolution cryo-SEM of block faces after cryo-sectioning. *J. Microsc. (Oxford, U. K.)*, **196**, pp. 279–287.

38. Provder, T. (2006). *Film Formation: Process and Morphology* (Amer Chemical Society, USA).

39. Hrubanova, K., Nebesarova, J., Ruzicka, F., and Krzyzanek, V. (2018). The innovation of cryo-SEM freeze-fracturing methodology demonstrated on high pressure frozen biofilm. *Micron*, **110**, pp. 28–35.

40. Samaj, J., Salaj, T., Matusova, R., Salaj, J., Takac, T., Samajova, O., and Volkmann, D. (2008). Arabinogalactan-protein epitope Gal4 is differentially regulated and localized in cell lines of hybrid fir (*Abies alba* × *Abies cephalonica*) with different embryogenic and regeneration potential. *Plant Cell Rep.*, **27**, pp. 221–229.

41. Bobák, M., Šamaj, J., Pre ová, A., Blehová, A., Hlinková, E., Ove ka, M., Hlava ka, A., and Kutar ová, Z. (2004). The histological analysis of indirect somatic embryogenesis on *Drosera spatulata* Labill. *Acta Physiol. Plant.*, **26**, pp. 353–361.

42. Nedela, V., Hrib, J., Havel, L., Hudec, J., and Runstuk, J. (2016). Imaging of Norway spruce early somatic embryos with the ESEM, cryo-SEM and laser scanning microscope. *Micron*, **84**, pp. 67–71.

43. Popielarska-Konieczna, M., Bohdanowicz, J., and Starnawska, E. (2010). Extracellular matrix of plant callus tissue visualized by ESEM and SEM. *Protoplasma*, **247**, pp. 121–125.

44. Planchon, O., Quénol, H., Irimia, L., and Patriche, C. (2014). European cold wave during February 2012 and impacts in wine growing regions of Moldavia (Romania). *Theor. Appl. Climatol.*, **120**, pp. 469–478.

45. Filho, J. L. M., Allebrandt, R., de Bem, B., Mudrei, P. I., Macedo, T. A., Schlemper, C., Lerin, S., Outemane, M., Kretzschmar, A. A., and Rufato, L. (2016). Damage to 'Cabernet Sauvignon' after late frost in the southern Brazilian highlands. *Acta Hortic.*, pp. 211–216.

46. Pierquet, P., Stushnoff, C., and Burke, M. J. (1977). Low-temperature exotherms in stem and bud tissues of *Vitis riparia* Michx. *J. Am. Soc. Hortic. Sci.*, **102**, pp. 54–55.

47. Fujikawa, S., Kuroda, K., and Ohtani, J. (1996). Seasonal changes in the low-temperature behaviour of xylem ray parenchyma cells in red osier dogwood (*Cornus sericea* L.) with respect to extracellular freezing and supercooling. *Micron*, **27**, pp. 181–191.

48. Pearce, R. S. (1988). Extracellular ice and cell shape in frost-stressed cereal leaves: A low-temperature scanning-electron-microscopy study. *Planta*, **175**, pp. 313–324.

49. Fujikawa, S. and Kuroda, K. (2000). Cryo-scanning electron microscopic study on freezing behavior of xylem ray parenchyma cells in hardwood species. *Micron*, **31**, pp. 669–686.

50. Ball, M. C., Canny, M. J., Huang, C. X., and Heady, R. D. (2004). Structural changes in acclimated and unacclimated leaves during freezing and thawing. *Funct. Plant Biol.*, **31**, pp. 29–40.

51. Kasuga, J., Tsumura, Y., Kondoh, D., Jitsuyama, Y., Horiuchi, R., and Arakawa, K. (2020). Cryo-scanning electron microscopy reveals that supercooling of overwintering buds of freezing-resistant interspecific hybrid grape 'Yamasachi' is accompanied by partial dehydration. *J. Plant Physiol.*, **253**, p. 153248.

52. Fujikawa, S., Kuroda, K., and Fukazawa, K. (1994). Ultrastructural study of deep supercooling of xylem ray parenchyma cells from *Styrax obassia*. *Micron*, **25**, pp. 241–252.

53. Bwanga, C. O., Einarsson, S., and Rodriguez-Martinez, H. (1991). Ii: Effect of cooling rate and duration of freezing point plateau on boar semen frozen in mini- and maxi-straws and plastic bags. *Acta Vet. Scand.*, **32**, pp. 455–461.

54. Eriksson, B. M. and Rodriguez-Martinez, H. (2000). Effect of freezing and thawing rates on the post-thaw viability of boar spermatozoa frozen in flatpacks and maxi-straws. *Anim. Reprod. Sci.*, **63**, pp. 205–220.

55. Eriksson, B. M., Vazquez, J. M., Martinez, E. A., Roca, J., Lucas, X., and Rodriguez-Martinez, H. (2001). Effects of holding time during cooling and of type of package on plasma membrane integrity, motility and in vitro oocyte penetration ability of frozen-thawed boar spermatozoa. *Theriogenology*, **55**, pp. 1593–1605.

56. Ekwall, H., Hernandez, M., Saravia, F., and Rodriguez-Martinez, H. (2007). Cryo-scanning electron microscopy (cryo-SEM) of boar semen frozen in medium-straws and MiniFlatPacks. *Theriogenology*, **67**, pp. 1463–1472.

57. Ekwall, H., Ploen, L., and Courtens, J. L. (1995). Embedding resin space, water contents and chromatin compaction in rabbit sperm nuclei: Electron microscopic X-ray spectrophotometry of a brominated probe. *Andrologia*, **27**, pp. 175–184.

58. Raess, B. U. and Tunnicliff, G. (1989). *The Red Cell Membrane: A Model for Solute Transport* (Humana Press, USA).

59. Naftalin, R. (1991). The red cell membrane: A model for solute transport. *Trends Biochem. Sci.*, **16**, pp. 45–48.

60. Wang, H., Hao, X., Shan, Y., Jiang, J., Cai, M., and Shang, X. (2010). Preparation of cell membranes for high resolution imaging by AFM. *Ultramicroscopy*, **110**, pp. 305–312.

61. Rash, J. E., Yasumura, T., Hudson, C. S., Agre, P., and Nielsen, S. (1998). Direct immunogold labeling of aquaporin-4 in square arrays of astrocyte and ependymocyte plasma membranes in rat brain and spinal cord. *Proc Natl Acad Sci U S A*, **95**, pp. 11981–11986.

62. Hermann, R., Walther, P., and Müller, M. (1996). Immunogold labeling in scanning electron microscopy. *Histochem. Cell Biol.*, **106**, pp. 356–356.

63. Edwards, H., Thompson, R. C., Koh, W. H., and Clode, P. L. (2012). Labeling surface epitopes to identify cryptosporidium life stages using

a scanning electron microscopy-based immunogold approach. *Mol. Cell. Probes*, **26**, pp. 21–28.

64. Treuel, L. and Nienhaus, G. U. (2012). Toward a molecular understanding of nanoparticle-protein interactions. *Biophys. Rev.*, **4**, pp. 137–147.

65. Cedervall, T., Lynch, I., Foy, M., Berggård, T., Donnelly, S. C., Cagney, G., Linse, S., and Dawson, K. A. (2007). Detailed identification of plasma proteins adsorbed on copolymer nanoparticles. *Angew. Chem.*, **119**, pp. 5856–5858.

66. Cedervall, T., Lynch, I., Lindman, S., Berggard, T., Thulin, E., Nilsson, H., Dawson, K. A., and Linse, S. (2007). Understanding the nanoparticle-protein corona using methods to quantify exchange rates and affinities of proteins for nanoparticles. *Proc Natl Acad Sci U S A*, **104**, pp. 2050–2055.

67. Niidome, T., Yamagata, M., Okamoto, Y., Akiyama, Y., Takahashi, H., Kawano, T., Katayama, Y., and Niidome, Y. (2006). PEG-modified gold nanorods with a stealth character for in vivo applications. *J. Controlled Release*, **114**, pp. 343–347.

68. Otsuka, H. (2003). *Electrical Phenomena at Interfaces and Biointerfaces* (John Wiley & Sons, USA).

69. Wang, C., Yan, Q., Liu, H. B., Zhou, X. H., and Xiao, S. J. (2011). Different EDC/NHS activation mechanisms between PAA and PMAA brushes and the following amidation reactions. *Langmuir*, **27**, pp. 12058–12068.

70. Stanković, V., Đurđić, S., Ognjanović, M., Antić, B., Kalcher, K., Mutić, J., and Stanković, D. M. (2020). Anti-human albumin monoclonal antibody immobilized on EDC-NHS functionalized carboxylic graphene/AuNPs composite as promising electrochemical HSA immunosensor. *J. Electroanal. Chem.*, **860**, pp. 113928.

71. Menger, F. M., Caran, K. L., and Apkarian, R. P. (1999). In-lens cryo-high-resolution scanning electron microscopy of large vesicles. *Langmuir*, **16**, pp. 98–101.

Chapter 9

Infrared Spectroscopy Studying Plasma Membrane

Lie Wu and Xiue Jiang
State Key Laboratory of Electroanalytical Chemistry,
Research Center of Biomembranomics,
Changchun Institute of Applied Chemistry,
Chinese Academy of Sciences, Changchun, Jilin, China
jiangxiue@ciac.ac.cn

Infrared (IR) spectroscopy is a powerful tool to elucidate the structure and conformation/orientation of water, lipids, and proteins and is widely applied in membrane-related studies. Surface-enhanced infrared absorption (SEIRA) spectroscopy, a booming variant of IR spectroscopy, is a surface-sensitive tool that can provide atomic/molecular information of membranes at the monolayer level. In this chapter, we first introduce the principles of IR spectroscopy with an emphasis on SEIRA spectroscopy, then discuss the application of IR spectroscopy in plasma membrane studies with a focus on the recently developed SEIRA spectroscopy, and finally describe the protocols.

Biomembranomics: Structure, Techniques, and Applications
Edited by Hongda Wang

ISBN 978-981-4968-61-4 (Hardcover), 978-1-003-45635-3 (eBook)
www.jennystanford.com

9.1 Introduction

Infrared (IR) spectroscopy, which relies on the interaction of low-energy IR light and matter, is a powerful non-destructive, label-free biophysical technology for in situ and real-time exploration of plasma membranes. Based on the "fingerprint" of vibrational modes of related molecules of plasma membranes, detailed molecular orientation and structure could be elucidated. Variations in spectral characteristics, including peak frequency, band intensity, and bandwidth, are closely related to structural and conformational changes in both lipid and protein components, as well as their environments, especially water molecules. And the vibrations in individual groups of specific molecules provide structural and environmental information of highly localized regions (down to chemical bands) of the membrane.

Water, the most important and abundant medium in cells, which is essential for the structure and function of biological membranes, has great absorption in the IR region. This property both aids and hinders the application of IR spectroscopy in membrane-related studies. On one hand, the sensitivity to water is a great advantage in studying the hydration water of membranes. On the other hand, the strong absorption of water overwhelms the whole spectrum, making the characteristic peaks of lipids and proteins hard to be distinguished, and thus limits the analysis of membranes in their natural state. Attenuated total reflection Fourier transform infrared spectroscopy (ATR-FTIR) and infrared reflection absorption spectroscopy (IRRAS) are two approaches that are suitable for biological membrane study in aqueous solution, and great progress has been achieved. In ATR-FTIR, the spectral information of the sample is detected by the evanescent wave generated at the surface of the internal reflection element with a penetration depth of several micrometers (~0.5–2 μm). Only molecules within this range are able to interact with IR light, and the contribution from the solution is largely eliminated. However, stacked multilayer membranes and/or proteins are required to generate a meaningful spectrum, and heterogeneity of samples is an inevitable issue. IRRAS, especially with polarization modulation (PM-IRRAS), can characterize the adsorbed matter on a metal surface at the monolayer level. But ultrathin layer cell (several μm) and D_2O solvent are necessary to reduce the contribution from the solution. Also, the signal-to-noise (S/N) ratio is limited by the extremely small number of molecules at the surface.

Surface-enhanced infrared absorption (SEIRA) spectroscopy using a nanostructured metal film as a substrate to enhance the IR absorption of molecules on the surface is a booming IR spectroscopy with surface sensitivity at the (sub-) monolayer level [1–3]. In the ATR mode (ATR-SEIRA spectroscopy), the interference of bulk solution could be effectively eliminated, realizing selective detection at the membrane/aqueous interface [4, 5]. Furthermore, compatibility with external modulations, such as light, electricity, and chemical reagent, makes the in situ monitor of spectral changes under external stimuli readily. Such a property endows ATR-SEIRA spectroscopy with the ability to investigate membranes in physiological conditions. Thus, ATR-SEIRA spectroscopy is an in situ label-free, surface-sensitive tool for biological membrane-related studies at the monolayer level.

In this chapter, we introduce the principles of IR spectroscopy with an emphasis on SEIRA spectroscopy, discuss the applications of IR spectroscopy in plasma membrane studies with a focus on the recently developed SEIRA spectroscopy through several representative research and strategies used, and finally describe the protocols.

9.2 Principles

9.2.1 Molecular Vibrations

In a given molecule, atoms are connected by covalent bonds in which the electron pairs are shared between them. The harmonic oscillator model is used to simply describe covalent bonds in a molecule. Massless springs are used to represent the intramolecular interactions (covalent bonds) between the atoms. And the molecule could be simply considered as an array of point masses connected by such springs. The frequency (ν) of a two-atomic oscillator (m_a and m_b) is determined by the nature of the covalent bond:

$$\nu = \frac{\sqrt{\frac{k}{\mu}}}{2\pi} \tag{9.1}$$

where k is the force constant between the two atoms, and μ is the reduced mass defined by

$$\mu = \frac{m_a \cdot m_b}{m_a + m_b} \quad (9.2)$$

According to the formula, two factors govern the vibration frequency. (1) The frequency of the vibration increases when the rigidity of the spring (or the strength of the bond) increases. This means any inter- or intramolecular factor, as well as environmental factors, that alters the electron density in a specific bond will affect the vibrational frequency. (2) The frequency of the vibration decreases with increasing masses of the atoms, which means the larger the mass of vibrating atoms, the slower the vibrational frequency. Thus, each individual vibration has its own inherent frequency (v), and chemical bonds in different chemical environments exhibit varying vibrational frequencies.

At temperatures above absolute zero, covalent bonds in molecules are not rigid sticks; all the atoms in molecules are in continuous motion with respect to each other, as their covalent bonds stretch, bend, and twist. There are $3n - 6$ and $3n - 5$ degrees of freedom for motions, which are fundamental vibrations for nonlinear molecules and linear molecules, respectively. These individual fundamental vibrations (also known as normal modes of vibration), a few of which are illustrated in Fig. 9.1, constitute the complex vibrations of a molecule, which provide a "fingerprint" of the molecular structure and chemical environments.

9.2.2 Infrared Absorption

During the motion of a molecule, a covalent bond oscillates due to the oscillation of the dipole of the molecule, producing a varying electromagnetic (EM) field. The greater the dipole moment changes through the vibration, the more intense the EM field generated. When a wave of incident light encounters this oscillating EM field generated by the oscillating dipole of the same frequency, the two waves couple, and a photon is absorbed. The molecule acquires a clearly defined amount of energy from the radiation and moves up to the next vibrational level, which matches the energy difference between two vibrational energy levels—the initial (e.g., ground state) and the final (e.g., first excited state) vibrational states. As these energy differences are in the order of 0.5 and 0.005 eV, light with wavelengths longer than 2.5 μm (IR light) is sufficient to excite the vibrational transitions. Thus, vibrational spectroscopy that is based on the direct absorption of light quanta is denoted as IR absorption.

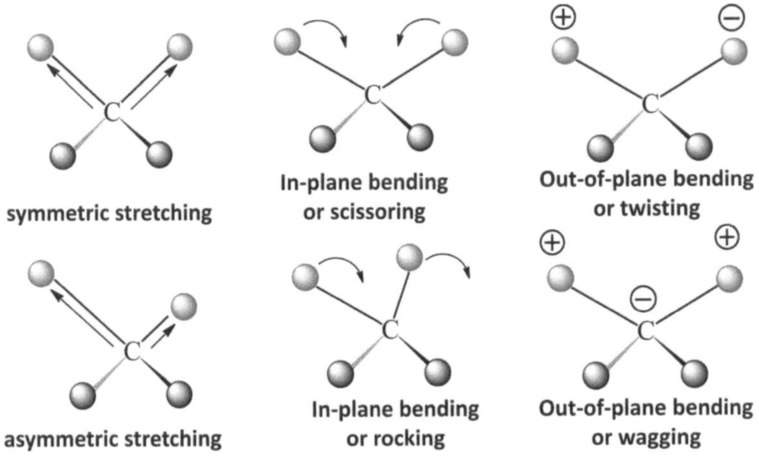

Figure 9.1 Typical fundamental vibration modes.

The total number of observed absorption bands is generally less than the predicted total number of fundamental vibrations ($3n - 6$ for nonlinear molecules and $3n - 5$ for linear molecules). There are several reasons that account for this phenomenon.

1. Some modes are not IR active. For a fundamental vibration to be an IR-active vibration, two prerequisites must be satisfied: (a) has the same frequency as in light radiation and (b) produces a net change in the dipole moment as a result of the vibration. Thus, only the vibrational mode with nonzero transition dipole moment is IR active. The greater the change in the dipole moment, the stronger the corresponding IR absorption. Homonuclear diatomic molecules, such as H_2, O_2, and N_2, do not have any IR absorption. Also, for a given molecule, some fundamental vibration modes could be IR active and other modes could be IR inactive.

2. The energies of some vibrations are identical, which means that a single frequency of incident light can activate more than one mode of vibrations to occur.
3. **Limitations of analytical techniques:** The absorption intensity is too low to detect, and/or the frequency is beyond the detecting range of the instrument. On the other hand, additional bands could be generated by the appearance of overtones (integral multiples of the fundamental vibrations), combinations or differences of fundamental frequencies, and coupling interactions of the above-mentioned frequencies, whose intensities are less than those of the fundamental bands. The combination and blending of all the factors thus create a unique IR spectrum for each compound. It should be noted that for most cases, IR absorption spectroscopy deals with the fundamental bands. The major types of molecular vibrations are stretching and bending, as shown in Fig. 9.1. The absorption involves discrete, quantized energy levels. However, the individual vibrational motion is usually accompanied by other rotational motions. These combinations lead to the absorption bands, not the discrete lines, commonly observed in the mid-IR region. It also should be noted that Raman scattering occurs if the polarizability of the bond changes during the vibration. Thus, IR-inactive vibrations are Raman active if the polarizability changes. Raman and IR spectra complement each other. And some vibrations can be both IR active and Raman active.

9.2.3 What Infrared Absorption Tells

Each individual vibration corresponds to a specific covalent bond and has its inherent frequency determined by the nature of the covalent bond. Any inter- or intramolecular factor that alters the force constant (electron density) of the covalent bond will produce variations in its IR absorption characteristics (peak frequency, band intensity, and bandwidth). Thus, IR absorption provides "fingerprint" information of both the molecular structure and its chemical environments.

Chemical structure: The chemical structure of a molecule determines the strengths of the vibrating bonds and the masses of the vibrating atoms, thus representing the dominating factor of the frequency and intensity of the IR absorption band. It often provides a fingerprint by which molecules can be identified. IR absorption is valuable and informative for the qualitative elucidation of molecular structures.

Hydrogen bonding: The formation of hydrogen bonds has a dramatic impact on the force constant of a covalent bond. Generally, hydrogen bonding lowers the restoring force, lowering the frequency of stretching vibrations, while increases the frequency of bending vibrations due to the production of an additional restoring force. IR absorption is one of the few methods that directly report on the strength of hydrogen bonds.

Conformational freedom: For a specific vibrational mode, several conformations may exist under a dynamic equilibrium. The absorption band position of each conformer is usually slightly different, leading to a heterogeneous band broadening. Structures with more conformational freedom will exhibit broader bands than rigid structures, resulting in a shift in the position of absorption maximum. Thus, the bandwidth and position of absorption maximum could be taken as a measure of conformational freedom.

Bond length and bond strength: Bond length and bond strength are two main dominating factors of a number of physicochemical parameters that determine the force constants of a specific vibration bond. Vibrational frequencies of stretching vibrations are a direct function of the force constants of the vibrating bonds, thus providing information on bond lengths and bond strength, which are very helpful for the interpretation of detailed structure changes.

Bond angles and conformation: Individual vibration is often coupled with nearby vibrations, and such coupling depends greatly on the details of the molecular geometry. Vibrational coupling, therefore, often provides insights into the three-dimensional (3D) structure of molecules due to the frequency–geometry correlations.

Electric fields: An electric field applied from the external environment or produced by a local environment could change the vibrational energy levels, which is known as the vibrational Stark effect (VSE) [6–8]. Infrared absorption probes the energy gap between the ground state and the first excited vibrational state. Since these states have (slightly) different dipoles, they will be (de) stabilized (slightly) differently by an external field, depending on the orientation between the dipole and the electric field. Therefore, the applied electric field will produce a shift in the vibrational transition energy that is linear with the difference in the dipole moments of the two states. Based on the vibrational Stark effect, the local electric field (where the covalent bond is located) could be analyzed with the position shift in IR absorption.

9.2.4 Infrared Spectrometers

Infrared spectrometers are analytic instruments used to obtain the IR absorption spectra, which in simple terms are obtained by detecting changes in transmittance (or absorption) intensity as a function of the frequency of incident light. Most commercial instruments separate and measure IR radiation using dispersive spectrometers or Fourier transform spectrometers.

9.2.4.1 Dispersive spectrometer

A dispersive spectrometer, introduced in the mid-1940s, is the first generation of infrared spectrometers. It consists of three basic components: radiation source, monochromator, and detector (Fig. 9.2A). IR radiation from a broadband source (typically SiC Globar) passes through the sample and is dispersed by a monochromator (prism or grating) into component frequencies. Then the beams focus on the detector, which generates an electrical signal and results in a recorder response. Thus, in a dispersive IR spectrometer, IR spectroscopy is obtained by successively measuring how much IR radiation is absorbed from a specific monochromatic beam. And IR spectrum is obtained by the frequency scanning of the monochromator. Most dispersive spectrometers have a double-beam design, and the difference between the IR radiation intensities of these two beams due to absorption by the sample is measured as an off-null signal recorded as IR absorption.

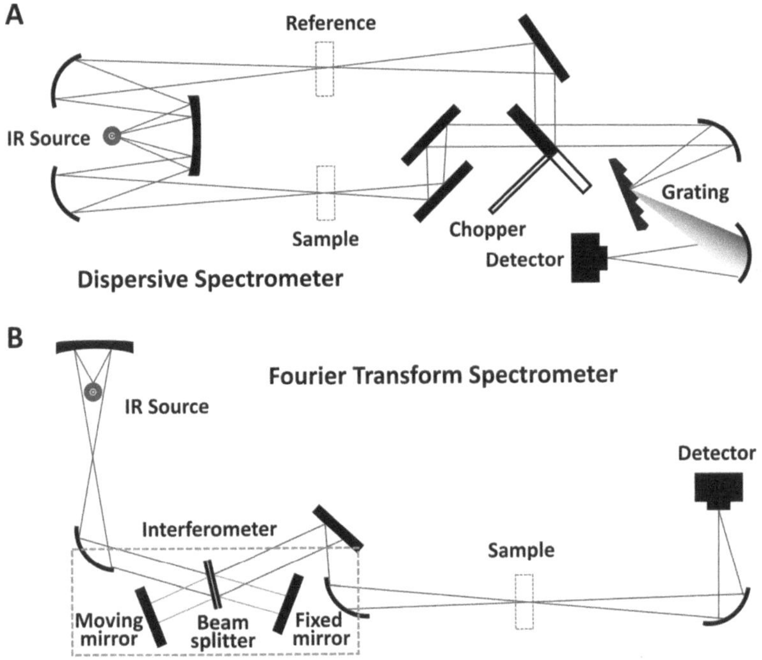

Figure 9.2 Simplified optical layout of typical dispersive spectrometer (A) and Fourier transform spectrometer (B).

9.2.4.2 Fourier transform spectrometer

A Fourier transform spectrometer with an integrated interferometer is the second generation of infrared spectrometers. Since its inception in mid-1960s, dispersive instruments have been widely replaced for most applications due to significant advantages of Fourier transform spectrometers. They have greatly extended the capabilities of infrared spectroscopy and have been applied to many fields that are very difficult or nearly impossible for dispersive instruments. The main difference between these two types of spectrometers lies in the way in which a spectrum is obtained. Instead of a frequency scan by the monochromator in dispersive IR spectrometers, the intensities of all frequencies are examined simultaneously in Fourier transform infrared (FTIR) spectrometers.

A simplified optical layout of a typical FTIR spectrometer is illustrated in Fig. 9.2B. An interferometer (mostly Michelson interferometer), which equally divides the incident beam, generates an optical path difference between the two beams, and then recombines them to produce repetitive interference signals, is applied to replace the monochromator. The recorded interferogram is a time domain spectrum that records the changes in detector response versus time within the mirror scan (difference in light path). When this IR beam is directed through the sample that absorbs light at a set of characteristic frequencies, the amplitudes of corresponding frequencies are reduced, resulting in a change in the output interference patterns. After Fourier transformation calculation, the interferogram (a time domain spectrum displaying intensity versus time within the mirror scan) is converted to the final IR spectrum, which is a frequency domain spectrum showing intensity versus frequency. Most benchtop FTIR spectrometers are single-beam instruments, and a background spectrum is necessary for measurement. The background spectrum not only contains the absorption contributions of the environment of optical bench, but also includes the performance of the spectrometer.

The FTIR spectrometer exhibits superior performance in many aspects, and almost every spectrometer used in mid-IR spectroscopy is of the Fourier transform type nowadays.

1. **Superior speed and sensitivity:** In an FTIR spectrometer, a full range spectrum can be obtained in a single scan of the interferometer, whose scanning speed (from several kHz to more than 100 kHz) is much faster than the grating in the monochromator. Thus, an FTIR instrument can achieve the same S/N ratio of a dispersive spectrometer in a fraction of the time (1 s or less versus 10 to 15 min).
2. **Higher optical throughput:** Energy-wasting slits used for shaping and filtering the dispersed beams are not required, and a circular aperture is commonly applied to adjust the optical throughput in the interferometer spectrometer. The beam area of an FTIR instrument is usually 75 to 100 times larger than that of a dispersive spectrometer, making more radiation energy available.
3. **Higher frequency accuracy and spectral resolution:** In an FTIR spectrometer, the frequency accuracy and spectral resolution are mainly determined by the position accuracy and

moving distance of the moving mirror in the interferometer. The interferometer has a highly stable HeNe laser as an internal reference to monitor and control the position of the moving mirror, providing an automatic calibration and a frequency accuracy better than 0.01 cm^{-1}. The spectral resolution could be improved by increasing the moving distance of the moving mirror. Generally, a spectral resolution of 2 cm^{-1} could be readily obtained, and 0.06 cm^{-1} for a research spectrometer and <0.0009 cm^{-1} for a high-resolution instrument could be achieved.

9.2.5 Surface-Enhanced Infrared Absorption Spectroscopy

In 1980, Hartstein et al. discovered that the IR absorption of organic compounds was effectively enhanced when adored on a deposited rough thin metal film [9], and defined it as the surface-enhanced infrared absorption (SEIRA) effect, opening the prelude to SEIRA spectroscopy. After the pioneer work of Osawa and Nishikawa et al. [1, 10–12], SEIRA spectroscopy has drawn more and more scientific attention and intensive research activities have been initiated in related fields. Taking advantage of the enhancement effect of nanostructured metal films, SEIRA spectroscopy is capable of detecting minute absorption changes (as small as 10^{-5}) of the adsorbed molecule at the (sub-) monolayer film level, even when immersed in water [13–15], providing a powerful tool for in situ interfacial analysis. Recently, the booming of patterned antenna enhancement substrates [16, 17] with an enhancement factor up to 10^5 has greatly promoted the development and application of SEIRA spectroscopy. With improvement in the enhancement mechanisms and development of enhanced substrates, SEIRA spectroscopy would be widely used in various fields.

9.2.5.1 Enhancement mechanisms for SEIRA

The theory of enhancement effect of SEIRA is basically derived from earlier established surface-enhanced Raman scattering, since a nanostructured metal film plays a crucial role in both enhancement phenomenon. Generally, it is believed that the observed total enhancement is the sum of the following mechanisms, which could be classified into electromagnetic (EM) mechanism and chemical mechanism [3, 18, 19]. It should be noted that the detailed enhancement mechanism of SEIRA is far from perfect and cannot explain the SEIRA from polar dielectric nanoparticles [20, 21] (such as SiC, Al_2O_3, AgCl), and new effects and models are emerging.

Surface plasmon resonance (electromagnetic mechanism): Island-like metal films crucial for SEIRA can be approximately represented as continuous or discontinuous ellipsoidal metal particles shown in Fig. 9.3. The EM mechanism of SEIRA includes two related theories.

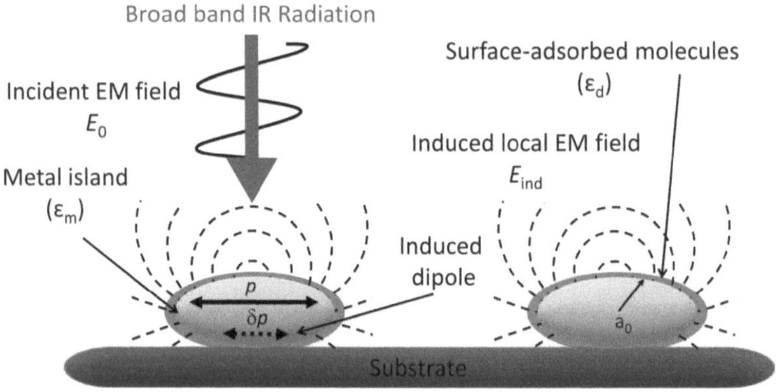

Figure 9.3 Schematic representation of the EM mechanism of SEIRA on metal island films.

1. **Surface plasmon resonance:** When the lateral size of the metal particles is small with respect to the wavelength of the incident light, the electromagnetic field of the incident light can effectively couple with the collective vibrations of the "free electrons" of the metal, which are denoted as surface plasmons. The incident electric field (E_0) induces an oscillating electric dipole p in the metal island and excites the surface plasmons, generating an additional electric field (E_{ind}) stronger than the incident photon field. Vibrational dipole moments in the molecule interact with this enhanced local EM field and result in enhanced absorption by the molecules located near the film. The vector of the induced local enhanced EM field is parallel to the surface normal of the islands, and its

intensity is determined by the shape and size of the islands and depends on the distance *d* from the metal surface:

$$\left(E_{\text{local}}\right)^2 \propto \frac{4p^2}{d^6} \tag{9.3}$$

And the decay in enhancement can be expressed as:

$$F_{\text{SEIRA}}(\text{d}) = F_{\text{SEIRA}}(0) \times \left(\frac{a_0}{a_0 + d}\right)^6 \tag{9.4}$$

where a_0 is the local radius of curvature of the island. This EM mechanism endows SEIRA spectroscopy with two important properties:

(a) **Optical near-field effect:** Due to the sharply decayed intensity of the local EM field, the effective enhancement distance is estimated to be in the range of 8–10 nm [11]. Signal contribution from species located beyond this distance is barely observed, endowing SERIA spectroscopy with surface sensitivity.

(b) **Surface selection rule:** As the vector of the induced local EM field is parallel to surface normal, only the dipole moment component perpendicular to the metal surface is enhanced. Thus, SEIRA signals are not detectable for vibrations associated with dipole moment changes parallel to the surface. This makes SERIA spectroscopy to readily analyze the orientation of the adsorbed molecules.

2. **Perturbation of optical properties of the metal film:** The vibrating dipoles of the adsorbed molecules can induce additional dipoles δp in the metal island, changing the dielectric function of the metal. The perturbation is most significant at frequencies where the adsorbed molecules are vibrating, making the spectrum of the metal nearly identical to the spectrum of the adsorbed molecules. Both the absorptivity and the volume fraction of the metal are much larger than that of the surface-adsorbed molecules (sub-monolayer level). Thus, the induced perturbation of the optical properties of the metal film would dominate the observed spectrum, causing an effective enhancement of the apparent IR absorption of the adsorbed molecules.

Chemical mechanism: It has been observed that the enhancement effect for molecules that bind to the metal surface by chemical adsorption is larger than that for the physically adsorbed ones. This indicates that surface chemistry also plays an extremely important role in the SEIRA effect, leading to the proposal of chemical mechanisms, which also include two parts.

(1) **Charge transfer and donor/acceptor interaction:** When the adsorbed molecule has a chemical interaction with an atom or atomic clusters of the metal surface, the incident light could induce charge transfer between the energy level of the molecule and the Fermi energy level of the metal. The electron is transferred from an occupied molecular orbital of the adsorbed molecule to an unfilled energy level of the metal or from a filled energy level of the metal to an excited molecular orbital of the adsorbed molecule. If the energy of the incident photon matches the electron transfer energy between the adsorbate and the metal substrate, the resultant oscillation increases the effective dipole moment of the adsorbate, enlarging the absorption coefficients of the adsorbed molecule by "intensity borrowing" from the charge transfer. In addition, the donor–acceptor interaction between the chemisorbed adsorbate and metals may also change the vibrational polarizability of the adsorbed molecules, thereby enhancing their infrared absorption.

(2) **Uniform orientation:** Chemisorption aligns the surface molecules with a preferential orientation (dipole moment along the surface normal of the metal) on the metal surface. This uniform orientation, on one hand, improves the adsorption efficiency and allows more molecules to be adsorbed to the metal surface. One the other hand, perpendicularly oriented vibrational modes give rise to IR absorption three times higher than that of the randomly oriented ones (vide infra) due to the surface selection rule. Thus, uniform orientation of the adsorbed molecules contributes to the apparent enhancement effect.

9.2.5.2 Enhancement substrate for SEIRA

Surface-enhancing substrate is a key factor in SERIA spectroscopy, and the development of a new enhancing substrate with a high enhancement factor and good reproducibility is of significant importance for SERIA spectroscopy and is a booming topic in SEIRA [16, 22]. The morphology of the metal island film, such as size, shape, particle density, and thickness, is the dominant factor of the enhancement effect of SEIRA, and the conductivity and electrochemical stability are also needed to be considered in some specific applications such as surface electrochemical research. In addition to its refractive index, the chemical properties of infrared window materials also affect the morphology of the metal island film, influencing the enhancement effect. Therefore, both SEIRA-active metals and infrared window materials have impact on the overall enhancement effect. SEIRA-active metals are not limited to Au and Ag; metals in which the real part of the dielectric function is larger than the imaginary component are all SERIA active [23], such as Al, Pt, Ni, Pd, Pb, Ru, Rh, Sn, etc. [24].

Methods used for the preparation of SEIRA substrates include dry and wet processes [18]. Dry processes mainly include evaporation and sputtering in vacuum: SEIRA-active metals are deposited on the infrared window materials in the vapor phase at large scale by electron beam heating, resistance heating, argon ion sputtering, etc. The disadvantages of the dry process include the requirement for high-cost equipment, time-consuming operation, sample contamination by organic species, poor performance reproducibility, and severely distorted spectra. Wet methods prepare SEIRA substrates by electrodeposition and/or chemical (or electroless) deposition, which are simple to operate and have low cost. In electrodeposition, metal ions in the plating solution are deposited on the support material by electrochemical reduction, in which good electrical conductivity of the support material is a prerequisite. In chemical deposition, the metal film is deposited by chemical reduction, which could be classified into autocatalytic, substrate-catalyzed, and galvanic displacement mechanisms, as shown in Fig. 9.4.

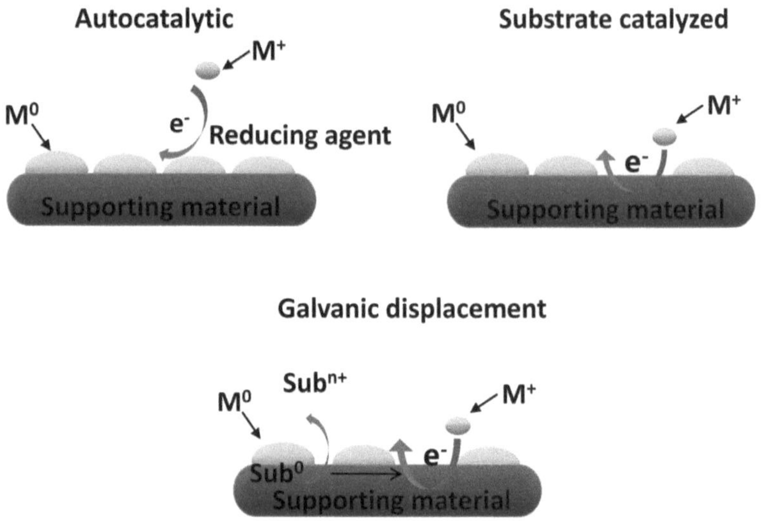

Figure 9.4 Three mechanisms in wet chemical deposition processes.

In the autocatalytic mechanism, the reduced metal serves as the catalyst for further reduction of the metal salt by an external reducing agent. For the substrate-catalyzed mechanism, the supporting material surface catalyzes the reduction of the metal salt by the reducing agent. Once a complete film is formed, reduction ceases because the substrate is no longer exposed. The difference with the autocatalytic mechanism is that metal reduction is facilitated on the substrate surface. Galvanic displacement proceeds in an entirely different manner, where deposition is carried out in the absence of an external reducing agent. The surface serves as the reducing agent and electron source for the reduction of the metal salt. Deposition continues as long as ions can permeate and electrons get transferred across the film. The adhesion strength of the chemically deposited metal film to the supporting material is stronger than that obtained from the dry process, endowing the enhancing substrate with good long-term stability. "Two-step wet chemical deposition" and "seed-mediated growth method" have been developed to extend the types of metals that could be chemically deposited with strengthened adhesive and electrochemical stability [25]. The colloidal metal nanoparticles are also SEIRA-active nanostructures, and the SEIRA spectrum could be obtained by collecting the signal of the preformed sample–colloidal nanoparticle complex through the transmission or attenuation total reflection (ATR) mode. The most commonly used enhancement substrate is the island-like gold films on silicon windows prepared by chemical deposition, as proposed by Osawa's group [26].

Recently, patterned periodic antenna-based enhancement substrate has drawn a lot of attention [16, 17]. Periodic nanostructured antennas feature pronounced plasmonic resonances and/or antenna resonance (mixed state of collective electronic excitations and photons localized at a metallic surface) in the IR region caused by collective oscillations of electrons, dramatically enhancing the IR absorption at corresponding frequencies. Resonant nano-antennas with tailored structure could be prepared by a variety of nanofabrication methods [16] such as electron beam lithography (EBL). SEIRA spectroscopy using resonant metal nano-antennas is called "resonant SEIRA" or "plasmon-enhanced infrared spectroscopy." The advantage of resonant SEIRA lies in the great enhancement factor (up to 10^5) and widely tunable resonant frequency. The former endows resonant SEIRA with high sensitivity, while the latter makes selective detection of specific vibration mode possible. Thus, resonant antenna enhancing substrates are quite promising, and resonant SEIRA is believed to be the next generation of SEIRA spectroscopy.

9.2.5.3 Accessory of SEIRA spectroscopy

SEIRA spectroscopy has several geometric and optical configurations, as shown in Fig. 9.5A. In transmission geometry, low-refractive index IR-transparent materials such as CaF_2 and BaF_2 are used as a supporting substrate. And the preformed sample–metal nanoparticle complexes directly deposited on the IR windows or sample adsorbed on very thin metal films (less than 10 nm to keep the metal film transmissive) are employed. In external reflection geometry, two optical configurations are used, and non-IR-transparent materials, such as glass, polymers, carbon, and metal, can be used as a supporting substrate. The IR beam is directed through the IR windows and is reflected at the supporting substrate or metal film surface. For the metal overlayer configuration (metal/sample/supporting substrate), the thin metal layer (~4–7 nm) is directly evaporated on the sample/support surface. This configuration is advantageous for detection in the trace analysis of a sample adsorbed on glass or polymer (in most cases in air or non-aqueous solvents). For the metal underlayer configuration (sample/metal film/supporting substrate), the enhancing substrate is prepared by depositing a relatively thick metal film (in the range of ten to several hundred nanometers) on the supporting material or roughening the bulk supporting material. The probing IR beam passes through the solution layer between the IR window and the supporting substrate twice. Thus, the surface sensitivity is relatively low in this configuration (surface signal is enhanced 5–10 times [27]) but is superior in the detection of solution phase species. To minimize the optical path through the solution, the metal surface is mounted in μm distance to the IR window.

The most widely employed optical arrangement of SEIRA spectroscopy is internal reflection (Kretschmann-attenuated total reflection) geometry, which is known as ATR-SEIRA spectroscopy. High-refractive-index IR windows are suitable to serve as supporting substrates in this configuration, including Si, Ge, and ZnSe, among which Si is most commonly used due to its high chemical stability. The evanescent wave generated during the total reflection of the incident IR beam could reach the surface of the metal film and couple to the enhancing field, resulting in enhanced vibrational signals of sample molecules.

In reflection configurations (both external and internal), additional accessories are needed to direct the IR beam onto the supporting substrate, which are placed in the sample chamber of the spectrometer. Several representative optical layouts of accessories are shown in Fig. 9.5B. The first one is the most complex one, which uses several couples of mirrors to control the incident angle of IR beam. Thus, the adjustment of incident angel is very easy, but the optical throughput is relatively low due to the reflection loss. The second one simplifies the optical path, making the accessory relatively easy to build. And a condenser lens is further used to focus the beam, improving the optical coupling of the reflection element and increasing the optical throughput. The third one has the simplest optical design with satisfiable performance for most applications and is the most commonly used design for homemade SERIA spectroscopy accessories. Half-cylindrical and half-spherical reflection elements or prisms are commonly used. The former two types are superior since the position of the focus on the metal film is not affected by the incident angle of IR beam.

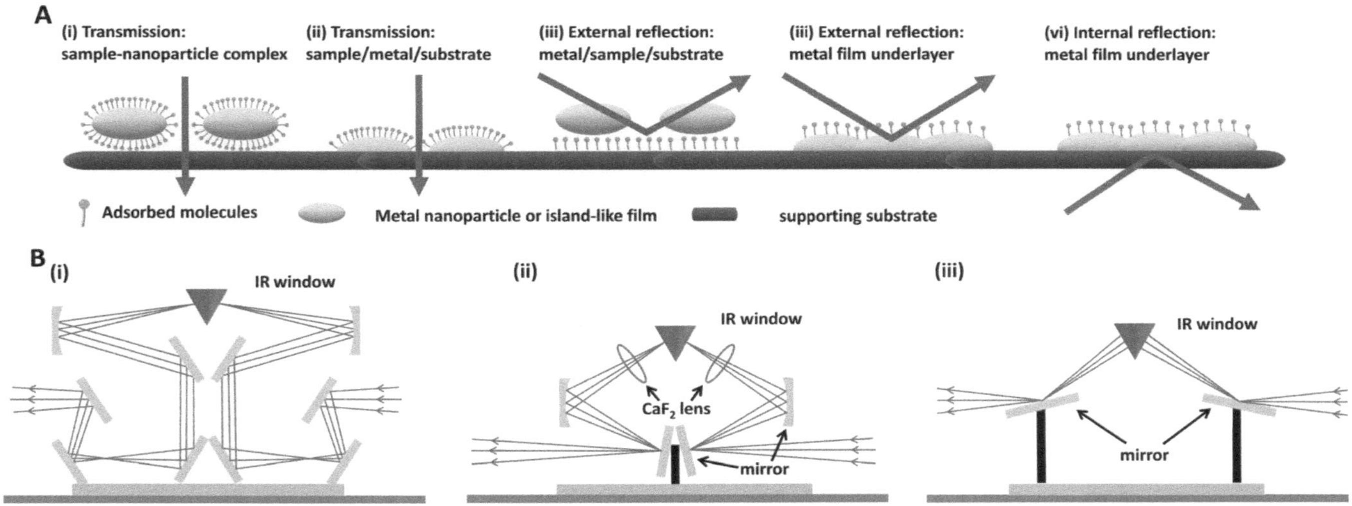

Figure 9.5 Schematic representation of geometric and optical configurations (A) and optical layout of accessories (B) for SEIRA spectroscopy.

9.2.5.4 Advantages of ATR-SEIRA spectroscopy

The inherent properties of ATR-SEIRA spectroscopy make it a powerful tool for interfacial analysis, among which several representative advantages are discussed below.

Ultra-sensitivity: ATR-SEIRA spectroscopy can detect minute changes (as small as 10^{-5}) in the absorption of the sample, making infrared detection at the (sub-) monolayer level possible. Taking advantage of patterned nano-antenna substrates, a signal enhancement in the order of 10^4–10^5 can be readily achieved, enabling the probe of adsorbates down to zeptomole concentrations (10^{-21} M) [28].

Surface selection rule: The vector of the locally enhanced EM field is parallel to the surface normal of the metal film. Thus, only vibrations associated with the dipole moment component perpendicular to the metal surface are enhanced. Based on this selection rule, the bonding and orientation of groups or molecules can be analyzed [11, 29].

Surface selectivity: In the ATR configuration, the coupling between the evanescent wave at the substrate surface and the sharply decayed locally enhanced EM field (optical near-field) at the metal surface nearly perfectly eliminates the signal contribution from the bulk solution, achieving the selective detection of surface species [1, 2, 13].

High stability: Under the internal reflection configuration, the IR beam propagates under the enhanced metal film and supporting substrate. Thus, the probing IR beam could be protected from atmospheric absorption by purging with a dry inert gas or even vacuum. Detection free from environmental fluctuations both dramatically improves the performance toward very weak spectral signals and endows with long-term stability.

In situ monitoring and compatibility with external modulation: As shown in Fig. 9.5B, the enhancing substrate is accessible to manipulation during measurements. Thus, ATR-SEIRA spectroscopy is readily capable of in situ monitor changes under external stimuli, such as light, chemical reagent, etc. More importantly, metal films (enhancing substrate) can serve as the work electrode, enabling the compatibility of SEIRA spectroscopy with electrochemical modulation. In the ATR configuration, the electrode surface is in direct contact with the bulk solution, eliminating the limitations of mass transfer and slow potential response, and large ohmic potential drop originating from thin layer cells required in the external reflection mode.

These unique characteristics make ATR-SEIRA spectroscopy a surface-sensitive in situ label-free spectro-electrochemical technique, showing great potential in various research fields.

9.3 Applications in Plasma Membrane Studies

Infrared spectroscopy with different technical variants, especially the emerging ATR-SEIRA spectroscopy, not only serves as a chemical characterization tool, but also has been extended deeply to many important aspects of plasma-membrane-related studies. In this part, we would like to briefly introduce the application of IR spectroscopy in the physicochemical characterization of membranes (Section 9.3.1), then focus on the newly developed application fields with ATR-SERIA spectroscopy (Sections 9.3.2–9.3.5) with an emphasis on interfacial-water-related studies (Section 9.3.2) by demonstrating several representative examples and strategies applied.

9.3.1 Physicochemical Characterization of Lipid Membranes

The lipid membrane is the structural basis of a real biological membrane and is widely used as artificial models of biological membranes. In this section, the application of IR spectroscopy in the characterization of the physicochemical properties of lipid membranes is briefly summarized.

9.3.1.1 Structure and orientation of lipids in membranes

PM-IRRAS has been proven as a useful tool for investigating the absolute orientation of lipid membranes. In PM-IRRAS, the direction of a transition dipole with respect to the surface normal can be determined from the experimental spectrum and the simulated spectrum of randomly oriented molecules using the following formula:

$$\cos^2\theta = \frac{1}{3}\left(\frac{A_{\text{exp}}}{A_{\text{sim}}}\right) \tag{9.5}$$

where A_{exp} is the intensity of the absorption band experimentally determined with bilayer at the metal surface and A_{sim} is the intensity of corresponding band in the hypothetical bilayer of randomly oriented molecules, which can be calculated from the spectra simulation. The intensities of the CH_2 stretching bands can provide useful information on the orientation of the acyl chains of lipid. The vector of the transition dipole for the asymmetric methylene stretch lies along the line joining the two hydrogen atoms of the CH_2 group, while the vector of the transition dipole for the symmetric stretch lies along the bisector of the CH_2 plane. Both vectors are perpendicular to each other and are also perpendicular to the line of the fully extended all-trans hydrocarbon chain. Therefore, the tilt angle of the hydrocarbon chain (θ_{chain}) in the bilayer can be calculated according to the following equation:

$$\cos^2\theta_{\text{chain}} = 1 - \cos^2\theta_{\text{s}} - \cos^2\theta_{\text{as}} \tag{9.6}$$

It should be noted that this equation can only be applied if the hydrocarbon chains are fully extended in the all-trans conformation. However, if part of the chains is melted, then gauche conformations are formed and the equation can be used only as a rough approximation.

The Jacek Lipkowski group in the University of Guelph has done excellent work in applying PM-IRRAS to investigate the structure and orientation of lipids in various biomimetic membranes supported on a metal under potential control and induced by interacting with other bioactive molecules. For detailed information, please refer to their published work.

9.3.1.2 Ordering and phase transition of membrane

The CH_2 stretching band of lipid molecules can provide useful information on the conformation of the membrane due to its sensitivity to environments. Both the peak frequency and the bandwidth are direct indicators of the ordering of membranes. The higher frequency and the broader band of CH_2 stretching indicate the more disorder of the acyl chain of membranes and *vice versa.*

Thus, the phase transition temperature (T_m) can be derived from the temperature-dependent frequency change of CH_2 stretching. Thus, the effect of biochemical and biophysical factors on the ordering and phase transition of membranes can be evaluated. For more information on the technical and scientific aspect, please refer to the specific reviews [30, 31].

9.3.1.3 Structural dynamics of membrane

Biological membranes hold a wide variety of proteins and other biomolecules that are essential to cell functions and serve as "solvent bath." Thus, the structural dynamics of membranes is an importance property for membrane function. The structural dynamics of membrane lipids in membranes can be investigated with nonlinear two-dimensional (2D) IR spectroscopy at the time scale of fs to ps. The nonlinear nature of ultrafast 2D IR requires a vibrational probe to have a strong intrinsic infrared transition strength in an uncluttered region of the spectrum while being highly sensitive to the local environment. The antisymmetric CO stretching of a nonpolar and water-insoluble tungsten hexacarbonyl ($W(CO)_6$) was successfully applied as a vibrational probe to study the dynamics in the interior region of membranes [32–34].

9.3.2 Structure and Functions of Interfacial Water of Membranes

Water molecules, the most important medium in cells, could associate with membranes through hydrogen bonding and dipole interactions, regulating the electrostatic potential of the membrane surface and participating in membrane-related interfacial interactions. Revealing the structure and functions of the interfacial water of membranes is crucial for deeply understanding the functions of biological membranes. IR spectroscopy is a powerful tool for studying the structure and orientation of water molecules owing to its sensitivity to hydrogen bonding. Water molecules exhibit a complex broad ν (OH) band in the range of 2800–3800 cm^{-1} and a δ (OH) band in the range of 1600–1700 cm^{-1}, whose shape and position are greatly influenced by the hydrogen banding state of the OH group. The frequency of the ν (OH) band shifts to lower wave numbers with increasing strength of the hydrogen bond, while the δ (OH) band exhibits a reverse tendency. Due to the complex hydrogen bond networks among water molecules, a broad ν (OH) band of water is often observed in IR spectroscopy. A deep analysis of the ν (OH) band with lipid-specific bands could reveal the change in water population species with membranes.

Using a stacked multilayer membrane as a model, water sorption kinetics, structure and hydrogen-bonding characteristics of hydration water, and the mutual interactions with lipid molecules during the hydration process with controlled relative humidity and temperature are systematically studied with ATR-FTIR [35–39]. However, the heterogeneous nature of the stacked multilayer membrane limits further in-depth research. Taking the phosphate vibration of lipid molecules [40], the ν (OH) of water in D_2O [41], vibrational labeled cholesterol derivative as probe [42], and the dynamics of interfacial water in vesicles or multilayers in solution could be determined. For more information and recent progress in the dynamics of interfacial water revealed by 2D IR spectroscopy, please refer to the specific reviews [43–45].

Taking advantage of surface sensitivity and compatibility with electrochemistry, ATR-SEIRA spectroscopy has been successfully applied to investigate the hydrogen-bonding network of the interfacial water in various bilayer model membranes under electrochemical modulation and its impact on membrane properties [46–49]. Thus, in the following section, we briefly summarize the recent progress achieved by ATR-SEIRA spectroscopy, especially the strategies developed.

9.3.2.1 Probing the structure and functions of local water bonded to phosphate groups at membrane/aqueous interface

Water's unique properties are given by hydrogen-bonding networks and dipoles, whose local structure and orientation greatly affect the electrostatic potential distribution at the interface [50, 51]. The presence of lipid membranes at the interface adds more complexity to the local water structure, greatly challenging the probe of the distinctive properties of water owing to an additional electric dipole originating from the polar groups of the membrane [52]. Through "ion and potential" double perturbation, the structure and functions of local water bonded to phosphate groups at zwitterionic phospholipid membranes/aqueous interface were successfully revealed [5].

Series polarization potentials (−0.3~0.7 V vs Ag/AgCl) were applied on a zwitterionic phospholipid (L-α-phosphatidylcholine, PC) hybrid bilayer membrane (constructed by PC vesicle fusion on 1-dodecane thiol (DT) monolayer) on Au film, and Ca^{2+} was further added to specifically disturb the hydration shell of PO^{2-} of PC membrane (Fig. 9.6A), taking advantage of its high affinity to PO^{2-} group (the binding constant of Ca^{2+} towards PO^{2-} is about 12–40 M^{-1} [53]). A broad negative band around 3000 cm^{-1} and a pair of coupled peaks at 1108(+)/1092(−) cm^{-1}, which indicate the dehydration of the PO^{2-} group [54], were simultaneously observed in the Ca^{2+}-induced difference spectra of the PC membrane (Fig. 9.6C). And almost completely reverse difference spectra can be induced by the further addition of EDTA·2Na to chelate and desorb Ca^{2+} (Fig. 9.6D). At the same time, current changes corresponding to the adsorption and desorption of Ca^{2+} were observed in the I-T curve (Fig. 9.6B). These results indicate that spectral changes around 3000 cm^{-1} are directly related to the adsorption and desorption of Ca^{2+} and originated from water molecules strongly hydrogen bonded to PO^{2-}. This is highly consistent with the previous results from sum frequency generation and molecular dynamics simulation. These results identified the spectral changes at about 3000 cm^{-1} to be originated from the strongly hydrogen-bonded water molecules with hydrogen (H) points to the PO^{2-} as a donor, which is restricted by the P^--N^+ dipole [55, 56].

The negative band around 3000 cm^{-1} was present in the Ca^{2+}-induced difference spectra at all applied potentials (−0.3~0.7 V vs Ag/AgCl) with a slight red-shift originating from the enhanced alignment of interfacial water under negative potential (Fig. 9.6C). At the phospholipid membranes/aqueous interface, water molecules exhibit a "H-down" orientation owing to the negative charge of the deprotonated PO^{2-}, significantly enhancing the hydrogen-bonding interaction. It is understandable that the reinforcement of the negative potential to the inherent electrostatics could enhance the alignment of interfacial water molecules. However, it is extraordinary that the structure and orientation of water molecules strongly hydrogen-bonded to the PO^{2-} group are maintained even under a positive potential (0.7 V vs Ag/AgCl), which is much higher than the potential of zero charge (0.05 V vs Ag/AgCl). And the expected water dipole flipping with the external electric field was not observed. These results indicate that the structure and orientation of water molecules located around the PO^{2-} group are not affected by the transmembrane potential. Further analysis of the effect of Ca^{2+} on the open circuit potential of PC membrane/aqueous interface suggests that these water molecules have dramatic impact on the electrostatic properties of the membranes, which should be closely related to the highly ordered orientation and strong dipole of water molecule. A detailed discussion and conclusion can be found in the original publication (Ref. [5]).

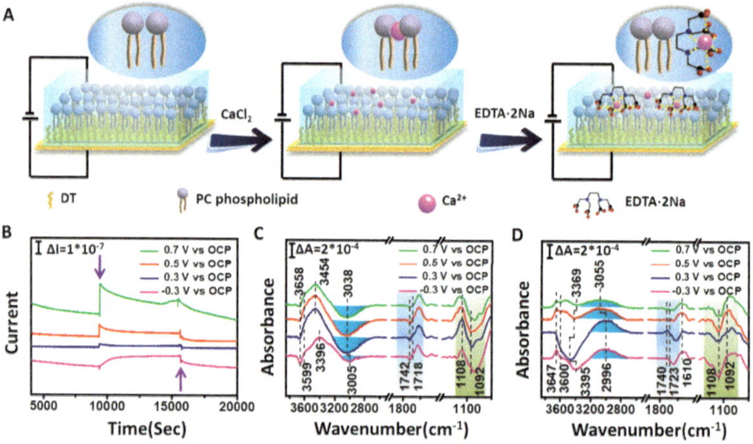

Figure 9.6 (A) Representation of the binding of Ca^{2+} with the phosphate group of membranes and the chelation of Ca^{2+} by EDTA·2Na in the presence of external electric fields. (B) I-T curves of membranes upon the addition of Ca^{2+} and EDTA·2Na salts at different applied potentials. The arrows indicate the addition of Ca^{2+} or EDTA·2Na. (C) Ca^{2+}-induced difference spectra with the lipid membranes under corresponding potentials as the reference. (D) Difference spectra recorded after further addition of EDTA·2Na to desorb Ca^{2+} with (C) as reference. Reproduced from Ref. [5] by permission of John Wiley & Sons Ltd.

The present study identified the strongly hydrogen-bonded water in the vicinity of the phosphate groups (PO^{2-}) of lipid membranes and revealed the resistance of this water to the disturbance of externally applied potentials, which is of significant influence on the electrostatic properties of the membranes, and further explored an effective way to regulate the electrical properties of phospholipid membranes. The "double disturbance" strategy of potential and ion establishes an effective approach for selectively studying the structure of the local water of membranes.

9.3.2.2 Revealing the molecular nature of structured water in the light-induced membrane capacitance changes

In recent years, the light modulation technology has received widespread attention in the field of neuroscience [57, 58] due to the capability of remotely controlling neuronal activity by affecting the action potential, even if only the phospholipid membranes served as the structural basis [59]. The change in membrane capacitance is believed to be the basis for the light modulation technology [59]. The total capacitance can be regarded as the intrinsic capacitance of phospholipid membranes in series with the electric double layer capacitance. Research on light-induced changes in membrane capacitance mainly focuses on the ionic thermal motion, the surface charge density of phospholipid bilayers, and the hydrophobic core dielectric constant [60]. However, the role of interfacial water, a major component in electric double layer, is still ambiguous. Nanostructured gold film employed as enhancing substrate in ATR-SEIRA spectroscopy could serves as both work electrode and heating substrate under light irradiation. Thus, ATR-SEIRA spectroscopy was applied to investigate the role of structured interfacial water in membrane capacitance change induced by near IR (NIR) light irradiation [61].

The sole contribution from the lipid bilayer was eliminated by potential clamping (−0.1 V vs Ag/AgCl), suggested by the absence of structural and conformational changes in lipid molecules induced by light irradiation (Fig. 9.7A). Spectral changes in the ν (OH) region with a broad negative band centered at 3363 cm^{-1} and a positive band at 3626 cm^{-1} (Fig. 9.7A) assigned to the loosely hydrogen-bonded tetrahedral water molecules and donor-bonded O-H stretch of three-coordinated water molecules at the interface, respectively [62–64], were simultaneously observed with the depolarizing current (Fig. 9.7B) of the model membranes upon light irradiation.

These results indicate that the rupture of the tetrahedral water also contributes to the light-induced increase in membrane capacitance. Furthermore, when specific Ca^{2+} was artificially introduced to perturb the interfacial water organization by desorbing the water closely bound to the PO^{2-} group of the phospholipid, both light-induced spectral changes in interfacial water and the change in

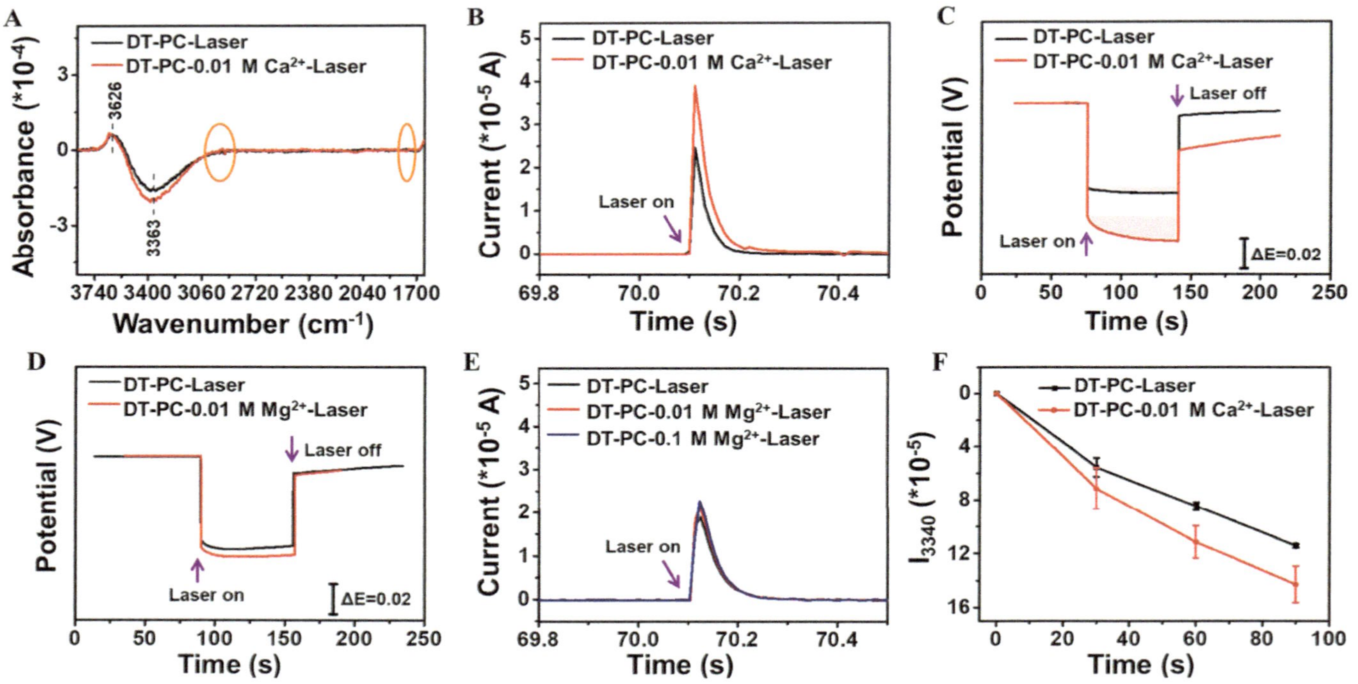

Figure 9.7 (A, B) Light-induced difference spectra of phospholipid membranes with and without Ca^{2+} under a clamping potential of −0.1 V (vs Ag/AgCl) (A) and the corresponding current jumps (B). (C, D) The changes in the open circuit potential of phospholipid membranes before and after light irradiation with and without Ca^{2+} (C) or Mg^{2+} (D). (E) Light-induced current jumps of phospholipid membranes with and without Mg^{2+} under a clamping potential of −0.1 V (vs Ag/AgCl). (F) Temporal profiles of the ν (OH) of interfacial water at 3340 cm^{-1} under light irradiation with and without Ca^{2+} (irradiation times of 0, 30, 60, and 90 s). Reprinted with permission from Ref. [61]. Copyright 2021 American Chemical Society.

the current amplitude were obviously enhanced. However, in the case of addition of Mg^{2+} with nonspecific adsorption instead of specific Ca^{2+}, only a slight enhancement was observed in the light-induced depolarizing current, even if the concentration of Mg^{2+} introduced was increased to 10-fold greater than that of Ca^{2+} (Fig. 9.7E).

The same phenomenon was also observed in the light-induced negative shift of open circuit potential (Fig. 9.7C,D). All these data indicate that the strongly hydrogen-bonded water bound to the PO^{2-} plays an important role in the light-induced membrane capacitance change. Its rupture restructures the hydrogen-bonding environment at the membrane/aqueous interface, leading to an enhanced change in loosely bonded tetrahedral water and thus an increase in the intensity of the negative peak at 3340 cm^{-1} upon irradiation (Fig. 9.7F). The introduction of Ca^{2+} simultaneously accelerated the trend of light-induced potential changes and the light-induced weakening of the hydrogen-bonding network, which indicated that the light-induced changes in membrane potential and/or capacitance were mediated by the interfacial water molecules, and the more disordered hydrogen-bonding network enhances the change in membrane capacitance. A detailed discussion and conclusion can be found in the original publication (Ref. [61]).

In this case study, using ATR-SEIRA spectroscopy combined with electrochemistry and ionic perturbation, the important role of the interfacial water structure and the local hydrogen-bonding environment in light-induced membrane capacitance increase is revealed. And it is demonstrated that the cleavage of loosely bonded tetrahedral water at the interface is an important source for converting the optical signals to electrical signals at the membrane interface. This work reported the significant regulation mechanism between the capacitance behavior and the molecular structure of interfacial water, laying the foundation for developing a new strategy for regulating interfacial capacitance in bioelectronic processes.

9.3.2.3 Studying the role of interface water controlled by transmembrane potentials in modulating the interfacial interaction

Cytochrome c (cyt c) is an important electron transfer protein in mitochondrion and is involved in apoptosis by its interaction with cardiolipin (CL) in the mitochondrion membrane [65]. Under physiological conditions, cyt c is exposed to a large electric field at membrane-bound state due to the transmembrane potential across the inner mitochondrial membrane. Revealing how transmembrane potential influences the interaction of cyt c with CL would yield new insights into the self-regulation mechanisms in apoptosis. Using the electrode potential to modulate the transmembrane potential, the adsorption of cyt c on CL membranes under different transmembrane potentials and resultant effect was successfully investigated by ATR-SEIRA spectroscopy [66].

Under an electrode potential of −0.2 V (vs Ag/AgCl), which is slightly higher than that of open circuit potential (OCP, −0.23 V vs Ag/AgCl), no change in lipid molecule was observed, while a broad positive peak with maximal peak intensity at 3250 cm^{-1} in the ν (OH) region was noted (Fig. 9.8A). This indicates that the applied electric field reorients the strongly hydrogen-bonded interfacial water molecules with dipoles more perpendicular to the surface without affecting the conformation of the membrane. Surprisingly, the adsorption of cyt c to CL membranes is significantly delayed compared with that under OCP (Fig. 9.8B). This suggests that the changes in interfacial water should be accounted for the resistance to protein adsorption. Besides, the adsorption kinetics of cyt c correlates well with the decrease in intensity at 3250 cm^{-1} caused by the replacement of interfacial water molecules by the adsorbed protein in all applied potentials. Such a relationship suggests that the applied potential could induce an additional repulsive force due to the rearrangement of interfacial water, while the additional repulsive force could be overcome in varying degrees by attractive forces originating from the potential-dependent structural changes in the CL membranes.

Based on the previously established relationship of interaction forces between cyt c and CL membrane and characteristics peaks in the redox difference spectra of CL-membrane-bonded cyt c [67], the delayed adsorption kinetics is identified to be originated from the decrease in hydrogen-bonding and hydrophobic interactions. Under apoptotic stimuli, the transmembrane potential of inner mitochondrial membrane (IMM) is reported to be decreased, accompanied by the detachment of cyt c [68]. Thus, the decreased hydrogen bond and hydrophobic interactions and the increase in the structured-water-induced repulsive force upon a drop in

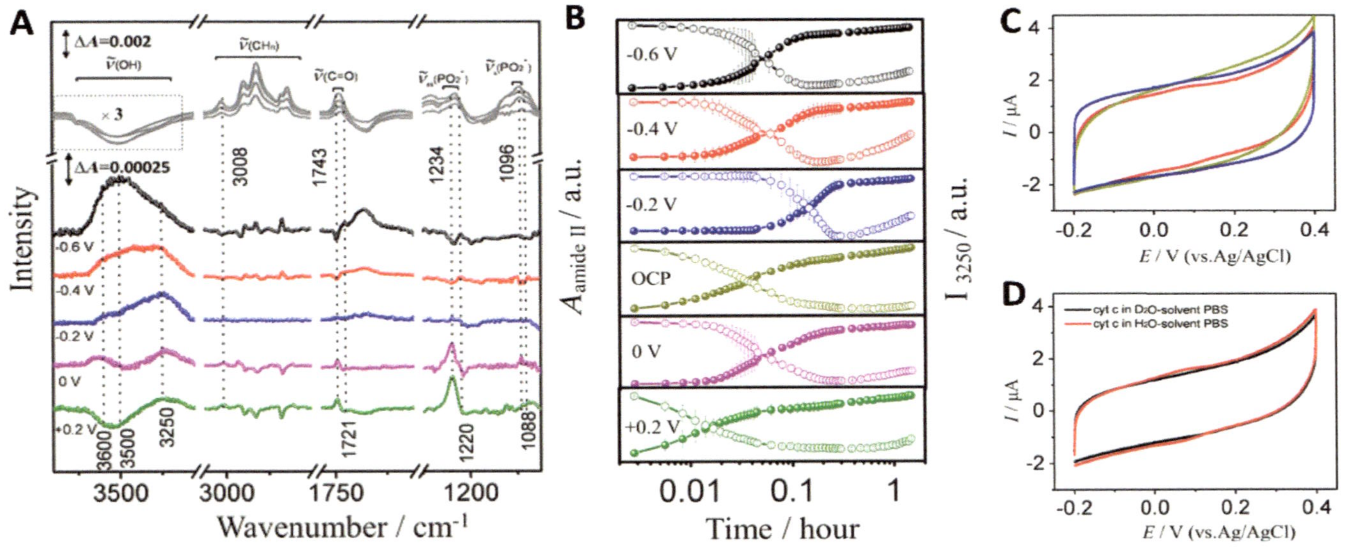

Figure 9.8 (A) Potential-induced difference spectra of the CL membranes at different potentials with the background spectra taken at the OCP (−0.23 V vs Ag/AgCl). (B) Adsorption kinetics of cyt c by calculating the time-dependent area of the amide II band (A_{amide} II) as the protein amount (full symbols) and time-dependent water absorption intensities at ν = 3250 cm^{-1} (I_{3250}, open symbols) at different potentials. (C) Cyclic voltammetric response of cyt c adsorbed on CL membranes under different potentials (OCP, yellow line; −0.4 V vs Ag/AgCl, red line; −0.2 V vs Ag/AgCl, blue line). (D) Cyclic voltammetric response of cyt c adsorbed on CL membrane at −0.4 V (vs Ag/AgCl) in deuteroxide solvent (black line) and then in deuteroxide solvent (red line). The scan rate for cyclic voltammetry was 50 mV/s. OCP: open circuit potential. Reproduced from Ref. [66] by permission of John Wiley & Sons Ltd.

the transmembrane potential across the IMM could be a new mechanism that facilitates the detachment of CL-bound cyt c, resulting in the redistribution of cyt c in the intermembrane space of the mitochondrion.

It was further found out that the electrons were successfully transferred to the electrode when cyt c was adsorbed on the highly hydrated CL membrane at potentials below −0.2 V (vs Ag/AgCl), whereas forbidden at potentials higher than −0.2 V (vs Ag/AgCl), where the CL membrane was dehydrated (Fig. 9.8C). This indicates that the potential-induced hydration state of the CL membranes determines the efficiency of the electron transfer between cyt c and the electrode, which might be originated from the water-mediated proton-coupled electron transfer process. Deuteroxide replacement was applied to estimate the proton effect in the electron transfer. No obvious redox peak can be observed in the cyclic voltammetry of cyt c adsorbed on CL membrane in deuteroxide solvent, while a pair of redox peaks with the formal potential at around 0.10 V (vs Ag/AgCl) was noted after the replacement of D_2O with water (Fig. 9.8D). This result confirms the important role of water in the electron transfer of protein, where water molecules around and in the CL membranes could act as sources of proton to assist the transfer of electrons between cyt c and the electrode. A detailed discussion and conclusion can be found in the original publication (Ref. [66]).

This present work revealed how transmembrane potential modulates the interaction between cyt c and CL membrane and regulates the redox behavior of CL-bound cyt c through the rearrangement of membrane interfacial water, which is significant to deeply understand the functions of mitochondrial membranes and may point to an underlying mechanism for the early apoptosis.

9.3.2.4 Revealing the interfacial-water-mediated proton transfer at membrane interface

Proton transfer plays a central role in chemistry and biology [69] and is the key process in many biological phenomena. Phosphatidylserine (PS) is one of the main components of neural cell membrane and crucial for neuron function. Its ultrahigh affinity toward copper ion is reported to be closely related to the deprotonation of PS [70]. Thus, the (de-) protonation state of carboxylate and amino moieties of PS dramatically affects its interactions with copper ion, which not only affects the physical properties of the membrane but also disrupts the PS-dependent protein-membrane binding. Probing the potential proton transfer and/or local pH change resulting from external stimuli is of great significance. The structural information on proton donor and/or acceptor, especially information on water structure, is significant for the proton transfer process. Using the carboxyl-terminated self-assembled monolayer (SAM) as the biomimetic model for PS membrane, the interfacial-water-mediated proton transfer of graphene oxide (GO), a representative of two-dimensional nanomaterials, was revealed by monitoring the IR-active vibrational marker modes of the protonated carboxyl group, deprotonated carboxylate group, and interfacial water [71].

Taking 11-mercaptoundecanoic acid (MUA)-modified Au surface immersed in buffer as reference, an obvious positive peak at the ν (C=O) region (orange box) and negative peaks in the ν (COO—) region (green box) were simultaneously observed after the addition of GO (Fig. 9.9A), which are basically consistent with that of pH-induced protonation (Fig. 9.9C). This indicates that the MUA nanolayer is protonated by the addition of GO. However, a dramatic difference was observed in the interfacial water region (Fig. 9.9B,D). This inconsistency is believed to be originated from the influence of surface-adsorbed GO, as suggested by a highly sensitive quartz crystal microbalance with dissipation (QCM-D) monitor, and atomic force microscopy (AFM) images. Interestingly, it is found that GO is still capable of inducing robust protonation of MUA SAM when the concentration of the buffer solution was increased to 50 mM to enhance the buffering capability (Fig. 9.9E,F), even if the pH of the bulk solution is barely changed. This is dramatically different from organic acid HCOOH, a simplified acidic group analogue of GO under the same conditions (Fig. 9.9G,H). These results indicate that the mechanism of GO-induced protonation of MUA SAM is different from that induced by HCOOH, which is simply from a change in the solution pH. Considering the unique structure of GO and the weak hydrogen-bonding network of interfacial water on both GO surface (Fig. 9.9I) and GO/MUA SAM interaction interface (Fig. 9.9B and F), proton transfer at the interaction interface was proposed (Fig. 9.9J) and further confirmed by the effect of the interfacial proton density and proton conductivity of GO on the protonation effect. In conclusion, in addition to acting as a macromolecular buffer and proton carrier,

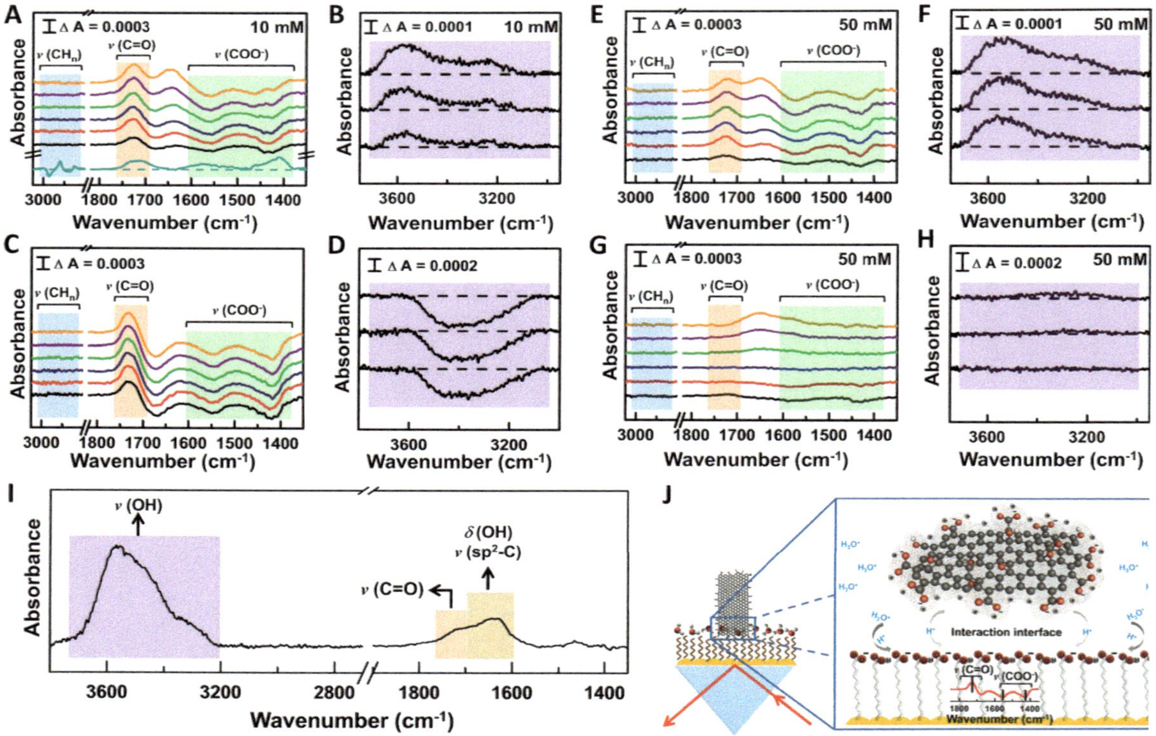

Figure 9.9 (A,B) GO-induced SEIRA difference spectra of MUA SAM (A) and interfacial water (B) with MUA SAM immersed in 10 mM MOPS buffer as background. (C,D) HCl-induced SEIRA difference spectra of MUA SAM (C) and interfacial water (D) with MUA SAM immersed in H_2O as background spectrum. (E–H) GO (E,F) and HCOOH (G,H) induced SEIRA difference spectra of MUA SAM (E,G) and interfacial water (F,H) with MUA SAM immersed in 50 mM MOPS buffer as background spectrum. (I) SEIRA spectrum of GO sheet in solution obtained by subtracting the SEIRA spectrum of GO solution and pure H_2O. (J) Schematic diagram of proton transfer at membrane interface. Reprinted with the permission from Ref. [71]. Copyright 2018 American Chemical Society.

the unique two-dimensional structure and oxygenated functional groups of GO can be co-assembled with the dissociated protons and interface water into an exchangeable proton pool, which is capable of dissociating and transferring protons at the membrane interface since the biological membrane is the first barrier that GO encounters in a biological environment [4]. A detailed discussion and conclusion can be found in the original publication (Ref. [71]).

9.3.3 Structure and Function Study of Membrane-Related Proteins

Membrane-related proteins (peripheral membrane proteins, integral membrane proteins, and lipid-anchored proteins) are important components in biological membranes and are the undertakers of numerous functions crucial for life. Any change in the structures and functions of membrane proteins may be fatal to living cells. And studying the mutual effects between membranes and related proteins is the foundation for deeply understanding the cellular function of biological membranes. IR spectroscopy is a powerful tool for the study of membrane-related proteins due to its sensitivity to the conformation and structure of both proteins and lipids. Great achievement has been made by applying IR spectroscopy in the structure and function study of membrane-related proteins, as well as their mutual interactions. Many reviews and chapters on this topic are available [72–78]. Besides, a special issue entitled "FTIR in membrane proteins and peptide studies" is available in *Biochimica et Biophysica Acta (BBA)-Biomembranes*. In this section, we focus on the studies on the monolayer level by ATR-SEIRA spectroscopy.

Cytochrome c (cyt c), a peripheral membrane protein of mitochondrion, is the first membrane-related protein that ATR-SEIRA spectroscopy studied. Ataka et al., for the first time, succeeded in the in situ monitoring of the adsorption of cyt c monolayer on mercaptopropionic acid (MPA)-modified gold films, and the detailed structural changes during the redox process were recorded by the potential-induced difference spectra in 2003 [2]. And the effect of surface chemical properties on the adsorbed cyt c was systematically investigated and a detailed band assignment was published [79]. It has been found that the chemical nature of the underlayer monolayer

[79, 80], solution pH [81], and adsorption orientation [82] have a dramatic influence on electroactivity. The ultrafast kinetics and rate constant of conformational change and electron transfer of surface-adsorbed cyt c monolayer were revealed by time-resolved SEIRA (tr-SEIRA) spectroscopy [83–85]. Recently, the interaction between cyt c and cardiolipin (CL) in the mitochondrion membrane has drawn much attention due to its important role in early apoptosis. Jiang et al. systematically investigated the interaction mode between cyt c and CL membrane and the resultant effect on the structure and function of cyt c by constructing a CL-containing hybrid bilayer membrane to mimic the inner membrane of mitochondrion [66, 67, 86, 87]. Results suggest that the adsorption of cyt c to CL is heterogeneous under the regulation of electrostatic, hydrogen-boding, and hydrophobic interactions, which is further modulated through the interfacial water molecules of membranes by the transmembrane potential. And the structure transition from α-helices to β-sheet and the hydration change of cyt c upon oxidization and reduction account for the electroactivity upon the adsorption on CL. Besides, they also studied the effect of nanomaterials on the structure and function of cyt c, which shows that interaction with nanomaterials could either positively or negatively affect the function of cyt c, and strong heterogeneity in induced effect is observed although interacting with one kind of nanomaterial [88, 89]. The structure and function of even proteins that desorbed after weakly interacting with nanomaterials have undergone great changes and thus affect the interaction with downstream ligands.

The integral membrane protein is another kind of membrane-related protein that has strict vector characteristics. Therefore, the key problem in applying ATR-SEIRA spectroscopy to integral membrane proteins is to construct a lipid bilayer membrane containing uniformly oriented protein on the enhancing substrate surface. Using the specific affinity recognition between the polyhistine tag (His-tag) and the Ni^{2+} chelated nitrilotriacetic acid (Ni-NTA), cytochrome c oxide (CcO), which is a large transmembrane protein complex with four redox-active metal centers, was uniformly oriented and assembled on the metal film and then reconstructed into the native phospholipid bilayer environment through in situ dialysis by Ataka et al. and in situ monitored by ATR-SEIRA spectroscopy

[27]. The orientation of the protein can be controlled by changes in the position of the genetically introduced His-tag [90]. This method can also be extended to other affinity recognition tags [91]. This strategy not only ensures the correct orientation of the adsorbed proteins, but also reconstructs the natural phospholipid membrane environment, which is extremely important for maintaining the activity and stability of membrane proteins, laying the foundation for ATR-SEIRA spectroscopy to study the structure and functions of integral membrane proteins. It is further found that the packing density of the bilayer reconstituted CcO has a significant impact on their activity [92], and direct electron transfer and potential-induced difference spectra were obtained through optimizing the surface packing density [93].

The conformational transitions and molecular hysteresis of CcO were investigated [94], and two-dimensional SEIRA and tr-SEIRA spectroscopy were further applied to reveal the different conformational states [95] and substantial conformational changes in secondary structures coupled to redox transitions [96] and enzymatic turnover [97] of the protein. Besides, based on this membrane-reconstituted strategy, the photocycle of the membrane photoreceptor sensory rhodopsin II of halobacteria was investigated [15, 98]. Considerable change in both the protein structure and the photocycle was observed between the detergent-solubilized and lipid-bilayer-reconstituted proteins, which is the first time that the effect of the lipid on the structure of integral membrane proteins has been revealed as a benefit of ATR-SEIRA spectroscopy. Further combined with electrochemistry, a molecular model has been established for the physiological regulation of the photocycle by modulating the potential drop across the membrane, which acts through the interplay between the change in local pH at the membrane surface and the external electric field.

In addition, ATR-SEIRA spectroscopy has also been widely used in the structure and function study of other proteins, including hydrogenase [99, 100], laccase [101], photosynthetic reaction centers (RCs) [102, 103], ion channels and ion-binding proteins [104–106], endonuclease [107], nitric oxide reductases [108], NADH:quinone oxidoreductase [109], and respiratory chain protein and complex [49, 110, 111].

9.3.4 Interfacial Interaction at Membrane/Water Interface

The biological membrane is the physical barrier that isolates the environment inside and outside a cell, and the realization of many cellular functions involves the interaction and passing through the membrane. It is of great importance to understand the interfacial interaction at the membrane/water interface. Surface-sensitive ATR-SEIRA spectroscopy can provide the fingerprint information of interface molecules and is a powerful tool for studying the interaction at the interface. By constructing an artificial biomimetic membrane on a metal film, the basic properties of the membrane and its interaction with external substances (proteins, bioactive small molecules, nanomaterials, etc.) can be studied in this model system.

Levin et al. combined surface-enhanced Raman spectroscopy and SEIRA spectroscopy to study the interaction of ibuprofen with lipid bilayer membranes and found that ibuprofen can be inserted into the membranes where the overall hydrophobicity plays an important role [112]. Quirk et al. explored the location of ubiquinone (UQ) in a biomimetic membrane both during assembly and as a function of transmembrane potential, and the isotope substitution method was used to extract the effect of UQ on the upper and lower leaflets of the membrane [113]. Forbrig et al. studied the adsorption of an antimicrobial peptide (alamethicin, ALM), which also serves as a voltage-gated ion channel model on lipid membrane and the potential-induced ion channel formation, and proposed a three-step potential-controlled interaction model [114]. Furthermore, Li et al. reported the importance of membrane dipole potential in regulating the orientation of adsorbed peptides [115]. Besides, Jiang et al. investigated the heterogeneous interaction mode and its regulation by transmembrane potential in the interaction between cyt c and the cardiolipin membrane [66, 67].

The booming of nanotechnology has accelerated the transformation of nanomaterials in living organisms. Since a biomembrane is the first barrier that nanomaterials may encounter in biological environments, its interaction with nanomaterials is of great significance. Jiang et al. systematically investigated the interaction between nanomaterials and biomembranes by

ATR-SEIRA spectroscopy [4, 71, 116–119]. Taking a lipid bilayer membrane constructed on a metal film as a reference, nanomaterial-induced changes in the characteristic peaks of both lipid molecules and nanomaterials are recorded, based on which the interaction mechanism could be deduced. This strategy was first applied to the interaction between zwitterionic D-penicillamine (DPA)-capped quantum dots (QDs) and red blood cell (RBC) membranes [116]. Results show that the DPA ligands on the QDs are in their zwitterionic form during interaction and both ν_{as} (CH_2) and ν_s (CH_2) of lipid molecules are blue shifted in the membrane, which is mimetic to either the inner or the outer leaflet of the RBC membrane. This indicates that interaction with zwitterionic DPA-QDs led to a softening of the lipid and resulted in a more flexible conformation, which facilitates the penetration of membrane without pore formation. Based on this model system, the interactions of RBC membranes with nanomaterials having different characteristics were further studied [117, 118]. It was found that nitrogen-doped graphene quantum dots (GQD) and mercaptosuccinic acid (MSA)-modified CdTe QDs with diameter less than 5 nm only slightly disturbed the conformation of membranes, while QDs with diameter larger than 5 nm and graphene oxide (GO) were able to induce hematolysis, but with different mechanisms.

QDs with larger particle size can coordinate with the PO^{2-} group of lipid molecules and break the phosphodiester bond, resulting in severe hemolysis, and GO can extract the lipid molecules from the membrane and thus damage the integrity of the membrane. Interestingly, it was observed that before the extraction occurs, in the ν (OH) region, a weak but distinguishable negative peak was observed immediately after the addition of GO (Fig. 9.10A), which is caused by the expelling and reorganization of interfacial water by the adsorption of GO at the membrane/water interface (Fig. 9.10B). The adsorption of substances at the membrane/water interface would desorb multiple water molecules resulting in a 1:N amplification, which further combined with the extremely large extinction coefficient of water molecules, resulting in a very sensitive change in the absorption of interfacial water before and after the adsorption of substances. A benefit of ATR-SEIRA spectroscopy is that this interfacial molecular event could be observed in situ. Thus, the initial replacement of interfacial water suggests the initial interaction between membranes and GO and is taken as an indicator of the adsorption of GO. Based on this sensitive label-free "water" probe, combined with the analysis of the characteristic peaks of interfacial water and lipid membranes induced by GO in the presence of different intermolecular force shielding agents, four noncovalent interaction forces were clearly identified: electrostatic replusion, electrostatic attraction, hydrogen bond interaction, and hydrophobic interaction (Fig. 9.10B) [4]. And it is revealed that long-range electrostatic attraction facilitates the short-range hydrogen-bonding interaction. Furthermore, by modulating the interaction forces via the content and type of oxygen-containing groups on GO, the synergic mechanism of noncovalent interactions and resultant biological effect (extraction of lipids from membranes) is established [119]. Besides, it has been proposed that due to its ultrathin structure and oxygen-containing groups, GO can be regarded as a two-dimensional exchangeable proton pool, transferring protons at the interaction interface [71].

9.3.5 Living Cell Membrane

The direct monitoring of cell activities in living cells is one of the directions that IR spectroscopy has been pursuing for a long term. ATR-SEIRA spectroscopy has been demonstrated to be capable of selectively monitoring biological events near the cell membrane in real time, which excludes the spectroscopic interference of complex components in cells.

Busalmen et al. monitored the adoption of living bacteria *Pseudomonas fluorescens* on gold electrodes for the first time and observed both the negative peak of ν (OH) caused by cell adsorption and the specific absorption peak of the cell component. It was further found that the oxidation potential made the cell get closer to the gold electrode and could oxidize the lipopolysaccharides outside the cell membrane under polarization [120]. Subsequent studies focused on the electrochemical properties of *Geobacter sulfurreducens* in microbial fuel cells (MFCs) [121–123]. The difference spectra induced by live-bacteria redox revealed that C-type cytochrome protein, which was in direct contact with the electrode surface at the outermost part of the cell, mediates the conduction of electrons from the bacteria to the electrode surface.

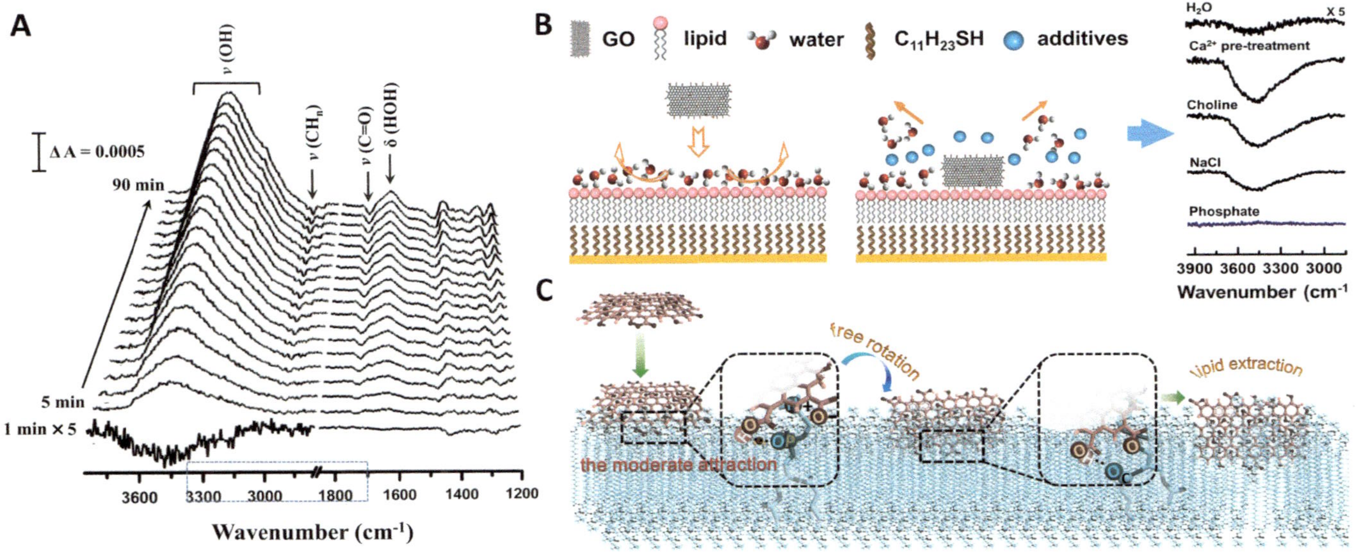

Figure 9.10 (A) 3D plot of GO-induced SEIRA spectral changes in zwitterionic lipid membrane in the presence of 50 µg/mL GO in water at 5 min interval. Reprinted with the permission from Ref. [4]. Copyright 2015 American Chemical Society. (B) Schematic diagram of using "water" probe to identify the noncovalent interaction forces. (C) Schematic diagram of synergic mechanism of noncovalent interaction forces at membrane/water interface. Reprinted with permission from Ref. [119]. Copyright 2019 American Chemical Society.

Cao et al. constructed a biocompatibility surface by coating the enhancing substrate with poly-L-lysine (PLL) and successfully monitored in situ the interaction between PEG-coated CuS nanoparticles with living cell membrane by ATR-SEIRA spectroscopy [124]. An obvious difference was observed in spectral changes induced by freshly prepared (CuS-new NPs) and aged (CuS-old NPs) nanoparticles, while the spectral pattern induced by the latter is nearly consistent with that induced by the addition of free Cu^{2+} (Fig. 9.11A). Cu^{2+} interacted with and is reduced to Cu^{+} by metalloreductases Fre1p/2p on the cell surface via the coordination of Cu^{2+} and nitrogen or oxygen donors, and is subsequently transported across the cell membrane by transmembrane copper transport protein 1 (Ctr1) by coordinating Cu^{+} with sulfur donor ligands [125]. Thus, the spectral changes at 1601 cm^{-1} and 1673 cm^{-1} originated from the coordination of copper ions with cell membranes and resultant conformational changes in related membrane proteins. Since a similar pattern was not observed in the spectra induced by the supernatant of CuS-old NPs, the potential effect of free copper ions in the CuS-old NPs stock solution was eliminated. These results suggest that the difference in spectra induced by CuS-new and CuS-old NPs originated from the difference in their surface properties and have a relationship with copper ions. Further characterization reveals that copper ions released from the degradation of nanoparticle coordinate with PEG on the surface, resulting in a dynamic surface property of nanoparticles.

Furthermore, in the ν (OH) region, positive peaks at 3255 cm^{-1} and 3683 cm^{-1} with different intensities were observed upon interaction with copper ion and CuS NPs caused by the changes in the local potential of membranes. These spectral changes might be served as an indicator of the interaction process of NPs. Dramatically different patterns were observed for CuS-new and CuS-old NPs; the growth of the peak intensity of ~3255 cm^{-1} reaches the plateau at 30 min for CuS-old NPs while continued increase for CuS-new NPs (Fig. 9.11B,C). The coordination of Cu^{2+} with PEG would decrease the electronegativity of the NPs, thus greatly weakening the electrostatic repulsion forces between NPs and the negatively charged cell plasma membrane, accelerating and increasing the adsorption of NPs on the cell membrane relative to that of CuS-new NPs.

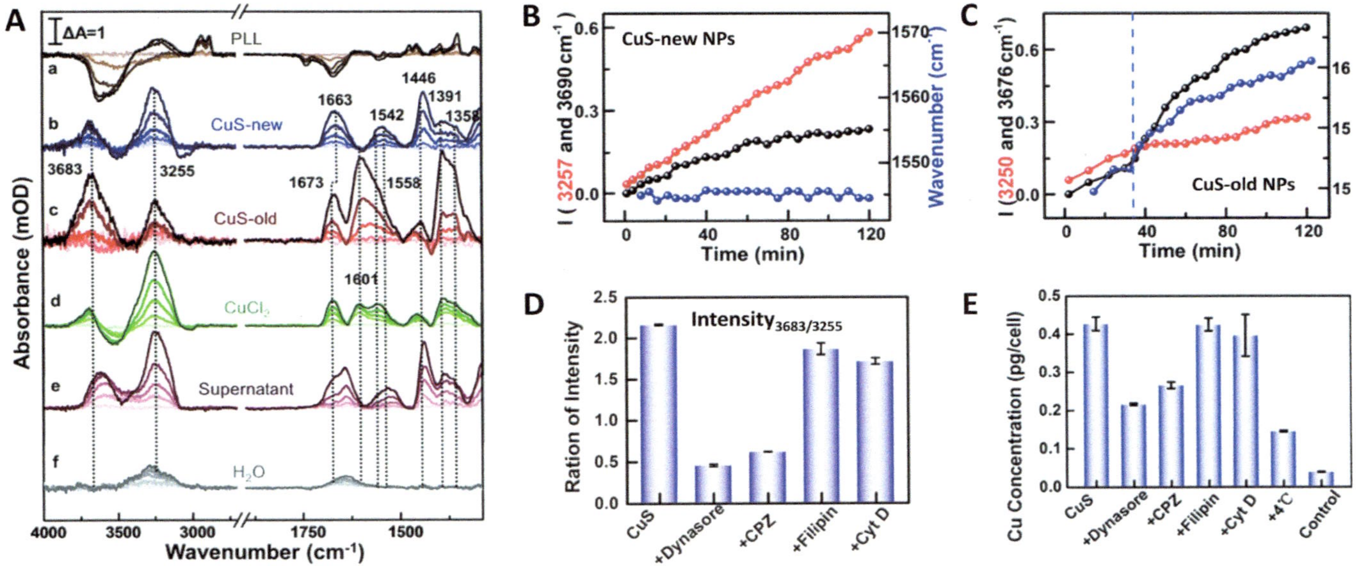

Figure 9.11 (A) SEIRA difference spectra of (a) CuS NP induced PLL by taking the PLL layer modified substrate as reference, and the difference spectra of HeLa cells induced by addition of (b) CuS-new NPs, (c) CuS-old NPs, (d) $CuCl_2$, (e) the supernatant of CuS-old NPs and (f) only water as a control by taking the cell incubated substrate as reference. (B, C) Intensity of interfacial water and the shifts of the maximum peak position at ~1542 cm^{-1} in CuS-new NPs (B) and CuS-old NPs (C) induced difference spectra of HeLa cells, respectively. (D, E) Cell uptake degree of CuS-old NPs quantified by the ratio of the peak intensity at ~3683 cm^{-1} to ~3255 cm^{-1} (D) and inductively coupled plasma mass spectroscopy (E) under different inhibitors. CPZ: chlorpromazine; Cyt D: Cytochalasin D. Reproduced from ref. [124] with permission from of the Royal Society of Chemistry.

More interestingly, it was noted that the gradual shift in the peak at 1577 cm^{-1} to 1601 cm^{-1} induced by the coordination of Cu^{2+} (Fig. 9.11A) showed similar tendency with the change in the peak intensity at 3683 cm^{-1}, while this phenomenon is dramatically different from that induced by pure free Cu^{2+}. Thus, the significant enhancement in the absorption at 3683 cm^{-1} is believed to be related to the endocytosis of NPs. This conclusion was further supported by the quite similar endocytosis inhibition results obtained from the ratio of the peak intensity at ~3683 cm^{-1} to ~3255 cm^{-1} and standard inductively coupled plasma mass spectroscopy (Fig. 9.11D,E). Both types of endocytosis inhibition results indicate that CuS-new and CuS-old NPs have different endocytosis pathways. Further, the in situ monitoring of photothermal-induced cell death revealed that the combination of phosphatidylserine flipping (from the inner to the outer membrane) and increasing in content of hydrophobic phenylalanine induced apoptosis, and the phosphatidylserine flipping further changed the arrangement of interfacial water. A detailed discussion and conclusion can be found in the original publication (Ref. [124]).

This case study demonstrated the great potential of ATR-SEIRA spectroscopy in monitoring the cellular activity, especially cell-membrane-related processes in living cells. With the valuable fingerprint information of molecular structure and related chemical environments, the underlying molecular mechanism of biological events might be revealed.

9.4 Protocols

The detailed protocols for preparing nanostructured metal films as enhancing substrate, preparing various vesicles (single/multiple component lipid vesicles, as well as membrane-protein-reconstituted vesicles), and constructing three types of biomimetic membranes (hybrid lipid bilayer, lipid-tethered bilayer, and protein-tethered bilayer) are already available in our earlier chapter [126]. In this section, we describe the procedure for combining ATR-SEIRA spectroscopy with electrochemistry and investigating living cell membranes with ATR-SEIRA spectroscopy.

9.4.1 Constructing ATR-SEIRA Spectroelectrochemistry

9.4.1.1 Materials

- Work electrode: surface-modified gold-nanostructured film deposited on Si prism (e.g., gold-film-supported hybrid L-α-phosphatidylcholine lipid bilayer membrane).
- Reference electrode (e.g., Ag/AgCl, 3M KCl).
- Counter electrode (e.g., Pt mesh).
- Electrolyte (e.g., 10 mM NaCl).
- Spectroelectrochemistry cell.
- ATR-SEIRA spectroscopy accessory.
- Fourier transform infrared (FTIR) spectrometer with mercury cadmium telluride (MCT) detector (e.g., VERTEX 80v).
- Electrochemical (EC) workstation (e.g., Autolab PGSTAT302N).
- Nitrogen gas.

9.4.1.2 Methods

1. Assemble the spectroelectrochemical cell (Fig. 9.12) and make sure good electrical contact between the gold layer deposited on the Si prism and Cu plate in the spectroelectrochemical cell (Note 1).

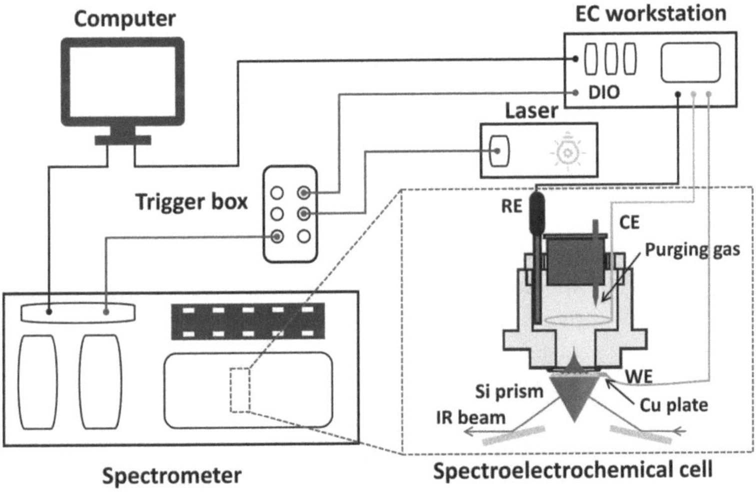

Figure 9.12 Schematic diagram of the configuration of ATR-SEIRA spectroelectrochemistry and external stimuli synchronous triggering.

2. Place the cell in the ATR accessory located in the sample chamber of the FTIR spectrometer. Adjust the incident angle if necessary.
3. Vacuum the spectrometer or purge with dry nitrogen gas and fill the MCT detector with liquid nitrogen.
4. Modify the gold film (e.g., hybrid lipid bilayer membrane) to construct the work electrode (Note 2).
5. Fill the cell with ~2–3 mL pre-degassed electrolyte (e.g., 10 mM NaCl).
6. Connect the electrodes to the electrochemical workstation (Fig. 9.12), purge the cell with nitrogen, and seal the cell (Note 3).
7. Set the spectrometer parameters, including spectral resolution (typically 4 cm^{-1}), wavelength range (typically 4000 to 800 cm^{-1}), number of scans (depend on the time interval and S/N ratio of spectrum) (Note 4).
8. Collect a background spectrum, and after a certain time, collect a sample spectrum (Note 5).
9. The potential modulated spectra can be collected either manually or automatically:
   a. **Manual collection**
      i. Set the potential at which the background spectrum is expected and collect the background spectrum when the current reaches its plateau.
      ii. Change the potential to which the sample spectrum is expected and collect the sample spectrum when the current reaches its plateau.
      iii. Repeat steps (i) and (ii) at the interested potentials.
      iv. Alternatively, set the potential at which the background spectrum is expected and recorded background spectrum and then successively change the potentials to which the sample spectrum is expected, and collect series sample spectra with the same background spectrum.
   b. **Automatic collection (Note 6)**
      i. Connect the EC workstation and spectrometer with the trigger box (Fig. 9.12), through which the two

devices communicate with each other by a voltage-level signal based on TTL serial communications.

ii. Edit the script that defines at which potentials the spectra should be collected in the spectrometer and edit the program that defines how the potential should be applied in the EC workstation.

iii. Run the script in the spectrometer and start the program in the EC workstation; the potential modulated spectra can be recorded automatically and synchronously with electrochemistry.

10. For stimuli-induced difference spectra under potential polarization:
    a. Set the potential at which the background spectrum is expected and collect the background spectrum when the current reaches its plateau.
    b. Record the sample spectra immediately after applying external stimuli (e.g., ion, protein, light, etc.) while maintaining the potential clamping. The current changes and spectral changes are recorded by the EC workstation and the spectrometer, respectively. Spectral collection can also be synchronously triggered by the application of stimuli (Fig. 9.12).

9.4.1.3 Notes

1. The ATR-SEIRA spectroscopy accessory and spectroelectrochemistry cell are necessary, which could be homemade or commercial (e.g., VeeMAX III ATR accessory and Jackfish spectroelectrochemical cell from PIKE Technologies).
2. The modification process could be in situ monitored by ATR-SEIRA spectroscopy if the measurement is started before modification.
3. Seal the cell with parafilm and make some pin holes for purging nitrogen gas to escape.
4. Increasing of scan number can effectively improve the S/N ratio of the spectrum, while the time required to acquire a spectrum can be controlled by the scanning rate.
5. A blank spectrum can be used to monitor the effectiveness of the water vapor and carbon dioxide removal. The longer

the time delay, the better; but it depends on individual experiments.

6. Whether the spectrometer and the EC workstation support synchronous triggering, and instructions on how to implement it need to be answered by their respective manufacturers.

9.4.2 Constructing Living Cell ATR-SEIRA Spectroscopy

9.4.2.1 Materials

1. Culture medium (e.g., Dulbecco's modified eagle medium (DMEM)).
2. 0.25% trypsin-EDTA solution.
3. Fetal bovine serum (FBS).
4. 100 × penicillin–streptomycin solution (10,000 U/mL).
5. 0.01% poly-L-lysine (PLL) solution (molecule weight 150~300 kD).
6. Phosphate-buffered saline (PBS) solution.

9.4.2.2 Methods

1. Culture living cells (e.g., HeLa cell) in DMEM medium supplemented with 10% (v/v) FBS, penicillin (100 U/mL), and streptomycin (100 ug/mL) (named complete medium) at 37°C under 5% CO_2 for 24 h in cell culture flasks (Note 1).
2. Remove the culture medium and rinse the cultured cells with 5 mL PBS by gently shaking the culture flask thrice (Note 2).
3. Add 1 mL 0.25% trypsin-EDTA solution, which covers the bottom of the culture flask to dissociate the cell monolayer for 30~60 s.
4. Add 5 mL complete medium to stop the dissociation process (Note 3).
5. Suspend the dissociated cells by gently pipetting and transfer the cell suspension to a sterile centrifuge tube of appropriate size and centrifuge for 10 min at 800 × g (Note 4).
6. Carefully remove the supernatant without disturbing the cell pellet and add an appropriate amount of complete medium to suspend the cells.

7. Prepare the enhancing substrate (e.g., gold-nanostructured films) and assemble the spectroelectrochemical cell.
8. Immerse the spectroelectrochemical cell and other necessary tools in 75% ethanol overnight and place them further under UV irradiation for 30 min to sterilize on the super clean bench.
9. Add 2 mL 0.01% PLL solution to the cell and incubate for ~6 h, and then thoroughly wash the modified surface with sterile water (Note 5).
10. Seed 2 mL of the as-prepared cell suspension (5×10^5 cells per mL) on the PLL-modified enhancing substrate and incubate at 37°C under 5% CO_2 for 24 h. Then, gently wash the cell three times with 5 mL PBS and leave 1 mL as incubating solution.
11. Place the cell in the ATR accessory and then ready it for ATR-SEIRA spectroscopy measurement (Note 6).

9.4.2.3 Notes

1. Cell-related operations need to be done on the super clean bench.
2. Pre-heating the related solution to 37°C is recommended.
3. The end point of dissociation by trypsin can be precisely monitored under a microscope: One should observe a large number of round cells, and when gently shake the flask, the cells should exhibit no obvious movement.
4. Certain cell lines are very sensitive to centrifugal force.
5. This modification process can be in situ monitored by ATR-SEIRA spectroscopy.
6. The time for the whole ATR-SEIRA spectroscopy measurement should be carefully considered to maintain the cell activity.

Acknowledgments

This work was supported by the Youth Innovation Promotion Association of CAS (Grant No. 2020233), and the National Natural Science Foundation of China (22074138), and the National Science Fund for Distinguished Young Scholars (22025406).

References

1. Osawa, M., Ataka, K., Yoshii, K., and Yotsuyanagi, T. (1993). Surface-enhanced infrared ATR spectroscopy for in situ studies of electrode/electrolyte interfaces, *J. Electron Spectrosc. Relat. Phenom.*, **64–65**, pp. 371–379.
2. Ataka, K. and Heberle, J. (2003). Electrochemically induced surface-enhanced infrared difference absorption (SEIDA) spectroscopy of a protein monolayer, *J. Am. Chem. Soc.*, **125**, pp. 4986–4987.
3. Osawa, M. (2006). *Handbook of Vibrational Spectroscopy*, eds. Chalmers, J.M. and Griffiths, P.R., Chapter 1 "Surface-enhanced infrared absorption spectroscopy," (John Wiley & Sons, Ltd, Chichester). pp. 785–799.
4. Wu, L., Zeng, L., and Jiang, X. E. (2015). Revealing the nature of interaction between graphene oxide and lipid membrane by surface-enhanced infrared absorption spectroscopy, *J. Am. Chem. Soc.*, **137**, pp. 10052–10055.
5. Li, S. S., Wu, L., Zhang, X. F., and Jiang, X. E. (2020). The structure of water bonded to phosphate groups at the electrified zwitterionic phospholipid membranes/aqueous interface, *Angew. Chem., Int. Ed.*, **59**, pp. 6627–6630.
6. Chattopadhyay, A. and Boxer, S. G. (1995). Vibrational stark-effect spectroscopy, *J. Am. Chem. Soc.*, **117**, pp. 1449–1450.
7. Boxer, S. G. (2009). Stark realities, *J. Phys. Chem. B*, **113**, pp. 2972–2983.
8. Fried, S. D. and Boxer, S. G. (2015). Measuring electric fields and noncovalent interactions using the vibrational Stark effect, *Acc. Chem. Res.*, **48**, pp. 998–1006.
9. Hartstein, A., Kirtley, J. R., and Tsang, J. C. (1980). Enhancement of the infrared absorption from molecular monolayers with thin metal overlayers, *Phys. Rev. Lett.*, **45**, pp. 201–204.
10. Nishikawa, Y., Fujiwara, K., Ataka, K., and Osawa, M. (1993). Surface-enhanced infrared external reflection spectroscopy at low reflective surfaces and its application to surface analysis of semiconductors, glasses, and polymers, *Anal. Chem.*, **65**, pp. 556–562.
11. Osawa, M., Ataka, K., Yoshii, K., and Nishikawa, Y. (1993). Surface-enhanced infrared spectroscopy: The origin of the absorption enhancement and band selection rule in the infrared spectra of molecules adsorbed on fine metal particles, *Appl. Spectrosc.*, **47**, pp. 1497–1502.

12. Osawa, M. (1997). Dynamic processes in electrochemical reactions studied by surface-enhanced infrared absorption spectroscopy (SEIRAS), *Bull. Chem. Soc. Jpn.*, **70**, pp. 2861–2880.

13. Ataka, K., Yotsuyanagi, T., and Osawa, M. (1996). Potential-dependent reorientation of water molecules at an electrode/electrolyte interface studied by surface-enhanced infrared absorption spectroscopy, *J. Phys. Chem.*, **100**, pp. 10664–10672.

14. Jiang, X., Zaitseva, E., Schmidt, M., Siebert, F., Engelhard, M., Schlesinger, R., Ataka, K., Vogel, R., and Heberle, J. (2008). Resolving voltage-dependent structural changes of a membrane photoreceptor by surface-enhanced IR difference spectroscopy, *Proc. Natl. Acad. Sci. U. S. A*, **105**, pp. 12113–12117.

15. Jiang, X. E., Engelhard, M., Ataka, K., and Heberle, J. (2010). Molecular impact of the membrane potential on the regulatory mechanism of proton transfer in sensory rhodopsin II, *J. Am. Chem. Soc.*, **132**, pp. 10808–10815.

16. Neubrech, F., Huck, C., Weber, K., Pucci, A., and Giessen, H. (2017). Surface-enhanced infrared spectroscopy using resonant nanoantennas, *Chem. Rev.*, **117**, pp. 5110–5145.

17. Yang, X. X., Sun, Z. P., Low, T., Hu, H., Guo, X. D., de Abajo, F. J. G., Avouris, P., and Dai, Q. (2018). Nanomaterial-based plasmon-enhanced infrared spectroscopy, *Adv. Mater.*, **30**, pp. 23.

18. Aroca, R. F., Ross, D. J., and Domingo, C. (2004). Surface-enhanced infrared spectroscopy, *Appl. Spectrosc.*, **58**, pp. 324A–338A.

19. Osawa, M. (2001). *Near-Field Optics and Surface Plasmon Polaritons*, eds. Kawata, S., Chapter 9 "Surface-enhanced infrared absorption," (Springer Berlin Heidelberg, Germany). pp. 163–187.

20. Anderson, M. S. (2003). Enhanced infrared absorption with dielectric nanoparticles, *Appl. Phys. Lett.*, **83**, pp. 2964–2966.

21. Kosower, E. M., Markovich, G., and Borz, G. (2004). Surface-enhanced infrared absorption of p-nitrobenzoic acid on planar silver halide fiber, *J. Phys. Chem. B*, **108**, pp. 12873–12876.

22. Adato, R., Aksu, S., and Altug, H. (2015). Engineering mid-infrared nanoantennas for surface enhanced infrared absorption spectroscopy, *Mater. Today*, **18**, pp. 436–446.

23. Larmour, I. A. and Graham, D. (2011). Surface enhanced optical spectroscopies for bioanalysis, *Analyst*, **136**, pp. 3831–3853.

24. Li, J., Zheng, B., Zhang, Q. W., Liu, Y., Shi, C. F., Wang, F. B., Wang, K., and Xia, X. H. (2017). Attenuated total reflection surface-enhanced infrared

absorption spectroscopy: A powerful technique for bioanalysis, *J. Anal. Test.*, **1**, pp. 8.

25. Yang, Y. Y., Zhang, H. X., and Cai, W. B. (2013). Recent experimental progresses on electrochemical ATR-SEIRAS, *J. Electrochem.*, **19**, pp. 6–16.
26. Miyake, H., Ye, S., and Osawa, M. (2002). Electroless deposition of gold thin films on silicon for surface-enhanced infrared spectroelectrochemistry, *Electrochem. Commun.*, **4**, pp. 973–977.
27. Ataka, K., Giess, F., Knoll, W., Naumann, R., Haber-Pohlmeier, S., Richter, B., and Heberle, J. (2004). Oriented attachment and membrane reconstitution of His-tagged cytochrome c oxidase to a gold electrode: In situ monitoring by surface-enhanced infrared absorption spectroscopy, *J. Am. Chem. Soc.*, **126**, pp. 16199–16206.
28. Adato, R., Yanik Ahmet, A., Amsden Jason, J., Kaplan David, L., Omenetto Fiorenzo, G., Hong Mi, K., Erramilli, S., and Altug, H. (2009). Ultra-sensitive vibrational spectroscopy of protein monolayers with plasmonic nanoantenna arrays, *Proc. Natl. Acad. Sci. U. S. A*, **106**, pp. 19227–19232.
29. Johnson, E. and Aroca, R. (1995). Surface-enhanced infrared spectroscopy of monolayers, *J. Phys. Chem.*, **99**, pp. 9325–9330.
30. Lewis, R. N. A. H. and McElhaney, R. N. (2007) *Methods in Membrane Lipids*, eds. Dopico, A.M., Chapter 14 "Fourier transform infrared spectroscopy in the study of lipid phase transitions in model and biological membranes," (Humana Press, Totowa, New Jersey). pp. 207–226.
31. Lewis, R. N. A. H. and McElhaney, R. N. (2013). Membrane lipid phase transitions and phase organization studied by Fourier transform infrared spectroscopy, *Biochim. Biophys. Acta, Biomembr.*, **1828**, pp. 2347–2358.
32. Kel, O., Tamimi, A., Thielges, M. C., and Fayer, M. D. (2013). Ultrafast structural dynamics inside planar phospholipid multibilayer model cell membranes measured with 2D IR spectroscopy, *J. Am. Chem. Soc.*, **135**, pp. 11063–11074.
33. Kel, O., Tamimi, A., and Fayer, M. D. (2014). Size-dependent ultrafast structural dynamics inside phospholipid vesicle bilayers measured with 2D IR vibrational echoes, *Proc. Natl. Acad. Sci. U. S. A*, **111**, pp. 918–923.
34. Kel, O., Tamimi, A., and Fayer, M. D. (2015). The influence of cholesterol on fast dynamics inside of vesicle and planar phospholipid bilayers

measured with 2D IR spectroscopy, *J. Phys. Chem. B*, **119**, pp. 8852–8862.

35. Binder, H. (2007). Water near lipid membranes as seen by infrared spectroscopy, *Eur Biophys J*, **36**, pp. 265–279.
36. Bridelli, M. G., Capelletti, R., and Mora, C. (2013). Structural features and functional properties of water in model DMPC membranes: Thermally stimulated depolarization currents (TSDCs) and Fourier transform infrared (FTIR) studies, *J. Phys. D: Appl. Phys.*, **46**, pp. 485401.
37. Disalvo, E. A., Lairion, F., Martini, F., Tymczyszyn, E., Frías, M., Almaleck, H., and Gordillo, G. J. (2008). Structural and functional properties of hydration and confined water in membrane interfaces, *Biochim. Biophys. Acta, Biomembr.*, **1778**, pp. 2655–2670.
38. Disalvo, E. A. and Frias, M. A. (2013). Water state and carbonyl distribution populations in confined regions of lipid bilayers observed by FTIR spectroscopy, *Langmuir*, **29**, pp. 6969–6974.
39. Disalvo, E. A., Bouchet, A. M., and Frias, M. A. (2013). Connected and isolated CH_2 populations in acyl chains and its relation to pockets of confined water in lipid membranes as observed by FTIR spectrometry, *Biochim. Biophys. Acta, Biomembr.*, **1828**, pp. 1683–1689.
40. Levinger, N. E., Costard, R., Nibbering, E. T. J., and Elsaesser, T. (2011). Ultrafast energy migration pathways in self-assembled phospholipids interacting with confined water, *J. Phys. Chem. A*, **115**, pp. 11952–11959.
41. Costard, R., Levinger, N. E., Nibbering, E. T. J., and Elsaesser, T. (2012). Ultrafast vibrational dynamics of water confined in phospholipid reverse micelles, *J. Phys. Chem. B*, **116**, pp. 5752–5759.
42. Osborne, D. G., Dunbar, J. A., Lapping, J. G., White, A. M., and Kubarych, K. J. (2013). Site-specific measurements of lipid membrane interfacial water dynamics with multidimensional infrared spectroscopy, *J. Phys. Chem. B*, **117**, pp. 15407–15414.
43. Flanagan, J. C., Valentine, M. L., and Baiz, C. R. (2020). Ultrafast dynamics at lipid-water interfaces, *Acc. Chem. Res.*, **53**, pp. 1860–1868.
44. Fayer, M. D. (2012). Dynamics of water interacting with interfaces, molecules, and ions, *Acc. Chem. Res.*, **45**, pp. 3–14.
45. Bredenbeck, J., Ghosh, A., Nienhuys, H. K., and Bonn, M. (2009). Interface-specific ultrafast two-dimensional vibrational spectroscopy, *Acc. Chem. Res.*, **42**, pp. 1332–1342.
46. Juhaniewicz-Dębińska, J., Konarzewska, D., and Sęk, S. (2019). Effect of interfacial water on the nanomechanical properties of negatively

charged floating bilayers supported on gold electrodes, *Langmuir*, **35**, pp. 9422–9429.

47. Su, Z. F., Juhaniewicz-Debinska, J., Sek, S., and Lipkowski, J. (2020). Water structure in the submembrane region of a floating lipid bilayer: The effect of an ion channel formation and the channel blocker, *Langmuir*, **36**, pp. 409–418.
48. Dziubak, D. and Sek, S. (2021). Physicochemical characterization of sparsely tethered bilayer lipid membranes: Structure of submembrane water and nanomechanical properties, *ChemElectroChem*, **8**, pp. 2564–2571.
49. Kriegel, S., Uchida, T., Osawa, M., Friedrich, T., and Hellwig, P. (2014). Biomimetic environment to study *E-coli* complex I through surface-enhanced IR absorption spectroscopy, *Biochemistry*, **53**, pp. 6340–6347.
50. Das, S., Chakraborty, S., and Mitra, S. K. (2012). Redefining electrical double layer thickness in narrow confinements: Effect of solvent polarization, *Phys. Rev. E*, **85**, pp. 051508.
51. Sin, J. S., Im, S. J., and Kim, K. (2015). Asymmetric electrostatic properties of an electric double layer: A generalized Poisson–Boltzmann approach taking into account non-uniform size effects and water polarization, *Electrochim. Acta*, **153**, pp. 531–539.
52. Berkowitz, M. L., Bostick, D. L., and Pandit, S. (2006). Aqueous solutions next to phospholipid membrane surfaces: Insights from simulations, *Chem. Rev.*, **106**, pp. 1527–1539.
53. Hua, W., Verreault, D., and Allen, H. C. (2015). Solvation of calcium-phosphate headgroup complexes at the DPPC/aqueous interface, *ChemPhysChem*, **16**, pp. 3910–3915.
54. Chen, X. K., Hua, W., Huang, Z. S., and Allen, H. C. (2010). Interfacial water structure associated with phospholipid membranes studied by phase-sensitive vibrational sum frequency generation spectroscopy, *J. Am. Chem. Soc.*, **132**, pp. 11336–11342.
55. Dogangun, M., Ohno, P. E., Liang, D., McGeachy, A. C., Bé, A. G., Dalchand, N., Li, T. Z., Cui, Q., and Geiger, F. M. (2018). Hydrogen-bond networks near supported lipid bilayers from vibrational sum frequency generation experiments and atomistic simulations, *J. Phys. Chem. B*, **122**, pp. 4870–4879.
56. Ishiyama, T., Terada, D., and Morita, A. (2015). Hydrogen-bonding structure at zwitterionic lipid/water interface, *J. Phys. Chem. Lett.*, **7**, pp. 216–220.

57. Yun, S. H. and Kwok, S. J. J. (2017). Light in diagnosis, therapy and surgery, *Nat. Biomed. Eng.*, **1**, pp. 8.

58. Jiang, Y., Carvalho-de-Souza, J. L., Wong, R. C. S., Luo, Z. Q., Isheim, D., Zuo, X. B., Nicholls, A. W., Jung, I. W., Yue, J. P., Liu, D. J., Wang, Y., De Andrade, V., Xiao, X. H., Navrazhnykh, L., Weiss, D. E., Wu, X. Y., Seidman, D. N., Bezanilla, F., and Tian, B. (2016). Heterogeneous silicon mesostructures for lipid-supported bioelectric interfaces, *Nat. Mater.*, **15**, pp. 1023–1030.

59. Shapiro, M. G., Homma, K., Villarreal, S., Richter, C. P., and Bezanilla, F. (2012). Infrared light excites cells by changing their electrical capacitance, *Nat. Commun.*, **3**, pp. 736.

60. Ebtehaj, Z., Hatef, A., Malekmohammad, M., and Soltanolkotabi, M. (2018). Computational modeling and validation of thermally induced electrical capacitance changes for lipid bilayer membranes irradiated by pulsed lasers, *J. Phys. Chem. B*, **122**, pp. 7319–7331.

61. Li, S. S., Wu, L., Zhen, W. Y., Zhu, M. Y., Cheng, X. W., and Jiang, X. E. (2021). Molecular nature of structured water in the light-induced interfacial capacitance changes at the bioelectric interface, *J. Phys. Chem. Lett.*, **12**, pp. 9982–9988.

62. Tian, C., Ji, N., Waychunas, G. A., and Shen, Y. R. (2008). Interfacial structures of acidic and basic aqueous solutions, *J. Am. Chem. Soc.*, **130**, pp. 13033–13039.

63. Tian, C. S. and Shen, Y. R. (2009). Structure and charging of hydrophobic material/water interfaces studied by phase-sensitive sum-frequency vibrational spectroscopy, *Proc. Natl. Acad. Sci. U. S. A*, **106**, pp. 15148–15153.

64. Nagata, Y. and Mukamel, S. (2010). Vibrational sum-frequency generation spectroscopy at the water/lipid interface: Molecular dynamics simulation study, *J. Am. Chem. Soc.*, **132**, pp. 6434–6442.

65. Kagan, V. E., Tyurin, V. A., Jiang, J. F., Tyurina, Y. Y., Ritov, V. B., Amoscato, A. A., Osipov, A. N., Belikova, N. A., Kapralov, A. A., Kini, V., Vlasova, I. I., Zhao, Q., Zou, M. M., Di, P., Svistunenko, D. A., Kurnikov, I. V., and Borisenko, G. G. (2005). Cytochrome c acts as a cardiolipin oxygenase required for release of proapoptotic factors, *Nat. Chem. Biol.*, **1**, pp. 223–232.

66. Zeng, L., Wu, L., Liu, L., and Jiang, X. E. (2017). The role of water distribution controlled by transmembrane potentials in the cytochrome c–cardiolipin interaction: Revealing from surface-enhanced infrared absorption spectroscopy, *Chem. Eur. J.*, **23**, pp. 15491–15497.

67. Zeng, L., Wu, L., Liu, L., and Jiang, X. E. (2016). Analyzing structural properties of heterogeneous cardiolipin-bound cytochrome c and their regulation by surface-enhanced infrared absorption spectroscopy, *Anal. Chem.*, **88**, pp. 11727–11733.

68. Koya, R. C., Fujita, H., Shimizu, S., Ohtsu, M., Takimoto, M., Tsujimoto, Y., and Kuzumaki, N. (2000). Gelsolin inhibits apoptosis by blocking mitochondrial membrane potential loss and cytochrome c release, *J. Biol. Chem.*, **275**, pp. 15343–15349.

69. Hynes, J. T. (1999). The protean proton in water, *Nature*, **397**, pp. 565–567.

70. Cong, X., Poyton, M. F., Baxter, A. J., Pullanchery, S., and Cremer, P. S. (2015). Unquenchable surface potential dramatically enhances Cu^{2+} binding to phosphatidylserine lipids, *J. Am. Chem. Soc.*, **137**, pp. 7785–7792.

71. Wu, L. and Jiang, X. E. (2018). Proton transfer at the interaction interface of graphene oxide, *Anal. Chem.*, **90**, pp. 10223–10230.

72. Barth, A. and Zscherp, C. (2003). What vibrations tell about proteins, *Q. Rev. Biophys.*, **35**, pp. 369–430.

73. Barth, A. (2007). Infrared spectroscopy of proteins, *Biochim. Biophys. Acta, Bioenerg.*, **1767**, pp. 1073–1101.

74. Baenziger, J. E. and Costa, C. J. B. (2008). *Biophysical Analysis of Membrane Proteins*, eds. Pebay-Peyroula, E., Chapter 11 "Membrane protein structure and conformational change probed using Fourier transform infrared spectroscopy," (Wiley-VCH Verlag GmbH & Co. KGaA, Darmstadt, Germany). pp. 259–288.

75. Tatulian, S. A. (2013). *Lipid–Protein Interactions: Methods and Protocols*, eds. Kleinschmidt, J.H., Chapter 9 "Structural characterization of membrane proteins and peptides by FTIR and ATR-FTIR spectroscopy," (Humana Press, Totowa, New Jersey). pp. 177–218.

76. Tatulian, S. A. (2019). *Lipid–Protein Interactions: Methods and Protocols*, eds. Kleinschmidt, J.H., Chapter 13 "FTIR analysis of proteins and proteinmembrane interactions," (Springer New York, New York, U.S.A.). pp. 281–325.

77. Lorenz-Fonfria, V. A. (2020). Infrared difference spectroscopy of proteins: From bands to bonds, *Chem. Rev.*, **120**, pp. 3466–3576.

78. Maiti, K. S. (2021). Two-dimensional infrared spectroscopy reveals better insights of structure and dynamics of protein, *Molecules*, **26**, pp. 6893.

79. Ataka, K. and Heberle, J. (2004). Functional vibrational spectroscopy of a cytochrome c monolayer: SEIDAS probes the interaction with different surface-modified electrodes, *J. Am. Chem. Soc.*, **126**, pp. 9445–9457.

80. Jiang, X. E., Ataka, K., and Heberle, J. (2008). Influence of the molecular structure of carboxyl-terminated self-assembled monolayer on the electront transfer of cytochrome c adsorbed on an Au electrode: In situ observation by surface-enhanced infrared absorption spectroscopy, *J. Phys. Chem. C*, **112**, pp. 813–819.

81. Jin, B., Wang, G. X., Millo, D., Hildebrandt, P., and Xia, X. H. (2012). Electric-field control of the pH-dependent redox process of cytochrome c immobilized on a gold electrode, *J. Phys. Chem. C*, **116**, pp. 13038–13044.

82. Lin, S. R., Jiang, X. E., Wang, L. X., Li, G. G., and Guo, L. P. (2012). Adsorption orientation of horse heart cytochromecon a bare gold electrode hampers its electron transfer, *J. Phys. Chem. C*, **116**, pp. 637–642.

83. Wisitruangsakul, N., Zebger, I., Ly, K. H., Murgida, D. H., Ekgasit, S., and Hildebrandt, P. (2008). Redox-linked protein dynamics of cytochrome c probed by time-resolved surface enhanced infrared absorption spectroscopy, *Phys. Chem. Chem. Phys.*, **10**, pp. 5276–5286.

84. Nowak, C., Luening, C., Knoll, W., and Naumann, R. L. C. (2009). A two-layer gold surface with improved surface enhancement for spectro-electrochemistry using surface-enhanced infrared absorption spectroscopy, *Appl. Spectrosc.*, **63**, pp. 1068–1074.

85. Nowak, C., Luening, C., Schach, D., Baurecht, D., Knoll, W., and Naumann, R. L. C. (2009). Electron transfer kinetics of cytochrome c in the submillisecond time regime using time-resolved surface-enhanced infrared absorption spectroscopy, *J. Phys. Chem. C*, **113**, pp. 2256–2262.

86. Liu, L., Wu, L., Zeng, L., and Jiang, X. E. (2015). Label-free surface-enhanced infrared spectro-electro-chemical analysis of the Redox potential shift of cytochrome c complexed with a cardiolipin-containing lipid membrane of varied composition, *Chin. Phys. B*, **24**, pp. 128201.

87. Liu, L., Zeng, L., Wu, L., and Jiang, X. E. (2015). Label-free surface-enhanced infrared spectroelectrochemistry studies the interaction of cytochrome c with cardiolipin-containing membranes, *J. Phys. Chem. C*, **119**, pp. 3990–3999.

88. Zhang, X. F., Zeng, L., Liu, L., Ma, Z. F., and Jiang, X. E. (2016). Surface-enhanced infrared absorption spectroscopy and electrochemistry reveal the impact of nanoparticles on the function of protein

immobilized on mimic biointerface, *Electrochim. Acta*, **211**, pp. 148–155.

89. Liu, L., Zeng, L., Wu, L., and Jiang, X. E. (2017). Revealing the effect of protein weak adsorption to nanoparticles on the interaction between the desorbed protein and its binding partner by surface-enhanced infrared spectroelectrochemistry, *Anal. Chem.*, **89**, pp. 2724–2730.
90. Ataka, K., Richter, B., and Heberle, J. (2006). Orientational control of the physiological reaction of cytochrome c oxidase tethered to a gold electrode, *J. Phys. Chem. B*, **110**, pp. 9339–9347.
91. Jiang, X. E., Zuber, A., Heberle, J., and Ataka, K. (2008). In situ monitoring of the orientated assembly of strep-tagged membrane proteins on the gold surface by surface enhanced infrared absorption spectroscopy, *Phys. Chem. Chem. Phys.*, **10**, pp. 6381–6387.
92. Friedrich, M. G., Kirste, V. U., Zhu, J., Gennis, R. B., Knoll, W., and Naumann, R. L. C. (2008). Activity of membrane proteins immobilized on surfaces as a function of packing density, *J. Phys. Chem. B*, **112**, pp. 3193–3201.
93. Nowak, C., Schach, D., Gebert, J., Grosserueschkamp, M., Gennis, R. B., Ferguson-Miller, S., Knoll, W., Walz, D., and Naumann, R. L. C. (2011). Oriented immobilization and electron transfer to the cytochrome c oxidase, *J. Solid State Electrochem.*, **15**, pp. 105–114.
94. Nowak, C., Santonicola, M. G., Schach, D., Zhu, J. P., Gennis, R. B., Ferguson-Miller, S., Baurecht, D., Walz, D., Knoll, W., and Naumann, R. L. C. (2010). Conformational transitions and molecular hysteresis of cytochrome c oxidase: Varying the redox state by electronic wiring, *Soft Matter*, **6**, pp. 5523–5532.
95. Nowak, C., Laredo, T., Gebert, J., Lipkowski, J., Gennis, R. B., Ferguson-Miller, S., Knoll, W., and Naumann, R. L. C. (2011). 2D-SEIRA spectroscopy to highlight conformational changes of the cytochrome c oxidase induced by direct electron transfer, *Metallomics*, **3**, pp. 619–627.
96. Schwaighofer, A., Steininger, C., Hildenbrandt, D. M., Srajer, J., Nowak, C., Knoll, W., and Naumann, R. L. C. (2013). Time-resolved surface-enhanced IR-absorption spectroscopy of direct electron transfer to cytochrome c oxidase from *R. sphaeroides*, *Biophys. J.*, **105**, pp. 2706–2713.
97. Steininger, C., Reiner-Rozman, C., Schwaighofer, A., Knoll, W., and Naumann, R. L. C. (2016). Kinetics of cytochrome c oxidase from R. sphaeroides initiated by direct electron transfer followed by tr-SEIRAS, *Bioelectrochemistry*, **112**, pp. 1–8.

98. Jiang, X. E., Zaitseva, E., Schmidt, M., Siebert, F., Engelhard, M., Schlesinger, R., Ataka, K., Vogel, R., and Heberle, J. (2008). Resolving voltage-dependent structural changes of a membrane photoreceptor by surface-enhanced IR difference spectroscopy, *Proc. Natl. Acad. Sci. U. S. A*, **105**, pp. 12113–12117.

99. Millo, D., Hildebrandt, P., Pandelia, M. E., Lubitz, W., and Zebger, I. (2011). SEIRA spectroscopy of the electrochemical activation of an immobilized [NiFe] hydrogenase under turnover and non-turnover conditions, *Angew. Chem., Int. Ed.*, **50**, pp. 2632–2634.

100. Gutierrez-Sanz, O., Marques, M., Pereira, I. A. C., De Lacey, A. L., Lubitz, W., and Ruediger, O. (2013). Orientation and function of a membrane-bound enzyme monitored by electrochemical surface-enhanced infrared absorption spectroscopy, *J. Phys. Chem. Lett.*, **4**, pp. 2794–2798.

101. Vaz-Domínguez, C., Pita, M., de Lacey, A. L., Shleev, S., and Cuesta, A. (2012). Combined ATR-SEIRAS and EC-STM study of the immobilization of laccase on chemically modified Au electrodes, *J. Phys. Chem. C*, **116**, pp. 16532–16540.

102. Gebert, J., Reiner-Rozman, C., Steininger, C., Nedelkovski, V., Nowak, C., Wraight, C. A., and Naumann, R. L. C. (2015). Electron transfer to light-activated photosynthetic reaction centers from *Rhodobacter sphaeroides* reconstituted in a biomimetic membrane system, *J. Phys. Chem. C*, **119**, pp. 890–895.

103. Nedelkovski, V., Schwaighofer, A., Wraight, C. A., Nowak, C., and Naumann, R. L. C. (2013). Surface-enhanced infrared absorption spectroscopy (SEIRAS) of light-activated photosynthetic reaction centers from *Rhodobacter sphaeroides* reconstituted in a biomimetic membrane system, *J. Phys. Chem. C*, **117**, pp. 16357–16363.

104. Kozuch, J., Steinem, C., Hildebrandt, P., and Millo, D. (2012). Combined electrochemistry and surface-enhanced infrared absorption spectroscopy of gramicidin A incorporated into tethered bilayer lipid membranes, *Angew. Chem., Int. Ed.*, **51**, pp. 8114–8117.

105. Kozuch, J., Weichbrodt, C., Millo, D., Giller, K., Becker, S., Hildebrandt, P., and Steinem, C. (2014). Voltage-dependent structural changes of the membrane-bound anion channel hVDAC1 probed by SEIRA and electrochemical impedance spectroscopy, *Phys. Chem. Chem. Phys.*, **16**, pp. 9546–9555.

106. Noguchi, H., Adachi, T., Nakatomi, A., Yazawa, M., and Uosaki, K. (2016). Biofunctionality of calmodulin immobilized on gold surface studied by surface enhanced infrared absorption spectroscopy-Ca^{2+} induced

conformational change and binding to a target peptide, *J. Phys. Chem. C*, **120**, pp. 16035–16041.

107. Moe, E., Sezer, M., Hildebrandt, P., and Todorovic, S. (2015). Surface enhanced vibrational spectroscopic evidence for an alternative DNA-independent redox activation of Endonuclease III, *Chem. Commun.*, **51**, pp. 3255–3257.

108. Kato, M., Nakagawa, S., Tosha, T., Shiro, Y., Masuda, Y., Nakata, K., and Yagi, I. (2018). Surface-enhanced infrared absorption spectroscopy of bacterial nitric oxide reductase under electrochemical control using a vibrational probe of carbon monoxide, *J. Phys. Chem. Lett.*, **9**, pp. 5196–5200.

109. Salewski, J., Batista, A. P., Sena, F. V., Millo, D., Zebger, I., Pereira, M. M., and Hildebrandt, P. (2016). Substrate-protein interactions of type II NADH:quinone oxidoreductase from *Escherichia coli*, *Biochemistry*, **55**, pp. 2722–2734.

110. Wiebalck, S., Kozuch, J., Forbrig, E., Tzschucke, C. C., Jeuken, L. J. C., and Hildebrandt, P. (2016). Monitoring the transmembrane proton gradient generated by cytochrome bo($_3$) in tethered bilayer lipid membranes using SEIRA spectroscopy, *J. Phys. Chem. B*, **120**, pp. 2249–2256.

111. Gutierrez-Sanz, O., Forbig, E., Batista, A. P., Pereira, M. M., Salewski, J., Mroginski, M. A., Gotz, R., De Lacey, A. L., Kozuch, J., and Zebger, I. (2018). Catalytic activity and proton translocation of reconstituted respiratory complex I monitored by surface-enhanced infrared absorption spectroscopy, *Langmuir*, **34**, pp. 5703–5711.

112. Levin, C. S., Kundu, J., Janesko, B. G., Scuseria, G. E., Raphael, R. M., and Halas, N. J. (2008). Interactions of ibuprofen with hybrid lipid bilayers probed by complementary surface-enhanced vibrational spectroscopies, *J. Phys. Chem. B*, **112**, pp. 14168–14175.

113. Quirk, A., Lardner, M. J., Tun, Z., and Burgess, I. J. (2016). Surface-enhanced infrared spectroscopy and neutron reflectivity studies of ubiquinone in hybrid bilayer membranes under potential control, *Langmuir*, **32**, pp. 2225–2235.

114. Forbrig, E., Staffa, J. K., Salewski, J., Mroginski, M. A., Hildebrandt, P., and Kozuch, J. (2018). Monitoring the orientational changes of alamethicin during incorporation into bilayer lipid membranes, *Langmuir*, **34**, pp. 2373–2385.

115. Li, S. S., Wu, L., Zhu, M. Y., Cheng, X. W., and Jiang, X. E. (2022). Effect of dipole potential on the orientation of voltage-gated alamethicin

peptides regulated by chaotropic anions, *J. Electroanal. Chem.*, **904**, pp. 115880.

116. Wang, T. T., Bai, J., Jiang, X. E., and Nienhaus, G. U. (2012). Cellular uptake of nanoparticles by membrane penetration: A study combining confocal microscopy with FTIR spectroelectrochemistry, *ACS Nano*, **6**, pp. 1251–1259.
117. Wang, T. T. and Jiang, X. E. (2015). The broken of phosphodiester bond: A key factor induced hemolysis, *ACS Appl. Mater. Interfaces*, **7**, pp. 129–136.
118. Wang, T. T., Zhu, S. J., and Jiang, X. E. (2015). Toxicity mechanism of graphene oxide and nitrogen-doped graphene quantum dots in RBCs revealed by surface-enhanced infrared absorption spectroscopy, *Toxicol. Res.*, **4**, pp. 885–894.
119. Zhang, X. F., Cao, F. J., Wu, L., and Jiang, X. E. (2019). Understanding the synergic mechanism of weak interactions between graphene oxide and lipid membrane leading to the extraction of lipids, *Langmuir*, **35**, pp. 14098–14107.
120. Busalmen, J. P., Berná, A., and Feliu, J. M. (2007). Spectroelectrochemical examination of the interaction between bacterial cells and gold electrodes, *Langmuir*, **23**, pp. 6459–6466.
121. Busalmen, J. P., Esteve-Núñez, A., Berná, A., and Feliu, J. M. (2008). C-type cytochromes wire electricity-producing bacteria to electrodes, *Angew. Chem., Int. Ed.*, **120**, pp. 4952–4955.
122. Busalmen, J. P., Esteve-Nunez, A., Berna, A., and Feliu, J. M. (2010). ATR-SEIRAs characterization of surface redox processes in *G. sulfurreducens*, *Bioelectrochemistry*, **78**, pp. 25–29.
123. Kuzume, A., Zhumaev, U., Li, J. F., Fu, Y. C., Füeg, M., Estévez, M., Borjas, Z., Wandlowski, T., and Esteve-Nuñez, A. (2014). An in situ surface electrochemistry approach towards whole-cell studies: The structure and reactivity of a *Geobacter sulfurreducens* submonolayer on electrified metal/electrolyte interfaces, *Phys. Chem. Chem. Phys.*, **16**, pp. 22229–22236.
124. Cao, F. J., Wu, L., Zhang, X. F., Li, S. S., Wang, C., Zhen, W. Y., and Jiang, X. E. (2019). Dynamic surface properties of PEG-coated CuS nanoparticles alter their interaction with cells as revealed by surface-enhanced infrared spectroscopy, *Nanoscale Adv.*, **1**, pp. 4268–4276.
125. Kim, B. E., Nevitt, T., and Thiele, D. J. (2008). Mechanisms for copper acquisition, distribution and regulation, *Nat. Chem. Biol.*, **4**, pp. 176–185.

126. Wu, L. and Jiang, X. E. (2018). *Membrane Biophysics: New Insights and Methods*, eds. Wang, H.D. and Li, G.H., Chapter 11 "Infrared spectroscopy for studying plasma membranes," (Springer Singapore, Singapore). pp. 319–354.

Chapter 10

Application of Single-Molecule Force Spectroscopy in Membrane Receptors Dynamics and Signal Transduction

Rui Qin[a] **and Wei Chen**[a,b,c]
[a]*Department of Cell Biology and Department of Cardiology of the Second Affiliated Hospital, Zhejiang University School of Medicine, Zhejiang University, Hangzhou 310058, China*
[b]*Liangzhu Laboratory, Zhejiang University Medical Center, 1369 West Wenyi Road, Hangzhou 311121, China*
[c]*Collaborative Innovation Center for Diagnosis and Treatment of Infectious Diseases, the MOE Frontier Science Center for Brain Science & Brain-Machine Integration, State Key Laboratory for Modern Optical Instrumentation, Key Laboratory for Biomedical Engineering of the Ministry of Education, College of Biomedical Engineering and Instrument Science, Zhejiang University, Hangzhou 310058, Zhejiang, China*
jackweichen@zju.edu.cn

Mechanical cues dynamically modulate interactions between membrane receptors and ligands to trigger various physiological and pathological processes, such as cell adhesion, viral invasion,

Biomembranomics: Structure, Techniques, and Applications
Edited by Hongda Wang

ISBN 978-981-4968-61-4 (Hardcover), 978-1-003-45635-3 (eBook)
www.jennystanford.com

and immune defense. Single-molecule force spectroscopy (SMFS) techniques have been widely used to reveal unprecedented dynamic biophysical regulatory mechanisms that are unable to be disclosed by qualitative biochemical analysis. Here, we summarize the recent advances and applications of these biophysical methodologies and novel regulatory mechanisms of the critical ligand recognition of viral and immune receptors, including SARS-CoV-2 spike protein, T-cell receptor (TCR), and other co-stimulatory immune receptors.

10.1 Introduction

Ligand recognition of membrane receptors has been resolved with high sensitivity, high specificity, and high spatial and temporal resolutions by single-molecule force spectroscopy (SMFS) techniques, which mainly contain biomembrane force probes (BFP), optical tweezers (OT), magnetic tweezers (MT), and atomic force microscopy (AFM), which works like soft and mechanically sensitive springs, transducing piconewton forces to the displacements of a bead or a cantilever [1–4]. Benefited from precise force manipulations on single-molecule bonds, these SMFS techniques have been applied to characterize force-induced conformational changes, and to characterize the force-dependent receptor–ligand binding kinetics under different types of dynamic piconewton forces, revealing crucial biophysical regulatory mechanisms, such as adhesion receptors (e.g., integrin, PSGL-1, and LFA-1 [5–13]), viral (e.g., severe acute respiratory syndrome coronavirus 2, SARS-CoV-2 spike [14]), and immune receptors (e.g., T-cell receptor (TCR) [15, 16] and Natural Killer Group 2, member D, NKG2D [17]). Mechanical cues exerted on the receptor–ligand binding complex mainly include the dynamic tensile or traction forces and the spatial confinement of the cellular plasma membrane [18–22]. The mechanical tensile or traction force generated by cytoskeletal actomyosin contraction, cell membrane bending, and cell migration could deform or change the conformations of receptors and/or ligands, differentially modulating the dissociation pathway of receptor–ligand interactions [23, 24]. The cellular plasma membrane provides a unique microenvironment that biomechanically restricts the orientation of the ectodomain of membrane receptors and the spatial diffusion or movement only within a two-dimensional (2D) membrane, thus inevitably impacting the association and dissociation processes of receptor–ligand interactions and their binding kinetics [19–21, 25, 26]. Furthermore, the mechanical regulation of receptor–ligand interactions executes crucial effects on the signal transduction of membrane receptors [27–29]. The underlying mechanisms have been dissected by biophysical methodologies that simultaneously record the binding kinetics and binding-triggered cell signaling, such as fBFP (fluorescent BFP [5, 15, 30]) and BATTLES (biomechanically assisted T-cell triggering for large-scale exogenous-pMHC screening [31, 32]).

Here, we summarize recent processes in the development of SMFS techniques and crucial functional mechanisms underlying receptor–ligand interactions for triggering cell adhesion, viral infection, and immune defense. Understanding the biomechanical characteristics of how mechanical and chemical cues couple in the regulation of membrane receptor–ligand interactions may help optimize the designs of neutralizing antibodies against SARS-CoVs infection and of TCR-T or neoantigen vaccines for T-cell-based immunotherapies.

10.2 Single-Molecule Force Spectroscopy

10.2.1 Biomembrane Force Probe

The biomembrane force probe (BFP) was pioneered by E. Evans to study the dynamics of membrane protein-protein interaction, such as P-selectin/PSGL-1 (P-selectin glycoprotein ligand-1), E-cadherin and ICAM-1 (intercellular adhesion molecule-1), and ICAM-1 and β2 integrin [33–35]. After that, Cheng Zhu's laboratory in the United States and related domestic laboratories (Long and Lou laboratories of the Chinese Academy of Sciences and our laboratory) widely used BFP to study the dynamic ligand recognition of cell surface membrane proteins, such as viral and immune receptors, including SARS-CoV-2 spike protein [14], TCR [15, 16], and other co-stimulatory immune receptors (NKG2D) [17]. BFP uses red blood cells as an ultrasensitive mechanical sensor, and its elastic coefficient can reach 0.1 pN/nm [36, 37]. The binding kinetics of receptor–ligand interactions can be derived from different methods, mainly containing dynamic force spectroscopy for measuring force-free dissociation rates, force-clamp assay for characterizing force-dependent dissociation rates, adhesion frequency assay for binding affinities and force-free association and dissociation rates, and thermal-fluctuation assay for association rates and force-free dissociation rates [2–4]. Taking TCR as an example, the binding kinetics measurements of TCR–pMHC interactions better match their ligand potency than surface plasmon resonance (SPR) measurements under physiologically relevant biophysical conditions [26, 38]. The integration of fluorescent spectroscopy into the BFP system enabled the measurement of TCR–pMHC binding kinetics and the recording of intracellular Ca^{2+} signaling simultaneously, digitalizing TCR triggering mechanisms more directly and revolutionizing canonical methodologies for studying the mechanisms of membrane receptors mechanosensing and triggering [15]. Our group further improved the clamping force stability and accuracy of BFP in 2020, enabling the lifetime measurements of single-molecule bonds on live cells with ultra-slow dissociation kinetics, such as the interaction between anti-PD-1 mAb and PD-1 on live T cells [36]. Furthermore, different cycle force loading can be integrated into the BFP system to get the fluctuated force-regulated kinetics to mimic the physiological force fluctuations [6, 39]. BFP is also divided into conventional BFP and dual-load BFP. Compared with the conventional BFP with only one set of force measurement devices, the dual-loaded BFP has two sets of force measurement devices, which can mechanically stimulate a cell and monitor the activation state of the cell, respectively [40].

10.2.2 Magnetic Tweezers

In recent years, magnetic tweezers (MT) have been gradually applied to detect the dynamic conformational changes in proteins by repeatable mechanical stretching, which can directly observe the unfolding or folding of different intermediate states of protein structures, such as Tilin, cytoskeleton-associated proteins Talin, and α-catenin [41–44]. By repeatedly stretching, the binding site of protein–protein interactions is continuously exposed or hidden, and force-dependent associated and dissociated rates are continuously measured. By integrating fluorescent spectroscopy into the MT system, the force-induced dynamic conformational change in protein during structural folding and unfolding and the biochemical function can be obtained simultaneously [45]. With the development of software and hardware technologies, the spatial measurement accuracy of protein conformation has reached 5 nm, and force range can be extended to within 1 pN, reaching 10^{-3} to 10^{4} pN. The stiffness of the system is very small, reaching 10^{-6}–10^{-3} pN/nm, and the sampling frequency is generally 60–200 Hz [46]. This makes magnetic tweezers a great advantage for more precise detection, such as the analysis of DNA/RNA helices [47]. In 2016, the Li laboratory in the Chinese Academy of Sciences used single-molecule magnetic tweezers to analyze the establishment process of force-regulated chromatin fiber structure and reveal the biophysical molecular regulatory mechanism, providing novel evidence that force plays a direct role in regulating gene transcription [48].

10.2.3 Atomic Force Microscopy and Optical Tweezers

Atomic force microscopy (AFM) and optical tweezers (OT) are also currently widely used SMFS technologies besides MT and AFM. Similar to BFP, AFM is a contact measurement method, which is especially suitable for the in situ detection of dynamic interactions between cell surface membrane proteins and their ligands [49, 50]. Different from AFM, an optical tweezer is a non-contact operation, which can detect intracellular protein–protein interactions by introducing microspheres into cells or observe conformational changes in protein dynamics by combining the fluorescent spectroscopy or FRET system (for details, see published reviews) [51, 52]. For example, in 2016, Hu et al. integrated the fluorescence spectrum into the AFM system to simultaneously monitor the force of the cantilever beam and the fluorescence spectrum information acquisition of calcium signal in T cells, revealing the mechanical–chemical feedback process of T-cell activation [53]. In 2017, Ellis's laboratory used optical tweezers to study the mechanical dissociation characteristics between TCR and pMHC interactions and found that mechanical force allosterically drives the conformational change in the FG loop in the constant region of the TCR β-subunit [52, 54, 55]. Thus, the rapidly developed SMFS techniques have become more efficient in dissecting the dynamic biophysical mechanisms of membrane receptor functions.

10.3 Mechanical Force

A growing number of studies suggest that T cells enforce piconewton forces (e.g., 12~19 pN for naïve CD8+ T cell) to TCR–pMHC or integrin with its ligand bonds and dynamically modulate their binding kinetics, as well as their conformations to accordingly transduce signals across cell membranes [56–59]. Mechanical force exerted by membrane bending has been reported to be involved in viral endocytosis [60]. When virions attach to the epithelial layer of the lung airways, the bent cell membrane exerts tensile forces (e.g., 0~30 pN) on the viral spike/ACE2 (angiotensin-converting enzyme 2, ACE2) complex, regulating their binding kinetics and accordingly mediating viral–host recognition, attachment, and invasion [14]. During the aforementioned processes, mechanical force regulates the dissociation rates of receptor–ligand interactions, exhibiting catch, slip, or ideal bonds. Catch bonds slow down the dissociation of receptor–ligand interactions and prolong the bond lifetimes in a specialized force range due to newly formed interactions between residues on receptor and ligand during the force-regulated dissociation pathway (Fig. 10.1A). In contrast, slip bonds, as Bell predicted, accelerate bond dissociation as mechanical force increases (Fig. 10.1B). Ideal bonds exhibit no effect on dissociation, which is independent of changes in mechanical forces (Fig. 10.1C) [2–4]. However, the detailed molecular mechanisms remain largely unknown.

10.3.1 Mechanical Force in Cell Adhesion

Conformational changes in force-regulated integrins execute an important effect on precisely controlling integrin-mediated cell adhesion and signaling. Integrins composed of 18 α and 8 β subunits in mammals are transmembrane receptors that mainly facilitate cell–cell and cell–extracellular matrix (ECM) adhesion [61]. Upon ligand binding, such as fibronectin, vitronectin, collagen, and laminin, integrins activate signal transduction pathways that mediate intracellular signal such as receptor tyrosine kinase signaling [62]. During this process, numerous studies have shown that mechanical force induces different conformational changes in integrins to prolong or decrease integrin–ligand bond lifetimes, thereby modulating integrin-mediated cell adhesion [2, 4, 7, 12, 63]. Here, we briefly describe the force-dependent dissociation pathway of the interaction between LFA-1 and ICAM-1 (for details, see published reviews and books) [2]. There are three different conformations, corresponding to short-lived, intermediate-lived, and long-lived force-dependent bond times when integrin LFA-1 interacts with ICAM-1 [7, 12]. As the force gradually increased, the proportion of short-lived bond times gradually decreases, while the proportion of intermediate-lived and long-lived bond times gradually increases, and the dynamic conformational change (~20 nm) in LFA-1 from the bent state to the upright state was also observed in situ [7, 12]. The effect of the dynamic conformational changes in LAF-1 reveals the biophysical and biochemical coupling regulatory mechanism of allosteric adhesion of leukocyte integrins. Furthermore, studies have found other adhesion molecules such as the interaction between L-selectin/PSGL-1 also shared similar mechanical–chemical coupling regulatory mechanisms (for details, see published reviews) [10, 64].

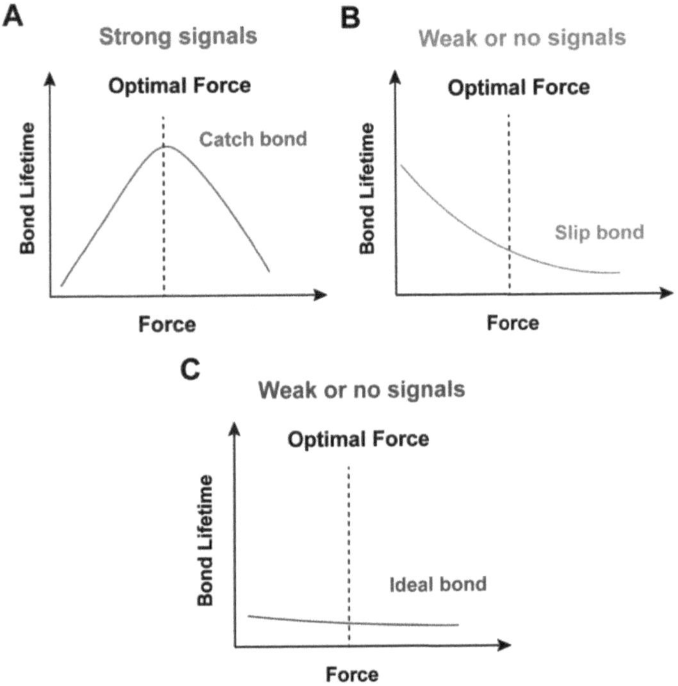

Figure 10.1 Mechanical forces regulate catch–slip bond formation between receptors and ligands. (A–C) Bond lifetime versus force curves showing that mechanical forces induce the formation of catch–slip bonds of receptor–ligand interactions (red) (A) but slip bonds (blue) (B) or ideal bonds (gray) (C).

10.3.2 Mechanical Force in Immune Defense

Force-regulated conformational changes in immune receptors (e.g., TCR) or their ligands (e.g., pMHC, MICA) have been reported to uncover the detailed molecular mechanisms of receptor–ligand interactions during immune defense. Single-molecule BFP technique revealed that mechanical force strengthens agonistic pMHC–TCR interactions through catch bonds (Fig. 10.1A). In contrast, antagonistic pMHC–TCR interactions only exhibit slip bonds under mechanical force (Fig. 10.1B) [15, 65].

To allosterically activate TCR–pMHC catch bonds, mechanical force induces conformational changes in pMHC to promote new interactions (hydrogen bonds) at the peptide–TCR or MHC–TCR binding interface [16]. Mechanical stretching of a single pMHC molecule results in an about 13 nm extension by magnetic tweezers, and locking such extension attenuates TCR–pMHC catch bonds, demonstrating that force-induced conformational changes in pMHC contribute to the catch bonds formation [16]. Mechanical force mainly drives three sequential steps of conformational changes in an agonistic pMHC/TCR complex to activate catch bonds (Fig. 10.2A). First, TCR exploits its CDRβ (complementarity determining region) to establish physical contact with the functional hotspot (e.g., the fourth or sixth residue) of the agonistic peptide. Second, force induces the MHC α1 and α2 domains to rotate toward TCR CDR loops, allosterically promoting new interactions at the TCR–MHC binding interface. Finally, force boosts the separation of β2m from the MHC α chain, resulting in greater rotation and extension of pMHC [16]. In addition to the force-induced conformational changes in pMHC, the TCR–pMHC catch-bond formation also involves moderate TCRαβ conformation changes [52, 55, 66]. Mechanical force allosterically drives the conformational change in the FG loop in the constant region of the TCR β-subunit, controlling the catch-bond formation and modulating bond lifetimes of TCR–pMHC interactions [52]. On the contrary, mechanical forces cannot induce the conformational change antagonistic pMHC and form the slip bonds (Fig. 10.2B) [16]. Collectively, the force-dependent conformational changes in pMHC and TCR allosterically regulate TCR–pMHC interactions to determine TCR antigen discriminatory power.

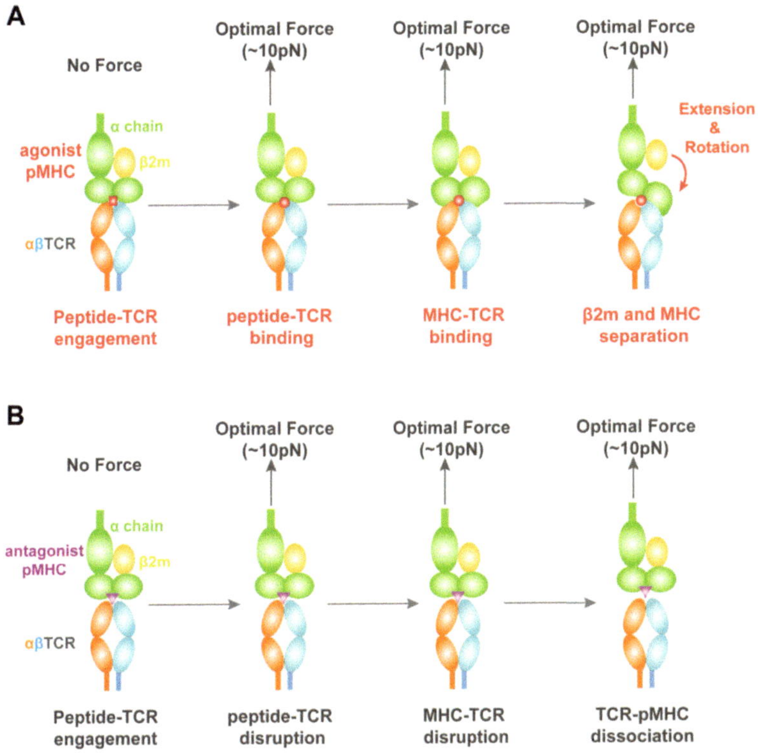

Figure 10.2 The dynamic structural model of force-induced conformational changes in agonistic pMHC-activating TCR antigen discrimination. (A and B) Dynamic sequential steps in the mechanical regulation of agonistic pMHC conformation (A) and antagonistic pMHC (B). First, TCR exploits its CDRβ to establish a physical contact with the functional hotspot of the agonistic peptide. Second, force induces the MHC α1 and α2 domains to rotate toward TCR CDR loops, allosterically promoting new interactions at the TCR–MHC binding interface. Finally, force boosts the separation of β2m from MHC α chain, resulting in greater rotation and extension of pMHC. On the contrary, a mechanical force cannot induce the conformational change in the antagonistic pMHC.

The biophysical regulatory mechanisms described above also prompt researches on other immune receptors in the innate immune system (e.g., NKG2D on lymphocyte cells). NKG2D receptors, as co-stimulatory receptors commonly expressed on the surface of natural killer (NK) or T cells, recognize autologous ligands from the MIC (MHC class I chain-associated, MICA and MICB) and ULBP

(UL16 binding protein, ULBP1-6) families expressed on stressed, transformed, or infected cells, to mediate the killing process of virus-infected or tumor cells [17, 67–70]. NK cells exert tensile forces through NKG2D receptors, which interact dynamically with their ligands (e.g., MICA, MICB, ULBP1, and ULBP3), converting mechanical stimuli into biochemical signals [17]. MICA is the most pronounced ligand to exploit mechanical force to form the catch–slip bonds and has the longest bond lifetimes with NKG2D at the optimal force of ~10 pN compared with MICB, ULBP1, and ULBP3 ligands [17]. Thus, mechanical force aids NKG2D the power to discriminate different ligands, similar to but less potent than TCRs, through differential force-induced ligand conformational change in the NKG2D/ligand complex.

Recent studies have demonstrated that catch-bond engineering can serve as a promising biophysical strategy targeting cancer immunotherapy, e.g., TCR engineering in T-cell-based adoptive cell therapy (ACT). In 2022, Zhao et al. exploited the biophysical parameters of TCR–pMHC interactions to develop an optimization strategy based on the force-induced conformational changes in the TCR/pMHC complex and screen high-potency TCR variants through modifying low-affinity and non-reactive or affinity-matured and off-target TCRs (Fig. 10.3B) [16, 32]. They specifically mutated polar or charged amino acids in the TCR CDR region (defined as "hotspot" on the TCR) that would potentially enhance/impede new H-bonds and/or salt bridge interactions at the TCR–pMHC binding surface under force, instead of mutating the residues that directly contact with pMHC, to ensure their affinities with catch-bond enhancement/inhibition, potentially boosting T-cell responses/avoiding excessive T-cell cross-reactivity, respectively [71, 72]. As a validation of the newly developed strategy, they selected a tumor-associated MAGE-A3-specific TCR (A3A-TCR) that was previously used in clinical trials with high binding affinities to its antigens and with severe off-target toxicity due to high cross-reactivity [73]. With the catch bonds engineering strategy, engineered analogs of A3A-TCR maintain binding affinity in the natural physiological range and high-efficiency activation and reduce TCR's cross-reactivity [32]. It is the milestone that catch bond is utilized to optimize TCR for immunotherapy, providing a new biophysical angle to revolutionize T-cell-mediated ACT. On the other hand, cancer-associated somatic mutations in pHLA-A2 were found to limit conformational extension in agonistic pHLA and impede TCR–pMHC catch bonds (Fig. 10.3A), which is also a promising target for the catch-bond engineering strategy [16].

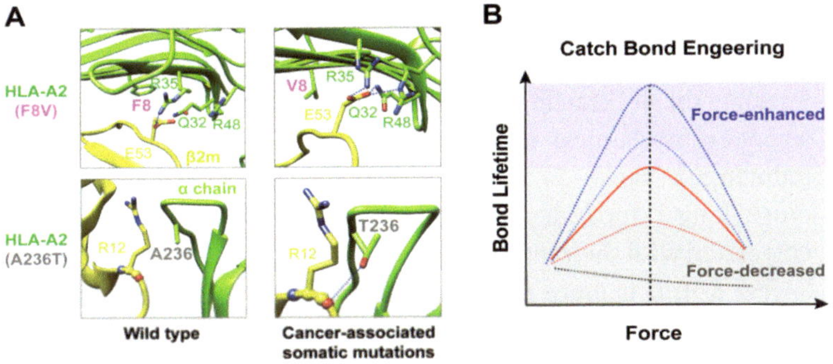

Figure 10.3 Catch-bond engineering between TCRs and pMHCs. (A) Cancer-associated somatic mutations (A236T and F8V in HLA-A2) induce new H-bonds between α and β2 m subunits. H-bonds are indicated as blue dashed lines. (B) Mechanical forces modulate the enhancement or reduction in the catch bond between TCRs and pMHCs.

10.3.3 Mechanical Force in Viral Invasion

COVID-19 pandemic has caused immeasurable damage worldwide, and new variants are constantly emerging, such as beta, gamma, delta, omicron, and omicron variants (https://www.gisaid.org) [74]. The initial step of viral invasion is that the spike protein (including S1 and S2 subunit) of SARS-CoV-2 (SARS2-S) recognizes host-cell receptors (targeting mainly ACE2 and also tyrosine-protein kinase receptor UFO (AXL), and immune system receptors, such as, toll-like receptors (TLR), C-lectin type receptors (CLR), neuropilin-1 (NRP1), and DPP4/CD26) [75–82]. The receptor-binding domain (RBD) in S1 subunit is the major domain for binding host receptor ACE2, while the S2 subunit forms fusion machinery to target host-cell plasma membrane after S1/S2 detachment. Both processes are potentially regulated by mechanical forces, and revealing the underlying mechanisms from the biomechanical angle would possibly provide novel strategies for preventing novel mutated SARS-CoV-2 strains that can evade therapeutic antibodies or vaccine protection [14, 83].

Studies have shown that mechanical forces generated by cell membrane bending boost ACE2-dependent SARS-CoV-2 invasion. First, in the viral recognition stage, mechanical force exerted on a single spike/ACE2 bond is approximately between 0 and 30 pN according to theoretical calculation [14]. Compared with SARS-S-RBD (~6% amino acid sequence difference compared to SARS2-S-RBD), the interactions of SARS2-S-RBD binding to ACE2 have stronger mechanical stability to resist stretching force through inducing a more pronounced catch–slip bond behavior with much longer force-dependent bond lifetimes, which is positively correlated with their infectivity (Fig. 10.4A) [14, 84]. Mechanistically, SARS2-S-RBD is more prone to adopt an open conformation under mechanical force loading, thereby promoting the formation of more H-bonds and hydrophobic interactions on the SARS2-S-RBD/ACE2 binding surface, compared to that in SARS-S-RBD/ACE2 binding. Second, in the fusion stage, mechanical force accelerates S1/S2 detachment to promote S2 structural rearrangement and fusion machinery formation. In this regard, our group for the first time revealed that mechanical force dramatically speeds up SARS2-S S1/S2 detachment by up to $\sim 10^3$ times faster than that in the force-free condition (Fig. 10.4B) [14].

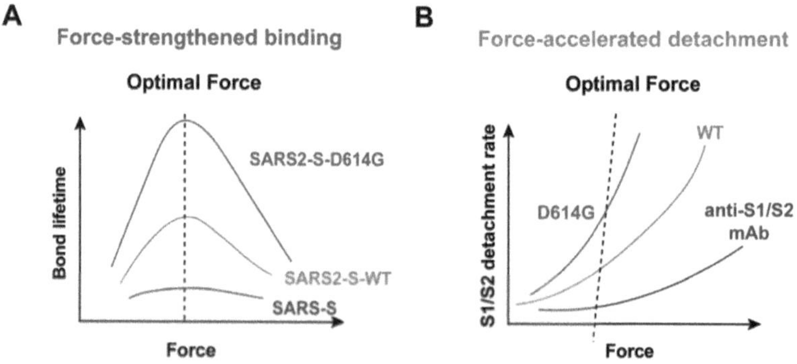

Figure 10.4 Mechanical forces strengthen the spike/ACE2 binding and accelerate S1/S2 detachment. (A) Schematics of the force-dependent bond lifetimes of the interactions of host ACE2 receptors interacting with SARS2-S-D614G (red), SARS2-S-WT (blue), and SARS-S (gray), respectively. (B) S1/S2 detachment rate versus force curves of SARS2-S-D614G (red), SARS2-S-WT (blue), and SARS2-S-WT in the presence of neutralizing antibody targeting S1/S2 (purple), respectively.

These mechanical regulations during viral invasion were further validated by the more infectious D614G mutation of SARS2-S. D614G mutation, located outside of the RBD of the S1 subunit, causes more fatality and is inherited in alpha, beta, delta, and recent omicron SARS-CoV-2 virus [85–90]. A recent study showed that D614G converts the S1 protein conformation (RBD and N-terminal domains of S1 subunit) to an ACE2-binding fusion-competent state and thus may increase viral infectivity [91]. Our group further found that mechanical force prolong the bond lifetimes of SARS2-S-D614G-ACE2 binding by nearly 4-fold and sped up the S1-S2 detachment by 35-fold compared to the wild-type SARS2-S-ACE2 binding (Fig. 10.4B) [14]. Thus, the more plausible molecular mechanism by which SARS2-S D614G mutation causes a higher infection rate from biophysical standpoint is that D614G mutation fosters stronger mechanical stability of the SARS2-S/ACE2 complex and better couples with mechanical force to induce much faster S1/S2 detachment.

The widely adopted treatment strategy to prevent viral invasion is to block the engagement of the spike-RBD and ACE2 interactions with neutralizing antibodies [92–95]. However, due to the high mutagenicity of SARS2-S1 (especially RBD), it is very likely that SARS2 evades neutralizing antibodies blockade through abolishing these mAb bindings with RBD or that SARS2 may adopt other entry pathways by binding with other host membrane proteins (e.g., AXL) to maintain viral infectivity, which are the intrinsic limitations for these RBD-blockade designs of neutralizing antibodies [96–102]. We speculate that the aforementioned mechanical regulatory mechanisms could provide another novel intervention strategy to prevent virus infection, that is, mechanically locking the S1 subunit to S2 subunit to impede their detachment (Fig. 10.5) [103–107]. In this way, revealing the underlying biophysical mechanisms of viral invasion, especially biomechanical activation of viral spikes, would potentially optimize the design of therapeutic antibodies.

Collectively, mechanical regulations on membrane receptors during cell adhesion, viral invasion, and immune defenses are important determinants for the precise recognition of different ligands, providing non-canonical strategies for designing novel neutralizing antibodies targeting SARS-CoV-2 and other infectious viruses and for engineering highly sensitive TCR-T in T-cell-based

ACT against tumors. Furthermore, we believe that revealing mechanical regulations on other immune receptors, as well as the popular immune checkpoint receptors (e.g., PD-1 and CTLA4), would also promote the optimization of immunotherapy strategies from a biomechanical angle to treat not only infectious diseases and tumors but also autoimmune diseases.

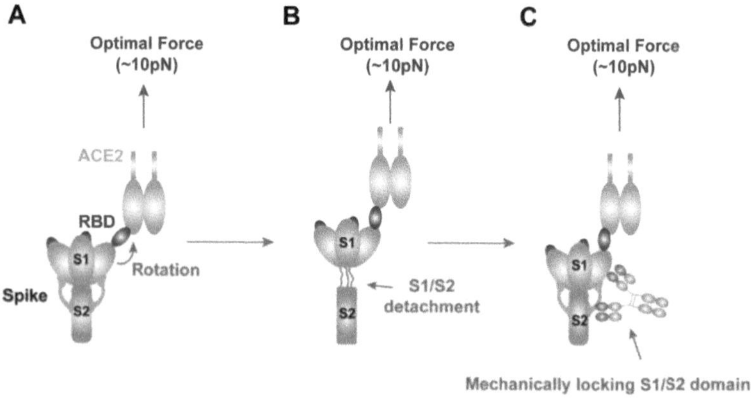

Figure 10.5 The dynamic structural model of forces regulating the spike/ACE2 binding and accelerating S1/S2 detachment, providing novel intervention strategy. (A–C) The dynamic structural model of force-regulated SARS2-S/ACE2 binding and conformational change in SARS2-S, force-induced S1/S2 detachment, and S1/S2-locking antibodies to impede SARS2-S/ACE2 binding.

10.3.4 Force Fluctuations

In addition, the force exerted on cell adhesion or receptor–ligand bonds in vivo potentially acts in a cyclically fluctuating manner and often repeats with multiple cycles, depending on cytoskeletal actin velocity at the immune synapse [53, 108]. During cell adhesion, the integrin-based focal adhesions (FAs) exhibit dynamic fluctuating traction through the movement of extracellular matrix cytoskeleton [109]. Kong et al. revealed that cyclical force could reinforce cell adhesion by prolonging the bond lifetimes of integrin and its ligand interactions, which will give a promising method for going further into the process of physiological fluctuations [6, 39]. TCR could exploit the cyclic force, which is exerted by actin movement, to significantly increase the T-cell signaling strength (Ca^{2+} flux), potentially providing a novel mechanical reinforcement possibility of T-cell triggering. When the cells are treated with Latrunculin A (LatA), which inhibits actin polymerization, the cyclical force could compensate the Ca^{2+} signals [53, 110]. But how TCR exploits force fluctuations to convert into biochemical signals is unknown. Mechanically induced dynamic catch bonds of TCR–pMHC interactions may be easier to buffer force fluctuations than slip bonds when the bonds experience fluctuating forces [111]. The detailed molecular mechanism needs to be further explored. At present, most studies are only based on the prediction of mathematical models [111]. Besides, the frequency and amplitude of force fluctuations play critical roles in determining receptor–ligand binding and membrane receptor functions [112, 113]. Different frequencies, amplitudes, and directions of dynamic force control different signal patterns and determine cell fates [3, 114, 115]. Therefore, gaining insights into how dynamic forces of different frequencies, amplitudes, and directions are functionally crucial to membrane receptors functions, worth for deeper exploration. We speculate these mechanisms could be shared by other mechanically sensing membrane receptors (e.g., integrins, NKG2D, and PD-1). Collectively, we anticipate that more experimental advances will enhance our understanding of the mechanical modulation of membrane receptors, which will provide more novel mechanisms of receptor–ligand interactions.

10.4 Spatial Confinement of Cell Plasma Membrane

In addition to detecting the effects of mechanical force, the spatial confinement by plasma membrane could also enforce significant impacts on the association and dissociation processes, especially their kinetic rates of receptor–ligand interactions and accordingly regulate membrane receptors functions [25].

The measurement of receptor–ligand interactions in a two-dimensional (2D) manner (e.g., micropipette assay), where the binding kinetics is detected based on two opposing cell membranes that reconstitute the physiological conditions of membrane proteins, is significantly different from those in three-dimensional (3D) manners (e.g., SPR) using purified ligands in soluble states

[19, 38, 116]. The existing regulatory mechanisms of how spatial confinement of plasma membrane regulates receptor–ligand interactions can be roughly divided into two categories. First, the 2D cell plasma membrane physically restricts the orientation of membrane receptors and/or ligands, affecting 2D binding affinities through changing association rates, such as interactions between human Fcγ receptor III (CD16, stimulatory receptor expressed mainly on NK cells as well as neutrophils, monocytes, macrophages, and T cells) or FcγRIIB (inhibitory receptor expressed on B lymphocytes) and their IgG ligands [2, 19, 21, 22, 26, 117, 118]. The changes in molecule orientation arise primarily from differences in the length or inclination of the TM regions of membrane receptors inside the membrane, altering their membrane-confined lateral mobility [119, 120]. Second, the cell plasma membrane influences the spatial diffusion or movement of surface molecules with long ectodomains (e.g., CD45), regulating the binding kinetics of receptor–ligand interactions at cell–cell interfaces [65, 121, 122]. When the receptor diffuses on the plasma membrane, both the distance and the local concentration of membrane receptors and/or ligands would change in the contact zone. The diffusion or movement within 2D cell plasma membrane can be driven by many possible mechanisms according to the kinetic segregation model [123], such as molecule length changes, spatial reorganization, or functional clustering. However, the mechanism by which molecular diffusion facilitates or prevents receptor–ligand interactions is still undefined. Collectively, the physical regulation of the spatial confinement of cell plasma membranes reveals a novel mechanism for the regulation of physical–chemical coupling on receptor–ligand interactions.

Recent studies found that single nucleotide polymorphisms (SNPs) in transmembrane regions of membrane receptors can allosterically regulate receptor–ligand interactions through coupling the TM–lipid bilayer biochemical interaction with the spatial confinement of cell plasma membrane. An SNP variant of FcγRIIB, I232T, in systemic lupus erythematosus (SLE), impedes its 2D binding affinities through association rates with its ligands (IgG1, IgG2, and IgG3) and suppresses the activation of immune cells [119]. Considering that the I232T variant is located in the transmembrane regions of the FcγRIIB receptor, and that the tilted transmembrane regions lead to a bent ectodomain conformation that impedes ligands association, SNP potentially couples with membrane confinement to affect the receptor–ligand recognition and signaling functions (Fig. 10.6). SNP mutations have also been identified in immune checkpoint molecules, such as LAG-3 (lymphocyte-activation gene 3 [124]); we speculate that this novel SNP–membrane confinement coupling mechanism would potentially inspire new strategies for the treatment of immune checkpoint blockade (ICB) therapy.

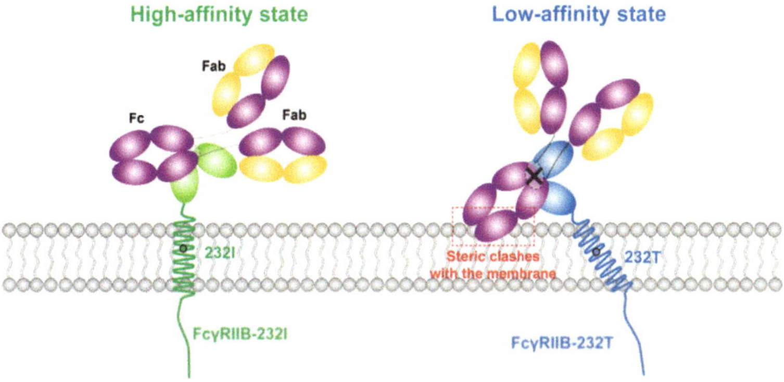

Figure 10.6 SNPs in the TM regions of membrane receptors exert allosteric regulatory effects on their binding affinity under membrane confinement. The binding affinity (*K*a) is defined by the ratio of k_{on}/k_{off}, reflecting receptor–ligand binding strength at equilibrium states; the association rates (k_{on}) characterize how fast the receptor–ligand bond forms, while the dissociation rates characterize how long the bond lasts (k_{off}). The SNP I232T impedes its two-dimensional binding affinities through association rates with its ligands (IgG1, IgG2, and IgG3) and suppresses immune cells activation.

10.5 Biophysical Regulation in Transmembrane Signaling

The triggering of the transmembrane signaling of membrane receptors also depends on the effects of mechanical cues. How cells convert biomechanical stimuli into intracellular biochemical signals (mechanotransduction) is a crucial yet unresolved question in mechanobiology. To directly observe membrane receptors-mediated mechanotransduction on living cells, fluorescence SMFS has been developed, which can concurrently analyze the receptor–ligand binding kinetics and intracellular signaling (e.g., calcium flux)

[15, 30]. Here, we take platelet glycoprotein Ib (GPIb) and TCRs as examples to illustrate recent advances.

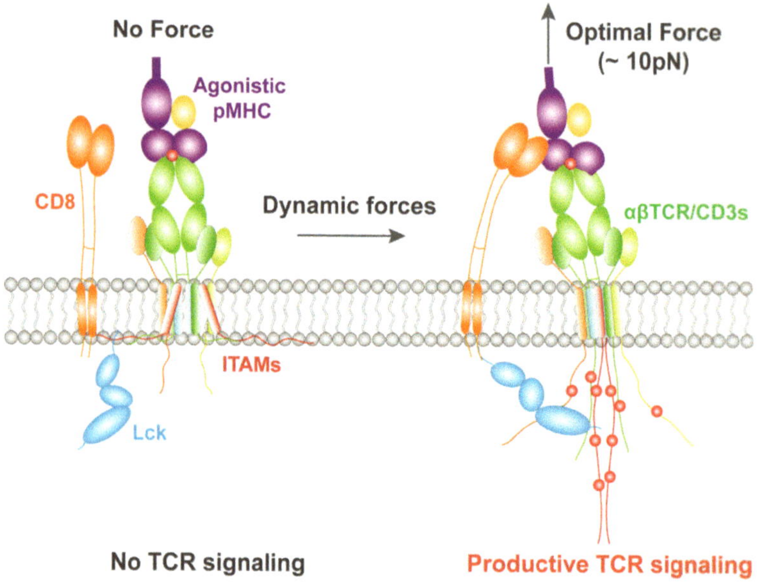

Figure 10.7 A dynamic model of mechanochemical coupling for receptor–ligand interactions and transmembrane signaling. When TCR recognizes agonistic pMHC, force induces conformational changes in pMHC and the extracellular domains of TCR, which sequentially propagates across the cell membrane to induce intracellular kinase activation (Lck) and structural changes in the cytoplasmic domains (CD3). Meanwhile, CD8-associated Lck binds to phosphorylated CD3 ITAMs inside the cell membrane and the TCR-CD3 complex outside the cell membrane, forming a positive feedback mechanical loop that amplifies productive TCR signaling.

The glycoprotein Ib (GPIb) membrane receptor on the surface of platelets, specifically one of its subunits (GPIba), recognizes the ligand of von Willebrand factor (VWF) on the surface of damaged arteries to counteract blood flow [115]. Ju et al. found that mechanical force induces the formation of catch bonds of VWF–GPIba interactions to slow their dissociation rates through the unfolding of the extracellular leucine-rich repeat domain (LRRD), which can concurrently enhance the strength of the calcium signal response in platelets aggregation [30]. More recently, the dual biomembrane force probe (dBFP) has been developed, which allows step-by-step receptor–ligand pre-stimulation, recognition, and signal transduction from one receptor to another on a single cell [40]. Chen et al. used the fluorescence dBFP, microfluidics, and cone-and-plate rheometry to initially control the ligand engagement of GPIbα via a mechano-signaling pathway, then sequentially stimulate $\alpha_{IIb}\beta_3$ on another platelet for further transition to the active state during integrin mechanical affinity maturation [5].

At least three models have been proposed to reveal the mechanotransduction mechanisms of T cells based on the aforementioned fluorescence SMFS (Fig. 10.7). First, force-induced conformational changes in the extracellular domains of membrane receptors (e.g., TCR) sequentially propagate across the cell membrane, inducing structural changes in the cytoplasmic domains and downstream signaling [52, 54, 55, 125]. In detail, when a TCR recognizes an agonistic pMHC, TCR's extracellular domains (e.g., FG loop) probably transmit forces to the TCR transmembrane (TM) regions that form a compact and precisely organized structure with the CD3 subunits ($\varepsilon\delta$, $\varepsilon\gamma$, $\zeta\zeta$) within the membrane [52, 55]. The juxtamembrane region of the CD3ζζ signaling module likely acts as a mechanical pivot connecting TCRα chain, thereby transmitting biochemical and biomechanical information across the cell membrane [126]. Second, the accumulation of dynamic catch bonds between TCR and pMHC triggers T-cell signaling [15]. By simultaneously measuring the TCR–pMHC force-dependent binding kinetics and binding-triggered intracellular signaling, Zhu et al. demonstrated that the accumulative bond lifetimes within the initial 60 seconds between TCR and agonist pMHC are the most relevant to intracellular signaling (e.g., Ca^{2+} flux), rather than the other parameters (e.g., binding affinities, peak bond lifetimes, average lifetimes, or longest lifetimes) [15]. On the contrary, for antagonistic pMHC ligands and TCR interactions, the accumulative bond lifetimes are very short, and the Ca^{2+} signals are not induced within the specific time window [15]. This indicates that the synchronization between extracellular receptor–ligand interactions and intracellular kinase activation is critical for triggering T cells. Third, mechanosensing through membrane receptors induces a mechanical feedback loop that affects T-cell activation [57]. Mechanically induced "dynamic catch" in trimolecular or multi-molecular interactions (e.g., TCR–pMHC–CD8 or TCR–pMHC–CD8–PD-1) regulates the linkage and cis-interactions of intracellular signaling molecules (e.g., linking CD8 and TCR–CD3 via lymphocyte-specific protein tyrosine kinase, Lck) to form a mechanotransduction loop, also known as inside-out signaling, further amplifying the ligand discriminative ability of membrane receptors [57]. In addition, PD-1 specifically blocks the mechanotransduction loop during T-cell antigen recognition [127]. However, the detailed mechanisms about how mechanical forces coordinate the mechanical signaling from these co-receptor, co-stimulatory, and co-inhibitory molecules to delicately tune the intracellular signaling cascades remain unclear.

Collectively, biomechanical parameters are key determinants of membrane receptors transmembrane signaling, providing novel physical transmission mechanisms for receptor's ligand recognition and signal transmission.

10.6 Protocols

10.6.1 Single-Molecule Biomembrane Force Probe

In this section, we introduce the method of single-molecule BFP, mainly containing force-clamp assay for characterizing force-dependent dissociation rates under constant forces or cyclic forces, adhesion frequency assay for binding affinities and force-free association and dissociation rates and force-dependent downstream signaling.

10.6.1.1 Ultra-stable force-clamp assay

A BFP uses red cells as mechanical sensors, which can optically detect the deformation of the red cell and observe the magnitude of the force acting on the molecules in real time by a high-speed camera [7, 15]. Within its force measurement range (e.g., 0 ~ 30 pN), the spring constant of red blood cells can be approximately regarded as obeying Hooke's law (Equation 10.1),

$$F = kx \tag{10.1}$$

and its stiffness k can be estimated by Equation 10.2:

$$k \approx \frac{\pi R\mathrm{p}\Delta\mathrm{p}}{\left(1 - \frac{R_\mathrm{p}}{R_0}\right)\mathrm{Ln}\left[\frac{4R_0^2}{R\mathrm{p}R_\mathrm{c}}\right]} \tag{10.2}$$

where R_p is the inner diameter of the micropipette for absorbing the red cell, Δ_p is the negative pressure for absorbing the red cell, R_0 is the radius of the red cell, and R_C is the radius of the connecting part between the red cell and the globules, and the red cell can be absorbed by changing the radius. The negative pressure can adjust the stiffness of the red blood cells at any time [36].

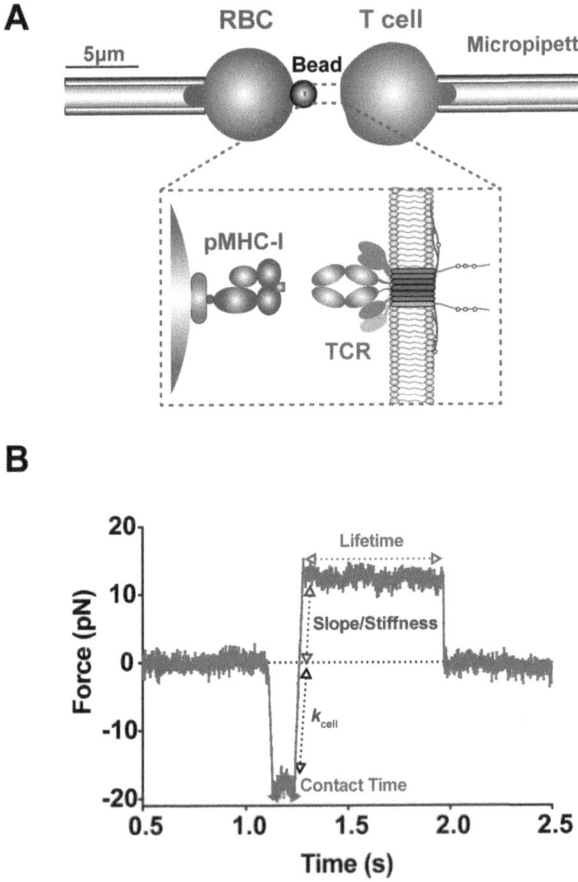

Figure 10.8 (A) Schematic diagram of the biomembrane force probe (BFP) experiment. (B) Force versus times curve of receptor–ligand interactions (e.g., TCR–pMHC) based on the force-clamp mode.

In brief, an SA bead coated with biotinylated proteins or pMHC was attached to the apex of RBCs, which was held by the micropipette on one side. The RBC functioned as an ultrasensitive biomechanical probe, the deformation of which was tracked in real time by a high-speed camera (Prosilica GE680, Allied Vision, Germany) and transformed into mechanical forces according to its adjustable spring constant (0.3–3 pN/nm) (Fig. 10.8A). A target cell was held by an opposing micropipette and was controlled by the LABVIEW program to contact the ultrasensitive biomechanical probe for 0.1 s (it depends on your need) and then retract until the stretching force reached preset clamp force (Fig. 10.8A,B). The stretching force on the single-molecule bond was precisely clamped until the bond ruptured. The duration sustaining the clamping force was recorded as a bond lifetime. Through collecting bond lifetimes under different force values, the force-regulated dissociation kinetics was dissected (Fig. 10.8B). Note that the adhesion frequency of each ligand–receptor pair was controlled less than 20% by adjusting the ligand density to ensure that most recorded adhesive events were mediated by single-molecule bonds. The experiments were performed in a cell chamber filled with complete Dulbecco's Modified Eagle Medium (DMEM, Shanghai Basal Media Technology Co., Ltd.) supplemented with 0.5% BSA.

10.6.1.2 Cyclic force-clamp assay

The cyclic force-clamp assay was similar to the ultra-stable force-clamp assay except that oscillating or cyclic forces (e.g., 0, 1, 3 cycles) were applied before the force-clamping stages (Fig. 10.9) [6, 39]. For example, the target cell was program controlled to overload the force to 15 pN and immediately unload to 5 pN to complete a forced cycle. The force was clamped at 10 pN after the preset cycles were finished, the duration of which was recorded as a bond lifetime after cyclic force loading.

10.6.1.3 Fluorescence biomembrane force probe

A fluorescence biomembrane force probe (fBFP) adds a fluorescent module based on the previous ultra-stable force-clamp mode (Fig. 10.10) [5, 15, 30]. Briefly, the biomechanical probe and the target cell was aligned in a temperature-controlled cell chamber filled with the imaging buffer (e.g., Hepes buffer) and observed under an inverted microscope (Nikon TiE, Nikon) through two cameras. One camera (GC1290, Procilica, MA) captured real-time images at a video rate (30 frames per second, fps), while the other (GE680,

Prosilica, MA) was able to record at a higher speed (1600 fps) as the images were limited to 100 pixels by 30 pixels rectangular region of interest across the contact interface between the bead and the RBCs. A customized image analysis LabView (National Instrument, TX) program tracked the bead position with a 3 nm (standard deviation, SD) displacement precision in real time.

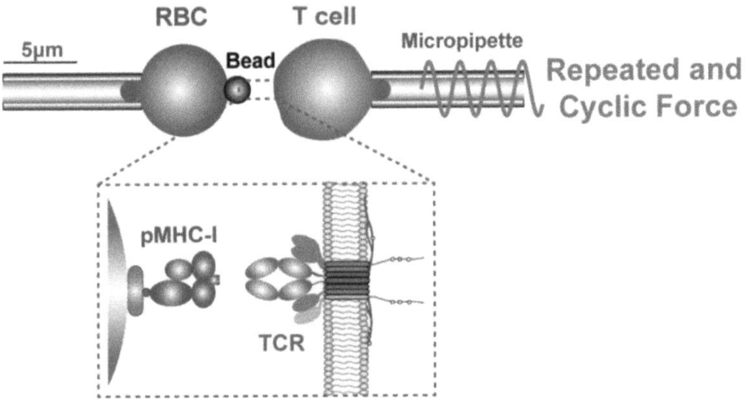

Figure 10.9 Schematic illustration of the interaction between receptors and ligands (e.g., TCR–pMHC) with different periodic cycles based on the conventional force-clamp mode.

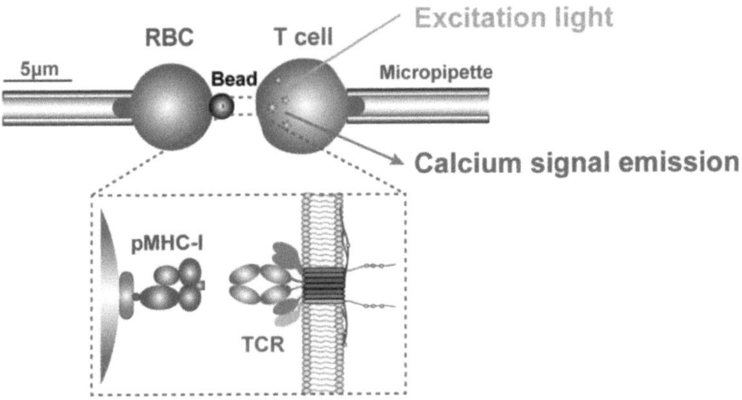

Figure 10.10 Conventional force-clamp assay combined with fluorescence module to dissect mechanochemical coupling of T-cell activation. The ultrasensitive biomechanical probe is coated with pMHCs to contact a TCR-expressing T cell. Ca^{2+} flux can be simultaneously detected by the fluorescence module when mechanical force is exerted on TCR/pMHC.

10.6.1.4 Dual biomembrane force probe (dBFP)

The dBFP is an upgrade from the previous conventional BFP [40]. The major improvement was made around the experiment stage, which consists of two components from the left to the right (Fig. 10.11): (1) The two micropipette sets in the left, Probe 1 and Probe 2, are anchored to a precision translational stage through a double-pipette holder set (HD-21, Narishgie), which can aspirate a functionalized biomechanical probe. (2) The other target micropipettes are placed on the right, which were driven by a computer program for nanometer-precision movement. By adjusting the reservoir height of a manometer, one can control the inner pressure of the micropipettes to aspirate or blow away a bead or a cell during experiments. The reaction chamber in the middle is similar to the previous experimental buffer, in which the micropipettes were inserted and beads and cells were injected.

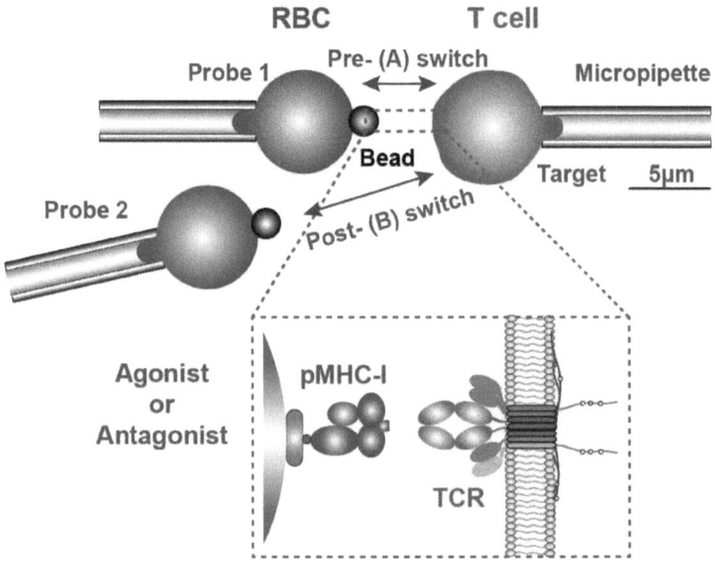

Figure 10.11 Schematic illustration of a dBFP setup between interactions of the target cell with two immobilized ligands (e.g., agonistic pMHC or antagonistic pMHC, respectively). Two beads functionalized with distinct proteins were attached to the apexes of two RBCs, respectively, aspirated by two micropipettes to form Probes 1 and Probes 2. TCR-expressing T cells are aligned with one probe at a time to sequentially engage two ligands, designated as the pre- (A) and post- (B) switch positions, respectively. Scale bars are 5 μm.

10.6.1.5 Micropipette adhesion assay

The fresh RBCs were biotinylated and then coated with streptavidin (SA), which were incubated with different concentrations of the biotinylated proteins aspirated by one micropipette [26, 38, 128]. A target cell or bead was aspirated by an apposed micropipette driven by a nanoscale PZT brake to impact the functional RBC, and then contacted and withdrew with a given time. When the RBCs were pulled and deformed, it is recorded as an "adhesion" event. If not, it is recorded as a "no adhesion" event. The experimental cycle was repeated 50 times at least and tested using 3–5 cell pairs to calculate the adhesion frequency (P_a). Change the contact time (t_c) and control the range of frequency at 20~80% and finally get the P_a versus t_c curve (Fig. 10.12) based on the adhesion frequency model:

$$P_a = 1 - \exp\{-m_r m_l A_c K_a [1 - \exp(k_{off} t_c)]\} \qquad (10.3)$$

m_r and m_1 are the densities of the receptor and ligand, respectively, which can be labeled and quantified by fluorescent antibody by flow cytometry corresponding to the standard fluorescence intensity bead (BD Biosciences). A_c is the contact area, K_a is the receptor–ligand binding affinity, and k_{off} is the force-free dissociation rate.

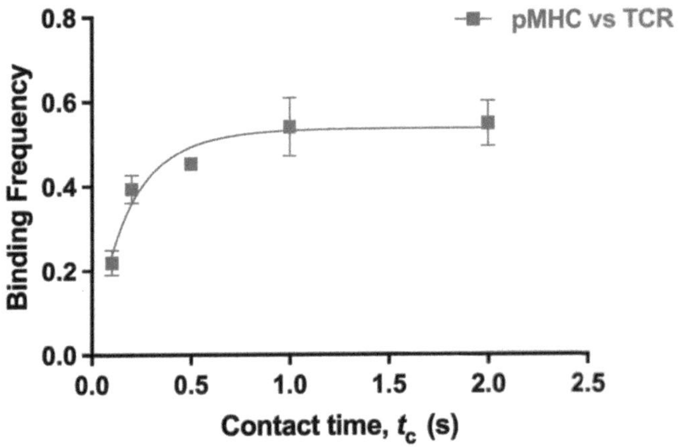

Figure 10.12 Example data from micropipette adhesion frequency experiments showing the fitting curve for the binding frequency of TCR–pMHC interactions under different t_c.

10.7 Conclusion

Over the past decade, a conceptual framework of physical–chemical regulation of membrane receptor dynamics by SMFS has emerged to address fundamental scientific questions of cell adhesion, viral invasion, and immune defense. These findings provide novel mechanisms by which mechanical cues modulate the conformation of membrane receptors or ligands, determining the receptor–ligand binding kinetics and signaling transduction. These dynamic mechanisms will very likely inspire non-canonical thoughts for developing novel clinically relevant immunotherapeutic treatments (e.g., CAR-T or TCR-T).

References

1. Zhu C: Kinetics and mechanics of cell adhesion. *J Biomech* **33**, 23–33 (2000).
2. Chen W, Zhu C: Mechanical regulation of T-cell functions. *Immunol Rev* **256**, 160–176 (2013).
3. Zhu C, Chen W, Lou J, Rittase W, and Li K: Mechanosensing through immunoreceptors. *Nat Immunol* **20**, 1269–1278 (2019).
4. Liu B, Chen W, and Zhu C: Molecular force spectroscopy on cells. *Annu Rev Phys Chem* **66**, 427–451 (2015).
5. Chen Y, Ju LA, Zhou F, Liao J, Xue L, Su QP, Jin D, Yuan Y, Lu H, Jackson SP, and Zhu C: An integrin αIIbβ3 intermediate affinity state mediates biomechanical platelet aggregation. *Nat Mater* **18**, 760–769 (2019).
6. Kong F, Li Z, Parks WM, Dumbauld DW, García AJ, Mould AP, Humphries MJ, and Zhu C: Cyclic mechanical reinforcement of integrin–ligand interactions. *Molecular Cell* **49**, 1060–1068 (2013).
7. Chen W, Lou J, Evans EA, and Zhu C: Observing force-regulated conformational changes and ligand dissociation from a single integrin on cells. *J Cell Biol* **199**, 497–512 (2012).
8. Ma VP-Y, Hu Y, Kellner AV, Brockman JM, Velusamy A, Blanchard AT, Evavold BD, Alon R, and Salaita K: The magnitude of LFA-1/ICAM-1 forces fine-tune TCR-triggered T cell activation. *Sci Adv* **8**, eabg4485 (2022).
9. Yago T, Wu J, Wey CD, Klopocki AG, Zhu C, and McEver RP: Catch bonds govern adhesion through L-selectin at threshold shear. *J Cell Biology* **166**, 913–923 (2004).

10. Sarangapani KK, Marshall BT, McEver RP, and Zhu C: Molecular stiffness of selectins. *J Biol Chem* **286**, 9567–9576 (2011).

11. Evans E, Leung A, Heinrich V, and Zhu C: Mechanical switching and coupling between two dissociation pathways in a P-selectin adhesion bond. *Proc National Acad Sci* **101**, 11281–11286 (2004).

12. Chen W, Lou J, and Zhu C: Forcing switch from short- to intermediate- and long-lived states of the ÃŽÂ±A domain generates LFA-1/ICAM-1 catch bonds. *J Biol Chem* **285**, 35967–35978 (2010).

13. Kinoshita K, Leung A, Simon S, and Evans E: Long-lived, high-strength states of ICAM-1 bonds to β 2 integrin, II: Lifetimes of LFA-1 bonds under force in leukocyte signaling. *Biophys J* **98**, 1467–1475 (2010).

14. Hu W, Zhang Y, Fei P, Zhang T, Yao D, Gao Y, Liu J, Chen H, Lu Q, Mudianto T, Zhang X, Xiao C, Ye Y, Sun Q, Zhang J, Xie Q, Wang P-H, Wang J, Li Z, Lou J, and Chen W: Mechanical activation of spike fosters SARS-CoV-2 viral infection. *Cell Res* 1–14 (2021).

15. Liu B, Chen W, Evavold BD, and Zhu C: Accumulation of dynamic catch bonds between TCR and agonist peptide-MHC triggers T cell signaling. *Cell* **157**, 357–368 (2014).

16. Wu P, Zhang T, Liu B, Fei P, Cui L, Qin R, Zhu H, Yao D, Martinez RJ, Hu W, An C, Zhang Y, Liu J, Shi J, Fan J, Yin W, Sun J, Zhou C, Zeng X, Xu C, Wang J, Evavold BD, Zhu C, Chen W, and Lou J: Mechano-regulation of peptide-MHC class I conformations determines TCR antigen recognition. *Mol Cell* **73**, 1015–1027.e7 (2019).

17. Fan J, Shi J, Zhang Y, Liu J, An C, Zhu H, Wu P, Hu W, Qin R, Yao D, Shou X, Xu Y, Tong Z, Wen X, Xu J, Zhang J, Fang W, Lou J, Yin W, and Chen W: NKG2D discriminates diverse ligands through selectively mechano-regulated ligand conformational changes. *Embo J* e107739 (2021).

18. Feng Y, Reinherz EL, and Lang MJ: αβ T cell receptor mechanosensing forces out serial engagement. *Trends Immunol* **39**, 596–609 (2018).

19. Chesla SE, Selvaraj P, and Zhu C: Measuring two-dimensional receptor–ligand binding kinetics by micropipette. *Biophys J* **75**, 1553–1572 (1998).

20. Huang J, Chen J, Chesla SE, Yago T, Mehta P, McEver RP, Zhu C, and Long M: Quantifying the effects of molecular orientation and length on two-dimensional receptor-ligand binding kinetics. *J Biol Chem* **279**, 44915–44923 (2004).

21. Chesla SE, Li P, Nagarajan S, Selvaraj P, and Zhu C: The membrane anchor influences ligand binding two-dimensional kinetic rates and three-dimensional affinity of FcγRIII (CD16). *J Biol Chem* **275**, 10235–10246 (2000).

22. Chen W, Zarnitsyna VI, Sarangapani KK, Huang J, and Zhu C: Measuring receptor–ligand binding kinetics on cell surfaces: From adhesion frequency to thermal fluctuation methods. *Cell Mol Bioeng* **1**, 276 (2008).

23. Charras G and Yap AS: Tensile forces and mechanotransduction at cell–cell junctions. *Curr Biol* **28**, R445–R457 (2018).

24. Hong J, Murugesan S, Betzig E, and Hammer JA: Contractile actomyosin arcs promote the activation of primary mouse T cells in a ligand-dependent manner. *Plos One* **12**, e0183174 (2017).

25. Zhang T, Hu W, and Chen W: Plasma membrane integrates biophysical and biochemical regulation to trigger immune receptor functions. *Front Immunol* **12**, 613185 (2021).

26. Huang J, Zarnitsyna VI, Liu B, Edwards LJ, Jiang N, Evavold BD, and Zhu C: The kinetics of two-dimensional TCR and pMHC interactions determine T-cell responsiveness. *Nature* **464**, 932–936 (2010).

27. Dupré L, Boztug K, and Pfajfer L: Actin dynamics at the T cell synapse as revealed by immune-related actinopathies. *Frontiers Cell Dev Biol* **9**, 665519 (2021).

28. Blumenthal D and Burkhardt JK: Multiple actin networks coordinate mechanotransduction at the immunological synapse. *J Cell Biol* **219**, 301–312 (2020).

29. Ma Z, Discher DE, and Finkel TH: Mechanical force in T cell receptor signal initiation. *Front Immunol* **3**, 217 (2012).

30. Ju L, Chen Y, Xue L, Du X, and Zhu C: Cooperative unfolding of distinctive mechanoreceptor domains transduces force into signals. *Elife* **5**, e15447 (2016).

31. Fordyce P, Feng Y, Zhao X, White A, and Garcia K: Structure-activity mapping of the peptide- and force-dependent landscape of T-cell activation. *BioRxiv* [Epub ahead of print], doi: 10.21203/rs.3.rs-636850/v1 (2021).

32. Zhao X, Kolawole EM, Chan W, Feng Y, Yang X, Gee MH, Jude KM, Sibener LV, Fordyce PM, Germain RN, Evavold BD, and Garcia KC: Tuning T cell receptor sensitivity through catch bond engineering. *Science* **376**, (2022).

33. Evans E, Heinrich V, Leung A, and Kinoshita K: Nano- to microscale dynamics of P-selectin detachment from leukocyte interfaces. I. Membrane separation from the cytoskeleton. *Biophys J* **88**, 2288–2298 (2005).

34. Merkel R, Nassoy P, Leung A, Ritchie K, and Evans E: Energy landscapes of receptor–ligand bonds explored with dynamic force spectroscopy. *Nature* **397**, 50–53 (1999).

35. Evans E, Kinoshita K, Simon S, and Leung A: Long-lived, high-strength states of ICAM-1 bonds to β 2 integrin, I: Lifetimes of bonds to recombinant α L β 2 under force. *Biophys J* **98**, 1458–1466 (2010).

36. An C, Hu W, and Chen W: Ultra-stable biomembrane force probe for accurately determining slow dissociation kinetics of PD-1 blockade antibodies on single living cells. *Nano Lett* 1–8 (2020).

37. Chenyi A, Wei C, China D of CB and D of C of the SAH Zhejiang University School of Medicine, Zhejiang University, Hangzhou 310058, China KL for BE of M of E and SKL for MOI Zhejiang University, Hangzhou 310058, China CIC for D and T of ID Zhejiang University, Hangzhou 310058: Multiplexed single-molecule force spectroscopy for dissecting biophysical regulation of membrane receptors functions on live cells. *Biophysics Rep* **7**, 377–383 (2021).

38. Liu B, Zhong S, Malecek K, Johnson LA, Rosenberg SA, and Krogsgaard CZM: 2D TCR–pMHC–CD8 kinetics determines T-cell responses in a self-antigen-specific TCR system. *Eur J Immunol* 1–12 (2013).

39. Li Z, Kong F, and Zhu C: A model for cyclic mechanical reinforcement. *Sci Rep-uk* **6**, 35954 (2016).

40. Ju L, Chen Y, Li K, Yuan Z, Liu B, Jackson SP, and Zhu C: Dual biomembrane force probe enables single-cell mechanical analysis of signal crosstalk between multiple molecular species. *Sci Rep* 1–12 (2017).

41. Rio A del, Perez-Jimenez R, Liu R, Roca-Cusachs P, Fernandez JM, and Sheetz MP: Stretching single talin rod molecules activates vinculin binding. *Science* **323**, 638–641 (2009).

42. Yao M, Qiu W, Liu R, Efremov AK, Cong P, Seddiki R, Payre M, Lim CT, Ladoux B, Mège R-M, and Yan J: Force-dependent conformational switch of α-catenin controls vinculin binding. *Nat Commun* **5**, 4525 (2014).

43. Yao M, Goult BT, Chen H, Cong P, Sheetz MP, and Yan J: Mechanical activation of vinculin binding to talin locks talin in an unfolded conformation. *Sci Rep-uk* **4**, 4610 (2014).

44. Yao M, Goult BT, Klapholz B, Hu X, Toseland CP, Guo Y, Cong P, Sheetz MP, and Yan J: The mechanical response of talin. *Nat Commun* **7**, 11966 (2016).

45. Guo Q, He Y, and Lu HP: Manipulating and probing enzymatic conformational fluctuations and enzyme–substrate interactions by

single-molecule FRET-magnetic tweezers microscopy. *Phys Chem Chem Phys* **16**, 13052–13058 (2014).

46. Neuman KC and Nagy A: Single-molecule force spectroscopy: Optical tweezers, magnetic tweezers and atomic force microscopy. *Nat Methods* **5**, 491–505 (2008).

47. Chen H, Fu H, Zhu X, Cong P, Nakamura F, and Yan J: Improved high-force magnetic tweezers for stretching and refolding of proteins and short DNA. *Biophys J* **100**, 517–523 (2011).

48. Li W, Chen P, Yu J, Dong L, Liang D, Feng J, Yan J, Wang P-Y, Li Q, Zhang Z, Li M, and Li G: FACT remodels the tetranucleosomal unit of chromatin fibers for gene transcription. *Mol Cell* **64**, 120–133 (2016).

49. Marshall BT, Long M, Piper JW, Yago T, McEver RP, and Zhu C: Direct observation of catch bonds involving cell-adhesion molecules. *Nature* **423**, 190–193 (2003).

50. Neuman KC, Lionnet T, and Allemand J-F: Single-molecule micromanipulation techniques. *Annu Rev Mater Res* **37**, 33–67 (2007).

51. Dufrêne YF, Evans E, Engel A, Helenius J, Gaub HE, and Müller DJ: Five challenges to bringing single-molecule force spectroscopy into living cells. *Nat Methods* **8**, 123–127 (2011).

52. Das DK, Feng Y, Mallis RJ, Li X, Keskin DB, Hussey RE, Brady SK, Wang J, Wagner G, Reinherz EL, and Lang MJ: Force-dependent transition in the T-cell receptor β-subunit allosterically regulates peptide discrimination and pMHC bond lifetime. *Proc Nat Acad Sci USA* **112**, 1517–1522 (2015).

53. Hu KH and Butte MJ: T cell activation requires force generation. *J Cell Biol* 1–11 (2016).

54. Feng Y, Brazin KN, Kobayashi E, Mallis RJ, Reinherz EL, and Lang MJ: Mechanosensing drives acuity of αβ T-cell recognition. *Proc Nat Acad Sci USA* **114**, E8204–E8213 (2017).

55. Hwang W, Mallis RJ, Lang MJ, and Reinherz EL: The αβTCR mechanosensor exploits dynamic ectodomain allostery to optimize its ligand recognition site. *Proc Nat Acad Sci USA* **117**, 21336–21345 (2020).

56. Li H, Zhang C, Hu Y, Liu P, Sun F, Chen W, Zhang X, Ma J, Wang W, Wang L, Wu P, and Liu Z: A reversible shearing DNA probe for visualizing mechanically strong receptors in living cells. *Nat Cell Biol* 1–24 (2021).

57. Hong J, Ge C, Jothikumar P, Yuan Z, Liu B, Bai K, Li K, Rittase W, Shinzawa M, Zhang Y, Palin A, Love P, Yu X, Salaita K, Evavold BD, Singer A, and

Zhu C: A TCR mechanotransduction signaling loop induces negative selection in the thymus. *Nat Immunol* **19**, 1379–1390 (2018).

58. Zhang Y, Ge C, Zhu C, and Salaita K: DNA-based digital tension probes reveal integrin forces during early cell adhesion. *Nat Commun* **5**, 1–10 (2014).
59. Liu Y, Blanchfield L, Ma VP-Y, Andargachew R, Galior K, Liu Z, Evavold B, and Salaita K: DNA-based nanoparticle tension sensors reveal that T-cell receptors transmit defined pN forces to their antigens for enhanced fidelity. *Proc Nat Acad Sci USA* **113**, 5610–5615 (2016).
60. Gao H, Shi W, and Freund LB: Mechanics of receptor-mediated endocytosis. *Proc National Acad Sci USA* **102**, 9469–9474 (2005).
61. Hynes RO: Integrins bidirectional, allosteric signaling machines. *Cell* **110**, 673–687 (2002).
62. Goel HL, Breen M, Zhang J, Das I, Aznavoorian-Cheshire S, Greenberg NM, Elgavish A, and Languino LR: β1A integrin expression is required for type 1 insulin-like growth factor receptor mitogenic and transforming activities and localization to focal contacts. *Cancer Res* **65**, 6692–6700 (2005).
63. Chen W, Lou J, Hsin J, Schulten K, Harvey SC, and Zhu C: Molecular dynamics simulations of forced unbending of integrin αVβ3. *Plos Comput Biol* **7**, e1001086 (2011).
64. Lou J and Zhu C: Flow induces loop-to-β-hairpin transition on the β-switch of platelet glycoprotein Ibα. *Proc Nat Acad Sci USA* **105**, 13847–13852 (2008).
65. Sibener LV, Fernandes RA, Kolawole EM, Carbone CB, Liu F, McAffee D, Birnbaum ME, Yang X, Su LF, Yu W, Dong S, Gee MH, Jude KM, Davis MM, Groves JT, Goddard WA, Heath JR, Evavold BD, Vale RD, and Garcia KC: Isolation of a structural mechanism for uncoupling T cell receptor signaling from peptide-MHC binding. *Cell* **174**, 672–687.e27 (2018).
66. Brazin KN, Mallis RJ, Boeszoermenyi A, Feng Y, Yoshizawa A, Reche PA, Kaur P, Bi K, Hussey RE, Duke-Cohan JS, Song L, Wagner G, Arthanari H, Lang MJ, and Reinherz EL: The T cell antigen receptor α transmembrane domain coordinates triggering through regulation of bilayer immersion and CD3 subunit associations. *Immunity* **49**, 829–841.e6 (2018).
67. Billadeau DD, Upshaw JL, Schoon RA, Dick CJ, and Leibson PJ: NKG2D-DAP10 triggers human NK cell–mediated killing via a Syk-independent regulatory pathway. *Nat Immunol* **4**, 557–564 (2003).

68. Raulet DH: Roles of the NKG2D immunoreceptor and its ligands. *Nat Rev Immunol* **3**, 781–790 (2003).

69. Raulet DH, Gasser S, Gowen BG, Deng W, and Jung H: Regulation of ligands for the NKG2D activating receptor. *Immunology* **31**, 413–441 (2013).

70. Andrade LF de, Tay RE, Pan D, Luoma AM, Ito Y, Badrinath S, Tsoucas D, Franz B, May KF, Harvey CJ, Kobold S, Pyrdol JW, Yoon C, Yuan G-C, Hodi FS, Dranoff G, and Wucherpfennig KW: Antibody-mediated inhibition of MICA and MICB shedding promotes NK cell–driven tumor immunity. *Science* **359**, 1537–1542 (2018).

71. Holler PD, Chlewicki LK, and Kranz DM: TCRs with high affinity for foreign pMHC show self-reactivity. *Nat Immunol* **4**, 55–62 (2002).

72. Chlewicki LK, Holler PD, Monti BC, Clutter MR, and Kranz DM: High-affinity, peptide-specific T cell receptors can be generated by mutations in CDR1, CDR2 or CDR3. *J Mol Biol* **346**, 223–239 (2005).

73. Cameron BJ, Gerry AB, Dukes J, Harper JV, Kannan V, Bianchi FC, Grand F, Brewer JE, Gupta M, Plesa G, Bossi G, Vuidepot A, Powlesland AS, Legg A, Adams KJ, Bennett AD, Pumphrey NJ, Williams DD, Binder-Scholl G, Kulikovskaya I, Levine BL, Riley JL, Varela-Rohena A, Stadtmauer EA, Rapoport AP, Linette GP, June CH, Hassan NJ, Kalos M, and Jakobsen BK: Identification of a titin-derived HLA-A1–presented peptide as a cross-reactive target for engineered MAGE A3–directed T cells. *Sci Transl Med* **5**, 197ra103 (2013).

74. Andreakos E, Abel L, Vinh DC, Kaja E, Drolet BA, Zhang Q, O'Farrelly C, Novelli G, Rodríguez-Gallego C, Haerynck F, Prando C, Pujol A, Effort CHG, Bastard P, Biggs CM, Bigio B, Boisson B, Bolze A, Bondarenko A, Brodin P, Chakravorty S, Christodoulou J, Cobat A, Condino-Neto A, Constantinescu SN, Feldman HB, Fellay J, Flores C, Halwani R, Jouanguy E, Lau Y-L, Meyts I, Mogensen TH, Okada S, Okamoto K, Ozcelik T, Pan-Hammarström Q, Diego RP de, Planas AM, Puel A, Quintana-Murci L, Renia L, Resnick I, Sediva A, Shcherbina A, Slaby O, Tancevski I, Turvey SE, Uddin KMF, Beek D van de, Zatz M, Zawadzki P, Zhang S-Y, Su HC, Casanova J-L, and Spaan AN: A global effort to dissect the human genetic basis of resistance to SARS-CoV-2 infection. *Nat Immunol* **23**, 159–164 (2022).

75. Hoffmann M, Kleine-Weber H, Schroeder S, Krüger N, Herrler T, Erichsen S, Schiergens TS, Herrler G, Wu N-H, Nitsche A, Müller MA, Drosten C, and Pöhlmann S: SARS-CoV-2 cell entry depends on ACE2 and TMPRSS2 and is blocked by a clinically proven protease inhibitor. *Cell* **181**, 271–280.e8 (2020).

76. Choudhury A and Mukherjee S: In silico studies on the comparative characterization of the interactions of SARS-CoV-2 spike glycoprotein with ACE-2 receptor homologs and human TLRs. *J Med Virol* **92**, 10.1002/jmv.25987 (2020).

77. Gadanec LK, McSweeney KR, Qaradakhi T, Ali B, Zulli A, and Apostolopoulos V: Can SARS-CoV-2 virus use multiple receptors to enter host cells? *Int J Mol Sci* **22**, 992 (2021).

78. Solerte SB, Sabatino AD, Galli M, and Fiorina P: Dipeptidyl peptidase-4 (DPP4) inhibition in COVID-19. *Acta Diabetol* **57**, 779–783 (2020).

79. Daly JL, Simonetti B, Klein K, Chen K-E, Williamson MK, Antón-Plágaro C, Shoemark DK, Simón-Gracia L, Bauer M, Hollandi R, Greber UF, Horvath P, Sessions RB, Helenius A, Hiscox JA, Teesalu T, Matthews DA, Davidson AD, Collins BM, Cullen PJ, and Yamauchi Y: Neuropilin-1 is a host factor for SARS-CoV-2 infection. *Science* **370**, 861–865 (2020).

80. Huang X, Ye Q, Chen M, Li A, Mi W, Fang Y, Zaytseva YY, O'Connor KL, Kooi CWV, Liu S, and She Q-B: N-glycosylation-defective splice variants of neuropilin-1 promote metastasis by activating endosomal signals. *Nat Commun* **10**, 3708 (2019).

81. Wang S, Qiu Z, Hou Y, Deng X, Xu W, Zheng T, Wu P, Xie S, Bian W, Zhang C, Sun Z, Liu K, Shan C, Lin A, Jiang S, Xie Y, Zhou Q, Lu L, Huang J, and Li X: AXL is a candidate receptor for SARS-CoV-2 that promotes infection of pulmonary and bronchial epithelial cells. *Cell Res* **31**, 126–140 (2021).

82. Diamond MS and Kanneganti T-D: Innate immunity: The first line of defense against SARS-CoV-2. *Nat Immunol* **23**, 165–176 (2022).

83. Wang J: Force boosts SARS-CoV-2 invasion. *Nat Cell Biol* **23**, 1051–1051 (2021).

84. Gorbalenya AE, Baker SC, Baric RS, Groot RJ de, Drosten C, Gulyaeva AA, Haagmans BL, Lauber C, Leontovich AM, Neuman BW, Penzar D, Perlman S, Poon LLM, Samborskiy D, Sidorov IA, Sola I, and Ziebuhr J: Severe acute respiratory syndrome-related coronavirus: The species and its viruses—a statement of the Coronavirus Study Group. *Biorxiv* 2020.02.07.937862 (2020).

85. Volz E, Hill V, McCrone JT, Price A, Jorgensen D, Robertson DL, Thomson EC, Rambaut A, and Connor TR: Evaluating the effects of SARS-CoV-2 spike mutation D614G on transmissibility and pathogenicity. *Cell* **184**, 64–75.e11 (2021).

86. Hou YJ, Chiba S, Halfmann P, Ehre C, Kuroda M, DinnonIII KH, Leist SR, Schäfer A, Nakajima N, Takahashi K, Lee RE, Mascenik TM, Graham

R, Edwards CE, Tse LV, Okuda K, Markmann AJ, Bartelt L, Silva A de, Margolis DM, Boucher RC, Randell SH, Suzuki T, Gralinski LE, Kawaoka Y, and Baric RS: SARS-CoV-2 D614G variant exhibits efficient replication ex vivo and transmission in vivo. *Science* **370**, 1464–1468 (2020).

87. Zhou B, Thao TTN, Hoffmann D, Taddeo A, Ebert N, Labroussaa F, Pohlmann A, King J, Steiner S, Kelly JN, Portmann J, Halwe NJ, Ulrich L, Trüeb BS, Fan X, Hoffmann B, Wang L, Thomann L, Lin X, Stalder H, Pozzi B, Brot S de, Jiang N, Cui D, Hossain J, Wilson MM, Keller MW, Stark TJ, Barnes JR, Dijkman R, Jores J, Benarafa C, Wentworth DE, Thiel V, and Beer M: SARS-CoV-2 spike D614G change enhances replication and transmission. *Nature* **592**, 122–127 (2021).

88. Plante JA, Liu Y, Liu J, Xia H, Johnson BA, Lokugamage KG, Zhang X, Muruato AE, Zou J, Fontes-Garfias CR, Mirchandani D, Scharton D, Bilello JP, Ku Z, An Z, Kalveram B, Freiberg AN, Menachery VD, Xie X, Plante KS, Weaver SC, and Shi P-Y: Spike mutation D614G alters SARS-CoV-2 fitness. *Nature* **592**, 116–121 (2021).

89. Shi AC and Xie X: Making sense of spike D614G in SARS-CoV-2 transmission. *Sci China Life Sci* **64**, 1062–1067 (2021).

90. Hewins B, Rahman M, Bermejo-Martin JF, Kelvin AA, Richardson CD, Rubino S, Kumar A, Ndishimye P, Ostadgavahi AT, Mahmud-Al-Rafat A, and Kelvin DJ: Alpha, beta, delta, omicron, and SARS-CoV-2 breakthrough cases: Defining immunological mechanisms for vaccine waning and vaccine-variant mismatch. *Frontiers Virology* **2**, 849936 (2022).

91. Yuan M, Liu H, Wu NC, Lee C-CD, Zhu X, Zhao F, Huang D, Yu W, Hua Y, Tien H, Rogers TF, Landais E, Sok D, Jardine JG, Burton DR, and Wilson IA: Structural basis of a shared antibody response to SARS-CoV-2. *Science* **369**, 1119–1123 (2020).

92. Cao Y, Su B, Guo X, Sun W, Deng Y, Bao L, Zhu Q, Zhang X, Zheng Y, Geng C, Chai X, He R, Li X, Lv Q, Zhu H, Deng W, Xu Y, Wang Y, Qiao L, Tan Y, Song L, Wang G, Du X, Gao N, Liu J, Xiao J, Su X, Du Z, Feng Y, Qin C, Qin C, Jin R, and Xie XS: Potent neutralizing antibodies against SARS-CoV-2 identified by high-throughput single-cell sequencing of convalescent patients' B cells. *Cell* **182**, 73–84.e16 (2020).

93. Dejnirattisai W, Huo J, Zhou D, Schreiber G, Stuart DI, and Screaton GR: SARS-CoV-2 omicron-B.1.1.529 leads to widespread escape from neutralizing antibody responses. *Cell* **185**, 467–484.e15 (2022).

94. Lan J, Ge J, Yu J, Shan S, Zhou H, Fan S, Zhang Q, Shi X, Wang Q, Zhang L, and Wang X: Structure of the SARS-CoV-2 spike receptor-binding domain bound to the ACE2 receptor. *Nature* **581**, 215–220 (2020).

95. McCallum M, Czudnochowski N, Rosen LE, Zepeda SK, Bowen JE, Walls AC, Hauser K, Joshi A, Stewart C, Dillen JR, Powell AE, Croll TI, Nix J, Virgin HW, Corti D, Snell G, and Veesler D: Structural basis of SARS-CoV-2 omicron immune evasion and receptor engagement. *Science* **375**, 864–868 (2022).

96. Tegally H, Wilkinson E, Giovanetti M, Iranzadeh A, Fonseca V, Giandhari J, Doolabh D, Pillay S, San EJ, Msomi N, Mlisana K, Gottberg A von, Walaza S, Allam M, Ismail A, Mohale T, Glass AJ, Engelbrecht S, Zyl GV, Preiser W, Petruccione F, Sigal A, Hardie D, Marais G, Hsiao M, Korsman S, Davies M-A, Tyers L, Mudau I, York D, Maslo C, Goedhals D, Abrahams S, Laguda-Akingba O, Alisoltani-Dehkordi A, Godzik A, Wibmer CK, Sewell BT, Lourenço J, Alcantara LCJ, Pond SLK, Weaver S, Martin D, Lessells RJ, Bhiman JN, Williamson C, and Oliveira T de: Emergence and rapid spread of a new severe acute respiratory syndrome-related coronavirus 2 (SARS-CoV-2) lineage with multiple spike mutations in South Africa. *Medrxiv* 2020.12.21.20248640 (2020).

97. Shi R, Shan C, Duan X, Chen Z, Liu P, Song J, Song T, Bi X, Han C, Wu L, Gao G, Hu X, Zhang Y, Tong Z, Huang W, Liu WJ, Wu G, Zhang B, Wang L, Qi J, Feng H, Wang F-S, Wang Q, Gao GF, Yuan Z, and Yan J: A human neutralizing antibody targets the receptor-binding site of SARS-CoV-2. *Nature* **584**, 120–124 (2020).

98. Iketani S, Liu L, Guo Y, Liu L, Chan JF-W, Huang Y, Wang M, Luo Y, Yu J, Chu H, Chik KK-H, Yuen TT-T, Yin MT, Sobieszczyk ME, Huang Y, Yuen K-Y, Wang HH, Sheng Z, and Ho DD: Antibody evasion properties of SARS-CoV-2 Omicron sublineages. *Nature* **604**, 553–556 (2022).

99. Costello SM, Shoemaker SR, Hobbs HT, Nguyen AW, Hsieh C-L, Maynard JA, McLellan JS, Pak JE, and Marqusee S: The SARS-CoV-2 spike reversibly samples an open-trimer conformation exposing novel epitopes. *Nat Struct Mol Biol* **29**, 229–238 (2022).

100. Yuan M, Wu NC, Zhu X, Lee C-CD, So RTY, Lv H, Mok CKP, and Wilson IA: A highly conserved cryptic epitope in the receptor binding domains of SARS-CoV-2 and SARS-CoV. *Sci New York N Y* **368**, 630–633 (2020).

101. Sui J, Li W, Murakami A, Tamin A, Matthews LJ, Wong SK, Moore MJ, Tallarico AStC, Olurinde M, Choe H, Anderson LJ, Bellini WJ, Farzan M, and Marasco WA: Potent neutralization of severe acute respiratory syndrome (SARS) coronavirus by a human mAb to S1 protein that blocks receptor association. *Proc National Acad Sci* **101**, 2536–2541 (2004).

102. Kim C, Ryu D-K, Lee J, Kim Y-I, Seo J-M, Kim Y-G, Jeong J-H, Kim M, Kim J-I, Kim P, Bae JS, Shim EY, Lee MS, Kim MS, Noh H, Park G-S, Park JS,

Son D, An Y, Lee JN, Kwon K-S, Lee J-Y, Lee H, Yang J-S, Kim K-C, Kim SS, Woo H-M, Kim J-W, Park M-S, Yu K-M, Kim S-M, Kim E-H, Park S-J, Jeong ST, Yu CH, Song Y, Gu SH, Oh H, Koo B-S, Hong JJ, Ryu C-M, Park WB, Oh M, Choi YK, and Lee S-Y: A therapeutic neutralizing antibody targeting receptor binding domain of SARS-CoV-2 spike protein. *Nat Commun* **12**, 288 (2021).

103. Weisblum Y, Schmidt F, Zhang F, DaSilva J, Poston D, Lorenzi JC, Muecksch F, Rutkowska M, Hoffmann H-H, Michailidis E, Gaebler C, Agudelo M, Cho A, Wang Z, Gazumyan A, Cipolla M, Luchsinger L, Hillyer CD, Caskey M, Robbiani DF, Rice CM, Nussenzweig MC, Hatziioannou T, and Bieniasz PD: Escape from neutralizing antibodies by SARS-CoV-2 spike protein variants. *Elife* **9**, e61312 (2020).

104. Chi X, Yan R, Zhang J, Zhang G, Zhang Y, Hao M, Zhang Z, Fan P, Dong Y, Yang Y, Chen Z, Guo Y, Zhang J, Li Y, Song X, Chen Y, Xia L, Fu L, Hou L, Xu J, Yu C, Li J, Zhou Q, and Chen W: A neutralizing human antibody binds to the N-terminal domain of the spike protein of SARS-CoV-2. *Sci New York N Y* **369**, 650–655 (2020).

105. Rogers TF, Zhao F, Huang D, Beutler N, Burns A, He W, Limbo O, Smith C, Song G, Woehl J, Yang L, Abbott RK, Callaghan S, Garcia E, Hurtado J, Parren M, Peng L, Ramirez S, Ricketts J, Ricciardi MJ, Rawlings SA, Wu NC, Yuan M, Smith DM, Nemazee D, Teijaro JR, Voss JE, Wilson IA, Andrabi R, Briney B, Landais E, Sok D, Jardine JG, and Burton DR: Isolation of potent SARS-CoV-2 neutralizing antibodies and protection from disease in a small animal model. *Science* **369**, 956–963 (2020).

106. Wajnberg A, Amanat F, Firpo A, Altman DR, Bailey MJ, Mansour M, McMahon M, Meade P, Mendu DR, Muellers K, Stadlbauer D, Stone K, Strohmeier S, Simon V, Aberg J, Reich DL, Krammer F, and Cordon-Cardo C: Robust neutralizing antibodies to SARS-CoV-2 infection persist for months. *Sci New York N Y* **370**, 1227–1230 (2020).

107. Liu L, Wang P, Nair MS, Yu J, Rapp M, Wang Q, Luo Y, Chan JF-W, Sahi V, Figueroa A, Guo XV, Cerutti G, Bimela J, Gorman J, Zhou T, Chen Z, Yuen K-Y, Kwong PD, Sodroski JG, Yin MT, Sheng Z, Huang Y, Shapiro L, and Ho DD: Potent neutralizing antibodies against multiple epitopes on SARS-CoV-2 spike. *Nature* **584**, 450–456 (2020).

108. Colin-York H, Javanmardi Y, Skamrahl M, Kumari S, Chang VT, Khuon S, Taylor A, Chew T-L, Betzig E, Moeendarbary E, Cerundolo V, Eggeling C, and Fritzsche M: Cytoskeletal control of antigen-dependent T cell activation. *Cell Rep* **26**, 3369–3379.e5 (2019).

109. Plotnikov SV, Pasapera AM, Sabass B, and Waterman CM: Force fluctuations within focal adhesions mediate ECM-rigidity sensing to guide directed cell migration. *Cell* **151**, 1513–1527 (2012).

110. Egan JR, Abu-Shah E, Dushek O, Elliott T, and MacArthur BD: Fluctuations in T cell receptor and pMHC interactions regulate T cell activation. *J Roy Soc Interface* **19**, 20210589 (2022).

111. Pullen RH and Abel SM: Catch bonds at T cell interfaces: Impact of surface reorganization and membrane fluctuations. *Biophys J* **113**, 120–131 (2017).

112. Utzny C, Faroudi M, and Valitutti S: Frequency encoding of T-cell receptor engagement dynamics in calcium time series. *Biophys J* **88**, 1–14 (2005).

113. Meldolesi J: Oscillation, activation, expression. *Nature* **392**, 863–865 (1998).

114. Yi J, Wu XS, Crites T, and Hammer JA: Actin retrograde flow and actomyosin II arc contraction drive receptor cluster dynamics at the immunological synapse in Jurkat T cells. *Mol Biol Cell* **23**, 834–852 (2012).

115. Zhu C, Chen Y, and Ju LA: Dynamic bonds and their roles in mechanosensing. *Curr Opin Chem Biol* **53**, 88–97 (2019).

116. Huang J, Edwards LJ, Evavold BD, and Zhu C: Kinetics of MHC-CD8 interaction at the T cell membrane. *J Immunol* **179**, 7653–7662 (2007).

117. Zhu C, Jiang N, Huang J, Zarnitsyna VI, and Evavold BD: Insights from in situ analysis of TCR–pMHC recognition: Response of an interaction network. *Immunol Rev* **251**, 49–64 (2013).

118. Zarnitsyna V and Zhu C: T cell triggering: Insights from 2D kinetics analysis of molecular interactions. *Phys Biol* **9**, 045005 (2012).

119. Hu W, Zhang Y, Sun X, Zhang T, Xu L, Xie H, Li Z, Liu W, Lou J, and Chen W: FcγRIIB-I232T polymorphic change allosterically suppresses ligand binding. *Elife* **8**, e46689 (2019).

120. Xu L, Xia M, Guo J, Sun X, Li H, Xu C, Gu X, Zhang H, Yi J, Fang Y, Xie H, Wang J, Shen Z, Xue B, Sun Y, Meckel T, Chen Y-H, Hu Z, Li Z, Xu C, Gong H, and Liu W: Impairment on the lateral mobility induced by structural changes underlies the functional deficiency of the lupus-associated polymorphism FcγRIIB-T232. *J Exp Med* **213**, 2707–2727 (2016).

121. Aramesh M, Stoycheva D, Sandu I, Ihle SJ, Zünd T, Shiu J-Y, Forró C, Asghari M, Bernero M, Lickert S, Kotowski M, Davis SJ, Oxenius A, Vogel V, and Klotzsch E: Nanoconfinement of microvilli alters gene expression and boosts T cell activation. *Proc Nat Acad Sci USA* **118**, e2107535118 (2021).

122. Chen KY, Jenkins E, Körbel M, Ponjavic A, Lippert AH, Santos AM, Ashman N, O'Brien-Ball C, McBride J, Klenerman D, and Davis SJ:

Trapping or slowing the diffusion of T cell receptors at close contacts initiates T cell signaling. *Proc Nat Acad Sci USA* **118**, e2024250118 (2021).

123. Davis SJ and Merwe PA van der: The kinetic-segregation model: TCR triggering and beyond. *Nat Immunol* **7**, 803–809 (2006).

124. Zhang Z, Duvefelt K, Svensson F, Masterman T, Jonasdottir G, Salter H, Emahazion T, Hellgren D, Falk G, Olsson T, Hillert J, and Anvret M: Two genes encoding immune-regulatory molecules (LAG3 and IL7R) confer susceptibility to multiple sclerosis. *Genes Immun* **6**, 145–152 (2005).

125. Wang J and Reinherz EL: The structural basis of αβ T-lineage immune recognition: TCR docking topologies, mechanotransduction, and co-receptor function. *Immunol Rev* **250**, 102–119 (2012).

126. Lee MS, Glassman CR, Deshpande NR, Badgandi HB, Parrish HL, Uttamapinant C, Stawski PS, Ting AY, and Kuhns MS: A mechanical switch couples T cell receptor triggering to the cytoplasmic juxtamembrane regions of CD3ζζ. *Immunity* **43**, 227–239 (2015).

127. Li K, Yuan Z, Lyu J, Ahn E, Davis SJ, Ahmed R, and Zhu C: PD-1 suppresses TCR-CD8 cooperativity during T-cell antigen recognition. *Nature Commun* 1–13 (2021).

128. Jiang N, Huang J, Edwards LJ, Liu B, Zhang Y, Beal CD, Evavold BD, and Zhu C: Two-stage cooperative T cell receptor-peptide major histocompatibility complex-CD8 trimolecular interactions amplify antigen discrimination. *Immunity* **34**, 13–23 (2011).

Chapter 11

Studying Membrane Dynamics Using Force Spectroscopy Based on Atomic Force Microscopy

Siying Li and Yuping Shan
School of Chemistry and Life Science, Advanced Institute of Materials Science, Changchun University of Technology, Changchun 130012, P.R. China
shanyp@ccut.edu.cn

The cell membrane is one of the most complicated biological complexes. Understanding the nature of the interaction dynamics between extracellular substances and cell membranes is required to both understand the transmembrane transporting mechanism of extracellular substances at work and for the treatment of membrane-related diseases. Force spectroscopy based on atomic force microscopy (AFM) is a powerful approach to studying the interaction dynamics between extracellular substances and cell membranes at the single-molecule level. In this chapter, single-molecule force spectroscopy (SMFS) and force-tracing techniques based on AFM are introduced, and the related experimental methods and applications

Biomembranomics: Structure, Techniques, and Applications
Edited by Hongda Wang

ISBN 978-981-4968-61-4 (Hardcover), 978-1-003-45635-3 (eBook)
www.jennystanford.com

are discussed, which will make several unprecedented advances in cell membrane research.

11.1 Introduction

Cell membranes are involved in various functions that contain thousands of types of lipids, proteins, and saccharides [1]. The structure region executes specific functions at the nanoscale and millisecond levels in the cell membrane. The investigation of membrane functions requires a technique with high spatiotemporal resolution under near-physiological conditions. AFM is capable of observing biological molecules at work and has become a powerful tool for exploring biological molecules, including proteins, DNAs, RNAs, enzymes and substrates, antigens, antibodies, and their complexes [2]. The force versus extension response of molecules can provide valuable insight into their mechanical stability, as well as details of the underlying energy landscape [3]. AFM shows great prospects for characterizing single molecules and thus offers many advantages in scientific discoveries.

Among AFM-based force spectroscopy technologies, SMFS and force tracing show unique advantages for detecting interactions between biological molecules due to the exquisite sensitivity of the AFM tip cantilever. They may be used as a force detector, obtaining information on the adhesion, elasticity, and interaction of molecules [4]. This knowledge can help us understand that we are interested in substances, such as viruses or nanodrugs during the dynamic process of cell entry. These techniques also can detect the structure, dynamics, and functions of single molecules, thus revealing their information that may be lost in ensemble averages. Here, we mainly discuss single-molecule techniques and their methods for measuring physicochemical parameters on cell membrane, and applications in the representative examples of revealing key information about cell membranes.

11.1.1 Single-Molecule Force Spectroscopy

SMFS based on AFM stands out among the fast-growing powerful approaches to analyze biological systems with the fine AFM tip cantilever [5, 6], including individual ligand–receptor pair unbinding, protein (un)-folding pathways, and DNA (un)-folding mechanics at the single-molecule level [7], yielding a radically new understanding of the mechanism for intermolecular binding and intramolecular elasticity [1, 8, 9]. Gaub et al. reported the first biological application of SMFS in detecting interactions of ligand–receptor pairs in 1994, and since then, the SMFS technique has especially attracted the attention of scientists [10]. In addition, detecting the mechanical properties and nanoscale forces occurring in the cellular dynamic process plays a remarkable role in understanding the physiological function of living cells and organisms [11, 12]. As mentioned above, SMFS measurements have been performed on a large variety of biological systems. The working principle of SMFS is based on monitoring the forces between the tip surface and the sample surface. Generally, these forces are measured via the vertical deflection of the AFM tip cantilever [13, 14]. Figure 11.1 shows a typical force–distance/time (FD/FT) curve obtained during force measurements [15].

In the force spectroscopy mode, the approach and retraction process reflect the interactions between the sample and the AFM tip in a controlled manner. The force signal is based on the detection of small shifts in the cantilever; cantilever deflection is then converted into FD or FT curves. When the tip approaches the sample surface, the attracting force can be detected, and the attracting force between the tip and the surface results in a downward deflection upon cantilever retraction; during the retraction of the tip that the deflection is increasing, the adhesive bonds are increasingly tensed until they break, and the unbinding of two interacted target molecules is detected [1, 15, 16]. The advantage of the single-molecule FD and FT curves is tracking individual dissociation events, intermediates, kinetics, and pathways so that the heterogeneity of coexisting biomolecule groups can be detected. It is often challenging or impossible to characterize such detailed understandings using biochemical batch analysis [17]. SMFS can detect nonspecific and specific biomolecular interactions [16, 18–20], including biomacromolecules (proteins, lipids, nucleic acids, sugars, viruses) or interactions between receptors and ligands [21]. The epidermal growth factor receptor (EGFR) is overexpressed in many malignancies.

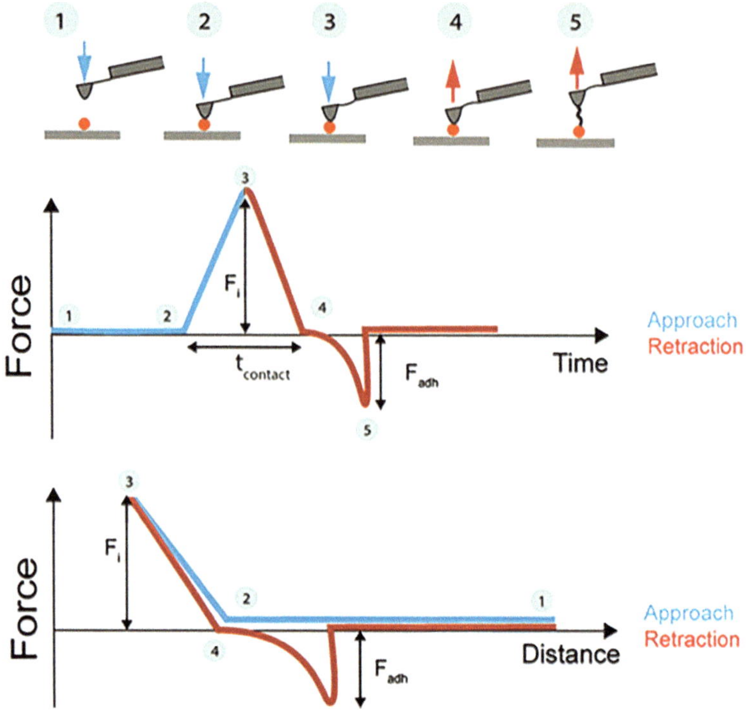

Figure 11.1 Examples of force–time (FT) and force–distance (FD) curves. The AFM tip, initially away from the surface (1), is brought into contact (approach curve in blue) with the surface (2) to which it is pressed until reaching a certain force threshold (3). Upon retraction (retraction curve in red) (4), adhesive forces (Fadh) are observed (5) at a separation distance depending on the presence of a polymeric linker tethering a molecular probe of interest (orange circle) to the AFM tip. Reprinted with the permission from Ref. [15]. Copyright 2021 American Chemical Society.

As shown in Fig. 11.2, Cai et al. using SMFS detected EGF–EGFR interactions in living esophageal cancer KYSE-150 cells to evaluate the anticancer activity of oridonin [22]. Oridonin shows strong anticancer activity, which is found to be related to specific inhibition effects on the EGFR tyrosine kinase activity [23]. The results indicated that oridonin could inhibit the binding between EGF and EGFR in KYSE-150 cells by decreasing the unbinding force and binding probability for EGF–EGFR complexes. Not only EGFR, Angiotensin II type 1 receptor (AT1R), a typical G protein-coupled receptor, plays a

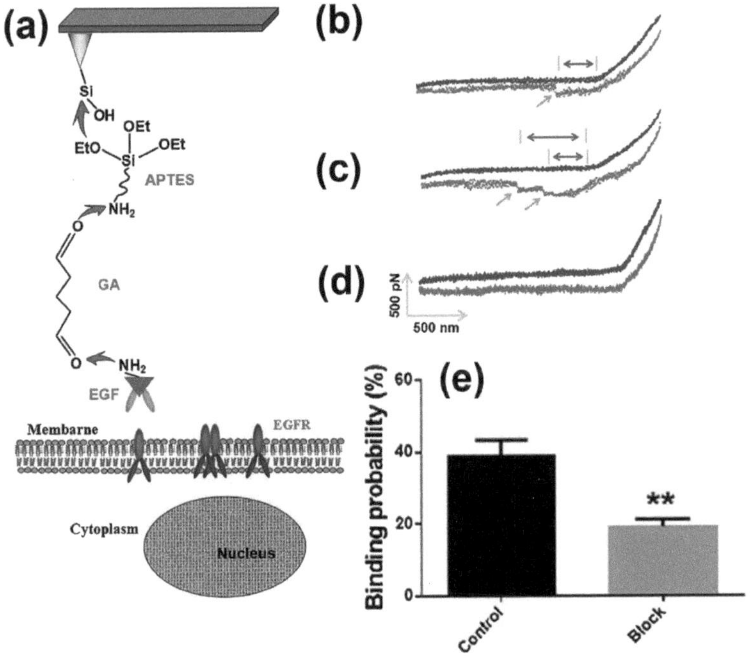

Figure 11.2 SMFS measurements with EGF-functionalized AFM tip on living KYSE-150 cells. (a) AFM tip functionalization procedure and schematic representation of the SMFS measurement between EGF-functionalized AFM tips and EGFR in KYSE-150 cells. (b, c) Representative force curves were obtained with EGF-modified AFM tips in KYSE-150 cells, and (d) after the system was blocked with free anti-EGFR antibody. (e) The binding probability of EGF-functionalized tips in KYSE-150 cells before and after blocking. Reprinted from Ref. [22]. Copyright 2017, with permission from Elsevier.

key role in regulating many cardiovascular functions [24]. Different ligands can bind with AT1R to selectively activate either G protein or β-arrestin pathway, or both pathways [25]. Fang et al. used SMFS to study the interactions of AT1R with three types of ligands, including balanced ligands [26]. The results revealed their difference in binding force and binding stability. This indicated that AT1R had different ligand-binding conformational substates and underwent different structural changes to activate downstream signaling pathways with variable agonist efficacies. These pioneering experiments established the framework for subsequent SMFS studies by defining the prerequisites for measuring specific binding forces between

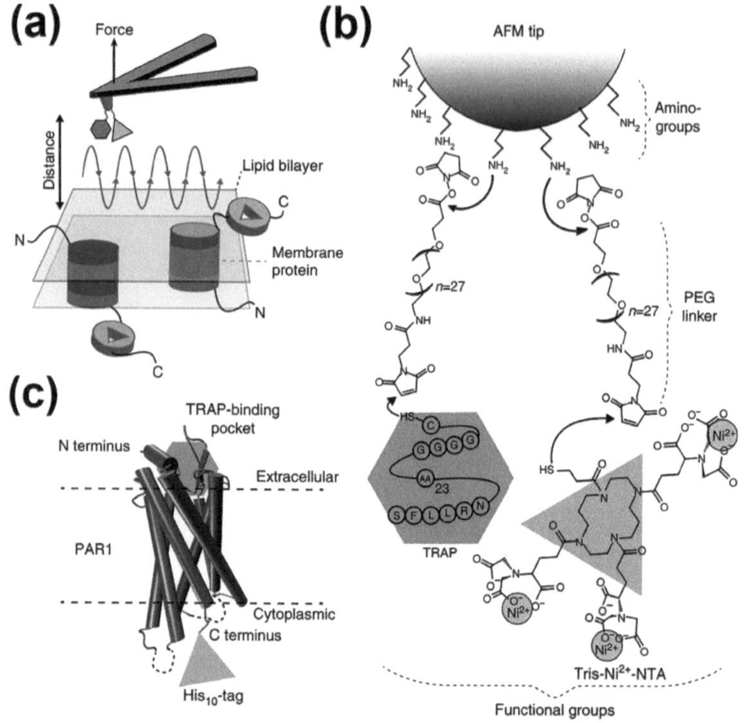

Figure 11.3 AFM mapping of two different ligand-binding events using a chemically bifunctionalized AFM tip. (a) Mapping adhesion forces on membrane proteins using FD-based AFM. When recording an AFM topograph, an approach (blue) and retraction (red) cycle between AFM tip and biological sample is performed for every pixel of the image. In each cycle, the cantilever deflection (for example, force) and the distance traveled by the AFM tip are monitored and transformed into an approach and retraction FD curve. (b) Bifunctionalization of the AFM tip. The amino-functionalized Si_3N_4 AFM tip is functionalized by hetero-bifunctional N-hydroxysuccinimide-PEG_{27}-maleimide linkers to which the thiol-bearing thrombin receptor-activating peptide (TRAP) and Ni^{2+}-loaded tris-NTA (tris-Ni^{2+}-NTA) are bound. (c) The membrane-embedded PAR1 reveals an intracellular C-terminal His10-tag (green) and an extracellular TRAP (here the native SFLLRN peptide ligand)-binding pocket (red), which can specifically bind the tris-Ni^{2+}-NTA and the SFLLRN ligands, respectively. Reprinted from Ref. [29] with permission from Springer Nature.

single ligand–receptor pairs [15, 27, 28]. A current challenge in life sciences is to image cell membrane receptors while characterizing their specific interactions with various ligands. Addressing this issue has been hampered by the lack of suitable nanoscopic methods. As shown in Fig. 11.3, Muller et al. addressed this challenge and introduced multifunctional high-resolution AFM to image human protease-activated receptors (PAR1) in the functionally important lipid membrane and to simultaneously localize and quantify their binding to two different ligands [29].

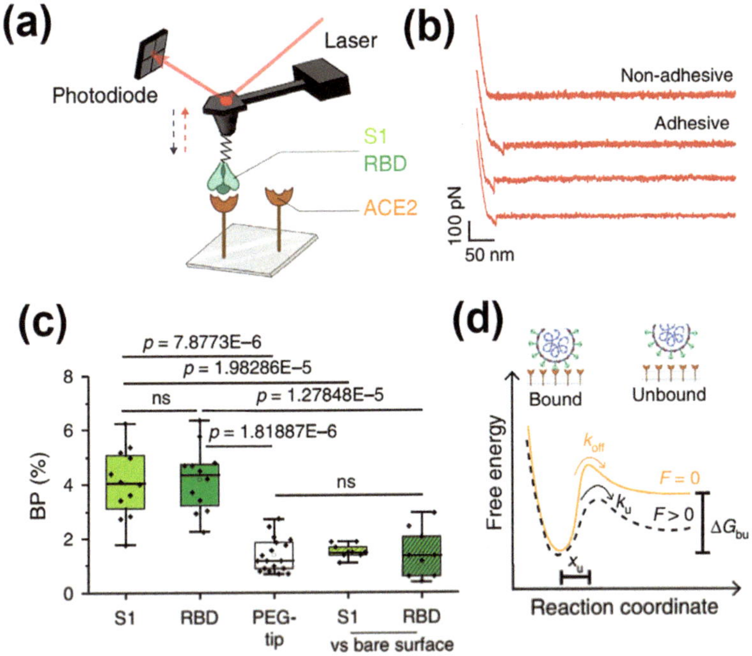

Figure 11.4 Probing S-glycoprotein binding to the ACE2 host receptor on the model surface. (a) The binding of S-glycoprotein subunit (S1 or RBD) is probed on an ACE2-coated surface. (b) Retraction part of four FD curves showing either nonadhesive or specific adhesive curves. (c) Box plot of specific binding probabilities (BP) measured by AFM between the functionalized tip (S1, RBD, or PEG) and the grafted surface (ACE2 or OH-/COOH-terminated alkanethiol (bare surface)). (d) The Bell–Evans model describing a virus–receptor bond as a two-state model. The bound state is separated from the unbound state by a single energy barrier located at distance x_u. k_{off} and k_{on} represent the dissociation and association rates, respectively. Reprinted from Ref. [9], with kind permission from Springer Nature.

Severe acute respiratory syndrome-related coronavirus-2 (SARS-CoV-2) has caused the coronavirus disease 2019 (COVID-19) pandemic worldwide. We require more investigation to understand SARS-CoV-2 [30]. As shown in Fig. 11.4, Alsteens et al. studied the interactions established between the SARS-CoV-2 glycoproteins and their host receptors by SMFS [9]. They demonstrated that on both model surfaces and living cells, the receptor-binding domain (RBD) serves as the binding interface within the S-glycoprotein with the ACE2 receptor and extract the kinetic and thermodynamic properties of this binding pocket.

11.1.2 Force Tracing

Cellular uptake is an ultrafast dynamic event; in these bulk experiments, the microenvironment effects of the heterogeneous cell surface on a single nanostructure were lost in the average outcome of an ensemble assay, and the results could not represent the behavior of single nanostructure internalization. According to the principle of SMFS, during the approaching process, the AFM tip moving in the same direction as cellular uptake may disturb the dynamic process of the cellular uptake. To capture the details of the dynamic process during cell entry in living cells and avoid disturbing from AFM tip cantilever moving, a static AFM tip cantilever to hold the molecule/nanoparticle is necessary. A technique based on SMFS, force tracing, was developed by Wang et al., which provides an unprecedented temporal–spatial resolution for the in situ observation of dynamic processes on living cell membrane. During force tracing, the functionalized AFM tip moves toward the cell surface and contacts living cells with a constant force/position mode. When using the constant position mode, changes in force can be monitored as substances pass through the cell membrane. When the cell undergoes endocytosis, it causes the cantilever of the AFM tip to bend downward. The laser light reflected by the cantilever also changes, and photodetectors detect and record the process. When using constant force mode, changes in the position of substances as they pass through the cell membrane can be monitored. When using the constant position mode, changes in the position of substances as they pass through the cell membrane can be monitored. The AFM tip contacts the cells with a constant force. When the cell undergoes endocytosis, the cantilever of the AFM tip bends downward, causing an instantaneous deflection, and the position of the piezoceramic is adjusted by the feedback system to keep the deflection of the cantilever at the setpoint. The force-tracing technique shows unique advantages for detecting the force, duration, average speed, and displacement during the dynamic process of transmembrane transporting at the single-molecules/particles level in living cells [31]. The principle of force tracing based on AFM is shown in Fig. 11.5 [32]. A laser beam is reflected by the AFM cantilever. The photodetector detects the laser position and records the change in the cantilever deflection. The AFM tip is approached to the cell surface with the contact mode of AFM; subsequently, FD curves are measured on the living cell to define the contact point between the AFM tip and the cell surface, where the AFM tip is slightly in contact with the cell membrane. In an FD cycle, the AFM tip stops at the position of the Z scan start, which is far from the contact point. Finally, the AFM tip is slowly moved to the contact point with the feedback system, and then the feedback is turned off; thus, the AFM tip is suspended on the cell surface. Upon the application of an external force (the cells uptake the molecule/particle tethered on the AFM tip and stretch the bifunctional crosslinker), the cantilever is deflected, and the change in vertical position of the laser on the photodetector is acquired using a PCI card that can readily monitor rapid processes down to 1 μs; this card is suitable for recording the ultrafast process of cell entry in living cells. The software is written in LabVIEW, and the sampling rate is set at a suitable value for recording the cellular uptake dynamic process of different nanoparticles and molecules [33, 34].

The typical FT curves for carbon dot (CD) entry cells are shown in Fig. 11.6. Initially, the CD attached onto the AFM tip is located on the cell membrane while contacting the cell membrane slightly, as shown on the left side of the typical FT curve. Once the CD is endocytosed, the AFM cantilever bends downward and generates a force–time signal in the FT curve (red arrow), meaning the CD entry cell process. Then, the piezoelectric ceramic drives the AFM tip cantilever to maintain a balanced position, and the FT curve returns flatten due to the principle of force tracing in a constant position mode. Collecting a large number of force-tracing curves is required for force-tracing experiments because transport events occur sporadically due to the early nature of cellular uptake activity. The duration and force for single CD nanoparticle transport could be measured straightforwardly from the force-tracing curves. The duration for single CD transport distribution varies from 5 to 65 ms with an average value of 13 ± 8.3 ms. The corresponding force distributes from 33 to 351 pN with an average value of 79 ± 50 pN [35]. The displacement of the molecule/nanoparticle during the transmembrane process includes two parts: the bending distance of the cantilever *d* and the extension length of the PEG linker *h*, as shown in Fig. 11.7 and Equation 11.1. The average speed of the single CD transported into the cell is calculated as 2.03 μm/s.

$$D = d + h \tag{11.1}$$

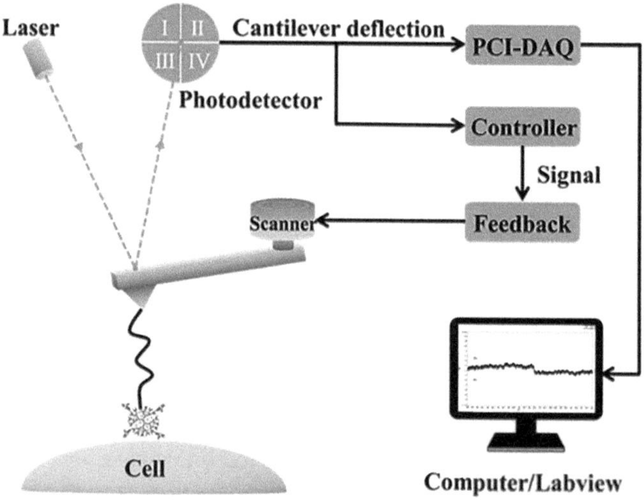

Figure 11.5 Schematic diagram of the force tracing technique workflow. Reproduced from Ref. [32] by permission of The Royal Society of Chemistry.

It is worth noting that irregular nanoparticles that enter into cells may show different FT curves and FT signals (Fig. 11.8) during their transmembrane transporting process, such as filamentous Ebola-virus-like particles [36]. The cell invasion process for Ebola-virus-like particles in vertical and horizontal directions was identified, and the interactions between Ebola-virus-like particles and cell membrane receptors were captured. In addition, we need to understand the difference between the measured force of SMFS and force tracing. SMFS measures the specific interaction force between a specific ligand and its receptor; however, the force measured by the force tracing is the force required by the cellular uptake of the substance.

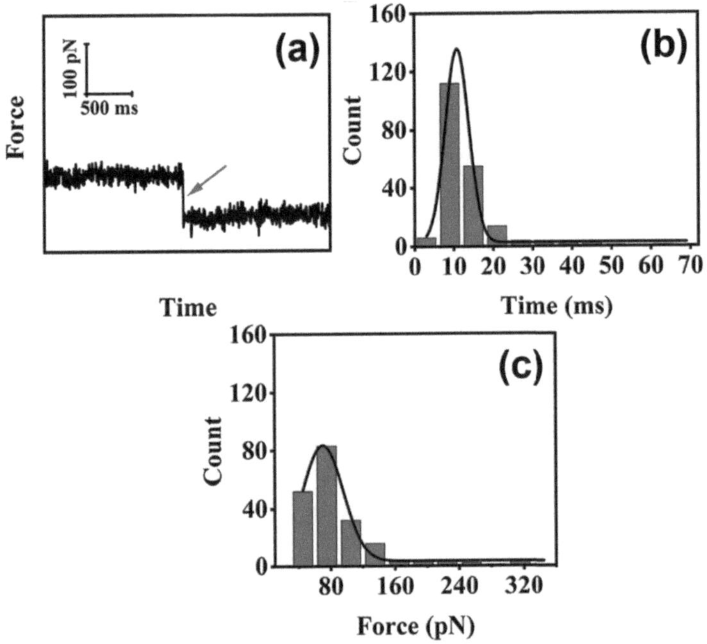

Figure 11.6 Force-tracing measurement results on Vero cells. (a) Typical force-tracing curve of transporting CDs into the Vero cells. (b) Duration distribution of transporting CDs through cell membranes. (c) Force distribution of transporting CDs through cell membranes (n = 200). Reprinted with permission from Ref. [35]. Copyright 2019 American Chemical Society.

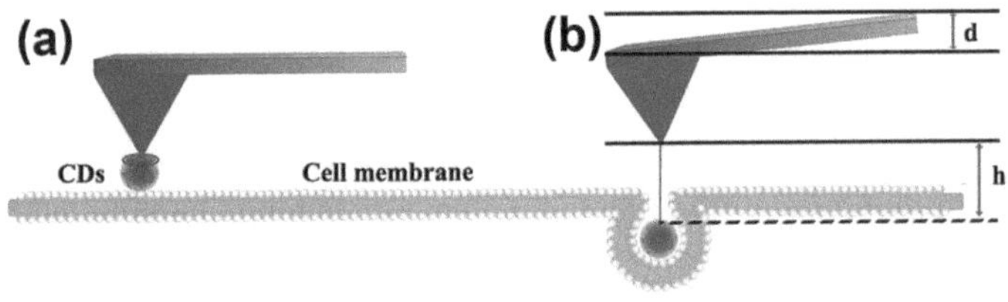

Figure 11.7 Schematic illustration of CD displacement during the transporting process. (a) CDs attached to the AFM tip were placed on the cell surface and the AFM tip cantilever maintained balance. (b) CD was internalized by the cell with the AFM tip cantilever bending (d) and the extension of PEG linker (h). Reprinted with permission from Ref. [35]. Copyright 2019, American Chemical Society.

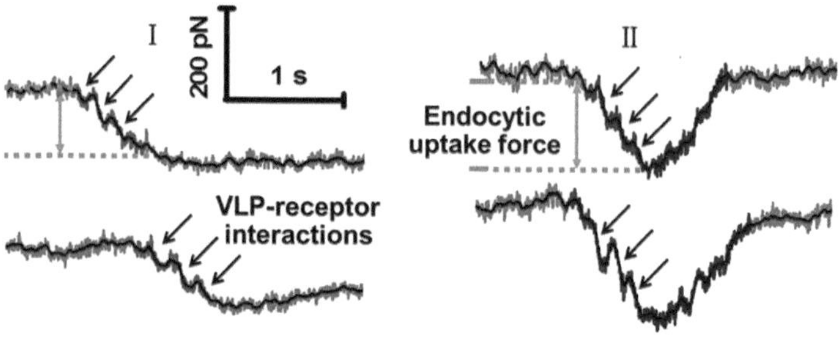

Figure 11.8 Two types of typical force tracing curves: EBO-VLP cell entry in the horizontal (force-tracing curve I) or vertical (force-tracing curve II) direction. The data are shown at full resolution (dark gray) and filtered smoothness (black). The green and red localized sections indicate the force signal of EBO-VLP endocytic uptake, and the arrow indicates the EBO-VLP–receptor interactions. The same number of peaks indicates the same EBO-VLP–receptor binding sites. Reprinted with permission from Ref. [36]. Copyright 2020 American Chemical Society.

11.2 Applications of Force Spectroscopy

11.2.1 Membrane Dynamics

Cell membrane defines the inside and outside boundaries of each cell. Cell membranes have dynamic functions for material transport, including endocytosis and membrane protein transport [1, 37]. In detail, it has the ability to allow the uptake of nutrients, small molecules, and ions; release waste products; bind ligands; transmit signals; convert energy; sense the environment; maintain cell adhesion; control cell migration; and much more while forming a tight barrier [38]. All of these are almost dynamic processes. Here, we provide an opportunity by use of force spectroscopy based on the AFM as a nanobiotechnological tool in modern membrane biology. The application of force spectroscopy to membrane dynamics will be described.

Single-molecule techniques, in particular, force spectroscopy studies, have yielded kinetic insights into the unfolding pathways of diverse macromolecules, including globular proteins [39], membrane proteins [40], multiprotein complexes [41], and ribozymes [42]. Proper identification of the intermediate states along the protein folding pathway is essential to describe the folding process accurately

[43]. The pioneering (un)folding study of Ribonuclease A by Nobel laureate Anfinsen takes advantage of the reductive denaturation of the stable S–S bonds inside the protein and provides the first picture of how the protein folds and unfolds [44]. As shown in Fig. 11.9 [45], recently, Zheng et al. successfully demonstrated the complex and multistep protein unfolding scenarios of the β-domain of a human protein metallothionein-3 (MT).

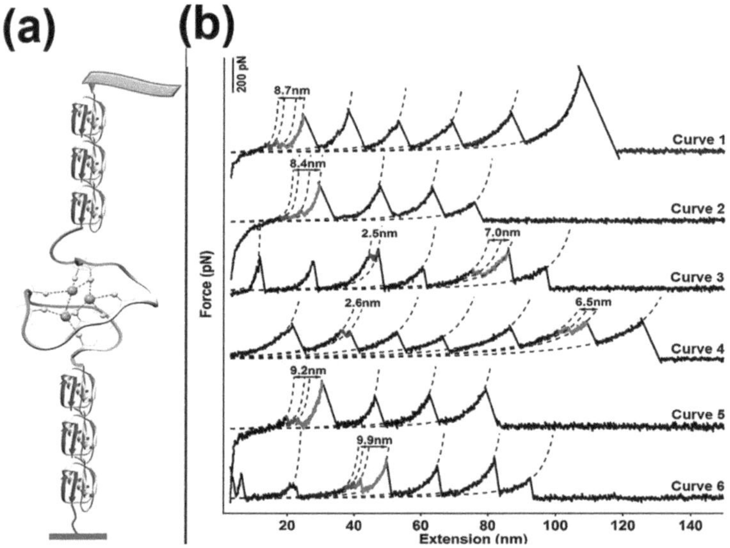

Figure 11.9 The three-step unfolding scenarios of the β-domain of human metallothionein III (βMT) from the rupture of the M_3S_9 cluster were revealed by single-molecule AFM. The setup of single-molecule AFM unfolding experiment for the gene of polyprotein $(GB1)_3$-βMT-$(GB1)_3$. (b) Representative force-extension curve of the polyprotein $(GB1)_3$-βMT-$(GB1)_3$ showed sawtooth-like unfolding force peaks. Reprinted from Ref. [45], with permission from Springer Nature.

The investigation for the kinetics of ligand–receptor complexes is usually performed using traditional surface plasmon resonance (SPR), which is based on ensemble methods, a technique inaccurate for very low-affinity interactions and also fails to identify subpopulations and different binding modes of multivalent energy landscapes [46]. SMFS is a powerful tool to address the mechanics of single proteins in their native environment [47]. Receptor–ligand recognition in cell membrane is a major means of communication with the outside world. Upon and during ligand occupancy, receptors elicit specific signaling cascades for cellular responses as diverse as migration, differentiation, secretion, survival, and even death. Defining the molecular characteristics of receptor–ligand bonds is necessary to understand the biology of a given receptor–ligand pair. Among the most important characteristics are the receptor–ligand binding kinetic rates and affinity, which define how fast the bond forms (the on rate, $k_{on)}$, how long it lasts (the off rate, k_{off}), and how many bonds exist in equilibrium (the affinity, $K_a = k_{on}/k_{off}$) [48]. It has long been recognized that force regulates molecular interactions by modulating k_{off} [49]. Many experiments demonstrate that the SMFS technique can be applied to study ligand–receptor interactions with dissociation constants, K_D, ranging from fM to μM with point mutant sensitivity. Several publications demonstrated the potential of this technique for affinity ranking applications [50, 51]. G protein-coupled receptors (GPCRs) are ubiquitously expressed membrane proteins and form one of the largest classes of therapeutic targets for drugs [52]. FD-based AFM was used for the functional development of GPCRs to simultaneously quantify the binding of two different ligands to the human protease-activated receptor 1 (PAR1). Such differentiation would be important to study how ligands compete during binding to the same receptor [53].

11.2.2 Single-Molecule Interaction and Transporting on Membrane

Small molecules (amino acids, sugars, nucleotides, ions, etc.) are transported by cellular uptake, which depends on the solute carrier superfamily, including a major class of cell membrane transporter [54, 55]. However, molecular transportation is dynamic; therefore, traditional methods (e.g., X-ray diffraction) to study the static structure of transporters cannot fully explain the dynamic mechanism of molecular transport. In general, the transport of a single molecule occurs at the micro to millisecond level. Wang et al. used a highly sensitive force-tracing technique to study the complicated process of single glucose transportation [56]. The force-tracing experimental results show that the force to transport a single D-glucose molecule across cell membranes via a glucose transporter is 37 ± 9 pN, and the transport event could be accomplished in approximately 20 ms with an average speed of 0.3 μm/s. Moreover, four residues (K38, E299, K300, and E426) on the extracellular side of the glucose transporter were identified to play critical roles in modulating the conformational transition, which helps develop potential therapeutic agents targeting glucose transporter 1.

SMFS allows the study of single-molecule interactions on cell membranes with slow dissociation dynamic processes that are inaccessible to other techniques. Virtually, the application of an external force on the ligand–receptor bond results in a loading rate-dependent deformation of the energy barrier separating the bound and unbound states, with a consequent reduction in the bond lifetime [57]. The force spectroscopy allows recording of the interaction forces in FD curves between the ligand on the tip and the receptor on the sample surface over a wide range of loading rates (i.e., speed at which the force is applied). By functionalizing the AFM tip with its endogenous ligand, the thrombin receptor-activating peptide (TRAP), it was possible to reconstruct the energy landscape of the ligand-activated receptor-1 (PAR1) interaction under physiological conditions [58]. Until now none of the SMFS experiments detected that a membrane protein establishes a new unfolding intermediate or stable structural segment. In the exploration of membrane proteins by SMFS, Muller et al. reconstituted the light-driven proton pump bacteriorhodopsin into lipid nanodiscs (Fig. 11.10).

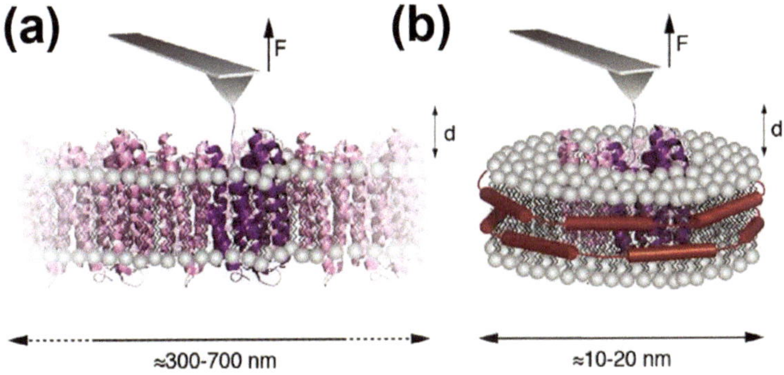

Figure 11.10 Schematic representation of SMFS of bacteriorhodopsin (BR) embedded in native purple membrane (BR_{PM}) and lipid nanodiscs (BR_{ND}). (a) and (b) are illustrations of BR trimers embedded in purple membrane (BR_{PM}) and in a lipid nanodisc (BR_{ND}), respectively. Reprinted with the permission from Ref. [59]. Copyright 2012 American Chemical Society.

The advantage of using nanodiscs is that membrane proteins can be handled like water-soluble proteins and characterized with similar ease. This demonstrates that lipid nanodiscs provide a unique approach for in vitro studies of native membrane proteins using SMFS [59].

11.2.3 Endocytosis of Nanoparticles

As the first step of the entry of nanoparticles into cells, the investigation of the transmembrane transporting process is particularly important. The transporting process dynamic parameters are influenced by many factors of nanoparticles involving size, shape, surface charge, surface functional groups, and hydrophobicity/hydrophilicity [60–63]. Using the ultrafast and sensitive force-tracing technique, the transporting process of a single fluorescent silicon nanoparticle (SiNP) was directly recorded on Vero and HeLa cells [64]. The results showed that the transport duration with a mean value of 16.80 ± 9.95 ms, and the corresponding force is 79.68 ± 40.80 pN in Vero cells. However, in HeLa cells, the force is 50.63 ± 17.9 pN, and the corresponding duration of the transport event is 9.98 ± 5.3 ms. The average speed (2.83 μm/s) in HeLa cells is higher than that in Vero cells (average speed 1.73 μm/s). The results could be attributed to the higher capability of the cellular uptake of cancer cells (HeLa) [35]. Meanwhile, it is found that the size of the nanoparticle will affect the endocytosis force, duration, and average speed during the endocytosis process. Wang's group detected the endocytosis of single gold nanoparticles (AuNPs) with different sizes by force tracing, revealing the size-dependent endocytosis dynamics. The average forces for the entry of AuNPs into living cells are 45 ± 14 pN, 67 ± 16 pN, and 126 ± 25 pN for 5 nm, 10 nm, and 20 nm AuNPs, respectively [34]. The shape of nanoparticles is another important factor that affect their endocytosis. Shan et al. found that transporting single tetrahedral DNA nanostructures (TDNs, 6 nm) into living HeLa cells needs 53 pN force and 25 ms duration with an average speed of 0.6 μm/s. Interestingly, the transporting dynamic parameters are irrelevant to the size of TDNs, while the larger TDNs rotated slightly during the transporting process [65]. Meanwhile, as shown in Fig. 11.11, in a recent study to map the detailed internalization behaviors of rod-shaped nanostructures with different aspect ratios, the aspect ratios are 1.01 (abbreviated as AR1), 2.00 (abbreviated as AR2), and 3.91 (abbreviated as AR4). The results indicated that the gold nanorod is endocytosed in a noncontinuous and force-rebound intermittent rotation manner and that the internalization force and time are positively correlated with the aspect ratios. However, internalization speed is negatively correlated with internalization time, irrespective of the aspect ratio [66].

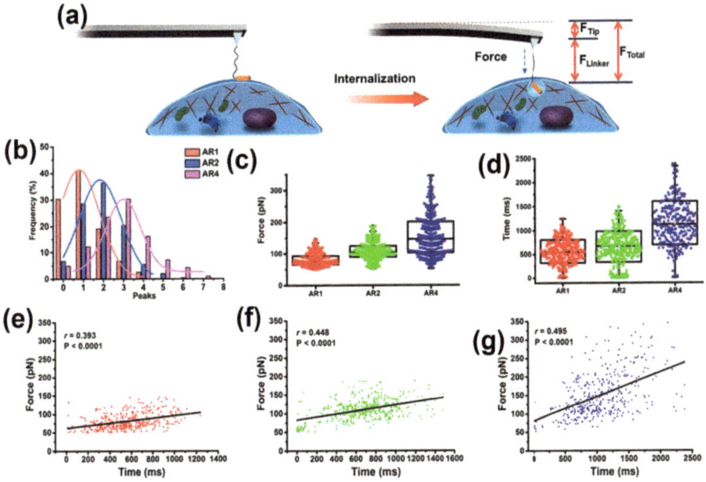

Figure 11.11 Capturing the dynamic internalization process of gold nanostructures. (a) Schematic diagram of nanostructure entry into a living cell. When the nanostructure enters the cell, the PEG linker is stretched, and the AFM cantilever bends downward. The internalization force and time are recorded in real time. (b) The distribution of the number of force-rebound events for AR1, AR2, and AR4 during the internalization process. The solid lines are a fit from the statistical results. (c) Boxplot of the distribution of the internalization force for three kinds of nanostructures. (d) Boxplot of the distribution of the internalization time for three kinds of nanostructures. (e–g) Distribution of the internalization force as a function of the internalization time for (d) AR1, (e) AR2, and (f) AR4. Linear regression (solid line) was used to model the related force trend with time. Reprinted with permission from Ref. [66]. Copyright 2022 American Chemical Society.

In addition, the effect of functional groups on the endocytosis of nanoparticles has also been discovered. It has been reported that the HAIYPRH (T7) peptide can target the transferrin receptor (TfR) and then enter cells easily with the help of transferrin (Tf) [67]. The dynamic mechanism by which transferrin promotes the entry of T7-conjugated nanostructures into cells remains unclear. Thus, the dynamic mechanism by which Tf enhances the cellular uptake of AuNP-T7 was explored by a force-tracing technique at the single-particle level [68]. The dynamic parameters, including duration, force, and average speed, were measured and compared in different cell lines. It was concluded that Tf enhances the uptake of AuNP-T7 by increasing the speed and decreasing the force of cellular uptake for A549 cells with a high level of TfR overexpression. For HeLa cells (lower TfR overexpression than A549 cells), only the cellular uptake speed is increased by Tf. In contrast, adding Tf decreases the speed of the AuNP-T7 entry into Vero cells without TfR overexpression.

11.2.4 Virus-Like Particles Invading Cells

Every so often news about a viral outbreak goes viral and catches widespread public attention in the media. Human immunodeficiency virus (HIV), West Nile virus, avian influenza (bird flu), Ebola, Middle East respiratory virus, and Zika virus in the past [69], and severe acute respiratory syndrome coronavirus 2 (SARS-CoV-2), which is still spreading [70], have each been in the media's spotlight [71]. Viruses rely on the host cell machinery in order to carry out an infectious life cycle and ultimately spread to new host cells [72]. Recognition and interactions with cellular receptors are the critical initial steps in the infectious viral life cycle and play a key regulatory role in the host cell viral pathogenesis [73]. Interactions with virus receptors are usually mediated by specific viral attachment proteins expressed on the surface of the virion [74]. Since the interaction between different specific viral attachment proteins is a dynamic process, viral binding events must be characterized at the molecular scale with a high temporal and spatial resolution to gain a comprehensive understanding of these complex multistep attachment processes [75–77]. The initial interactions between viruses and receptors are diverse and can range from nonspecific to highly specific. In most instances, viruses first engage a receptor in a low-affinity, high-avidity interaction and then have a subsequent interaction with a secondary or tertiary receptor in a high-affinity interaction event [78, 79]. Force spectroscopy based on AFM has proved to be extremely useful in the discovery of viral receptors, revealing the mechanisms by which virus receptors function to mediate attachment, entry, and signaling events [9, 33]. Virus entry into host cell is the first step of virus infection; however, its dynamic process via endocytosis remains largely elusive. As shown in Fig. 11.12, using SMFS to study at the single virion level, how enzymes trigger change at the virion surface and how it influences binding properties to cell surface receptors. The results elucidate that the binding of Ebola virus-like particle glycoprotein (GP) to phosphatidylserine (PS) receptors (TIM-1) and C-type lectin receptors (DC-SIGNR) acts as the initial high-affinity binding just before the cellular entry. This two-step entry mechanism, orchestrated by a single cellular receptor, demonstrates the complexity and elegance of the EBOV infection mechanism [80].

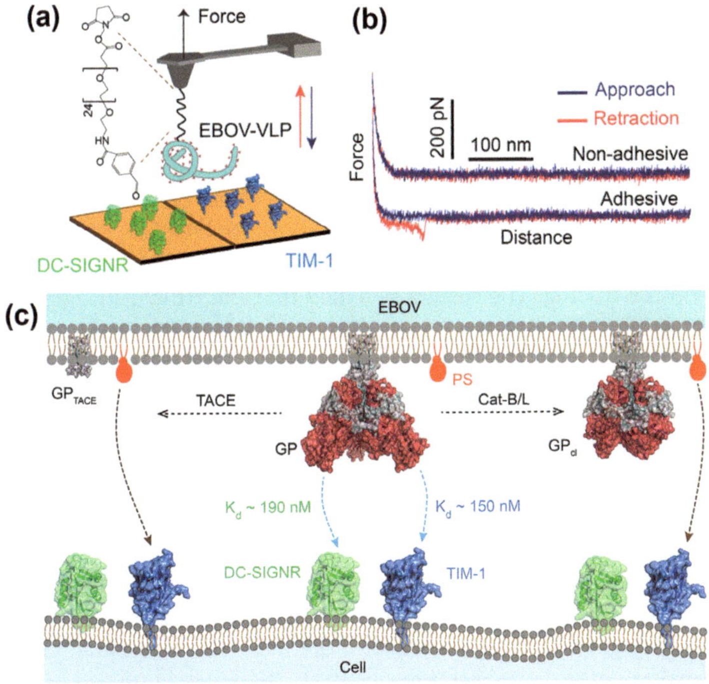

Figure 11.12 EBOV-VLPs specifically bind to purified receptors. (a) Schematic of the functionalized AFM tip with a PEG_{24} spacer grafted to the free amino group of EBOV-VLP. (b) Representative FD curves recorded on the model surface showing nonadhesion and specific adhesion event. (c) Multiple receptors mediate EBOV binding to host cells with high affinity. Models of EBOV binding to receptors: GP mediates EBOV high-affinity binding to DC-SIGNR and TIM-1. Binding to PS is sterically hindered during the early binding step. Reprinted with permission from Ref. [80]. Copyright 2022 American Chemical Society.

For millions of years of evolution, each virus particle carries the minimum number of genes and proteins to ensure their reproduction within host cells. However, the role of some viral proteins is not yet unraveled, with some appearing even redundant. For example, murid herpesvirus 4, the current model for human gammaherpesvirus infection, can bind to cell surface glycosaminoglycans using both glycoproteins gp70 and gH/gL. Alsteens et al. discriminated their relative contribution during virus binding to cell surface glycosaminoglycans. They demonstrated that gH/gL is the main actor in glycosaminoglycan binding, engaging in more numerous and stable interactions [81]. The virus must penetrate the cell membrane to complete the replication cycle; however, the true dynamic parameters of this step, such as the forces generated during endocytosis and the time for the virus to be encapsulated by the cell membrane, are not directly accessible by SMFS alone. The infection of a virus is directly related to its endocytosis, so the lack of understanding of the endocytosis dynamic parameters can seriously hinder the study of viruses. To solve this problem, Wang et al. investigated the invagination process of a single human enterovirus 71 (HEV71) via cell membranes during its host cell entry by force tracing. The results revealed that the HEV71 invaginates in membrane vesicles at a force of 58 ± 16 pN and in a duration of 278 ± 68 ms.

11.2.5 Nanodrug Delivery

Regarding the use of nanomaterials in drug delivery, the most reliable approach is to attach the drug to a designed nanocarrier and functionalize it with targeting ligands to target the drug to a specific site of action or specific cell. Nanodrug provides protection against rapid degradation or clearance of the drug and also enhances drug concentration in target tissues [82]. However, different types of nanodrugs have different therapeutic effects; therefore, assessing treatment efficacy by measuring the delivery process is necessary. When nanodrug nanoparticles reach the exterior membrane of a cell, they can interact with the components of the plasma membrane or extracellular matrix and enter the cell to take effect, mainly through endocytosis. As shown in Fig. 11.13, Shan et al. evaluated the therapeutic efficacy of nanodrugs targeting EGFR from the perspective of entry cell efficiency and induced cell mechanical properties using force-tracing and nanoindentation techniques at a single-particle/cell level in real time [83].

Nanodrug delivery offers great potential and benefits for cancer diagnosis and treatment, and revealing the kinetic mechanism of transmembrane transport in living cells will provide more understanding of the diagnosis and treatment of membrane-related diseases. However, the nonspecific administration of nanodrugs will also have adverse side effects on normal cells, thus influencing the therapeutic effect of the drugs. Therefore, it is very significant to improve the selectivity of nanodrugs to tumor cells and control the release of drugs. The entry of PAMAM-loading CPT with DNA aptamer AS1411 (PAMAM-CPT-AS1411) as target ligand into cancer and normal cells were also explored using the force-tracing technique by Shan's group. The forces of PAMAM-CPT-AS1411 entry into A549 cells and Vero cells are 85.86 ± 26.36 pN and 88.60 ± 17.92 pN, respectively. The durations of entry into A549 cells and Vero cells are 171.11 ± 63.42 ms and 81.20 ± 18.98 ms, respectively. The force of nanodrug entry into the two cell lines was not significantly different; however, the corresponding time was significantly distinguishable. The result provided the precise kinetic mechanism for targeted nanodrug delivery. In addition, they respectively used EGF and GE11 conjugated polyamindoamine (PAMAM)-CPT (camptothecin) nanodrugs to target cancer cells (Fig. 11.13a). It was found that the force of the PAMAM-CPT-GE11 entry cell is a little bit smaller and the duration is shorter than that of PAMAM-CPT-EGF, which may be attributed to the larger molecular structure of EGF and the more complex EGF–EGFR receptor-mediated endocytosis (Fig. 11.13b). PAMAM-CPT-EGF and PAMAM-CPT-GE11 cell entries had an average speed of 0.2 μm/s and 0.3 μm/s, respectively (Fig. 11.13c, inset). Thus, in terms of transporting, they can conclude the higher therapeutic efficacy of PAMAM-CPT-EGF with higher cell entry probability and prolonged therapeutic effect [35]. Next, the therapeutic efficacy of PAMAM-CPT-EGF and PAMAM-CPT-GE11 was studied by using the AFM-based nanoindentation to evaluate the therapeutic efficacy of drugs.

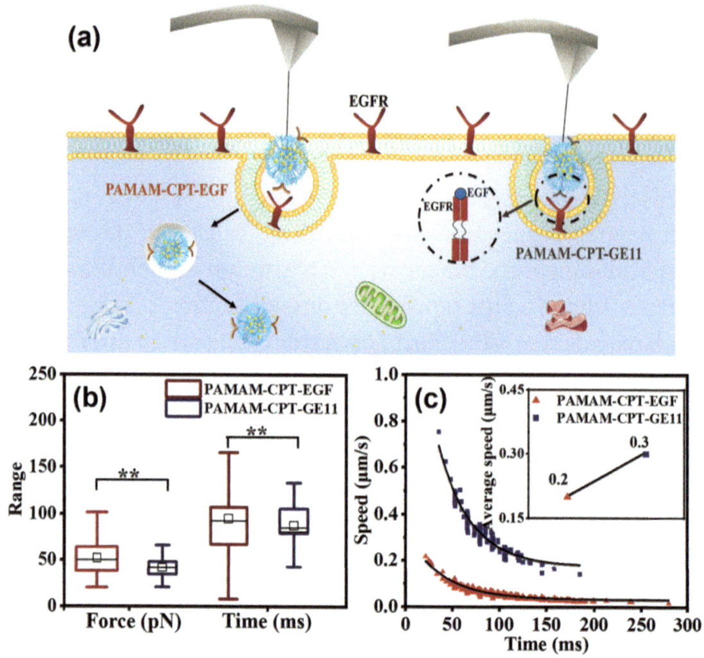

Figure 11.13 Schematic of the endocytosis of PAMAM-CPT-EGF and PAMAM-CPT-GE11 monitored by force tracing. (a) Diagram of cellular uptake of PAMAM-CPT-EGF and PAMAM-CPT-GE11 on the AFM tip. (b) Box plot graphs of the internalization force and duration for PAMAMCPT-EGF and PAMAM-CPT-GE11 cell entry. The boxes show the main distribution (50%) of values. The vertical lines through the boxes represent the 5% to 95% data range; the horizontal lines in each box represent the median; and the smaller square in each box is the mean value. (c) Scatter plot of the relationship between the duration and speed for PAMAM-CPT-EGF and PAMAM-CPT-GE11 cell entries. The inset shows the internalized average speed. Two-tailed Student's t-test: $*P < 0.05$, $**P < 0.01$. Reproduced from Ref. [83] by permission of The Royal Society of Chemistry.

11.3 Protocols

11.3.1 Spring Constant and Sensitivity Calibration of AFM Tip Cantilever

In the SMFS experiment, the binding force between the ligand functionalized on the AFM tip and the receptor on the cell membrane drives the tip to bend downward, and this deflection is converted into a voltage by the reflected signal of the laser, which is reflected in the photodetector. In order to convert the voltage into mechanical force, the AFM tip cantilever must be calibrated before each force spectroscopy experiment to obtain accurate spring constant and deflection sensitivity [84].

The deflection sensitivity measurement is performed by inserting the cantilever containing the substance-functionalized tip into the AFM scanner. Note that the cantilever never keeps drying out, as this may denature the substance modified on the tip. The AFM tip is used to performe classic FD curves on a hard substrate. In the inclined section where the tip contacts the substrate, the measured value on the photodiode (V) changes linearly with the displacement (nm) of the piezoelectric ceramic; therefore, the slope of the curve, the voltage signal can be converted into a displacement signal [85, 86]. To obtain the slope as accurately as possible, the down pressure valve we set should be as large as possible (usually greater than 2 V) to reduce the possible repulsive force between the tip and the substrate and the influence of the substrate contaminants. At the same time, the average value should be measured to exclude the influence of substrate unevenness. It is worth noting that due to the high pressure, there is a possibility of damaging the AFM tip or affecting the functionalized molecule on the tip. Researchers should deal with it as appropriate according to the type and function of the tip.

The spring constant can be measured in several ways, either using a calibrated reference [87], the added mass method [88], the thermal noise method [89], or the sadder method [90]. Generally, the thermal noise method is most widely used and commonly accepted. This method was first proposed by Hutter and Bechhoefer, who noted that the spring constant of a cantilever could be related to its thermal energy by the equipartition theorem [89, 91]. Following the equation [89]:

$$\frac{1}{2}kx^2 = \frac{1}{2}K_B T \qquad (11.2)$$

where k is the spring constant, K_B is the Boltzmann constant, T is the absolute temperature, and x^2 is the mean squared displacement of the cantilever's thermal motion. It was later pointed out that most AFM systems use the thermal noise method to calibrate the spring constant of the tip with the necessary corrections. Therefore, considering the relevant corrections, the following equation is applied:

$$k = \frac{0.971 K_{\mathrm{B}} T}{x^2} \tag{11.3}$$

It is worth mentioning that the commercial AFM software has a built-in part of the calculation with the development of AFM, and the user only needs to perform the fitting step by step according to the operation process. The spring constant of the tips used for SMFS measurement is in the range of 6–100 pN/nm. In addition, each of the other methods has advantages and drawbacks, and a good overview can be found in Ref. [92].

11.3.2 Functionalization of AFM Tip

For the SMFS and force-tracing experiments, the functionalization of AFM tips with relevant biomolecules is required, which is critical for successfully performing a single-molecule experiment [93, 94]. Because of the short lifetime of functionalized tips (due to tip contamination or damage) and samples, they should be freshly prepared just before the experiments. The most common method of tip functionalization is the covalent attachment of molecules/particles via PEG linkers. The biomolecules (e.g., proteins) are conjugated with silanized AFM tips through a bifunctional crosslinker (e.g., Acetal-PEG-NHS) via covalent bonds. The acetal end of the PEG linker is used to connect biomolecules/nanoparticles, and the NHS end is connected to the AFM tip, and the succinimide group attaches to the amino group on the tip [32]. The detailed steps are as follows: First, AFM tips are cleaned in a UV cleaner for 20 min to get rid of any organic contamination on the tips. The tips are functionalized by incubation with 50 μL of APTES (99%) and 20 μL of N,N-diisopropylethylamine (99%) for 2 h by a vapor deposition method. Subsequently, the unreacted APTES is removed by washing AFM tips with $CHCl_3$; then about 1 mg of bifunctional PEG crosslinker and 10 μL of triethylamine are mixed in 150 μL of $CHCl_3$, and the AFM tips are placed into the solution and treated for 2–3 h. The tips are washed with $CHCl_3$ and dried with argon. The PEG-functionalized AFM tips were then immersed in 1% citric acid solution for 10 min, with the conversion of the terminal acetal group into the aldehyde group, and the AFM tips were put into the mixture of nanoparticles/molecules with $-NH_2$ and 1 M $NaCNBH_3$ (freshly prepared by dissolving 32 mg of $NaCNBH_3$ in 500 μL of 10 mM NaOH) for 2 h. Finally, 5 mL of 1 M ethanolamine was added to the reaction solution to inactivate the unreacted aldehyde group. After reaction for 10 min, the tips were washed with phosphate buffer solution (PBS) (100 mM NaCl, 50 mM Na_2HPO_4, pH 7.5) and stored in PBS at 4°C until using (up to 7 days) [95, 96]. The functionalized tip is available for recognition imaging and force spectroscopy tests. Proper functionalization of the molecules/nanoparticles onto the AFM tip and the surface is thus crucial for the success of this type of measurement.

To probe the individual target molecules on the cell surface, AFM tips are functionalized with antibodies that can specifically bind to the target molecules on the cell surface [97]. Xu et al. used a PEG derivative functionalized with a thiol and azide ending group; this combination could be attached to a gold-coated AFM tip by a strong Au-thiol bond. The azide of the immobilized PEG was used for the attachment of an anti-ricin antibody modified with an alkyne group using a "click reaction" [98].

It should be noted that various biomolecules may require a PEG crosslinker with different functional groups. Take another AuNP that was functionalized onto the AFM tip as an example. After AFM tips are placed into the SH-PEG-NHS solution and treated, the AFM tips were immersed in a 200 μL solution containing AuNPs for 1 h. After functionalization, the AFM tips were washed with water three times and protected from light stored at 4°C in water until use [68]. The biomolecules should be strongly attached at a low surface density on the AFM tip to achieve single-molecule detection without unspecific adsorption. The methods mentioned above are also available for attaching cells and viruses onto AFM cantilevers [36]. The AFM tip linked molecules can be performed using different methods, mainly depending on the tip material and the ligand to be bound. As shown in Fig. 11.14, the PDP-PEG-NHS linker is the most suitable choice for bioligands containing a free thiol group [99]. The aldehyde-PEG-NHS (Fig. 11.14b) are preferred as linkers for thiol-free protein attachment, which can directly couple the protein through its lysine residues (using the primary amine) [100]. While both end groups of this linker are amino reactive, the NHS ester has a higher affinity for the amine in chloroform, which avoids loop formation. In other more specialized conjugation strategies, the conjugation strategy of small molecules (biotin [101], hormones [102], or sugars [103]), tethering of hexahistidine (His6)-tagged proteins is an important conjugation method [104, 105]. For this, the NTA-PEG-PDP linker (Fig. 11.14c) is coupled onto the AFM tip, resulting in outstanding nitrilotriacetic acid (NTA) groups. The NTA should be substituted by tris-NTA, which is known to form an even stronger triple complex.

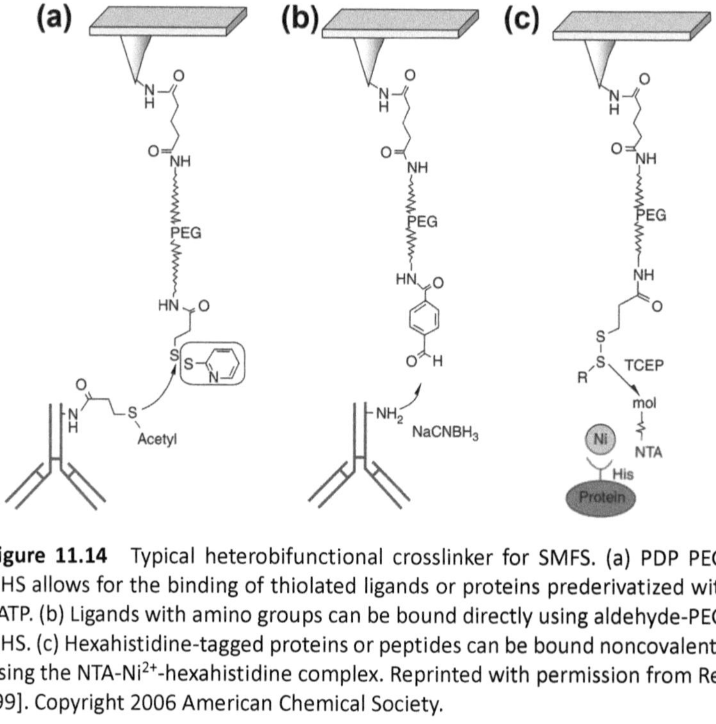

Figure 11.14 Typical heterobifunctional crosslinker for SMFS. (a) PDP PEG-NHS allows for the binding of thiolated ligands or proteins prederivatized with SATP. (b) Ligands with amino groups can be bound directly using aldehyde-PEG-NHS. (c) Hexahistidine-tagged proteins or peptides can be bound noncovalently using the NTA-Ni^{2+}-hexahistidine complex. Reprinted with permission from Ref. [99]. Copyright 2006 American Chemical Society.

11.3.3 Preparing Samples on Supporting Surfaces

SMFS requires the functionalization of AFM tips and samples with applicable biomolecules (or cells). It must be highlighted that the high quality and reproducibility of this preparation are the key factors that govern the success and the dependability of a single-molecule force experiment. Generally, mica, glass, and silicon could be used as supports for biomolecules and cell culture. For mica, the freshly pealed surface with a positive charge can be modified with APTES, and the function of glass and silicon is just the same as that of preparing an AFM tip cantilever [106, 107]. For force spectroscopy measurements on living cells, cells were grown on a 25 mm cover slip. Better results would be achieved when the cells covered ~75% of the cover slip. Before experiments, the adherent cells (e.g., Vero and A549) were washed with PBS three times and a serum-free medium one time to remove cell debris and unattached cells. As a rule, it is recommended to keep functionalized tips and samples hydrated and in conditions where their functionality remains intact [56, 108]. Distinct from cells, microbial cells (e.g., bacteria and yeast) cannot spread on solid supports. Therefore, immobilization through simple adsorption procedures is often unsuitable as this usually leads to cell detachment by the AFM tip. A stronger attachment may be performed either by pretreating the support with lectins [109] or polycations [110] or by binding the cells covalently to the support [111]. An alternative method is to immobilize the cells mechanically in a polymer membrane with a pore size comparable to the dimensions of the cell. This method allows one to perform repeated force spectroscopy on living cells without inducing cell separation or cell injury [112].

11.3.4 Recording Force Curves: Force–Distance Curve and Force–Time Curve

By performing SMFS and recording FD cycles to measure the interaction force of the complexes, the dissociation of the complexes is random. So many such FD cycles with signals need to be obtained to count the distribution of rupture forces. Another consequence of this process is that the rupture force distribution depends on the loading rate (LR). From this correlation, the kinetic off rate could be extracted:

1. The cells were washed thrice with PBS and once with a serum-free medium in sequence to remove cell debris and unattached cells before installing the cell sample in the AFM sample stage.

2. The functionalized tip was mounted and placed over a cell with the help of a CCD camera. During the approach of the AFM tip, it was taken care that the tip did not hit the cell to avoid striking the tip or nonspecific adhesion of the protein to the AFM tip. A "dirty" tip results in false force signals or no force signals at all.
3. Usually, several thousand force curves should be recorded for each set of experiments at the same time. The maximum force applied by the cantilever was kept as low as possible to prevent the contamination of the tip. More quantitative measurements at the single-molecule level in living cells, including bond association/dissociation kinetics, require further measurements and calculations.
4. The FD curve record also uses different sweep duration and different scan sizes. As shown in Fig. 11.15 [80], for very short durations (e.g., 0.1 s and 0.2 s), as many force curves as possible were recorded to compensate for the low binding probabilities due to the short contact times.

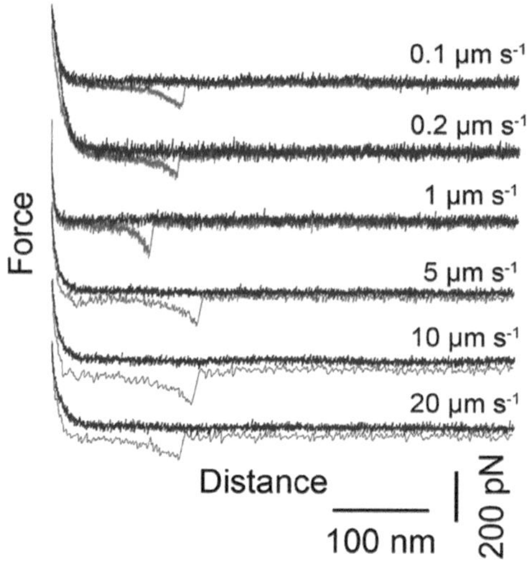

Figure 11.15 Representative FD curves were recorded at various retraction speeds (from 0.1 to 20 μm/s) on the model surface and showed specific adhesion events. Reprinted with permission from Ref. [80]. Copyright 2022 American Chemical Society.

However, for the FT curve collected by force tracing, steps 1 and 2 are the same as that of the SMFS experiment. The next steps are as follows:

Based on step 3 mentioned above, the contact point was defined. The typical FD curves were obtained to find the contact point (Fig. 11.16a) [113]. Subsequently, the functionalized AFM tip was slowly moved onto the cell membrane by turning on the proportional–integral (PI) control system (P = 0.001; I = 0.001), and then, the feedback system was closed. When the substances modified on the AFM tip (e.g., rabies virus particles) were internalized by the cell, the cantilever bent downward to generate a force signal, and the deflection of the AFM tip cantilever was recorded by a 16-bit DA/AD card (PCI-6361e, National Instruments, controlled by LabVIEW) to generate the FT curve (Fig. 11.16b). The 20 kS/s sampling rate (20k data points per second) of data acquisition was usually applied, and the high-frequency electronic noise was filtered by a 100-low pass filter. The collected FT curves were analyzed by LabVIEW software.

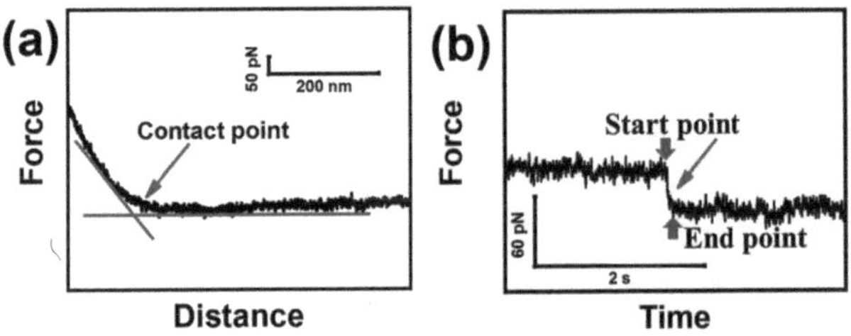

Figure 11.16 Force-tracing measurements of a single rabies virus particle. (a) The contact point between the functionalized AFM tip and the cell membrane surface. The contact point is the intersection of the slope and flat part in the trace curve. (b) Typical force-tracing curves showing the process of cell entry. Reprinted from Ref. [113], with permission from Springer Nature.

11.3.5 Data Analysis: FD Curve and FT Curve (Force, Time, Displacement, and Speed)

The FD curves are usually analyzed by the customized software written in MATLAB. By executing a command of "kspec19," the files containing the sensitivity of the AFM tip cantilever are loaded, and the spring constant of the AFM tip cantilever is inputted. The FD curves will be loaded by the software, after which, every FD curve will be shown successively, and the force signal would be selected and measured by the software, while the measuring and statistics of data are accomplished. In addition, a simple two-state binding model with a separating energy barrier can be used to describe dynamic force spectroscopy experiments. The details are shown as follows:

1. The loading rate (r) of a dataset was determined by changing the scan size and sweep time to be [91]:

$$r = K_{\text{eff}} \frac{2 \times \text{Scan size}}{\text{Sweep time}} \tag{11.4}$$

   where K_{eff} is the effective spring constant of the AFM tip cantilever used, which could be measured and calculated by the related software.

2. Obtaining the corresponding most probable unbinding force F_u under different loading rate r.
3. Plot a scatter of the unbinding force versus the logarithm of loading rate.
4. A straight line ($y = ax + b$) with slope a and Y-axis intersept b was fitted into the plot (Fig. 11.17) [81].

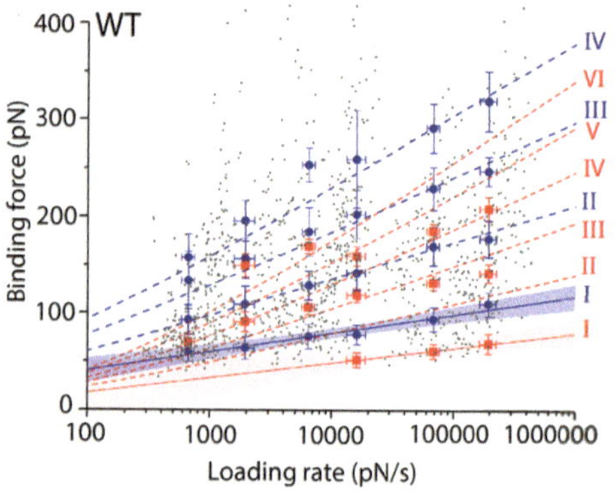

Figure 11.17 Rupture forces separating single murid herpesvirus 4 (MuHV-4)-the glycosaminoglycans (GAGs) bondsplotted against LR, enabling analysis with the Bell–Evans model (full lines). Dashed lines represent the predicted binding forces for multiple uncorrelated interactions breaking simultaneously, according to the Williams–Evans model. Gray dots represent the force and LR values extracted from each individual FD curve. Reprinted with permission from Ref. [81]. Copyright 2021 American Chemical Society.

Parameters were extracted according to the following equation:

$$F_u = \frac{K_B T}{X_\beta}\ln\left(\frac{rX_\beta}{K_B T K_{off}}\right) = \frac{K_B T}{X_\beta}\ln r + \frac{K_B T}{X_\beta}\ln\left(\frac{X_\beta}{K_B T K_{off}}\right) \quad (11.5)$$

where X_β is the separation distance from the equilibrium position, K_{off} is the dissociation kinetic rate constant at zero force, K_B is the Boltzmann constant, and T is the Kelvin temperature.

The kinetic on-rate constant K_{on} defines how fast the bond forms [48]. The receptor–ligand complex reaction can be approximated by pseudo-first-order kinetics; K_{on} for the ligand to the corresponding receptors was extracted from the binding probability (BP) measured at various contact time [18, 114–116]:

$$K_{on} = \frac{\frac{1}{2}\cdot 4\pi r^3{}_{eff}\cdot N_A}{3n_b\tau} \quad (11.6)$$

where N_A is the Avogadro constant (6.022×10^{23}/mol); an effective concentration was estimated from the number of binding partners, n_b (typically 1); the free volume was approximated by a sphere; r_{eff} is the effective radius of the sphere, being the sum of the equilibrium length of the crosslinker and the diameter of the substance modified on the AFM tip (e.g., virus, nanoparticle, and peptides); τ is the interaction time, which can be calculated from the binding probability at different dwell times by using the following equation:

$$\mathrm{BP} = A^*\left[1-\exp\left(\frac{-(t-t_0)}{\tau}\right)\right] \quad (11.7)$$

where A is the maximum BP, t_0 is the lag time, t is a certain holding time, and the AFM tip is in contact with the sample surface. As our software-controlled AFM does not directly support dwell time on the surface, we set a relative maximum deflection limit of 0.2 V. The dwell time was then determined from the force curve, i.e., the time needed to travel from the contact point to the maximum force limit and back again. Origin software is used to fit and extract τ as shown in Fig. 11.18 [117].

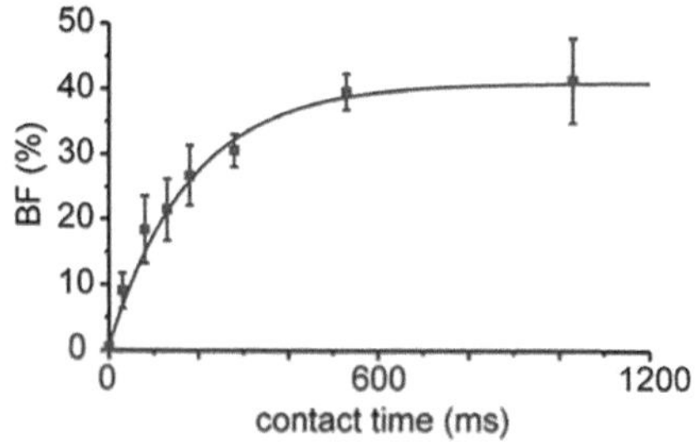

Figure 11.18 The binding frequency (BF) is plotted as a function of contact time for SARS-CoV-2 mutants (Alpha) and the ACE2 receptor-binding domain (RBD). The data points were fitted using a least-squares fit of a monoexponential growth. Reprinted from Ref. [117], with permission from Springer Nature.

The FT curve shows unique advantages for analyzing the force, duration, and displacement during the dynamic transmembrane transporting process at the single-molecule/particle level in living cells [64, 118]. They can be summarized as follows:

1. The FT curves were recorded by the PCI-DAQ card and analyzed by LabVIEW. A functionalized AFM tip (e.g., rabies virus) was located on a cell membrane, FT curves were collected for 20 min on one position of the cell membrane (200 FT curves), and the data were divided by cell into several files (~12 s/file). After ~20 min of detection, the AFM tip is retracted and changed to another cell. The typical force-tracing curve is shown in Fig. 11.16b. The red arrow is the detected force–time signal, the X-axis in the typical force-tracing curve shows time, and the distance between the start point and the end point of the X-axis represents the duration for the completion of a single-particle internalization. The Y-axis in the typical force-tracing curve represents the force converted from the AFM tip cantilever deflection, and the distance between the start point and the end point of the Y-axis represents the force for the completion of the internalization of a single particle. Generally, for an experimental condition such as the force and time required to detect the endocytosis of a single rabies virus, about 200–300 FT curves with force signals are obtained from 10–15 cells at different positions to be selected for statistics. For calculating the probability of observing force–time signals,

the number of FT curves with signal was divided by the total number of curves [113].

2. Calculate the average speed of endocytosis. The displacement D (Fig. 11.7, formula 11.1) of the molecule/nanoparticle during the transmembrane process includes two parts: the bending distance of the cantilever, d, and the extension length of the PEG linker, h. According to the relationship between PEG stretching behavior and the force, the value of h could be obtained by the extended worm-like chain (WLC) model [119]:

$$\frac{FL_\mathrm{p}}{K_\mathrm{B}T} = \frac{1}{4}\left(1 - \frac{h}{L_0} + \frac{F}{K_0}\right)^{-2} - \frac{1}{4} + \frac{h}{L_0} - \frac{F}{K_0} \tag{11.8}$$

where L_p is the persistence length, K_B stands for the Boltzmann constant [120], T refers to the absolute temperature, h is the extension of the PEG linker, L_0 is the contour length [121], and K_0 presents the enthalpic correction.

Based on Hooke's law, the bending distance d of the AFM tip cantilever could be obtained from the following equation:

$$F = k \times d \tag{11.9}$$

where F is the force of single-molecule/nanoparticle transporting into cells measured from the force-tracing curves, and k represents the spring constant of the AFM tip cantilever. Thus, d can be calculated by Equation 11.9, and the average speed is calculated as D/time.

11.4 Conclusion

In summary, we have illustrated the knowledge of using SMFS and force-tracing techniques based on AFM to detect the cell membrane dynamics, including the dynamics of the interaction between materials and cell membranes and the transmembrane tansport dynamics, through introducing AFM tip functionalization, substrate modification, and typical operation modes. In the case studies, we demonstrated the applications of SMFS and force-tracing techniques for studying the interactions between biomolecules, especially for research in virus and nanodrugs. In addition, the analysis of the interaction information obtained from single-molecule force experiments promoted the potential applications of the SMFS and force-tracing techniques for estimating the binding free energy, and the design of various biomolecule-based nanomaterials.

References

1. Shan, Y. P. and Wang, H. D. (2015) The structure and function of cell membranes examined by atomic force microscopy and single-molecule force spectroscopy, *Chem. Soc. Rev.*, **44**, pp. 3617–3638.
2. Zhang, Q. R., Li, S. Y., Yang, Y., Shan, Y. P., and Wang, H. D. (2021) Studying structure and functions of cell membranes by single molecule biophysical techniques, *Biophy. Rep.*, **7**, pp. 384–398.
3. Hughes, M. L. and Dougan, L. (2017) The physics of pulling polyproteins: A review of single molecule force spectroscopy using the AFM to study protein unfolding, *Rep. Prog. Phys.*, **79**, pp. 076601.
4. Hinterdorfer, P. and Dufrêne, Y. (2006) Detection and localization of single molecular recognition events using atomic force microscopy, *Nat. Methods.*, **3**, pp. 347–355.
5. Hamdani, N., Warren, C. M., Solaro, R. J., Linke, W. A., and Fernandez, J. M. (2014) S-glutathionylation of cryptic cysteines enhances titin elasticity by blocking protein folding, *Cell.*, **156**, pp. 1235–1246.
6. Popa, I., Kosuri, P., Alegre-Cebollada, J., Garcia-Manyes, S., and Fernandez, J. M. (2013) Force dependency of biochemical reactions measured by single-molecule force-clamp spectroscopy, *Nat. Protoc.*, **8**, pp. 1261–1276.
7. Alsteens, D., Gaub, H. E., Newton, R., Pfreundschuh, M., Gerber, C., and Muller, D. J. (2017) Atomic force microscopy-based characterization and design of biointerfaces, *Nat. Rev. Mater.*, **2**, pp. 17008.
8. Li, S. Y., Pang, X. L., Zhao, J., Zhang, Q. R., and Shan. Y. P. (2021) Evaluating the single-molecule interactions between targeted peptides and the receptors on living cell membrane, *Nanoscale*, **13**, pp. 17318–17324.
9. Yang, J. S., Petitjean, S. J. L., Koehler, M., Zhang, Q. R., Dumitru, A. C., Chen, W. Z., Derclaye, S., Vincent, S. P., Soumillion, P., and Alsteens, D. (2020) Molecular interaction and inhibition of SARS-CoV-2 binding to the ACE2 receptor, *Nat. Commun.*, **11**, pp. 4541.
10. Florin, E., Moy, V., and Gaub, H. (1994) Adhesion forces between individual ligand-receptor pairs, *Science.*, **264**, pp. 415–417.

11. Dufrene, Y. F. and Pelling, A. E. (2013) Force nanoscopy of cell mechanics and cell adhesion, *Nanoscale.*, **5**, pp. 4094–4104.
12. Sharma, S., Rasool, H. I., Palanisamy, V., Mathisen, C., Schmidt, M., Wong, D. T., and Gimzewski, J. K. (2010) Structural-mechanical characterization of nanoparticle exosomes in human saliva, using correlative AFM, FESEM, and force spectroscopy, *ACS. Nano.*, **4**, pp. 1921–1926.
13. Krieg, M., Fläschner, G., Alsteens, D., Gaub, B. M., Roos, W. H., Wuite, G. J. L., Gaub, H. E., Gerber, C., Dufrêne, Y. F., and Muller, D. J. (2018) Atomic force microscopy-based mechanobiology, *Nat. Rev. Phys.*, **1**, pp. 41–57.
14. Dufrene, Y. F., Ando, T., Garcia, R., Alsteens, D., Martinez-Martin, D., Engel, A., Gerber, C., and Muller, D. J. (2017) Imaging modes of atomic force microscopy for application in molecular and cell biology, *Nat. Nanotechnol.*, **12**, pp. 295–307.
15. Muller, D. J., Dumitru, A. C., Lo Giudice, C., Gaub, H. E., Hinterdorfer, P., Hummer, G., De Yoreo, J. J., Dufrene, Y. F., and Alsteens, D. (2021) Atomic force microscopy-based force spectroscopy and multiparametric imaging of biomolecular and cellular systems, *Chem. Rev.*, **121**, pp. 11701–11725.
16. Wang, L., Qian, Y. H., Sun, Y. T., Liu, B., and Wei, G. (2020) Single-molecule force spectroscopy: A facile technique for studying the interactions between biomolecules and materials interfaces, *Rev. Anal. Chem.*, **39**, pp. 116–129.
17. Yang, B., Liu, Z. W., Liu, H. P., and Nash, M. A. (2020) Next generation methods for single-molecule force spectroscopy on polyproteins and receptor-ligand complexes, *Fron.t Mol. Biosci.*, **7**, pp. 85.
18. Cuellar-Camacho, J. L., Bhatia, S., Reiter-Scherer, V., Lauster, D., Liese, S., Rabe, J. P., Herrmann, A., and Haag, R. (2020) Quantification of multivalent interactions between sialic acid and influenza a virus spike proteins by single-molecule force spectroscopy, *J. Am. Chem. Soc.*, **142**, pp. 12181–12192.
19. Zhao, W. D., Cai, M. J., Xu, H. J., Jiang, J. G., and Wang, H. D. (2013) A single-molecule force spectroscopy study of the interactions between lectins and carbohydrates on cancer and normal cells, *Nanoscale.*, **5**, pp. 3226–3229.
20. Yu, J. P., Jiang, Y. X., Ma, X. Y., Lin, Y., and Fang, X. H. (2007) Energy landscape of aptamer/protein complexes studied by single-molecule force spectroscopy, *Chem. Asian. J.*, **2**, pp. 284–289.

21. Hao, X., Zhu, N., Gschneidtner, T., Jonsson, E. O., Zhang, J. D., Moth-Poulsen, K., Wang, H. D., Thygesen, K. S., Jacobsen, K. W., Ulstrup, J., and Chi, Q. J. (2013) Direct measurement and modulation of single-molecule coordinative bonding forces in a transition metal complex, *Nat. Commun.*, **4**, pp. 2121.

22. Pi, J., Jin, H., Jiang, J. H., Yang, F., Cai, H. H., Yang, P. H., Cai, J. Y., and Chen, Z. W. (2017) Single molecule force spectroscopy for in-situ probing oridonin inhibited ROS-mediated EGF-EGFR interactions in living KYSE-150 cells, *Pharm. Res.*, **119**, pp. 479–489.

23. Kang, N., Zhang, J. H., Qiu, F., Tashiro, S. I., Onodera, S., and Ikejima, T. (2010) Inhibition of EGFR signaling augments oridonin-induced apoptosis in human laryngeal cancer cells via enhancing oxidative stress coincident with activation of both the intrinsic and extrinsic apoptotic pathways, *Cancer. Lett.*, **294**, pp. 147–158.

24. Balakumar, P. and Jagadeesh, G. (2014) A century old renin–angiotensin system still grows with endless possibilities: AT1 receptor signaling cascades in cardiovascular physiopathology, *Cell. Signal.*, **26**, pp. 2147–2160.

25. Strachan, R. T., Sun, J. P., Rominger, D. H., Violin, J. Thomsen, D., S. A. A. R. B., Zhu, X., Kleist, A., Rober, T. C., and Lefkowitz, R. J. (2014) Divergent transducer-specific molecular efficacies generate biased agonism at a G protein-coupled receptor (GPCR), *J. Biol. Chem.*, **289**, pp. 14211–14224.

26. Li, W. H., Xu, J. C., Kou, X. L., Zhao, R., Zhou, W., and Fang, X. H. (2018) Single-molecule force spectroscopy study of interactions between angiotensin II type 1 receptor and different biased ligands in living cells, *Anal. Bioanal. Chem.*, **410**, pp. 3275–3284.

27. Ma, C. D., Acevedo-Vélez, C., Wang, C. X., Gellman, S. H., and Abbott, N. L. (2016) Interaction of the hydrophobic tip of an atomic force microscope with oligopeptides immobilized using short and long tethers, *Langmuir*, **32**, pp. 2985–2995.

28. Shang, X., Shan, Y. P., Pan, Y. G., Cai, M. J., Jiang, J. G., and Wang, H. D. (2013) The force of transporting a single amino acid into the living cell measured using atomic force microscopy, *Chem. Commun. (Camb).*, **49**, pp. 8163–8165.

29. Pfreundschuh, M., Alsteens, D., Wieneke, R., Zhang, C., Coughlin, S. R., Tampé, R., Kobilka, B. K., and Muller, D. J. (2015) Identifying and quantifying two ligand-binding sites while imaging native human membrane receptors by AFM, *Nat. Comm.*, **6**, pp. 8857.

30. Shang, J., Wan, Y. S., Luo, C. M., Ye, G., Geng, Q. B., Auerbach, A., and Li, F. (2020) Cell entry mechanisms of SARS-CoV-2, *Proc. Natl. Acad. Sci. U S A*, **117**, pp. 11727–11734.

31. Zhang, Q. R., Li, S. Y., Yang, Y., Shan, Y. P., and Wang, H. D. (2021) Studying structure and functions of cell membranes by single molecule biophysical techniques, *Biophy. Rep.*, **7**, pp. 384–398.

32. Liu, Y. L, Yang, Y., Zhang, Q. R., Lu, D. H., Li, S. Y., Li, J. F., Yang, G. C., and Shan, Y. P. (2021) Dynamics of delivering aptamer targeted nano-drugs into cells, *J. Mater. Chem. B.*, **9**, pp. 952–957.

33. Pan, Y. G., Wang, S. W., Shan, Y. P., Zhang, D. L., Gao, J., Zhang, M., Liu, S. H., Cai, M. J., Xu, H. J., Li, G. H., Qin, Q. W., and Wang, H. D. (2015) Ultrafast tracking of a single live virion during the invagination of a cell membrane, *Small*, **11**, pp. 2782–2788.

34. Ding, B. H., Tian, Y. M., Pan, Y. G., Shan, Y. P., Cai, M. J., Xu, H. J., Sun, Y. C., and Wang, H. D. (2015) Recording the dynamic endocytosis of single gold nanoparticles by AFM-based force tracing, *Nanoscale*, **7**, pp. 7545–7549.

35. Lu, D. H., Yang, X. D., Zhang, Q. R., Wang, R. X., Zhou, S. Y., Yang, G. C., and Shan, Y. P. (2019) Tracking the single-carbon-dot transmembrane transport by force tracing based on atomic force microscopy, *ACS. Biomater. Sci. Eng.*, **5**, pp. 432–437.

36. Zhang, Q. R., Tian, F. L., Wang, F., Guo, Z. Y., Cai, M. J., Xu, H. J., Wang, H. D., Yang, G. C., Shi, X. H., Shan, Y. P., and Cui, Z. Q. (2020) Entry dynamics of single Ebola virus revealed by force tracing, *ACS Nano*, **14**, pp. 7046–7054

37. Goni, F. M. (2014) The basic structure and dynamics of cell membranes: An update of the Singer–Nicolson model, *Biochim. Biophys. Acta.*, **1838**, pp. 1467–1476.

38. Muller, D. J. (2008) AFM: A nanotool in membrane biology, *Biochemistry*, **47**, pp. 7986–7998.

39. Stigler, J., Ziegler, F., Gieseke, A., Gebhardt, J. C. M., and Rief, M. (2011) The complex folding network of single calmodulin molecules, *Science*, **334**, pp. 512–516.

40. Oesterhelt, F., Oesterhelt, D., Pfeiffer, M., Engel, A., Gaub, H., and Muller, D. J. (2000) Unfolding pathways of individual bacteriorhodopsins, *Science*, **288**, pp. 143–146.

41. Gao, Y., Zorman, S., Gundersen, G., Xi, Z. Q., Ma, L., Sirinakis, G., Rothman, J. E., and Zhang, Y. L. (2012) Single reconstituted neuronal SNARE complexes zipper in three distinct stages, *Science*, **337**, pp. 1340–1343.

42. Onoa, B., Dumont, S., Liphardt, J., Smith, S. B., Tinoco, I., and Bustamante, C. (2003) Identifying kinetic barriers to mechanical unfolding of the *T. thermophila* ribozyme, *Science*, **299**, pp. 1892–1895.

43. Dill, K. A. and Maccallum, J. L. (2012) The protein-folding problem, 50 years on, *Science*, **338**, 1042–1046.

44. Anfinsen, C. B. (1973) Principles that govern the folding of protein chains, *Science*, **181**, pp. 223–230.

45. Yuan, G. D., Ma, Q., Wu, T., Wang, M. D., Li, X., Zuo, J., and Zheng, P. (2019) Multistep protein unfolding scenarios from the rupture of a complex metal cluster Cd3S9, *Sci. Rep.*, **9**, pp. 10518.

46. Fuhrmann, A. and Ros, R. (2010) Single-molecule force spectroscopy: A method for quantitative analysis of ligand-receptor interactions, *Nanomedicine*, **5**, pp. 657–666.

47. Engel, A. and Muller, D. J. (2000) Observing single biomolecules at work with the atomic force microscope, *Nat. Struct. Biol.*, **7**, pp. 715–718.

48. Liu, B. Y., Chen, W., and Zhu, C. (2015) Molecular force spectroscopy on cells, *Annu. Rev. Phys. Chem.*, **66**, pp. 427–451.

49. Dembo, M., Torney, D. C., Saxman, K., and Hammer, D. (2018) The reaction-limited kinetics of membrane-to-surface adhesion and detachment, *Proc. R. Soc. Lond. B.*, **234**, pp. 55–83.

50. Sisquella, X., Pourcq, K., Alguacil, J., Robles, J., Sanz, F., Anselmetti, D., Imperial, S., and Fernandez-Busquets, X. (2010) A single-molecule force spectroscopy nanosensor for the identification of new antibiotics and antimalarials, *FASEB J.*, **24**, pp. 4203–4217.

51. Morfill, J., Blank, K., Zahnd, C., Luginbühl, B., Kühner, F., Gottschalk, K. E., Plückthun, A., and Gaub, H. E. (2007) Affinity-matured recombinant antibody fragments analyzed by single-molecule force spectroscopy, *Biophy. J.*, **93**, pp. 3583–3590.

52. Lundstrom, K. (2009) An overview on GPCRs and drug discovery: Structure-based drug design and structural biology on GPCRs, *Meth. Mol. Biol.*, **552**, pp. 51.

53. Pfreundschuh, M., Alsteens, D., Wieneke, R., Zhang, C., Coughlin, S. R., Tampe, R., Kobilka, B. K., and Muller, D. J. (2015) Identifying and quantifying two ligand-binding sites while imaging native human membrane receptors by AFM, *Nat. Commun.*, **6**, pp. 8857.

54. Krishnamurthy, H., Piscitelli, C. L., and Gouaux, E. (2009) Unlocking the molecular secrets of sodium-coupled transporters, *Nature*, **459**, pp. 347–355.

55. Braun, V. (1998) Pumping iron through cell membranes, *Science*, **282**, pp. 2202–2203.

56. Pan, Y. G., Zhang, Y. B., Gongpan, P. C., Zhang, Q. R., Huang, S. T., Wang, B., Xu, B. Q., Shan, Y. P., Xiong, W. Y., Li, G. H., and Wang, H. D. (2018) Single glucose molecule transport process revealed by force tracing and molecular dynamics simulations, *Nanoscale Horiz.*, **3**, pp. 517–524.

57. Lo Giudice, C., Dumitru, A. C., and Alsteens, D. (2019) Probing ligand-receptor bonds in physiologically relevant conditions using AFM, *Anal. Bioanal. Chem.*, **411**(25), pp. 6549–6559.

58. Alsteens, D., Pfreundschuh, M., Zhang, C., Spoerri, P. M., Coughlin, S. R., Kobilka, B. K., and Muller, D. J. (2015) Imaging G protein-coupled receptors while quantifying their ligand-binding free-energy landscape, *Nat. Methods*, **12**, pp. 845–851.

59. Zocher, M., Roos, C., Wegmann, S., Bosshart, P. D., Dötsch, V., Bernhard, F., and Muller, D. J. (2011) Single-molecule force spectroscopy from nanodiscs: An assay to quantify folding, stability, and interactions of native membrane proteins, *ACS Nano*, **6**, pp. 961–971.

60. Yang, L. and Monteiro-Riviere, N. A. (2016) Mechanisms of cell uptake, inflammatory potential and protein corona effects with gold nanoparticles, *Nanomedicine*, **11**, pp. 3185–3203.

61. Reifarth, M., Hoeppener, S., and Schubert, U. S. (2018) Uptake and intracellular fate of engineered nanoparticles in mammalian cells: Capabilities and limitations of transmission electron microscopy-polymer-based nanoparticles, *Adv. Mater.*, **30**, pp. 1703704.

62. Santos-Oliveira, R. (2019) Application of technetium 99 metastable radioactive nanosystems: Nanoparticles, liposomes, and nanoemulsion for biomedical application, *Curr. Pharm. Rep.*, **5**, pp. 281–302.

63. Zhang, Y. N., Wilson, P., Tavares, A. J., McGilvray, I. D., and Chan, W. C. W. (2016) Nanoparticle-liver interactions: Cellular uptake and hepatobiliary elimination, *J. Control. Release*, **240**, pp. 332–348.

64. Wang, R. X., Yang, X. D., Lu, D. H., Zhang, Q. R., Zhou, S. Y., Yang, Y., Yang, G. C., and Shan, Y. P. (2019) Monitoring the trans-membrane transport of single fluorescent silicon nanoparticles based on the force tracing technique, *Anal. Methods*, **11**, pp. 1724–1728.

65. Chen, X., Tian, F. L., Li, M., Xu, H. J., Cai, M. J., Li, Q., Zuo, X. L., Wang, H. D., Shi, X. H., Fan, C. H., Baigude, H., and Shan, Y. P. (2020) Size-independent transmembrane transporting of single tetrahedral DNA nanostructures, *Glob. Chall.*, **4**, pp. 1900075.

66. Wang, Y. F., Zhang, Q. R., Tian, F. L., Wang, H. D., Wang, Y. F., Ma, X. W., Huang, Q. Q., Cai, M. J., Ji, Y. L., Wu, X. C., Gan, Y. L., Yan,Y., Dawson, K. A., Guo, S. T., Zhang, J. C., Shi, X. H., Shan, Y. P., and Liang, X. J. (2022) Spatiotemporal tracing of the cellular internalization process of rod-shaped nanostructures, *ACS Nano*, **16**, pp. 4059–4071.

67. Han, L., Huang, R. Q., Liu, S. H., Huang, S. X., and Chen, J. (2010) Peptide-conjugated PAMAM for targeted doxorubicin delivery to transferrin receptor overexpressed tumors, *Mol. Pharm.*, **7**, pp. 2156–2165.

68. Li, S. Y., Wang, R. X., Li, J. F., Liu, Y. L., Fu, Y. F., Zhou, J., Yang, G. C., and Shan, Y. P. (2021) Revealing the dynamic mechanism by which transferrin promotes the cellular uptake of HAIYPRH peptide-conjugated nanostructures by force tracing, *Mol. Pharm.*, **18**, pp. 1480–1485.

69. Cohen, F. S. (2016) How viruses invade cells, *Biophys. J.*, **110**, pp. 1028–1032.

70. Xu, H., Liu, B., Xiao, Z., Zhou, M. L., Ge, L., Jia, F., Liu, Y. L., Jin, H. S., Zhu, X. L., Gao, J., Akhtar, J., Xiang, B., Tan, K., and Wang, G. Y. (2021) Computational and experimental studies reveal that thymoquinone blocks the entry of coronaviruses into in vitro cells, *Infect. Dis. Ther.*, **10**, pp. 483–494.

71. Krishnamoorthy, S., Swain, B., Verma, R. S., and Gunthe, S. S. (2020) SARS-CoV, MERS-CoV, and 2019-nCoV viruses: An overview of origin, evolution, and genetic variations, *Virusdisease*, pp. 1–13.

72. Maginnis, M. S. (2018) Virus-receptor interactions: The key to cellular invasion, *J. Mol. Biol.*, **430**, pp. 2590–2611.

73. Grove, J. and Marsh, M. (2011) The cell biology of receptor-mediated virus entry, *J. Cell. Biol.*, **195**, pp. 1071–1082.

74. Marsh, M. and Helenius, A. (2006) Virus entry: Open sesame, *Cell*, **124**, pp. 729–740.

75. Alsteens, D., Newton, R., Schubert, R., Martinez-Martin, D., Delguste, M., Roska, B., and Muller, D. J. (2016) Nanomechanical mapping of first binding steps of a virus to animal cells, *Nat. Nanotechnol.*, **12**, pp. 177–183.

76. Kukura, P., Ewers, H., Muller, C., Renn, A., Helenius, A., and Sandoghdar, V. (2009) High-speed nanoscopic tracking of the position and orientation of a single virus, *Nat. Methods*, **6**, pp. 923–927.

77. Yang, J., Petitjean, S., Derclaye, S., Koehler, M., and Alsteens, D. (2020) Molecular interaction and inhibition of SARS-CoV-2 binding to the ACE2 receptor, *Nat. Commun.*, **11**, pp. 4541.

78. Dermody, T. S., Kirchner, E., Guglielmi, K. M., Stehle, T., and Manchester, M. (2009) Immunoglobulin superfamily virus receptors and the evolution of adaptive immunity, *PLoS. Pathogens*, **5**, pp. 1–4.

79. Yamauchi, Y. and Helenius, A. (2013) Virus entry at a glance, *J. Cell. Sci.*, **15**, pp. 1289–1295.

80. Zhang, Q. R., Yang, J. S., Tillieux, S., Guo, Z. Y., Natividade, R. D. S., Koehler, M., Petitjean, S., Cui, Z. Q., and Alsteens, D. (2022) Stepwise enzymatic-dependent mechanism of Ebola virus binding to cell surface receptors monitored by AFM, *Nano Lett.*, **22**, pp. 1641–1648.

81. Delguste, M., Brun, G. L., Cotin, F., Machiels, B., Gillet, L., and Alsteens, D. (2021) Single-virus force spectroscopy discriminates the intrinsic role of two viral glycoproteins upon cell surface attachment, *Nano Lett.*, **21**, pp. 847–853.

82. Mohapatra, S., Ranjan, S., Dasgupta, N., Mishra, R. K., and Thomas, S. (2018). *Applications of Targeted Nano Drugs and Delivery Systems*, Rana, V. and Sharma, R. Chapter 5 "Recent advances in development of nano drug delivery," (Elsevier Press), pp. 93–131.

83. Zhao, J., Li, S. Y., Pang, X. L., and Shan, Y. P. (2022) Evaluating the therapeutic efficacy of nano-drugs targeting epidermal growth factor receptor, *Chem. Commun. (Camb).*, **58**, pp. 2726–2729.

84. Burnham, N. A., Chen, X., Hodges, C. S., Matei, G. A., Thoreson, E. J., Roberts, C. J., Davies, M. C., and Tendler, S. (2002) Comparison of calibration methods for atomic-force microscopy cantilevers, *Nanotechnology*, **14**, pp. 1.

85. Sumbul, F. and Rico, F. (2019) Single-molecule force spectroscopy: Experiments, analysis, and simulations. *Methods. Mol. Biol.*, **1886**, pp. 163–189.

86. Putman, C. A. J., De Grooth, B. G., Van Hulst, N. F., and Greve, J. (1992) A detailed analysis of the optical beam deflection technique for use in atomic force microscopy, *J. Appl. Phys.*, **72**, pp. 6–12.

87. Gibson, C. T., Watson, G. S., and Myhra, S. (1996) Determination of the spring constants of probes for force microscopy/spectroscopy, *Nanotechnology*, **7**, pp. 259–262.

88. Cleveland, J. P., Manne, S., Bocek, D., and Hansma, P. K. (1993) A nondestructive method for determining the spring constant of cantilevers for scanning force microscopy, *Rev. Sci. Instrum.*, **64**, pp. 403–405.

89. Hutter, J. L. and Bechhoefer, J. (1993) Calibration of atomic-force microscope tips, *Rev. Sci. Instrum.*, **64**, pp. 1868–1873.

90. Sader, J. E., Sanelli, J. A., Adamson, B. D., Monty, J. P., Wei, X., Crawford, S. A., Friend, J. R., Marusic, I., Mulvaney, P., and Bieske, E. J. (2012) Spring constant calibration of atomic force microscope cantilevers of arbitrary shape, *Rev. Sci. Instrum.*, **83**, pp. 103705.

91. Rankl, C., Wildling, L., Neundlinger, I., Kienberger, F., Gruber, H., Blaas, D., and Hinterdorfer, P. (2011) Determination of the kinetic on- and off-rate of single virus-cell interactions, *Methods Mol. Biol.*, **736**, pp. 197–210.

92. Ohler, B. (2007) Practical advice on the determination of cantilever spring constants. *Veeco Application Note.*

93. Yang, B. S., Liu, Z. W., Liu, H. P., and Nash, M. A. (2020) Next generation methods for single-molecule force spectroscopy on polyproteins and receptor-ligand complexes, *Front. Mol Biosci.*, **7**. pp. 85.

94. Kolberg, A., Wenzel, C., Hugel, T., Gallei, M., and Balzer, B. N. (2020) Covalent attachment of single molecules for AFM-based force spectroscopy, *J. Vis. Exp.*, **16**, pp. e60934.

95. Braga, P. C. (2016) *Atomic Force Microscopy in Biomedical Research: Methods and Protocols* (Springer-Verlag GmbH).

96. Wang, H. D. and Li, G. H. (2018) *Membrane Biophysics* Cai, M. J., Gao, J., and Wang. H. D. Chapter 3 "Imaging membranes by high-resolution tomic force microscopy" (Springer-Verlag GmbH), pp. 45–89.

97. Li, M., Dan, D., Liu, L. Q., Ning, X., and Wang, Y. C. (2017) Imaging and force recognition of single molecular behaviors using atomic force microscopy, *Sensors.*, **17**, pp. 200.

98. Chen, G. J., Ning, X. H., Park, B., Boons, G. J., and Xu, B. Q. (2009) Simple, clickable protocol for atomic force microscopy tip modification and its application for trace ricin detection by recognition imaging, *Langmuir*, **25**, pp. 2860–2864.

99. Kamruzzahan, A. S. M., Ebner, A., Wildling, L., Kienberger, F., Riener, C. K., Hahn, C. D., Pollheimer, P. D., Winklehner, P., Hölzl, M., Lackner, B., Schörkl, D. M., Hinterdorfer, P., and Gruber, H. J. (2006) Antibody linking to atomic force microscope tips via disulfide bond formation, *Bioconjugate Chem.*, **17**, pp. 1473–1481.

100. Ebner, A., Wildling, L., Kamruzzahan, A. S. M., Rankl, C., Wruss, J., Hahn, C. D., HoLzl, M., Zhu, R., Kienberger, F., and Blaas, D. (2007) A new, simple method for linking of antibodies to atomic force microscopy tips, *Bioconjugate Chem.*, **18**, pp. 1176–1184.

101. Stroh, C. M., Ebner, A., Klampfl, C., Gall, A. A., Romanin, C., Lyubchenko, Y. L., Hinterdorfer, P., and Gruber, H. J. (2003) Simple test system for

single molecule recognition force microscopy, *Anal. Chim. Acta.*, **1**, pp. 59–75.

102. Wildling, L., Hinterdorfer, P., Kusche-Vihrog, K., Treffner, Y., and Oberleithner, H. (2009) Aldosterone receptor sites on plasma membrane of human vascular endothelium detected by a mechanical nanosensor, *Pflug. Arch. Eur. J. Phy.*, **458**, pp. 223–230.
103. Puntheeranurak, T., Wimmer, B., Castaneda, F., Gruber, H. J., Hinterdorfer, P., and Kinne, R. (2007) Substrate specificity of sugar transport by rabbit SGLT1: Single-molecule atomic force microscopy versus transport studies, *Biochemistry*, **46**, pp. 2797–2804.
104. Rangl, M., Nevo, R., Liashkovich, I., Shahin, V., Reich, Z., Ebner, A., and Hinterdorfer, P. (2009) Stable, non-destructive immobilization of native nuclear membranes to micro-structured PDMS for single-molecule force spectroscopy, *Chem. Phys. Chem.*, **10**, pp. 1322.
105. Kienberger, F., Kada, G., Gruber, H. J., Pastushenko, V. P., Riener, C., Trieb, M., Knaus, H. G., Schindler, H., and Hinterdorfer, P. (2015) Recognition forces spectroscopy studies of the NTA-His6 bond, *Single. Mol.*, **1**, pp. 59–65.
106. Scholl, Z. N. and Marszalek, P. E. (2018) AFM-based single-molecule force spectroscopy of proteins, *Methods. Mol. Biol.*, **1814**, pp. 35–47.
107. Wang, H. D. and Li, G. H. (2018) *Membrane Biophysics*, Shan, Y. P. Chapter 4 "Detection of membrane mechanical properties and endocytosis by single molecule force spectroscopy" (Springer-Verlag GmbH), pp. 91–115.
108. Yang, B. Y., Xu, H. J., Wang, S. W., Cai, M. J., Shi, Y., Yang, G. C., Wang, H. D., and Shan, Y. P. (2016) Studying the dynamic mechanism of transporting a single drug carrier-polyamidoamine dendrimer through cell membranes by force tracing, *Nanoscale*, **8**, pp. 18027–18031.
109. Gad, M. and Itoh, A. (1997) Mapping cell wall polysaccharides of living microbial cells using atomic force microscopy, *Cell. Biol. Int.*, **21**, pp. 697–706.
110. Schaer-Zammaretti, P. and Ubbink, J. (2003) Imaging of lactic acid bacteria with AFM-elasticity and adhesion maps and their relationship to biological and structural data, *Ultramicroscopy*, **97**, pp. 199–208.
111. Camesano, T. A., Natan, M. J., and Logan, B. E. (2000) Observation of changes in bacterial cell morphology using tapping mode atomic force microscopy, *Langmuir*, **16**, pp. 4563–4572.
112. Kasas, S. and Ikai, A. (1995) A method for anchoring round shaped cells for atomic force microscope imaging, *Biophys. J.*, **68**, pp. 1678.

113. Li, S. Y., Pan, Y. G., Teng, H. D., Shan, Y. P., Yang, G. C., and Wang, H. D. (2022) Revealing the cell entry dynamic mechanism of single rabies virus particle, *Chem. Res. Chin. Univ.*, **38**, pp. 838–842.

114. Wildling, L., Rankl, C., Haselgrubler, T., Gruber, H. J., Holy, M., Newman, A. H., Zou, M. F., Zhu, R., Freissmuth, M., Sitte, H. H., and Hinterdorfer, P. (2012) Probing binding pocket of serotonin transporter by single molecular force spectroscopy on living cells, *J. Biol. Chem.*, **287**, pp. 105–113.

115. Rankl, C., Kienberger, F., Wildling, L., Wruss, J., Gruber, H. J., Blaas, D., and Hinterdorfer, P. (2008) Multiple receptors involved in human rhinovirus attachment to live cells, *P. Nat. Acad. Sci.*, **105**, pp. 17778–17783.

116. Baumgartner, W., Gruber, H. J., Hinterdorfer, P., and Drenckhahn, D. (2000) Affinity of trans-interacting VE-cadherin determined by atomic force microscopy, *Single. Mol.*, **1**, pp. 119–122.

117. Koehler, M., Ray, A., Moreira, R. A., Juniku, B., Poma, A. B., and Alsteens, D. (2021) Molecular insights into receptor binding energetics and neutralization of SARS-CoV-2 variants, *Nat. Commun.*, **12**, pp. 6977.

118. Yang, Y., Zhang, Q. R., Cai, M. J., Xu, H. J., Lu, D. H., Liu, Y. L., Fu, Y. F., Yang, G. C., and Shan, Y. P. (2020) Size-dependent transmembrane transport of gold nanocages, *ACS. Omega*, **5**, pp. 9864–9869.

119. Zhou, S. Y., Yang, B. Y., Chen, Y., Zhang, Q. R., Cai, M. J., Xu, H. J., Yang, G. C., Wang, H. D., and Shan, Y. P. (2018) Exploring the trans-membrane dynamic mechanisms of single polyamidoamine nano-drugs via a "force tracing" technique, *RSC. Adv.*, **8**, pp. 8626–8630.

120. Marantan, A. and Mahadevan, L. (2018) Mechanics and statistics of the worm-like chain, *Ame. J. Phy.*, **86**, pp. 86–94.

121. Fu, Y. B., Liu, Y. R., Wang, P. Y., and Xie, P. (2018) A revised worm-like chain model for elasticity of polypeptide chains, *J. Polym. Sci. Part B: Pol. Phys.*, **56**, pp. 297–307.

Chapter 12

Computer Simulations to Explore Membrane Organization and Transport

Tingting Fu,[a] Tao Zhang,[b] Guixuan Xing,[a] Shi Feng,[a] and Qingchuan Zheng[a]
[a]*Institute of Theoretical Chemistry, College of Chemistry, Jilin University, Changchun 130023, Jilin, China*
[b]*State Key Laboratory of Electroanalytical Chemistry, Research Center of Biomembranomics, Changchun Institute of Applied Chemistry, Chinese Academy of Sciences, Changchun 130022, Jilin, China*
zhengqc@jlu.edu.cn

The membrane is composed of phospholipids, proteins, and sugars in diverse amounts and proportions, and it undertakes many important biological activities, including material transport, signal transduction, energy transfer, and enzymatic reaction. Computer simulations play an increasingly important role in the study of membrane organization and transport at the atomic level, especially in providing dynamic information. As computing power increases, cell-scale molecular simulations are also applied to larger and more complex membrane systems. Computer simulations not only allow the verification of experimental findings or new hypotheses,

Biomembranomics: Structure, Techniques, and Applications
Edited by Hongda Wang

ISBN 978-981-4968-61-4 (Hardcover), 978-1-003-45635-3 (eBook)
www.jennystanford.com

but also provide data that cannot be obtained from experiments or become valuable complements to experiments. This chapter mainly introduces the technical principles and applications of computational simulation in membrane systems and also contains initial model construction, advantages and disadvantages of methods, and applicable scenarios. We aim to summarize the latest technological developments and provide references for the implementation of computational simulation in membrane systems. We believe that computational simulations will become more and more widely used in membrane systems with the rapid improvements in hardware and algorithms.

12.1 Introduction

The membrane is a complex, diverse, and dynamic supramolecular liquid crystalline structure, which consists of phospholipids, proteins, and sugars. Each component has a variety of types and functions, for example, membrane proteins [1]. Membrane proteins have been found to comprise approximately one-third of the human genome, and over half of these are known as drug targets [2]. Among them, most members are G protein-coupled receptors (GPCRs), transporters, ion channels, and membrane-bound enzymes that are vital to material transport, signal transduction, energy transfer, and enzymatic reaction [3]. Thus, the biological function of membrane proteins is an important focus in fundamental research.

Technological breakthroughs in structural biology, such as cryo-electron microscopy (cryo-EM), X-ray, and nuclear magnetic resonance (NMR), have led to the identification of an increasing number of membrane protein structures [4]. In fact, many structures of membrane protein are incomplete, and obtaining full-length protein structures with high resolution remains challenging due to their relatively hydrophobic surfaces [5]. A variety of molecular biology or biochemical methods, including directed evolution [3], high-throughput stability screening [6], genomic edits [7], and crystallization chaperone strategies [8], are applied to improve expression systems and improve the stabilities of membrane proteins for crystallization. Nevertheless, these protein structures only represent static snapshots obtained under specific experimental conditions, and they may miss important information on conformational dynamics. On the other hand, computational methods play an increasingly important role in the study of structure, function, and dynamic information of membrane proteins, such as homology modeling, docking, molecular dynamics (MD) simulations, and quantum mechanical (QM) calculations. These computational methods not only allow the verification of experimental findings or new hypotheses at the atomic level, but also provide data that cannot be obtained from experiments, such as the pressure profile of membranes [9, 10].

Currently, MD simulations have become an indispensable tool to obtain structural information and interaction details, probe the conformational landscapes, and uncover biomolecular mechanisms of membrane proteins. MD simulations predict how every atom in membrane protein systems will move over time based on molecular force fields [11]. In contrast to QM methods, electrons in molecular force fields are not treated explicitly but together with the nuclei as effective atoms in molecular force fields [10]. Therefore, MD simulations can be used to investigate the biochemical and biophysical systems that contain tens of thousands of atoms. Membrane proteins often undergo large-scale conformational transitions and visit distinct conformational states during their biological functions [12]. A complete understanding of their conformational transition pathways is essential for biology and pharmacology. Due to large energy barriers, the conformational transitions usually take place on timescales of milliseconds or even longer [13]. In addition to unbiased (or classical) MD simulation, many enhanced sampling techniques are also widely used in membrane protein systems allowing simulations to capture longer timescale events, such as accelerated molecular dynamics (aMD), Gaussian accelerated molecular dynamics (GaMD), steered molecular dynamics (SMD) simulations, metadynamics (MetaD), and umbrella sampling. Furthermore, the development of graphics processing units (GPUs) has increased computing power and allowed simulations to be run locally at a modest cost. As computing power increases, MD simulations are also applied to larger and more complex membrane systems, such as cell-scale or organelle-scale systems.

MD simulation is a powerful and accessible technique routinely used to investigate the dynamic motions of macromolecules at the atomic level [14–18]. We believe that computer simulations will become faster, cheaper, broader, and more accurate with the improvements in hardware and algorithms, which can be widely used to characterize and quantify rare events and become valuable complements to experimental and clinical approaches.

In the following sections, we introduce the general technical principle of computational simulations. Then, two tools, SWISS-MODEL [19] for constructing structures of membrane proteins and CHARMM-GUI [20] for generating membrane systems, are introduced. Finally, the applications of MD simulation in membrane proteins are presented.

12.2 Technical Principle

12.2.1 Force Fields

The success of computer simulation depends on the quality of force fields, which define the interaction parameters between the particles [21]. Force fields are mathematical expressions describing the dependence of the system energy on the coordinates of its particles, and they are highly simplifying assumptions [22]. In biological systems, there are many force fields. An important distinction between the force fields is the level of resolution considered. Depending on the level of resolution, force fields can be roughly divided into three categories: all-atom model, coarse-grained (CG) model, and ultra-coarse-grained (UCG) model. At the all-atom level, all atoms are considered explicitly. In the CG model, small groups of atoms and associated hydrogens are represented by CG beads. Additionally, at the UCG level, lipids and proteins are represented only qualitatively by few-bead models, and solvent is considered implicitly.

It is worth noting that each force field has its particular strengths and weaknesses related to the data and procedure used in its parametrization. The final choice of force field depends on the particular problem being considered [23]. Moreover, these force fields have been integrated into most modern simulation software or packages, for instance, AMBER [24], NAMD [25], GROMACS [26], and

OpenMM [27]. They perform similar computations but differ in how efficiently they map to various hardware and supported features (e.g., enhanced sampling methods, temperature and pressure control schemes, and support for CG simulations) [28].

12.2.1.1 All-atom models

At the all-atom level, all atoms are considered explicitly. The simplicity of the force field representation of molecular features—springs for bond length and angles, periodic functions for bond rotations and Lennard–Jones potentials, and Coulomb's law for van der Waals and electrostatic interactions, respectively—assures that energy and force calculations are fast [22]. The popular all-atom force fields include AMBER, CHARMM, GROMOS, and OPLS. Among them, AMBER, CHARMM, and GROMOS are commonly used in simulations of biomolecules, while OPLS is originally developed to simulate condensed matter and may have the most extensively optimized ligand parameters [29]. It is worth noting that these force fields are constantly evolving and different versions are available. The current AMBER force field for proteins is ff19SB [30]; Lipid17 [31] is used for lipids, and GAFF [32] is developed to model ligands interacting with proteins. The functional form of the AMBER force field can be shown as follow:

$$V = \sum_{\text{bonds}} \frac{k}{2}(l - l_0)^2 + \sum_{\text{angles}} \frac{k}{2}(\theta - \theta_0)^2 + \sum_{\text{torsions}} \frac{V_n}{2}[1 + \cos(n\omega - \gamma)] + \sum_{i<j} \left\{ 4\varepsilon_{ij} \left[\left(\frac{\sigma_{ij}}{r_{ij}} \right)^{12} - \left(\frac{\sigma_{ij}}{r_{ij}} \right)^{6} \right] + \frac{q_i q_j}{4\pi\varepsilon_0 r_{ij}} \right\} \tag{12.1}$$

where l, θ, ω are the bond length, angle, and dihedral angle, respectively. The fourth term represents the sum of the non-bonded interactions.

The CHARMM force field also forms several improved versions, such as CHARMM36 [33]. In the GROMACS format of CHARMM36, the CGenFF version 4.6 and the CHARMM36m protein force field are the latest revisions. The form of CHARMM force fields is as follow [10]:

$$V = \sum K_b (r - r_0)^2 + \sum K_\theta (\theta - \theta_0)^2 + \sum \left[|k_\varphi| - k_\varphi \cos(n\varphi) \right] +$$

$$\sum k_\chi (\chi - \chi_0)^2 + \sum_{i,j} \frac{q_i q_j}{4\pi\varepsilon_0 r_{ij}} + \sum_{i,j} \left(\frac{A_{ij}}{r_{ij}^{12}} - \frac{B_{ij}}{r_{ij}^{6}} \right) \quad (12.2)$$

For the OPLS force field, the newest is OPLS3e released in 2019, which is optimized for free-energy computations [34]. The OPLS force field can be represented as follow:

$$U(R) = \sum_b K_b (b - b_0)^2$$

$$+ \sum_\theta K_\theta (\theta - \theta_0)^2 + \sum_\varphi \frac{k_\varphi}{2} \left[1 + \cos(n\varphi - \varphi_0) \right]$$

$$+ \sum_\sigma \left\{ 4\varepsilon_{ij} \left[\left(\frac{\sigma_{ij}}{r_{ij}} \right)^{12} - \left(\frac{\sigma_{ij}}{r_{ij}} \right)^{6} \right] + \frac{q_i q_j}{r_{ij}} \right\} \quad (12.3)$$

12.2.1.2 Coarse-grained (CG) models

Although all-atom models can capture the atomic details of biomolecules in atomic detail, biological processes of general interest, which occur in milliseconds, are beyond the length of the typical all-atom simulation time [21]. To study cellular processes on large time and long scales, CG models have emerged, such as the MARTINI force field [35]. CG models generally use a four-to-one mapping scheme, that is, four heavy atoms correspond to one CG bead. Such force fields can greatly increase the spatial and temporal scales that can be accessed by simulation systems while retaining some chemical specificity of biomolecules. The CG models can be parameterized using a top-down or bottom-up strategy. In a bottom-up approach, efficient CG interactions are extracted from reference data, such as atomically detailed simulations or structural databases, aiming to faithfully reproduce the structural features of the reference data. In the top-down approach, the emphasis is on reproducing experimental data, especially thermodynamic properties such as density, heat of vaporization, and partitioning data [21]. The bottom-up approach produces highly accurate CG models whose accuracy depends on the reference data state points. The CG model produced by the top-down approach has limited accuracy but can reproduce the global division of CG molecules on a larger scale.

For CG models, widely used CG force fields for membrane systems are MARTINI [35, 36], SDK [37], SIRAH [38], ELBA [39], and a number of solvent-free models. Another advantage of the CG model over the all-atom model is that it does not require high quality of the initial structure. The ease of initial structure construction and the larger accessible timescales enable simultaneous exploration and running of hundreds of simulations through CG models. For example, CHARMM-GUI supports the model building of membrane systems based on the MARTINI force field.

12.2.1.3 Ultra-coarse-grained (UCG) models

To explore membrane-related processes on larger spatiotemporal scales, the researchers adopted a more aggressive strategy, namely the ultra-coarse-grained (UCG) model. The UCG models use fewer beads to describe molecules, such as proteins using several beads, where one bead represents a domain of the protein molecule. A more radical level of simplification is to integrate out lipids (and water) altogether, such that proteins are represented as particles interacting in a continuum membrane environment [21]. For example, Voth et al. developed a framework for lipids in which two or three particles are used for lipid molecules [40, 41]. This can be used to simulate lipid vesicles with a diameter of 200 nm containing about 500,000 lipids [41].

12.2.2 Molecular Dynamic Simulation

12.2.2.1 Model construction

The initial structure of a simulation system is usually obtained from either experimental structure or computational model [42]. Current protein-structure prediction methods can be broadly separated into two classes: homology modeling (or comparative modeling) and *de novo* methods [43]. Homology modeling has emerged as a valuable approach for generating full-length protein structures of membrane proteins, which relies on the similarities between target sequences and template structures [44, 45]. For example, SWISS-MODEL (https://swissmodel.expasy.org/) [19] is a fully automated online server that uses homology modeling to build structural models. *De novo* methods only use amino acid sequences to build structural models relying on effective conformation-searching algorithms and good energy functions [43, 46]. Based on protein structures, molecular docking is one of the most frequently used methods to obtain the initial poses of ligands binding in proteins [47]. Simulation systems for membrane proteins often require membrane-embedded models, and CHARMM-GUI is a famous tool for constructing the membrane environment, including lipid bilayers, water models, and solvent ions [20]. Moreover, cell-scale models have recently been proposed for molecular simulations of larger and more complex membrane systems, such as an entire virus particle and a small bacterial cell [48].

12.2.2.2 Unbiased molecular dynamics simulation

Unbiased (or classical) MD simulation is an effective computer simulation method for studying the physical movement of atoms and molecules [10]. Atoms and molecules are allowed to interact for a fixed period of time, giving a view of the dynamical evolution of the system. In the most common version, the trajectories of atoms and molecules are determined by numerically solving Newton's equations of motion for a system of interacting particles, where forces between the particles and their potential energies are calculated using interatomic potentials or molecular mechanics force fields. Currently, unbiased (or classical) MD simulation is widely used to study binding modes [15, 49], allosteric modulators [49], ligand selectivities [50], and effects of mutations [51] of membrane protein systems. The simulation trajectory is obtained by solving the different equations embodied in Newton's second law [10]:

$$\frac{d^2x}{dt^2}=\frac{F_i}{m_i} \tag{12.4}$$

The equation describes the motion of a particle of mass m_i along one coordinate (x_i) with Fx_i being the force on the particle in the corresponding direction, and can be shown as:

$$F_i=-\nabla U(r_i,\ldots,r_N) \tag{12.5}$$

where $U(r_i, \ldots, r_N)$ is the potential energy function of N particles, which contains the bonded and non-bonded interactions. The bonded interactions represent the interactions of the covalently bound atoms. The non-bonded interactions can be decomposed into four items: electrostatic interaction, polarization interaction, van der Waals potential, and short-range repulsion. The electrostatic interaction can be described as follow:

$$U_{\text{Coul}} = \frac{q_i q_j}{4\pi\varepsilon_0 r_{ij}} \tag{12.6}$$

And the van der Waals potential can be represented as:

$$U_{\text{LJ}} = 4\varepsilon\left[(\sigma / R)^{12} - (\sigma / R)^{6}\right] \tag{12.7}$$

where ε is the depth of potential at the minimum $r = 2^{1/6}\sigma$, and the potential vanishes at $r = \sigma$.

In the integration algorithms, giving the position and velocities of N particles at time t, a straightforward integration of Newton's equation of motion yields the following at $t + \Delta t$:

$$v_i(t+\Delta t) = r_i(t) + \frac{F_i(t)}{m_i}\Delta t \tag{12.8}$$

$$v_i(t+\Delta t) = r_i(t) + r_i(t)\Delta t + \frac{F_i(t)}{2m_i}\Delta t^2 \tag{12.9}$$

In the popular Verlet algorithm, one eliminates velocities by adding the time-reversed position at $t - \Delta t$:

$$v_i(t-\Delta t) = r_i(t) - r_i(t)\Delta t + \frac{F_i(t)}{2m_i}\Delta t^2 \tag{12.10}$$

The $v_i(t + \Delta t)$ can be written as:

$$v_i(t+\Delta t) = 2r_i(t) - r_i(t-\Delta t) + \frac{F_i(t)}{m_i}\Delta t^2 \tag{12.11}$$

If required, velocities can be calculated from:

$$v_i(t) = \frac{1}{2\Delta t}\left[r_i(t+\Delta t) - r_i(t-\Delta t)\right] \tag{12.12}$$

In the leap-frog algorithm, where the positions and velocities are calculated at different times separated by $\Delta t/2$:

$$v_i\left(t+\frac{\Delta t}{2}\right) = v_i\left(t-\frac{\Delta t}{2}\right) + \frac{F_i(t)}{m_i}\Delta t \tag{12.13}$$

$$r_i(t+\Delta t)=r_i(t)+v_i\left(t+\frac{\Delta t}{2}\right)\Delta t \tag{12.14}$$

The initial coordinate can be obtained from a structure file. After energy minimization, the coordinates give the t = 0 time atom positions. The initial velocities are sampled from a Maxwell–Boltzmann distribution:

$$P(v_{ix})=\left(\frac{m_i}{2\pi\text{kT}}\right)\exp\left[-m_i v_{ix}^2 / 2kT\right] \tag{12.15}$$

During MD simulation, the boundary effects should be considered [10]. The most common method is to use periodic boundary conditions, in which the simulation box is replicated in all directions just like in a crystal. The cube and the rectangular prism are the most common box shapes, though other shapes are also possible. Application of the periodic boundary conditions results in an infinite system, but it raises the question of accurate calculation of the long-range electrostatic interactions. Ewald's sum can be used to address this problem, where the long-range part is separately evaluated in the reciprocal Fourier space. During MD simulation, the NVE ensemble is the most used, where all three quantities (number of atoms, volume, and energy) are constant. Due to truncation errors, keeping the energy constant in long-timescale MD simulations can be problematic. The alternative NVT and NPT ensembles can be employed to avoid this problem. The temperature of the system is obtained from the average kinetic energy:

$$\langle K\rangle=\frac{3}{2}\,\text{NkT} \tag{12.16}$$

A method to keep the temperature constant at T_0 is to scale the velocities as follow:

$$v_i(t)\rightarrow\lambda v_i(t),\lambda=\sqrt{T_0 / T(t)} \tag{12.17}$$

This is a rough approach since kinetic energy has obvious fluctuations. A Berendsen thermostat is a better method that can achieve the same result more smoothly, in which the atoms are weakly coupled to an external heat bath with the desired temperature T_0:

$$m_i\frac{d^2}{dt^2}r_i=F_i+m_i\gamma_i\left[\frac{T_0}{T(t)}-1\right]\frac{dr_i}{dt} \tag{12.18}$$

12.2.2.3 Enhanced sampling molecular dynamics simulations

Membrane proteins often undergo large-scale conformational transitions and visit distinct conformational states during their biological functions [12]. Completely capturing their large-scale conformational transitions along the activation pathway is of critical importance for biology and pharmacology. The conformational transitions usually take place on timescales of milliseconds or even longer due to large energy barriers [13]. This has presented a challenge for unbiased MD simulations that are limited to typically hundreds of nanoseconds to tens of microseconds. To address this challenge, a variety of enhanced sampling techniques have been applied in membrane protein systems to capture longer timescale events. These methods include accelerated molecular dynamics (aMD) simulation altering the force field to reduce the height of energetic barriers, Gaussian accelerated MD (GaMD) simulation adding a Gaussian-distributed harmonic boost potential to smoothen the system potential energy surface, steered molecular dynamics (SMD) simulation and umbrella sampling pulling a biomolecule from a desired initial position to a desired final position, and Metadynamics (MetaD) simulation pushing a simulation away from regions of conformational space it has already visited [23].

Accelerated molecular dynamics (aMD) simulation: aMD simulation works often by adding a boost potential to smoothen the biomolecular potential energy surface [13, 52]. The method modifies the potential energy landscape by raising energy wells that are below a certain threshold level, while leaving those above this level unaffected. As a result, barriers separating adjacent energy basins are reduced, allowing the system to sample conformational space that cannot be easily accessed in an unbiased MD simulation [53]. In addition, aMD simulation does not require predefined reaction coordinate(s) or collective variables, which can be advantageous for exploring the biomolecular conformational space without a priori knowledge or restraints [13].

In aMD simulation, when the system's potential energy falls below a threshold energy E, a bias potential is added, such that the modified potential, $V^*(r)$, is related to the original potential $V(r)$ [53]:

$$V^*(r) = V(r) + \Delta V(r) \tag{12.19}$$

where $\Delta V(r)$ is the boost potential:

$$\Delta V(r) = \begin{cases} 0 & V(r) \geq E \\ \dfrac{(E - V(r))^2}{\alpha + E - V(r)} & V(r) < E \end{cases} \tag{12.20}$$

The threshold energy E controls the portion of the potential surface affected by the boost, while the acceleration factor α determines the shape of the modified potential. α cannot be set to zero; otherwise, the derivative of the modified potential is discontinuous.

From an aMD simulation, the ensemble average, $\langle A \rangle$, of an observable, $A(r)$, can be calculated using the following reweighting procedure:

$$\langle A \rangle = \frac{\langle A(r)\exp(\beta \Delta V(r))\rangle^*}{\langle \exp(\beta \Delta V(r))\rangle^*} \tag{12.21}$$

In which $\beta = 1/k_B T$, and $\langle ... \rangle$ and $\langle ... \rangle^*$ represent the ensemble average in the original and the aMD ensembles, respectively.

Gaussian accelerated MD (GaMD) simulation: GaMD simulation is an unconstrained enhanced sampling method that enhances the conformational sampling of biomolecules by adding a boost potential to smoothen the system's potential energy surface, without the need to set predefined reaction coordinates [13]. The boost potential follows Gaussian distribution, which allows for accurate reweighting using cumulant expansion to the second order [54]. Furthermore, the free-energy profiles obtained from the reweighting of the GaMD simulations help us to identify distinct low-energy states of the biomolecules and characterize protein-folding and ligand-binding pathways quantitatively [13].

When the system's potential energy $V(r)$ is lower than a threshold energy E, the following boost potential is added [12, 54]:

$$V^*(r) = V(r) + \Delta V(r) \tag{12.22}$$

where $\Delta V(r)$ is the boost potential:

$$\Delta V(r) = \begin{cases} 0 & V(r) \geq E \\ \dfrac{1}{2}K(E - V(r))^2 & V(r) < E \end{cases} \tag{12.23}$$

where K is the harmonic force constant. E and K are automatically determined by the following criteria. One is that the boost potential does not change the order of the biased potential value; the other is that the smoothened energy surface is smaller than that of the original. Thus, the range of E and K is:

$$V_{\max} \le E \le V_{\min} + \frac{1}{k} \tag{12.24}$$

where $V_{\max}$ and $V_{\min}$ are the system maximum and minimum potential energies, respectively. To make Equation 12.24 valid, K should be guaranteed:

$$k = k_0 \frac{1}{V_{\max} - V_{\min}} \tag{12.25}$$

The k_0 is determined by:

$$k_0 = \min\left(1.0, \frac{\sigma_0}{\sigma_v} \times \frac{V_{\max} - V_{\min}}{V_{\max} - V_{\text{avg}}}\right) \tag{12.26}$$

The V_{avg} and σ_v are the average and standard variation of potential energies, while σ_0 is a user-defined upper limit.

Metadynamics (MetaD) simulation: MetaD allows acceleration of rare events and estimation of the free energy of simulation systems. It is reconstructing the free-energy surface of complex systems as a function of a small number of suitably chosen collective variables (CVs), which has been informally described as "filling the free energy wells with computational sand" [55, 56]. In MetaD, sampling is accelerated by a history-dependent bias potential, which is adaptively constructed in the space of CVs [57, 58]. This potential can be written as the sum of Gaussians deposited along the system trajectory in the CVs space to discourage the system from revisiting configurations that have already been sampled. Choosing a good set of collective variables that fully describe the slow events of the system is crucial to the convergence of the calculations and the free-energy values to be trusted. The ratio between accuracy and computational cost can be optimized by adjusting the time interval between the addition of two Gaussian functions, as well as the Gaussian height and Gaussian width. MetaD can rapidly generate a rough energy landscape using large size Gaussians, or it can generate a more fine-grained description with a smaller size of Gaussian.

In MetaD, an additional history-dependent bias in the form of a Gaussian is added to the system potential [59]:

$$V_G(S,t)=w\sum_{t'=\tau_G,2\tau_G}^{t}\exp\left(-\frac{(S-s(t'))^2}{2\delta s^2}\right) \tag{12.27}$$

where w and δs are the Gaussian height and width, τ_G is the time interval for the Gaussian deposition, and $s(t) = S(x(t))$ is the value of the collective variable at time t. It is assumed in MetaD that for a sufficiently long simulation, $V_G(S, t)$ provides a calculation of the underling free energy [59]:

$$\lim_{t\to\infty} V_G(S,t)=-F(S)+C \tag{12.28}$$

Steered molecular dynamics (SMD) simulation: The basic idea of SMD simulation is applying an external force to one or more atoms, pulling them to another fixed group of atoms. In SMD simulation, a parabolic potential ΔU is applied to increase the probability of sampling a specific phase space region [42]. Additionally, the center of the parabola moves in time over the desired range of the reaction coordinate ξ. In detail:

$$\Delta U=\frac{1}{2}K\left(\xi-\xi(t)\right)^2 \tag{12.29}$$

where the center $\xi_0(t)$ often moves at constant velocity as in:

$$\xi_0(t)=\xi_0 0+vt \tag{12.30}$$

v is the value of the constant velocity in the collective variable space.

Umbrella sampling: Umbrella sampling is the progenitor of the family of enhanced sampling methods. In this technique, a bias is applied to the system to ensure efficient sampling along the whole reaction coordinate [42]. This can either be aimed at one simulation or different simulations (windows), the distributions of which overlap [60]. The effect of the bias potential to connect energetically separated regions in phase space gave rise to the name umbrella sampling. Umbrella sampling with the potential $U(r)$ [42]:

$$\hat{U}=U(r)+\frac{1}{2}K\left(\xi(r)-\xi 0\right)^2 \tag{12.31}$$

where the second term is a harmonic restraint centered on $\xi 0$. Umbrella sampling usually coupled with the weighted histogram analysis method (WHAM) to reconstruct the free-energy profile [61].

12.3 Applications of Computer Simulations

12.3.1 Unbiased Molecular Dynamic Simulation Case Study

12.3.1.1 Membrane lipid composition regulating conformation and hydration of influenza virus B M2 channel

Cellular membranes consist of phospholipids, proteins, and sugars in diverse amounts and proportions. Numerous pieces of evidence indicate that lipid composition affects membrane protein functions, such as ion channels [62–64]. The influenza virus B M2 (BM2) proton channel is an acid-activated transmembrane proton channel, responsible for the acidification of viral interior and transferring proton, which is an important step in the virus life cycle [65]. Histidine19 (His19) located in the conserved HxxxW motif of BM2 proton channel (Fig. 12.1A) plays an important role in acid activation and proton conduction [66, 67]. The function of BM2 channel is affected by the membrane lipid composition, but the detailed molecular mechanism is challenging to determine. We reveal that the mechanism of lipid composition (negatively charged lipid palmitoyloleoyl phosphatidylglycerol (POPG) and cholesterol) affects the function of BM2 proton channel by unbiased MD simulation [68], which provides a unique perspective for understanding lipid composition regulating conformation and function of membrane protein. In the next section, we briefly introduce the application of unbiased MD simulation in revealing the mechanism of negatively charged lipid membrane regulating the structure and function of BM2 proton channel [68], providing a reference for the system preparation of membrane protein and the implementation of unbiased MD simulation.

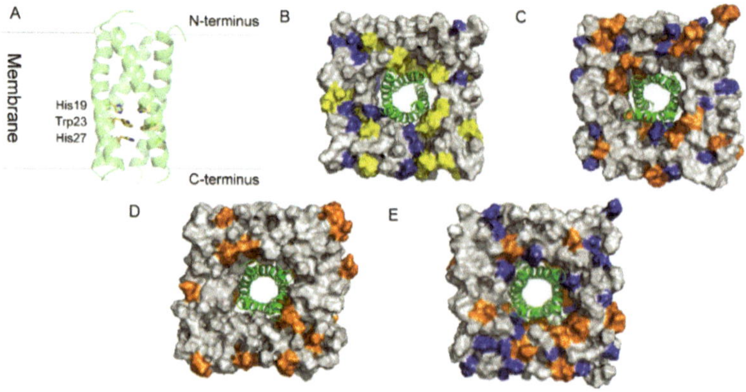

Figure 12.1 Structure of BM2 tetramers in different membrane environments. (A) The solution NMR structure of BM2 (PDB: 2KIX [69]) HxxxW motif. (B–E) Structure models of BM2 tetramers in the neutral membrane PC/PE/Chol, negatively charged membrane PC/PG/Chol, membrane without cholesterol PC/PG (4:1), and membrane containing cholesterol PC/PG/Chol (4:1:2). Cholesterol, palmitoyloleoyl phosphatidylcholine (POPC), palmitoyloleoyl phosphatidylethanolamine (POPE), and palmitoyloleoyl phosphatidylglycerol (POPG) are shown in blue, gray, yellow, orange, respectively. Reprinted with permission from Ref. [68]. Copyright 2020 American Chemical Society.

12.3.1.2 Methods

The high-resolution NMR structure of BM2 (PDB: 2KIX [69] shown in Fig. 12.1A) was obtained from the Protein Data Bank (https://www.rcsb.org/). According to previous work [70–73], the protonation states of residue His19 in BM2 were set to two types: $His19_4^0$ (four His19 residues were neutral representing the closed state of the BM2 channel) and $His19_4^{4+}$ (four His19 residues were all cationic representing the open state of the BM2 channel). Additionally, the His27 tetrad of BM2 channel should become protonated before the His19 tetrad [66, 70, 71, 74]. Four His27 residues were neutral at $His19_4^0$, and they were all cationic at $His19_4^{4+}$. Four different membrane models were constructed by CHARM-GUI (Fig. 12.1) [20, 75], including the negatively charged membrane POPC/POPG/cholesterol (PC/PG/Chol, at molar ratios of 60:20:20%), the neutral membrane POPC/POPE/cholesterol (PC/PE/Chol, at molar ratios of 60:20:20%), the cholesterol absent membrane POPC:POPG (PC/PG, at molar ratios of 4:1), and the membrane containing cholesterol

POPC/POPG/cholesterol (PC/PG/Chol, at molar ratios of 4:1:2). For each system, a 15 Å thick water layer with TIP3P model was added on each side of the lipid bilayer, and NaCl solutions (0.15 M) were used to mimic the physiological conditions.

All the simulations were performed by the AMBER16 package with GPU-accelerated PMEMD [24]. For each system, AMBER ff14SB [76] and Lipid11 [77] force fields were utilized to model the protein and lipids, respectively. To eliminate the bad contact, the minimization and equilibration were carried out before the production simulation. Three stages of minimization were applied, including 20000 steps of the steepest descent algorithm and 10000 steps of the conjugate gradient algorithm [68]. In the first stage, 500 kcal/mol $Å^2$ positional restraints were applied to the Cα atoms of the BM2 channel and all atoms of lipids. In the second stage, the restraints for lipids were removed—500 kcal/mol $Å^2$ restraints only for Cα atoms of the BM2 channel. In the third stage, all atoms were allowed to move freely. Then, each system was gradually heated to 310 K and performed 2 ns equilibration simulation. Finally, 500 ns production MD simulations were performed at 310 K with the NPT ensemble. The Particle Mesh Ewald (PME) method was utilized to deal with the long-range electrostatic interactions with a cutoff value of 10 Å, and all bonds with hydrogen atoms were constrained by the SHAKE algorithm. The time step was set to 2 fs. Moreover, the simulation was repeated five times.

12.3.1.3 Mechanism of negatively charged membrane regulating conformation and hydration of the BM2 channel

In the case study, two mixed lipid bilayers with negative charge and neutral charge were constructed to evaluate the effects of the anionic lipid bilayer on regulating the behaviors of the BM2 channel, including conformational transitions and hydration. In Fig. 12.2A, the probability density distributions of the conformation divided the curves into two regions: one is centered at 2.4–2.7 Å representing the closed state of the BM2 channel ($His19_4^0$), and the other has a wider distribution centered at 5.2–6.2 Å, adopting the open state ($His19_4^{4+}$). Compared to the neutral membrane, the root mean square deviation (RMSD) increased more in negatively charged membrane (Fig. 12.2A), especially in the open state ($His19_4^{4+}$). Thus, the negatively charged membrane may induce significant conformational transitions of the BM2 channel. To further verify the conformation transitions induced by negatively charged membrane, the perimeter formed by key residues (Phe5, Ser9, Ser12, Ser16, His19, Trp23, His27, and Arg33) and the pore radius along the channel axis were evaluated. As illustrated in Fig. 12.2B, the negatively charged membrane could lead to a sharply enhanced C-terminal perimeter of MB2 (His19, Trp23, His27, and Arg33), especially in the $His19_4^{4+}$ state. The data of pore radius along the channel axis calculated by CAVER [78] also indicated that the negatively charged membrane strengthened the dilation of the C-terminal portion (Fig. 12.2C).

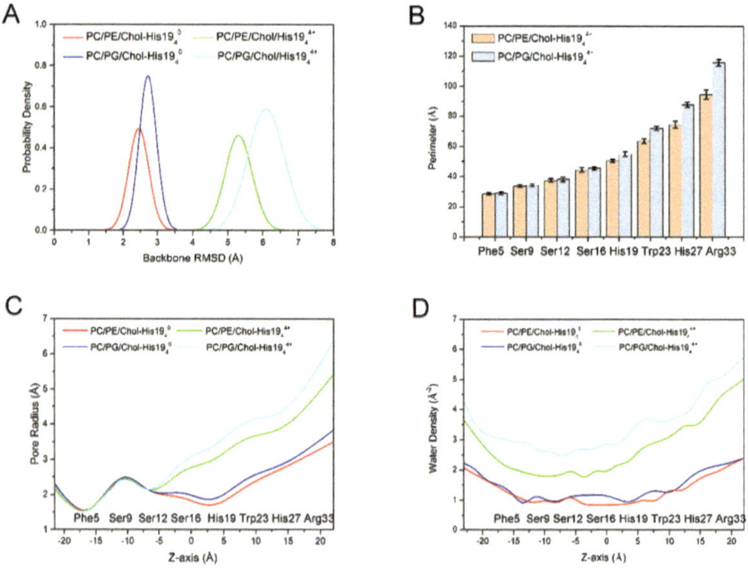

Figure 12.2 Conformational changes at $His19_4^0$ and $His19_4^{4+}$ in neutral membrane and negatively charged membrane. (A) Probability distribution of the backbone RMSD with respect to the starting structure. (B) Perimeter of the tetragon formed by the Cα atoms of the specified residue on the four helices at $His19_4^{4+}$. The residues are those lining the pore: Phe5, Ser9, Ser12, Ser16, His19, Trp23, His27, and Arg33. The perimeters are averaged over the simulation time. The error bars indicate the standard deviations. (C) Pore radius along the channel. The pore radii are averaged over the simulation time. (D) Water density profiles along the channel axis. Water oxygen atoms per $Å^2$ where the pore cross sections from CAVER were calculated. Reprinted with permission from Ref. [68]. Copyright 2020 American Chemical Society.

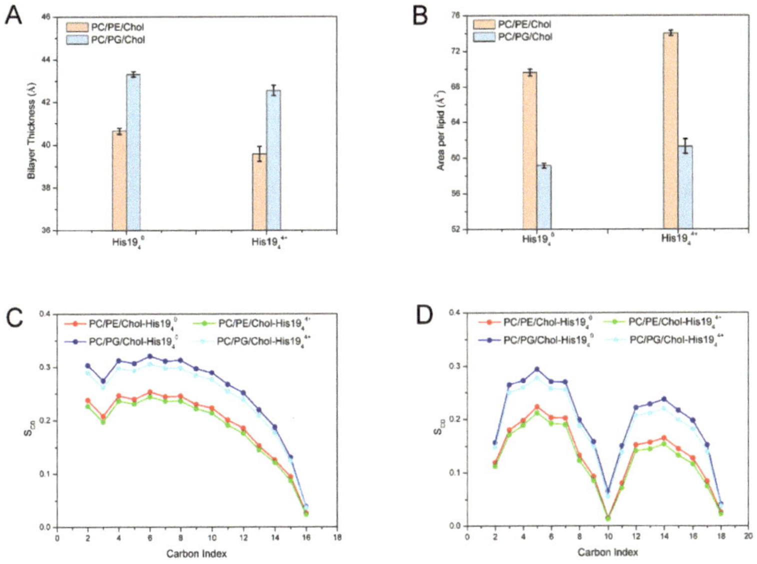

Figure 12.3 Comparison of membrane properties in neutral membrane and negatively charged membrane averaged over the five repeat simulations. (A) Bilayer thickness in neutral membrane and negatively charged membrane; the error bars indicate standard deviation. (B) Area per lipid in different membranes; the error bars indicate standard deviations. (C) SCD of lipid bilayer for sn-1 tails. (D) SCD of lipid bilayer for sn-2 tails. Reprinted with permission from Ref. [68]. Copyright 2020 American Chemical Society.

The water density along the channel axis was measured to explore the effect of the negatively charged membrane on channel hydration. As illustrated in Fig. 12.2D, the water density obviously increased in the negatively charged membrane in the $His19_4^{4+}$ state, suggesting that the negatively charged membrane remarkably facilitates channel hydration as the channel adopts the open state. To further explore how the negatively charged membrane affects the hydration of the BM2 channel, the membrane properties (membrane thickness, surface area per lipid, and deuterium order parameter) and the lipid–protein interactions were evaluated. In Fig. 12.3A, the thickness of the neutral membrane in the $His19_4^{4+}$ state was 39.6 ± 0.3 Å, and it increased to 42.7 ± 0.2 Å in the presence of anionic lipid POPG. The data demonstrate that the negatively charged membrane can increase the bilayer thickness. In Fig. 12.3B, the area per lipid decreased in the negatively charged membrane, which means the addition of the anionic lipid leads to close lipid packing. The deuterium order parameter (SCD) represents the order of the lipid tails. In Fig. 12.3C and D, the negatively charged lipid bilayer was more ordered than a neutral lipid bilayer. Furthermore, the negatively charged membrane significantly promotes channel hydration at the channel in the open state and is associated with the hydrogen-bond networks between anionic lipid POPG and the C-terminal residues. In the open state (PC/PG/Chol-His19_4^{4+}), the addition of anionic lipid POPG can form stable direct hydrogen-bond networks with the C-terminal residues (including full-cationic His27, positively charged Lys32, and Arg33) of BM2 tetramers (Fig. 12.4A–C), which could further stabilize the opening of the C-terminus, allowing more water molecules to enter the interior channel and promoting channel hydration.

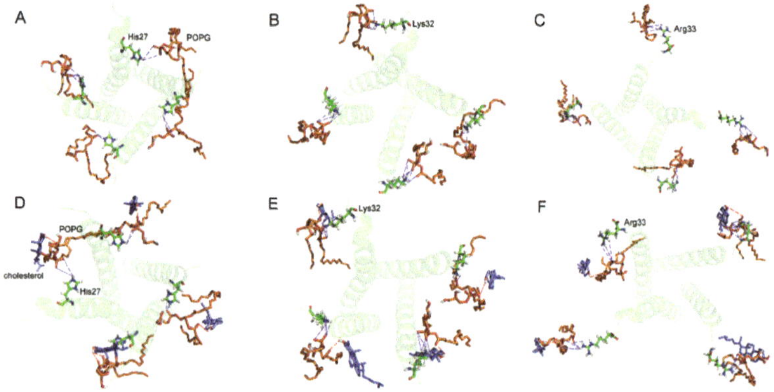

Figure 12.4 Hydrogen-bond networks are formed by negatively charged lipids POPG (orange), cholesterol (blue), and the C-terminal portion of the BM2 channel. The hydrogen bonds between cholesterol and POPG are colored red. The hydrogen bonds between POPG and the C-terminus are colored blue. The nonpolar hydrogen atoms of POPG and cholesterol are removed for clarity. (A) Hydrogen-bond networks between POPG and full-cationic His27 tetrad. (B) Hydrogen-bond networks between POPG and Lys32 tetrad. (C) Hydrogen-bond networks between POPG and Arg33 tetrad. (D) Hydrogen-bond networks formed by cholesterol, POPG, and full-cationic His27 tetrad. (E) Hydrogen-bond networks formed by cholesterol, POPG, and Lys32 tetrad. (F) Hydrogen-bond networks formed by cholesterol, POPG, and Arg33 tetrad. Reprinted with permission from Ref. [68]. Copyright 2020 American Chemical Society.

The negatively charged lipid POPG reduces membrane fluidity and augments membrane rigidity. A more rigid membrane environment is beneficial to expand the BM2 channel, allow more water to enter the channel, promote channel hydration, and then even affect the proton conduction facilitated by the hydrated channel. This case study reveals that the negatively charged lipid POPG plays a positive role in facilitating conformational transitions and channel hydration of the BM2 channel through unbiased MD simulation. The simulation results provide a unique perspective for understanding the lipid composition that regulates the structure and function of membrane proteins and are valuable complements to experimental approaches.

12.3.2 Enhanced Sampling Molecular Dynamic Simulation Case Study

12.3.2.1 Conformational transition mechanism of AT1 receptor activation

G protein-coupled receptors (GPCRs) constitute a large superfamily of membrane proteins, including approximately 800 members [79]. GPCRs represent the therapeutic targets of nearly one-third of all approved drugs, due to their substantial involvement in human pathophysiology and their pharmacological tractability [80]. A complete mechanistic understanding of large-scale conformational transitions of GPCRs activation is of critical importance for therapeutic drug development. Due to large energy barriers, the conformational transitions usually take place on timescales of milliseconds or even longer [13]. Introducing biases or potentials in a variety of enhanced sampling techniques usually allows simulations to capture longer timescale events. For example, GaMD simulation adds a Gaussian-distributed harmonic boost potential to smoothen the system potential energy surface, which can help us to identify distinct low-energy states of the biomolecules and characterize the protein-folding and ligand-binding pathways [13]. In 2021, Zhang's group applied GaMD simulation to verify the existence of an intermediate state for angiotensin II type 1 receptor (AT1 receptor) [12], which provides insights into the application of enhanced sampling simulations in membrane protein systems. The AT1 receptor is a prototypical class A GPCR, and it is a prominent therapeutic target for hypertension and cardiovascular diseases [81]. Currently, the high-resolution structures of AT1 receptor in active (agonist-bound, PDB ID: 6DO1) [82] and inactive (antagonist-bound, PDB ID: 4YAY) [83] states are available. A possible intermediate state was first proposed based on extensive unbiased MD simulations (300 μs) and Markov state models (MSMs), taking inactive structure (PDB ID: 4YAY) [83] as the initial state and active structure (PDB ID: 6DO1) [82] as the end state. Then, the agonist (AngII) and the inverse agonist (olmesartan) were docked into the orthosteric site of the AT1 receptor (in a possible intermediate state) to form two complex models. Subsequently, GaMD simulations were performed on the complex models to confirm the existence of intermediate state. This work identifies the intermediate state of the AT1 receptor through GaMD simulation [12], which is of great significance for a comprehensive understanding of the activation mechanism of GPCRs. The following is a brief introduction to the application of GaMD simulation in AT1 receptor conformational identification [12].

12.3.2.2 System setup

The natural agonist (AngII) and the inverse agonist (olmesartan) were docked into the orthosteric site of AT1 receptor (in a possible intermediate conformation), using the molecular operating environment (MOE) [84]. The best-scored poses were chosen for further simulation study. All complex models were oriented in the Orientations of Proteins in Membranes (OPM) server [85] and were inserted into a POPC membrane with the CHARMM36 additive force field using the CHARMM-GUI [20]. Furthermore, TIP3P water molecules with a length of 10 Å were added to the top and bottom of the systems, and an additional 0.15 mol/L KCl was solvated in the systems.

Before the GaMD simulation, 2 μs unbiased MD simulations were performed to relax the simulation systems. Two stages of minimization were applied. The systems were first minimized with a restraint of 500 kcal/mol Å^2 on the AT1 receptor and lipids, while water and counterions were minimized in 8000 steepest descent cycles, followed by 7000 conjugate gradient cycles. Second, all atoms encountered 1.5 × 104 cycles of steepest descent and 1.5 × 104 cycles of conjugate gradient minimization. Then, the systems were gradually heated from 0 to 300 K in 300 ps and equilibrated for 700 ps under canonical ensemble conditions, with a 10 kcal/mol Å^2 restraint on proteins and lipids. Finally, 2 μs unbiased MD simulations were performed to relax the systems. During the simulation, the PME method was applied to calculate the long-range electrostatic interactions, and a cutoff of 10 Å was used for short-range electrostatic and van der Waals interactions. The SHAKE algorithm was also employed for covalent bonds containing hydrogen. A temperature of 300 K was controlled by a Langevin thermostat, and the collision frequency was 1.0 ps^{-1}. The time step was set to 2 fs, and the snapshots were written out every 200 ps. Other simulation analyses were based on AMBER suite.

During the GaMD simulation, the total potential and dihedral energy were boosted. The average and standard variations of energy were calculated every 400 ps, while the upper limit was 6.0 kcal/mol. The principle of GaMD simulation can be found in "12.3.2.2 Enhanced sampling molecular dynamic simulations." For each system, three independent 1 μs production simulations were performed. In each simulation, a 26 ns short unbiased MD simulation was first applied to evaluate potential energies for acceleration parameters, and then a 50 ns equilibration run with the boost energy was employed. The sampling time step was set to 100 ps.

12.3.2.3 Confirming the existence of an intermediate state of AT1 receptor

The main differences in the active and inactive AT1 receptor structures are the conformations of the transmembrane domains (TMs) 5, 6, and 7. Therefore, two activation parameters were defined to capture the conformational landscape of AT1 receptor. One parameter is the distance between the Cα atoms of $L^{5.55}$ and $N^{7.46}$ (superscripts indicate the Ballesteros–Weinstein numbering for GPCR residues), which represents the conformational changes in TM5 and TM7. The other one is the angle among the Cα atoms of $F^{6.34}$, $S^{6.47}$, and $V^{2.41}$, which reflects the outward movement of TM6.

According to the activation parameters calculated from GaMD trajectories, free-energy landscapes were constructed and shown in

Fig. 12.5. In the AngII (agonist) binding system, the TM6 angle of the AT1 receptor largely adopted more than 55° (inactive state: ~35–42°; intermediate state: ~46–51°; active state: >55°), implying a tendency fitted the outward movement of TM6 (Fig. 12.5A–C), which resembles the active state of the AT1 receptor. In the olmesartan (inverse agonist) binding system, the angle is significantly reduced, making the free-energy landscape downward (Fig. 12.5D–F). On the other hand, the distances are around 15–17 Å in the AngII-binding system, which is close to the active conformation. In the olmesartan binding system, the distances are in the range of 18–20 Å, which are consistent with the feature of the inactive conformations. As mentioned earlier, agonist AngII binding results in an active state of AT1 receptor, whereas inverse agonist olmesartan binding results in an inactive state. The binding of inverse agonist and agonist shifts the possible intermediate state toward the inactive and active states, respectively, suggesting the existence of intermediate states in the inactive-to-active transition pathway of AT1 receptor. The AT1 receptor intermediate state identified by the GaMD simulation is essential for elucidating the dynamics and underlying steps of AT1 receptor cycle, and it is helpful in the development of new drugs. This case study provides insights into the application of enhanced sampling simulations in membrane protein systems.

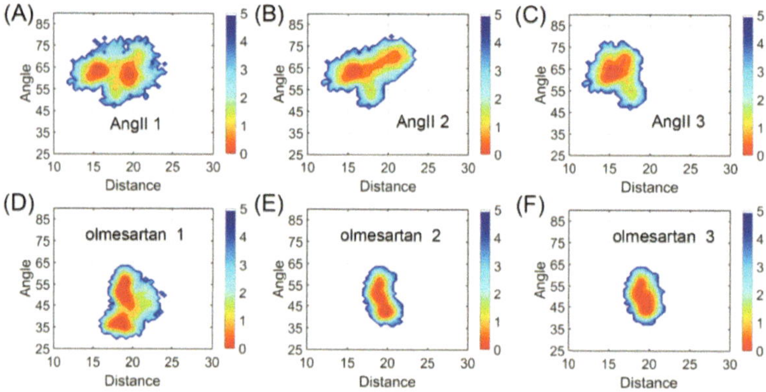

Figure 12.5 The conformational landscape of AT1 receptor generated using the Cα atom distance between $L^{5.55}$ and $N^{7.46}$, and the angle among the Cα atoms of $F^{6.34}$, $S^{6.47}$, and $V^{2.41}$ as the order parameters along the activation pathway [12]. (A–C) GaMD rounds 1–3 in the intermediate AT1 receptor with AngII bound. (D–F) GaMD rounds 1–3 in the intermediate AT1 receptor with olmesartan bound.

12.3.3 Large-Scale Molecular Dynamics Simulation Case Study

12.3.3.1 Assembly of cell-scale membrane envelopes

As computing power increases, MD simulations can be used for larger and more complex systems, such as an entire virus particle and a small bacterial cell. Efforts are currently under way to develop software tools, procedures, and workflows for constructing cell-scale models to reveal the interaction networks and couplings in important biological processes at subcellular and even cellular scales. Here, we introduce the workflow proposed by Vermaas et al. to provide ideas for model building at the cellular scale [48].

This section describes the efforts in cell-scale modeling that develop and implement a method for constructing a cellular envelope composed of lipids and embedded proteins surrounding the cytoplasm, which is called a "protocell" [48]. This method is based on the combination of subtractive assembly technology and structural splicing tools, and the workflow is shown in Fig. 12.6. Briefly, the method first separates proteins and lipids into two separate Fibonacci spherical shells, which are generated along two paths and finally merge into a single protocell.

The top half of Fig. 12.6 shows the protein path. First, the protein composition and content to be modeled are determined and a Fibonacci spherical shell corresponding to a uniform distribution of protein numbers is generated. The proteins were then randomly inserted into Fibonacci spherical shells. Due to the variety of proteins, protein shells generated by random insertion may have steric conflicts. To improve the nonphysical initial protein alignment, protein positions were optimized through a UCG simulation. For a single membrane protein, a maximum of four UCG particles are used, including the outer leaflet, the inner leaflet, the extracellular domain, and the intracellular domain, where the mass of the UCG particle is determined by the mass of all the particles that make it up. The sum is determined, and the radius of a UCG particle is determined by the maximum distance of the particles it is composed of to its center. After the UCG simulation was completed, the optimized protein positions were determined by the first and last frames of the simulation.

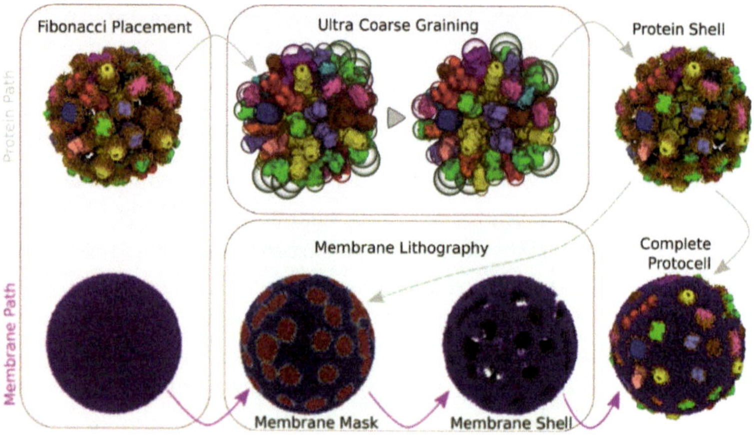

Figure 12.6 Protocell construction workflow. Protocell assembly across two overall paths. The protein path (gray arrows) prepares the protein components (multicolored surfaces) and their tightly associated lipids (orange), by optimizing protein position through a UCG simulation (gray triangle) where each protein is represented by up to four particles (semitransparent beads). The membrane path (purple arrows) prepares the bulk membrane lipids (purple) through membrane lithography using the optimized positions of the protein shell components (gray dashed line) to generate a membrane mask. The membrane mask evaluates bulk lipid overlap with the protein shell (red for high overlap, blue for low overlap, and white for intermediate) and selects as many lipids with minimal overlap as needed to complete the protocell. The complete protocell, thus, has optimally placed protein components (multicolored surfaces), lipids tightly associated with the proteins (orange), and bulk lipids (purple). Reprinted with permission from Ref. [48]. Copyright 2022 American Chemical Society.

The lower half of Fig. 12.6 shows the lipid path. Similarly, the lipid species and content in the lipid shell are first determined and a Fibonacci spherical shell corresponding to the amount of lipid is generated. The lipids were then randomly inserted into Fibonacci spheres. The lipids that produce the greatest physical overlap with the protein shell are removed by the "Computational Membrane Lithography" method. Then, protein shell and lipid shell were merged using the "fastmerge" tool, and the final protocell was generated. A detailed description of the "Computational Membrane Lithography" method and the "fastmerge" tool can be found in the original paper [48].

This workflow proposed by Vermaas et al. [48] has been shown to be efficient and robust in modeling subcellular-sized vesicles and could be of great help in moving toward cellular-scale modeling. This case study provides a cell-scale model construction idea, which is helpful for understanding and implementing molecular simulations of larger and more complex membrane systems.

12.3.3.2 Molecular dynamics simulation at the organelle scale

Currently, the computational simulation has gradually transformed from a simplified simulation model to a simulation model that reflects a more complex and more realistic cellular environment. This transition is critical for uncovering interaction networks and couplings between components in important biological processes. Cell-scale or organelle-scale models can resolve the physical and chemical events that underlie the emergent phenotypes and ultimately render a comprehensive picture of the cellular machinery in action [86]. The amazing work of Singharoy et al. developed a comprehensive organelle-scale model that describes the energy conversion processes in purple bacterial photosynthesis and accounts for the cooperation between the constituent proteins in terms of molecular interactions based on the model [86]. Here we briefly describe the application of computational simulations (unbiased MD simulation and CG simulation) at the organelle scale to reveal changes in the local curvature of the lipid membrane surrounding the proteins.

An all-atom model at the organelle scale for the low-light-adapted chromatophore containing 100 million atoms was first constructed based on earlier models and a lot of experimental data (Fig. 12.7) [86]. The structure model was solvated and ionized within a water box of 110 nm^3. Before the production MD simulation, a series of equilibrium simulations were performed to eliminate instability. All simulations were performed with the NAMD 2.12 using the CHARMM36 force field for proteins and lipids [87]. The force field parameters for the photosynthetic cofactors were obtained from previous work [88, 89]. The time step was set to 1 fs, and bonded interactions, short-range non-bonded interactions, and long-range electrostatic interactions were calculated every 1 fs, 2 fs, and 4 fs, respectively. A cutoff of 12 Å was used for van der Waals and short-range electrostatic interactions, and the PME method was used to calculate the long-range electrostatics with a grid point density of

1 Å^{-1}. Periodic boundary conditions were used for the full system. The systems were kept at constant temperature using Langevin dynamics for all non-hydrogen atoms.

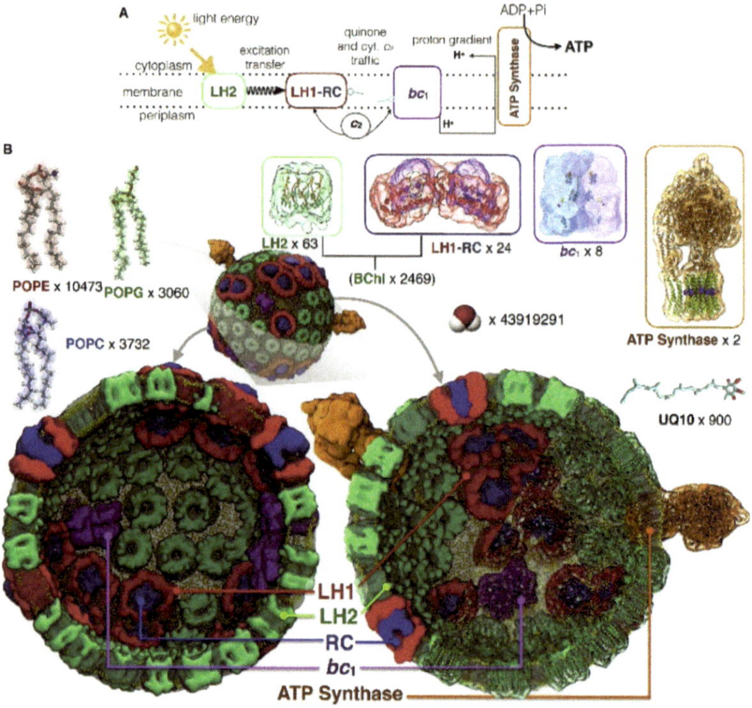

Figure 12.7 Physical model of a chromatophore. (A) Photosynthetic energy conversion in a purple bacterial chromatophore. Light energy absorbed by light-harvesting LH2 and LH1 complexes is channeled to the reaction center (RC), where a charge separation is stabilized as a reduced quinone, and the electron hole is filled by reduced cyt c_2. Subsequent turnovers of the bc_1 complex generate a proton-motive force that drives ATP synthesis. (B) All-atom model of the chromatophore from the purple bacterium *Rhodobacter sphaeroides*. Illustrated using two halves of a transverse section; the final configuration of the all-atom model was obtained after 0.5 ms of explicit solvent all-atom equilibrium MD simulation. The model features 82 bioenergetic complexes (63 light-harvesting LH2 complexes (green), 11 dimeric and 2 monomeric RC-LH1 complexes (LH1, red; RC, blue), 4 bc_1 dimers [magenta], and 2 ATP synthases (orange)), together with 4,011 light-absorbing antenna molecules (2,469 BChls and 1,542 carotenoids) embedded in a membrane of approximately 17,200 lipid molecules (lipid phosphate is indicated in yellow). Reprinted from Ref. [86], Copyright 2019, with permission from Elsevier.

CG simulations were performed using GROMACS 5.1.1. The Martinize script was used to convert the atomistic bc_1 complex to a CG MARTINI model [35]. The ElNeDyn elastic network model was applied to particles within a cutoff value of 0.9 nm using a force constant of 500 kJ/mol nm^2 [35]. Then the protein was embedded in the chromatophore membrane model using the Insane script. The system was solvated using the standard MARTINI water model along with ions used to neutralize the simulation system. Periodic boundary conditions and semi-isotropic pressure coupling were also applied to the simulation. In minimization simulation, 1000 steps of the steepest descent algorithm were performed. In equilibration steps, position restraints were first applied to all protein beads and then applied to the backbone beads. In the production simulation, the temperature at 310 K and pressure 1 bar were maintained using a v-rescale thermostat and a Berendsen barostat, respectively. The time step was set to 20 fs. Furthermore, electrostatic interactions were treated using a reaction field (cutoff was set to 1.1 nm); Lennard–Jones interactions were calculated (cutoff was set to 1.1 nm).

The content of each component of the all-atom model is shown in Fig. 12.7. The model contains 82 bioenergetic complexes (cytochrome bc_1, light-harvesting LH1, light-harvesting LH2, and others), 4011 photosynthetic cofactors, and approximately 17200 lipid molecules. Then, the initial all-atom model was simulated using unbiased MD for more than 500 ns. An analysis of the MD trajectories revealed changes in the local curvature of the lipid membrane surrounding the proteins.

In Fig. 12.8A and B, the membrane is nearly flat in the vicinity of the cytochrome bc_1 (bc_1) complex, but convex near light-harvesting LH1 (LH1) and light-harvesting LH2 (LH2) complexes. The LH1 dimers and the hexagonally packed LH2 complexes form three- to four-fold more hydrogen bonds on the cytoplasmic side of the chromatophore than on the periplasmic side. The difference in hydrogen bonds between the cytoplasmic side and the periplasmic side in the LH1 and LH2 complexes leads to a curvature in the membrane. For the bc_1 complex, the hydrogen-bond difference is minimal, eliciting a flatter topography in its vicinity. Furthermore, the results of the 40 μs CG MD simulation also confirmed that the bc_1 complexes elicit a flat membrane (Fig. 12.8C), which is consistent with the curvature

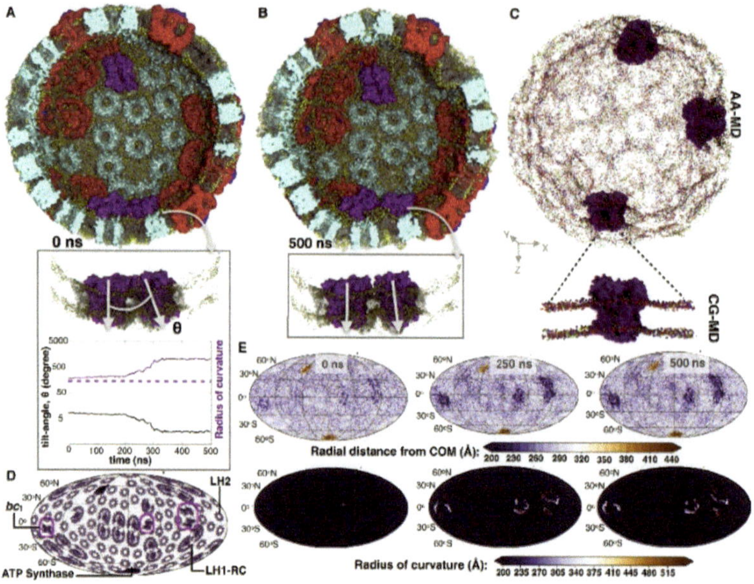

Figure 12.8 Snapshots of the chromatophore vesicle at the beginning (A) and after 500 ns (B) of the MD simulation. The membrane is nearly flat in the vicinity of the bc_1 complexes [magenta] but convex near other proteins. (Insets) Zoom-in view of two proximal bc_1 complexes that are found to induce the highest degree of local membrane curvature change. The tilt angle, θ, between these two proximal bc_1 complexes decreases during the simulation, reflecting a gradual increase in the local radius of curvature (tilt angle radius of curvature relationship discussed in Ref. [86]). (C) Flat-to-convex curvature in the vicinity of the bc_1 complexes observed in the all-atom (AA) and coarse-grained (CG) MD simulations; phosphate head groups of POPE, POPG, and POPC are labeled as red, green, and blue beads, respectively. (D) 2D Mollweide projection map of the chromatophore vesicle annotated with the locations of all four membrane protein types. (E) (Upper panels) Mollweide maps of the chromatophore membrane illustrating the radial distances of its atoms to the vesicle's center of mass at 0, 250, and 500 ns of the AA MD simulation. The distance is the highest for the ATP synthase motors that protrude the most, the smallest for the bc_1-rich areas, and almost uniform across the rest of the chromatophore membrane. (Lower panels) Local radius of curvature obtained at every point on the chromatophore membrane derived with a 2D curve fitting protocol at 0, 250, and 500 ns of AA MD. The radius of curvature is uniform across the chromatophore membrane, except for the bc_1-rich patches where the membrane is considerably flatter, and the radius exceeds 500 Å. Reprinted from Ref. [86], Copyright 2019, with permission from Elsevier.

modulation observed in the 500 ns unbiased MD simulation. The proteins can induce reshaping of the chromatophore membrane and changes in the local curvature of the lipid membrane. The uneven hydrogen bonds that LH1, LH2, and bc1 complexes form with the lipid head groups at the cytoplasmic and periplasmic sides of the chromatophore bilayer are responsible for differences in the local curvature of the membrane surrounding the proteins (Fig. 12.8D and E).

This work elucidates the changing mechanism of membrane curvature through MD simulation at the organelle scale. It also demonstrates the amazing potential of simulation to provide unprecedented detail of the structure and function of cellular processes at atomic resolution.

12.4 Protocols

In this section, we briefly introduce the processes of homology modeling using SWISS-MODEL [19] and the construction of the membrane protein system required for MD simulation using CHARMM-GUI [20].

12.4.1 SWISS-MODEL

SWISS-MODEL (https://swissmodel.expasy.org/) [19] is a fully automated online server for building protein structures. It predicts protein structures based on the homology modeling method, which relies on the similarities between target sequences and template structures [90]. Besides, SWISS-MODEL also contains a structural database: SWISS-MODEL Repository (SMR) [91]. Currently, SMR contains >2,200,000 models derived from SWISS-MODEL homology modeling and >170,000 experimental structures obtained from the Protein Data Bank (https://www.rcsb.org/). These data cover 12 core species, such as *Homo sapiens, Mus musculus, Caenorhabditis elegans, Escherichia coli,* and *Arabidopsis thaliana.* Here, we take the influenza virus B M2 (BM2) proton channel as an example to briefly introduce homology modeling and model evaluation using SWISS-MODEL, and the process can be divided into the following three steps.

12.4.1.1 Upload target sequence

First, the sequence of BM2 channel (target sequence) can be obtained from NCBI [92] or UniProt [93] database, and the target sequence needs to be provided to the server through the search box or upload.

12.4.1.2 Search for templates and build models

Second, multiple potential templates of the BM2 channel can be searched from the template library based on sequence similarities, in which BLAST [94] and HHblits [95] are used in parallel for identifying templates and obtaining target–template alignments. The combination of these two tools improves the efficiency of template searching and sequence alignment. Additionally, the template of the BM2 channel can also be uploaded from a local machine. After the "build model" command is issued, the structure of BM2 channel is automatically constructed by ProMod3 within a few minutes [96].

12.4.1.3 Structure assessment

Third, the constructed structural model can be downloaded on the results page. It is known that model evaluation is critical for protein-structure prediction. SWISS-MODEL can evaluate the structural model from multiple perspectives, and the results are displayed in the "Structure Assessment" menu, including the Ramachandran plot, MolProbity Results, Quality Estimate results, and Residue Quality.

It is worth noting that the QMEANBrane [97] module implemented in SWISS-MODEL is specifically designed to measure modeling qualities for membrane proteins, which determines the transmembrane regions of proteins by an implicit solvent model and later determines different regions by special statistical potentials. QMEANBrane [97] can be accessed from the QMEAN (qualitative model energy analysis) tool of SWISS-MODEL.

12.4.2 CHARMM-GUI

CHARMM-GUI (http://www.charmm-gui.org) [20] is a web-based graphical user interface to prepare complex biomolecular systems for MD simulations. CHARMM-GUI has been continuously expanded since its release in 2006. The advantage of CHARMM-GUI is that it can provide input files for most MD simulation software, such as CHARMM [98], AMBER [24], NAMD [25], GROMACS [26], and OpenMM [27], which helps us easily and interactively build complex membrane protein systems [10]. Here, we also take the BM2 channel (PDB: 2KIX [69]) as an example to show how to generate an MD simulation system for membrane protein using the Membrane Builder module of CHARMM-GUI.

12.4.2.1 Read protein coordinates

First, we need to give the coordinates of the BM2 channel (PDB: 2KIX [69]) to the server, which can be downloaded from Protein Data Bank (https://www.rcsb.org/) or Orientations of Proteins in Membranes (OPM) (http://opm.phar.umich.edu) database [85, 99, 100] or uploaded from our local machine. The OPM database is a curated web resource that provides spatial positions of membrane-bound peptides and proteins of known three-dimensional structures in the lipid bilayer, together with their structural classification, topology, and intracellular localization [100]. Then, we need to choose the model/chain of the structure according to our needs. The disulfide bonds, different protonation states of titratable residues, and other post-translational modifications can also be easily handled in this step.

12.4.2.2 Orient the protein structure

In this step, the orientation of the BM2 channel embedded in the membrane needs to be determined. If the PDB coordinates are from Protein Data Bank or a local machine, the orientation of the protein with respect to the membrane can be handled by VMD software, OPM database [20, 99]. In addition, the Membrane Builder can also adjust the orientation of the protein.

12.4.2.3 Determine the system size

In this step, we need to determine the system size, the type, and the ratio of lipids and sterols. There are three aspects of parameters that need to be set in the case of heterogeneous bilayer generation: (1) box type, (2) length along Z (water thickness and hydration number), and (3) length along with X and Y (ratios of lipid components or numbers of lipid components). After filling out the table, we need to click "show the system info" to calculate the lipid number and the size of the XY system, and then go to the next step. If the calculation results show that the system is unreasonable, the parameters need to be modified.

12.4.2.4 Build the components

According to the system size calculated in the previous step, ions can be added to the system in this step. If the ion type and ion concentration are specified, the numbers of ions are automatically calculated by the ion-accessible volume and the total charge of the system.

12.4.2.5 Assemble the components

In this step, all the components (proteins, lipid bilayers, sterols, additional water, and ions) are assembled together to form a membrane-embedded complex system, and this procedure will take minutes to hours depending on the system size [20]. If a problem is found in this step, the job can be reverted to previous steps to regenerate the system before quitting the browser. Additionally, the successful assembly of the initial structure in this step can be used to verify that the system is reasonable.

12.4.2.6 Equilibrate the system

Equilibration is necessary to relax the initial system before MD simulation. In this step, the force field parameters, input file formats, ensemble types, and other parameters can be defined. Subsequently, we can obtain the input files for MD simulation, such as the AMBER format coordinates and topology file.

Acknowledgments

This work was supported by the National Natural Science Foundation of China (No. 21773084).

References

1. Kang, M.K. and Tullman-Ercek, D. (2018). Engineering expression and function of membrane proteins, *Methods*, **147**, pp. 66–72.

2. Davis, J.E., Raharnan, O., and Patel, S. (2009). Molecular dynamics simulations of a DMPC bilayer using nonadditive interaction models, *Biophys. J.*, **96**, pp. 385–402.

3. Lluis, M.W., Godfroy, J.I., and Yin, H. (2013). Protein engineering methods applied to membrane protein targets, *Protein Eng. Des. Sel.*, **26**, pp. 91–100.

4. Thal, D.M., Vuckovic, Z., Draper-Joyce, C.J., Liang, Y.L., Glukhova, A., Christopoulos, A., and Sexton, P.M. (2018). Recent advances in the determination of G protein-coupled receptor structures, *Curr. Opin. Struct. Biol.*, **51**, pp. 28–34.

5. Carpenter, E.P., Beis, K., Cameron, A.D., and Iwata, S. (2008). Overcoming the challenges of membrane protein crystallography, *Curr. Opin. Struct. Biol.*, **18**, pp. 581–586.

6. Kotov, V., Bartels, K., Veith, K., Josts, I., Subhramanyam, U.K.T., Gunther, C., Labahn, J., Marlovits, T.C., Moraes, I., Tidow, H., Low, C., and Garcia-Alai, M.M. (2019). High-throughput stability screening for detergent-solubilized membrane proteins, *Sci. Rep.*, **9**, pp. 10379.

7. Jensen, H.M., Eng, T., Chubukov, V., Herbert, R.A., and Mukhopadhyay, A. (2017). Improving membrane protein expression and function using genomic edits, *Sci. Rep.*, **7**, pp. 13030.

8. Lieberman, R.L., Culver, J.A., Entzminger, K.C., Pai, J.C., and Maynard, J.A. (2011). Crystallization chaperone strategies for membrane proteins, *Methods*, **55**, pp. 293–302.

9. Gullingsrud, J. and Schulten, K. (2004). Lipid bilayer pressure profiles and mechanosensitive channel gating, *Biophys. J.*, **86**, pp. 3496–3509.

10. Chu, H., Zhang, Y., Li, Y., and Li, G. (2018) *Membrane Biophysics: New Insights and Methods*, eds. Wang, H. and Li, G., Chapter 12 "Computer Simulations to Explore Membrane Organization and Transport," (Springer Singapore, Singapore) pp. 355–392.

11. Karplus, M. and McCammon, J.A. (2002). Molecular dynamics simulations of biomolecules, *Nat. Struct. Biol.*, **9**, pp. 646–652.

12. Lu, S., He, X., Yang, Z., Chai, Z., Zhou, S., Wang, J., Rehman, A.U., Ni, D., Pu, J., Sun, J., and Zhang, J. (2021). Activation pathway of a G protein-coupled receptor uncovers conformational intermediates as targets for allosteric drug design, *Nat. Commun.*, **12**, pp. 4721.

13. Miao, Y., Feher, V.A., and McCammon, J.A. (2015). Gaussian accelerated molecular dynamics: Unconstrained enhanced sampling and free energy calculation, *J. Chem. Theory Comput.*, **11**, pp. 3584–3595.

14. Hildebrand, P.W., Rose, A.S., and Tiemann, J.K.S. (2019). Bringing molecular dynamics simulation data into view, *Trends Biochem. Sci*, **44**, pp. 902–913.

15. Zhang, Y. and Zheng, Q.C. (2021). In silico analysis revealed a unique binding but ineffective mode of amantadine to influenza virus B M2 channel, *J. Phys. Chem. Lett.*, **12**, pp. 1169–1174.

16. Perilla, J.R., Goh, B.C., Cassidy, C.K., Liu, B., Bernardi, R.C., Rudack, T., Yu, H., Wu, Z., and Schulten, K. (2015). Molecular dynamics simulations of large macromolecular complexes, *Curr. Opin. Struct. Biol.*, **31**, pp. 64–74.

17. Wang, Q., Shao, X., Leung, E.L.H., Chen, Y., and Yao, X. (2021). Selectively targeting individual bromodomain: Drug discovery and molecular mechanisms, *Pharmacol. Res.*, **172**, pp. 105804.

18. Fu, T., Zheng, Q., and Zhang, H. (2022). Investigation of the molecular and mechanistic basis for the regioselective metabolism of midazolam by cytochrome P450 3A4, *Phys. Chem. Chem. Phys.*, **24**, pp. 8104–8112.

19. Waterhouse, A., Bertoni, M., Bienert, S., Studer, G., Tauriello, G., Gumienny, R., Heer, F.T., de Beer, T.A.P., Rempfer, C., Bordoli, L., Lepore, R., and Schwede, T. (2018). SWISS-MODEL: Homology modelling of protein structures and complexes, *Nucleic Acids Res.*, **46**, pp. W296–W303.

20. Jo, S., Kim, T., Iyer, V.G., and Im, W. (2008). CHARMM-GUI: A web-based graphical user interface for CHARMM, *J. Comput. Chem.*, **29**, pp. 1859–1865.

21. Marrink, S.J., Corradi, V., Souza, P.C.T., Ingolfsson, H.I., Tieleman, D.P., and Sansom, M.S.P. (2019). Computational modeling of realistic cell membranes, *Chem. Rev.*, **119**, pp. 6184–6226.

22. Hospital, A., Goni, J.R., Orozco, M., and Gelpi, J.L. (2015). Molecular dynamics simulations: Advances and applications, *Adv. Appl. Bioinform. Chem.*, **8**, pp. 37–47.

23. Hollingsworth, S.A. and Dror, R.O. (2018). Molecular dynamics simulation for all, *Neuron*, **99**, pp. 1129–1143.

24. Case, D., Betz, R., Cerutti, D.S., Cheatham, T., Darden, T., Duke, R., Giese, T.J., Gohlke, H., Götz, A., Homeyer, N., Izadi, S., Janowski, P., Kaus, J., Kovalenko, A., Lee, T.-S., LeGrand, S., Li, P., Lin, C., Luchko, T., and Kollman, P.A. (2016). Amber 2016, *University of California, San Francisco*.

25. Phillips, J.C., Hardy, D.J., Maia, J.D.C., Stone, J.E., Ribeiro, J.V., Bernardi, R.C., Buch, R., Fiorin, G., Henin, J., Jiang, W., McGreevy, R., Melo, M.C.R.,

Radak, B.K., Skeel, R.D., Singharoy, A., Wang, Y., Roux, B., Aksimentiev, A., Luthey-Schulten, Z., Kale, L.V., Schulten, K., Chipot, C., and Tajkhorshid, E. (2020). Scalable molecular dynamics on CPU and GPU architectures with NAMD, *J. Chem. Phys.*, **153**, pp. 044130.

26. Hess, B., Kutzner, C., van der Spoel, D., and Lindahl, E. (2008). GROMACS 4: Algorithms for highly efficient, load-balanced, and scalable molecular simulation, *J. Chem. Theory Comput.*, **4**, pp. 435–447.
27. Eastman, P., Swails, J., Chodera, J.D., McGibbon, R.T., Zhao, Y., Beauchamp, K.A., Wang, L.P., Simmonett, A.C., Harrigan, M.P., Stern, C.D., Wiewiora, R.P., Brooks, B.R., and Pande, V.S. (2017). OpenMM 7: Rapid development of high performance algorithms for molecular dynamics, *PLoS Comput. Biol.*, **13**, pp. e1005659.
28. Collier, T.A., Piggot, T.J., and Allison, J.R. (2020). Molecular dynamics simulation of proteins, *Methods Mol. Biol.*, **2073**, pp. 311–327.
29. Harder, E., Damm, W., Maple, J., Wu, C., Reboul, M., Xiang, J.Y., Wang, L., Lupyan, D., Dahlgren, M.K., Knight, J.L., Kaus, J.W., Cerutti, D.S., Krilov, G., Jorgensen, W.L., Abel, R., and Friesner, R.A. (2016). OPLS3: A force field providing broad coverage of drug-like small molecules and proteins, *J. Chem. Theory Comput.*, **12**, pp. 281–296.
30. Tian, C., Kasavajhala, K., Belfon, K.A.A., Raguette, L., Huang, H., Migues, A.N., Bickel, J., Wang, Y., Pincay, J., Wu, Q., and Simmerling, C. (2020). ff19SB: Amino-acid-specific protein backbone parameters trained against quantum mechanics energy surfaces in solution, *J. Chem. Theory Comput.*, **16**, pp. 528–552.
31. Madej, B.D., Gould, I.R., and Walker, R.C. (2015). A parameterization of cholesterol for mixed lipid bilayer simulation within the Amber Lipid14 force field, *J. Phys. Chem. B*, **119**, pp. 12424–12435.
32. Wang, J.M., Wolf, R.M., Caldwell, J.W., Kollman, P.A., and Case, D.A. (2005). Development and testing of a general amber force field (vol 25, pp. 1157, 2004), *J. Comput. Chem.*, **26**, pp. 114–114.
33. Huang, J., Rauscher, S., Nawrocki, G., Ran, T., Feig, M., de Groot, B.L., Grubmuller, H., and MacKerell, A.D., Jr. (2017). CHARMM36m: An improved force field for folded and intrinsically disordered proteins, *Nat. Methods*, **14**, pp. 71–73.
34. Roos, K., Wu, C., Damm, W., Reboul, M., Stevenson, J.M., Lu, C., Dahlgren, M.K., Mondal, S., Chen, W., Wang, L., Abel, R., Friesner, R.A., and Harder, E.D. (2019). OPLS3e: Extending force field coverage for drug-like small molecules, *J. Chem. Theory Comput.*, **15**, pp. 1863–1874.

35. Marrink, S.J., Risselada, H.J., Yefimov, S., Tieleman, D.P., and de Vries, A.H. (2007). The MARTINI force field: Coarse grained model for biomolecular simulations, *J. Phys. Chem. B*, **111**, pp. 7812–7824.

36. Lopez, C.A., Sovova, Z., van Eerden, F.J., de Vries, A.H., and Marrink, S.J. (2013). Martini force field parameters for glycolipids, *J. Chem. Theory Comput.*, **9**, pp. 1694–708.

37. Shinoda, W., DeVane, R., and Klein, M.L. (2010). Zwitterionic lipid assemblies: Molecular dynamics studies of monolayers, bilayers, and vesicles using a new coarse grain force field, *J. Phys. Chem. B*, **114**, pp. 6836–6849.

38. Darre, L., Machado, M.R., Brandner, A.F., Gonzalez, H.C., Ferreira, S., and Pantano, S. (2015). SIRAH: A structurally unbiased coarse-grained force field for proteins with aqueous solvation and long-range electrostatics, *J. Chem. Theory Comput.*, **11**, pp. 723–739.

39. Orsi, M. and Essex, J.W. (2011). The ELBA force field for coarse-grain modeling of lipid membranes, *PLoS One*, **6**, pp. e28637.

40. Dama, J.F., Sinitskiy, A.V., McCullagh, M., Weare, J., Roux, B., Dinner, A.R., and Voth, G.A. (2013). The theory of ultra-coarse-graining. 1. General principles, *J. Chem. Theory Comput.*, **9**, pp. 2466–2480.

41. Ayton, G.S. and Voth, G.A. (2009). Hybrid coarse-graining approach for lipid bilayers at large length and time scales, *J. Phys. Chem. B*, **113**, pp. 4413–4424.

42. Decherchi, S. and Cavalli, A. (2020). Thermodynamics and kinetics of drug-target binding by molecular simulation, *Chem. Rev.*, **120**, pp. 12788–12833.

43. Shen, Y. and Bax, A. (2015). Homology modeling of larger proteins guided by chemical shifts, *Nat. Methods*, **12**, pp. 747–750.

44. Fu, T., Li, F., Zhang, Y., Yin, J., Qiu, W., Li, X., Liu, X., Xin, W., Wang, C., Yu, L., Gao, J., Zheng, Q., Zeng, S., and Zhu, F. (2022). VARIDT 2.0: Structural variability of drug transporter, *Nucleic Acids Res.*, **50**, pp. D1417–D1431.

45. Reichel, K., Fisette, O., Braun, T., Lange, O.F., Hummer, G., and Schafer, L.V. (2017). Systematic evaluation of CS-Rosetta for membrane protein structure prediction with sparse NOE restraints, *Proteins*, **85**, pp. 812–826.

46. Ovchinnikov, S., Kim, D.E., Wang, R.Y., Liu, Y., DiMaio, F., and Baker, D. (2016). Improved de novo structure prediction in CASP11 by incorporating coevolution information into Rosetta, *Proteins*, **84 Suppl 1**, pp. 67–75.

47. Meng, X.Y., Zhang, H.X., Mezei, M., and Cui, M. (2011). Molecular docking: A powerful approach for structure-based drug discovery, *Curr. Comput. Aided Drug Des.*, **7**, pp. 146–157.

48. Vermaas, J.V., Mayne, C.G., Shinn, E., and Tajkhorshid, E. (2022). Assembly and analysis of cell-scale membrane envelopes, *J. Chem. Inf. Model.*, **62**, pp. 602–617.

49. Xue, W., Fu, T., Deng, S., Yang, F., Yang, J., and Zhu, F. (2022). Molecular mechanism for the allosteric inhibition of the human serotonin transporter by antidepressant escitalopram, *ACS Chem. Neurosci.*, **13**, pp. 340–351.

50. Mattedi, G., Deflorian, F., Mason, J.S., de Graaf, C., and Gervasio, F.L. (2019). Understanding ligand binding selectivity in a prototypical GPCR family, *J. Chem. Inf. Model.*, **59**, pp. 2830–2836.

51. Lin, B.H., Zhang, H.X., and Zheng, Q.C. (2020). How do mutations affect the structural characteristics and substrate binding of CYP21A2? An investigation by molecular dynamics simulations, *Phys. Chem. Chem. Phys.*, **22**, pp. 8870–8877.

52. Hamelberg, D., Mongan, J., and McCammon, J.A. (2004). Accelerated molecular dynamics: A promising and efficient simulation method for biomolecules, *J. Chem. Phys.*, **120**, pp. 11919–11929.

53. Wang, Y., Harrison, C.B., Schulten, K., and McCammon, J.A. (2011). Implementation of accelerated molecular dynamics in NAMD, *Comput. Sci. Discov.*, **4**.

54. Pang, Y.T., Miao, Y., Wang, Y., and McCammon, J.A. (2017). Gaussian accelerated molecular dynamics in NAMD, *J. Chem. Theory Comput.*, **13**, pp. 9–19.

55. Ensing, B., De Vivo, M., Liu, Z., Moore, P., and Klein, M.L. (2006). Metadynamics as a tool for exploring free energy landscapes of chemical reactions, *Acc. Chem. Res.*, **39**, pp. 73–81.

56. Conces, D.J., Jr. and Tarver, R.D. (1991). Noninfectious and nonmalignant pulmonary disease in AIDS, *J. Thorac. Imaging*, **6**, pp. 53–59.

57. Barducci, A., Bonomi, M., and Parrinello, M. (2011). Metadynamics, *Wiley Interdiscip. Rev. Comput. Mol. Sci.*, **1**, pp. 826–843.

58. Laio, A. and Parrinello, M. (2002). Escaping free-energy minima, *Proc. Natl. Acad. Sci. U. S. A.*, **99**, pp. 12562–12566.

59. Laio, A. and Gervasio, F.L. (2008). Metadynamics: A method to simulate rare events and reconstruct the free energy in biophysics, chemistry and material science, *Rep. Prog. Phys.*, **71**.

60. Kastner, J. (2011). Umbrella sampling, *Wiley Interdiscip. Rev. Comput. Mol. Sci*, **1**, pp. 932–942.

61. Zhu, F. and Hummer, G. (2012). Convergence and error estimation in free energy calculations using the weighted histogram analysis method, *J. Comput. Chem.*, **33**, pp. 453–465.

62. Harayama, T. and Riezman, H. (2018). Understanding the diversity of membrane lipid composition, *Nat. Rev. Mol. Cell Biol.*, **19**, pp. 281–296.

63. Ng, H.W., Laughton, C.A., and Doughty, S.W. (2014). Molecular dynamics simulations of the adenosine A2a receptor in POPC and POPE lipid bilayers: Effects of membrane on protein behavior, *J. Chem. Inf. Model.*, **54**, pp. 573–581.

64. Cross, T.A., Sharma, M., Yi, M., and Zhou, H.X. (2011). Influence of solubilizing environments on membrane protein structures, *Trends Biochem. Sci*, **36**, pp. 117–125.

65. Pinto, L.H. and Lamb, R.A. (2006). The M2 proton channels of influenza A and B viruses, *J. Biol. Chem.*, **281**, pp. 8997–9000.

66. Williams, J.K., Tietze, D., Lee, M., Wang, J., and Hong, M. (2016). Solid-state NMR investigation of the conformation, proton conduction, and hydration of the influenza B virus M2 transmembrane proton channel, *J. Am. Chem. Soc.*, **138**, pp. 8143–8155.

67. Williams, J.K., Shcherbakov, A.A., Wang, J., and Hong, M. (2017). Protonation equilibria and pore-opening structure of the dual-histidine influenza B virus M2 transmembrane proton channel from solid-state NMR, *J. Biol. Chem.*, **292**, pp. 17876–17884.

68. Zhang, Y., Zhang, H.X., and Zheng, Q.C. (2020). In silico study of membrane lipid composition regulating conformation and hydration of influenza virus B M2 channel, *J. Chem. Inf. Model.*, **60**, pp. 3603–3615.

69. Wang, J., Pielak, R.M., McClintock, M.A., and Chou, J.J. (2009). Solution structure and functional analysis of the influenza B proton channel, *Nat. Struct. Mol. Biol.*, **16**, pp. 1267–1271.

70. Zhang, Y., Zhang, H., and Zheng, Q. (2019). A unique activation-promotion mechanism of the influenza B M2 proton channel uncovered by multiscale simulations, *Phys. Chem. Chem. Phys.*, **21**, pp. 2984–2991.

71. Zhang, Y. and Zheng, Q.C. (2019). What are the effects of the serine triad on proton conduction of an influenza B M2 channel? An investigation by molecular dynamics simulations, *Phys. Chem. Chem. Phys.*, **21**, pp. 8820–8826.

72. Sharma, M., Yi, M., Dong, H., Qin, H., Peterson, E., Busath, D.D., Zhou, H.X., and Cross, T.A. (2010). Insight into the mechanism of the

influenza A proton channel from a structure in a lipid bilayer, *Science*, **330**, pp. 509–512.

73. Hu, F., Luo, W., and Hong, M. (2010). Mechanisms of proton conduction and gating in influenza M2 proton channels from solid-state NMR, *Science*, **330**, pp. 505–508.

74. Williams, J.K., Tietze, D., Wang, J., Wu, Y.B., DeGrado, W.F., and Hong, M. (2013). Drug-induced conformational and dynamical changes of the S31N mutant of the influenza M2 proton channel investigated by solid-state NMR, *J. Am. Chem. Soc.*, **135**, pp. 9885–9897.

75. Lin, C.W., Mensa, B., Barniol-Xicota, M., DeGrado, W.F., and Gai, F. (2017). Activation pH and gating dynamics of influenza A M2 proton channel revealed by single-molecule spectroscopy, *Angew. Chem. Int. Ed. Engl.*, **56**, pp. 5283–5287.

76. Maier, J.A., Martinez, C., Kasavajhala, K., Wickstrom, L., Hauser, K.E., and Simmerling, C. (2015). ff14SB: Improving the accuracy of protein side chain and backbone parameters from ff99SB, *J. Chem. Theory Comput.*, **11**, pp. 3696–3713.

77. Skjevik, A.A., Madej, B.D., Walker, R.C., and Teigen, K. (2012). LIPID11: A modular framework for lipid simulations using amber, *J. Phys. Chem. B*, **116**, pp. 11124–11136.

78. Chovancova, E., Pavelka, A., Benes, P., Strnad, O., Brezovsky, J., Kozlikova, B., Gora, A., Sustr, V., Klvana, M., Medek, P., Biedermannova, L., Sochor, J., and Damborsky, J. (2012). CAVER 3.0: A tool for the analysis of transport pathways in dynamic protein structures, *PLoS Comput. Biol.*, **8**, pp. e1002708.

79. Stevens, R.C., Cherezov, V., Katritch, V., Abagyan, R., Kuhn, P., Rosen, H., and Wuthrich, K. (2013). The GPCR Network: A large-scale collaboration to determine human GPCR structure and function, *Nat. Rev. Drug Discov.*, **12**, pp. 25–34.

80. Hauser, A.S., Attwood, M.M., Rask-Andersen, M., Schioth, H.B., and Gloriam, D.E. (2017). Trends in GPCR drug discovery: New agents, targets and indications, *Nat. Rev. Drug Discov.*, **16**, pp. 829–842.

81. Maning, J., Negussie, S., Clark, M.A., and Lymperopoulos, A. (2017). Biased agonism/antagonism at the AngII-AT1 receptor: Implications for adrenal aldosterone production and cardiovascular therapy, *Pharmacol. Res.*, **125**, pp. 14–20.

82. Wingler, L.M., McMahon, C., Staus, D.P., Lefkowitz, R.J., and Kruse, A.C. (2019). Distinctive activation mechanism for angiotensin receptor revealed by a synthetic nanobody, *Cell*, **176**, pp. 479–490 e12.

83. Zhang, H., Unal, H., Gati, C., Han, G.W., Liu, W., Zatsepin, N.A., James, D., Wang, D., Nelson, G., Weierstall, U., Sawaya, M.R., Xu, Q., Messerschmidt, M., Williams, G.J., Boutet, S., Yefanov, O.M., White, T.A., Wang, C., Ishchenko, A., Tirupula, K.C., Desnoyer, R., Coe, J., Conrad, C.E., Fromme, P., Stevens, R.C., Katritch, V., Karnik, S.S., and Cherezov, V. (2015). Structure of the angiotensin receptor revealed by serial femtosecond crystallography, *Cell*, **161**, pp. 833–844.

84. Vilar, S., Cozza, G., and Moro, S. (2008). Medicinal chemistry and the molecular operating environment (MOE): Application of QSAR and molecular docking to drug discovery, *Curr. Top. Med. Chem.*, **8**, pp. 1555–1572.

85. Lomize, M.A., Pogozheva, I.D., Joo, H., Mosberg, H.I., and Lomize, A.L. (2012). OPM database and PPM web server: Resources for positioning of proteins in membranes, *Nucleic Acids Res*, **40**, pp. D370–D376.

86. Singharoy, A., Maffeo, C., Delgado-Magnero, K.H., Swainsbury, D.J.K., Sener, M., Kleinekathofer, U., Vant, J.W., Nguyen, J., Hitchcock, A., Isralewitz, B., Teo, I., Chandler, D.E., Stone, J.E., Phillips, J.C., Pogorelov, T.V., Mallus, M.I., Chipot, C., Luthey-Schulten, Z., Tieleman, D.P., Hunter, C.N., Tajkhorshid, E., Aksimentiev, A., and Schulten, K. (2019). Atoms to phenotypes: Molecular design principles of cellular energy metabolism, *Cell*, **179**, pp. 1098.

87. Phillips, J.C., Braun, R., Wang, W., Gumbart, J., Tajkhorshid, E., Villa, E., Chipot, C., Skeel, R.D., Kale, L., and Schulten, K. (2005). Scalable molecular dynamics with NAMD, *J. Comput. Chem.*, **26**, pp. 1781–802.

88. Chandler, D.E., Hsin, J., Harrison, C.B., Gumbart, J., and Schulten, K. (2008). Intrinsic curvature properties of photosynthetic proteins in chromatophores, *Biophys. J.*, **95**, pp. 2822–2836.

89. Chandler, D.E., Strumpfer, J., Sener, M., Scheuring, S., and Schulten, K. (2014). Light harvesting by lamellar chromatophores in Rhodospirillum photometricum, *Biophys. J.*, **106**, pp. 2503–2510.

90. Marks, D.S., Hopf, T.A., and Sander, C. (2012). Protein structure prediction from sequence variation, *Nat. Biotechnol.*, **30**, pp. 1072–1080.

91. Bienert, S., Waterhouse, A., de Beer, T.A.P., Tauriello, G., Studer, G., Bordoli, L., and Schwede, T. (2017). The SWISS-MODEL Repository: New features and functionality, *Nucleic Acids Res.*, **45**, pp. D313–D319.

92. Sayers, E.W., Bolton, E.E., Brister, J.R., Canese, K., Chan, J., Comeau, D.C., Connor, R., Funk, K., Kelly, C., Kim, S., Madej, T., Marchler-Bauer, A., Lanczycki, C., Lathrop, S., Lu, Z., Thibaud-Nissen, F., Murphy, T., Phan,

L., Skripchenko, Y., Tse, T., Wang, J., Williams, R., Trawick, B.W., Pruitt, K.D., and Sherry, S.T. (2022). Database resources of the national center for biotechnology information, *Nucleic Acids Res.*, **50**, pp. D20–D26.

93. UniProt, C. (2021). UniProt: The universal protein knowledgebase in 2021, *Nucleic Acids Res.*, **49**, pp. D480–D489.

94. Camacho, C., Coulouris, G., Avagyan, V., Ma, N., Papadopoulos, J., Bealer, K., and Madden, T.L. (2009). BLAST+: Architecture and applications, *BMC Bioinform.*, **10**, pp. 421.

95. Steinegger, M., Meier, M., Mirdita, M., Vohringer, H., Haunsberger, S.J., and Soding, J. (2019). HH-suite3 for fast remote homology detection and deep protein annotation, *BMC Bioinform.*, **20**.

96. Studer, G., Tauriello, G., Bienert, S., Biasini, M., Johner, N., and Schwede, T. (2021). ProMod3: A versatile homology modelling toolbox, *PLoS Comput. Biol.*, **17**.

97. Studer, G., Biasini, M., and Schwede, T. (2014). Assessing the local structural quality of transmembrane protein models using statistical potentials (QMEANBrane), *Bioinformatics*, **30**, pp. I505–I511.

98. Brooks, B.R., Brooks, C.L.III, Mackerell, A.D. Jr., Nilsson, L., Petrella, R.J., Roux, B., Won, Y., Archontis, G., Bartels, C., Boresch, S., Caflisch, A., Caves, L., Cui, Q., Dinner, A.R., Feig, M., Fischer, S., Gao, J., Hodoscek, M., Im, W., Kuczera, K., Lazaridis, T., Ma, J., Ovchinnikov, V., Paci, E., Pastor, R.W., Post, C.B., Pu, J.Z., Schaefer, M., Tidor, B., Venable, R.M., Woodcock, H.L., Wu, X., Yang, W., York, D.M., and Karplus, M. (2009). CHARMM: The biomolecular simulation program, *J. Comput. Chem.*, **30**, pp. 1545–1614.

99. Lomize, M.A., Lomize, A.L., Pogozheva, I.D., and Mosberg, H.I. (2006). OPM: Orientations of proteins in membranes database, *Bioinformatics*, **22**, pp. 623–625.

100. Lomize, M.A., Pogozheva, I.D., Joo, H., Mosberg, H.I., and Lomize, A.L. (2012). OPM database and PPM web server: Resources for positioning of proteins in membranes, *Nucleic Acids Res.*, **40**, pp. D370–D376.

Index